Lecture Notes in Computer Science   10118

Commenced Publication in 1973
Founding and Former Series Editors:
Gerhard Goos, Juris Hartmanis, and Jan van Leeuwen

More information about this series at http://www.springer.com/series/7412

Chu-Song Chen · Jiwen Lu
Kai-Kuang Ma (Eds.)

Computer Vision – ACCV 2016 Workshops

ACCV 2016 International Workshops
Taipei, Taiwan, November 20–24, 2016
Revised Selected Papers, Part III

 Springer

Editors
Chu-Song Chen
Institute of Information Science
Academia Sinica
Taipei
Taiwan

Jiwen Lu
Tsinghua University
Beijing
China

Kai-Kuang Ma
School of Electrical and Electronic
  Engineering
Nanyang Technological University
Singapore
Singapore

ISSN 0302-9743   ISSN 1611-3349   (electronic)
Lecture Notes in Computer Science
ISBN 978-3-319-54525-7   ISBN 978-3-319-54526-4   (eBook)
DOI 10.1007/978-3-319-54526-4

Library of Congress Control Number: 2017932642

LNCS Sublibrary: SL6 – Image Processing, Computer Vision, Pattern Recognition, and Graphics

Printed on acid-free paper

This Springer imprint is published by Springer Nature
The registered company is Springer International Publishing AG
The registered company address is: Gewerbestrasse 11, 6330 Cham, Switzerland

Preface

It is our great pleasure to present the workshop proceedings of three LNCS volumes, which contain the papers carefully reviewed and selected from the 17 workshops that were held in conjunction with the 13th Asian Conference on Computer Vision (ACCV), during November 20–24, 2016, in Taipei, Taiwan. There are 134 papers selected from 223 papers submitted to all the 17 workshops as listed below.

1. New Trends in Image Restoration and Enhancement (NTIRE): 14 papers
2. Workshop on Assistive Vision: 6 papers
3. ACCV 2016 Workshop on Hyperspectral Image and Signal Processing: 6 papers
4. Computer Vision Technologies for Smart Vehicle: 7 papers
5. Spontaneous Facial Behavior Analysis: 8 papers
6. 3D Modelling and Applications: 16 papers
7. 4th ACCV Workshop on e-Heritage: 4 papers
8. Multiview Lip-Reading Challenges: 5 papers
9. Workshop on Facial Informatics (WFI): 11 papers
10. Discrete Geometry and Mathematical Morphology for Computer Vision: 4 papers
11. Workshop on Mathematical and Computational Methods in Biomedical Imaging and Image Analysis: 15 papers
12. International Workshop on Driver Drowsiness Detection from Video: 6 papers
13. Workshop on Meeting HCI with CV: 6 papers
14. Workshop on Human Identification for Surveillance (HIS) Methods and Applications: 8 papers
15. Benchmark and Evaluation of Surveillance Task (BEST): 9 papers
16. The Third Workshop on Computer Vision for Affective Computing (CV4AC): 3 papers
17. Workshop on Interpretation and Visualization of Deep Neural Nets: 6 papers

The workshop topics are related to computer vision and its applications, interdisciplinary themes with other application areas, as well as challenges or competitions. Every workshop handles its own paper submission system, and each paper is reviewed by two to three reviewers. We thank all the workshop organizers for their great efforts in holding these successful workshops. We also thank the help of the publication chairs in making this publication possible.

November 2016

Chu-Song Chen
Jiwen Lu
Kai-Kuang Ma

Organization

W01: 3D Modelling and Applications

Chia-Yen Chen	National University of Kaohsiung, Taiwan
Min-Chun Hu	National Cheng Kung University, Taiwan
Li-Wei Kang	National Yunlin University of Science and Technology, Taiwan
Chih-Yang Lin	Asia University, Taiwan
Tang-Kai Yin	National University of Kaohsiung, Taiwan
Guo-Shiang Lin	Da-Yeh University, Taiwan
Chia-Hung Yeh	National Sun Yat-Sen University, Taiwan

W02: 4th ACCV Workshop on e-Heritage

Katsushi Ikeuchi	Microsoft Research Asia, China
El Mustapha Mouaddib	Université de Picardie Jules Verne, France
Takeshi Masuda	AIST, Japan
Takeshi Oishi	The University of Tokyo, Japan

W03: ACCV 2016 Workshop on Hyperspectral Image and Signal Processing

Keng-Hao Liu	National Sun Yat-sen University, Taiwan
Wei-Min Liu	National Chung Cheng University, Taiwan

W04: Benchmark and Evaluation of Surveillance Task (BEST)

Xiaokang Yang	Shanghai Jiao Tong University, China
Chong-Yang Zhang	Shanghai Jiao Tong University, China
Bingbing Ni	Shanghai Jiao Tong University, China
Lin Mei	The Third Research Institute of the Ministry of Public Security, China

W05: Computer Vision Technologies for Smart Vehicle

Li-Chen Fu	National Taiwan University, Taiwan
Pei-Yung Hsiao	National University of Kaohsiung, Taiwan
Shih-Shinh Huang	National Kaohsiung First University of Science and Technology, Taiwan

W06: Discrete Geometry and Mathematical Morphology for Computer Vision

Jean Cousty   Université Paris-Est, ESIEE Paris, France
Yukiko Kenmochi   Université Paris-Est, CNRS, France
Akihiro Sugimoto   National Institute of Informatics, Japan

W07: International Workshop on Driver Drowsiness Detection from Video

Chen-Kuo Chiang   National Chung Cheng University, Taiwan
Shang-Hong Lai   National Tsing Hua University, Taiwan
Michel Sarkis   Qualcomm Technologies Inc., USA

W08: Large-Scale 3D Human Activity Analysis Challenge in Depth Videos

Gang Wang   Nanyang Technological University, Singapore
Amir Shahroudy   Nanyang Technological University, Singapore
Jun Liu   Nanyang Technological University, Singapore

W09: Multiview Lip-Reading Challenges

Ziheng Zhou   University of Oulu, Finland
Guoying Zhao   University of Oulu, Finland
Takeshi Saitoh   Kyushu Institute of Technology, Japan
Richard Bowden   University of Surrey, UK

W10: New Trends in Image Restoration and Enhancement (NTIRE)

Radu Timofte   ETH Zurich, Switzerland
Luc Van Gool   ETH Zurich, Switzerland
Ming-Hsuan Yang   University of California at Merced, USA

W11: Spontaneous Facial Behavior Analysis

Xiaopeng Hong   University of Oulu, Finland
Guoying Zhao   University of Oulu, Finland
Stefanos Zafeiriou   Imperial College London, UK
Matti Pietikäinen   University of Oulu, Finland
Maja Pantic   Imperial College London, UK

W12: The Third Workshop on Computer Vision for Affective Computing (CV4AC)

Abhinav Dhall	Abhinav Dhall, University of Waterloo, Canada
Roland Goecke	University of Canberra/Australian National University, Australia
O.V. Ramana Murthy	Amrita University, India
Jesse Hoey	University of Waterloo, Canada
Nicu Sebe	University of Trento, Italy

W13: Workshop on Assistive Vision

Chetan Arora	Indraprastha Institute of Information Technology, Delhi, India
Vineeth N. Balasubmanian	Indian Institute of Technology, Hyderabad, India
C.V. Jawahar	International Institute of Information Technology, Hyderabad, India
Vinay P. Namboodiri	Indian Institute of Technology, Kanpur, India
Ramanathan Subramanian	International Institute of Information Technology, Hyderabad, India

W14: Workshop on Facial Informatics (WFI)

Gee-Sern (Jison) Hsu	National Taiwan University of Science and Technology, Taiwan
Moi Hoon Yap	Manchester Metropolitan University, UK
Xiaogang Wang	Chinese University of Hong Kong, Hong Kong, SAR China
Su-Jing Wang	Chinese Academy of Science, China
John See	Multimedia University, Malaysia

W15: Workshop on Meeting HCI with CV

Liwei Chan	National Chiao Tung University, Taiwan and Keio Media Design, Japan
Yi-Ping Hung	National Taiwan University, Taiwan

W16: Workshop on Human Identification for Surveillance (HIS): Methods and Applications

Wei-Shi Zheng	Sun Yat-sen University, China
Ruiping Wang	Institute of Computing Technology, Chinese Academy of Sciences, China

Weihong Deng   Beijing University of Posts and Telecommunications,
                                China
Shenghua Gao   ShanghaiTech University, China

W17: Workshop on Interpretation and Visualization of Deep Neural Nets

Alexander Binder   Singapore University of Technology and Design,
                                Singapore
Wojciech Samek   Fraunhofer Heinrich Hertz Institute, Germany

W18: Workshop on Mathematical and Computational Methods in Biomedical Imaging and Image Analysis

Atsushi Imiya   Chiba University, Japan
Xiaoyi Jiang   Universität Münster, Germany
Hidetaka Hontani   Nagoya Institute of Technology, Japan

Contents – Part III

Workshop on Meeting HCI with CV

Workshop on Human Identification for Surveillance (HIS) Methods and Applications

Benchmark and Evaluation of Surveillance Task (BEST)

**The Third Workshop on Computer Vision for Affective
Computing (CV4AC)**

Workshop on Interpretation and Visualization of Deep Neural Nets

Workshop on Mathematical and Computational Methods in Biomedical Imaging and Image Analysis

Segmentation of Trabecular Bone for In Vivo CT Imaging Using a Novel Approach of Computing Spatial Variation in Bone and Marrow Intensities

Cheng Chen[1(✉)], Dakai Jin[1], Xiaoliu Zhang[1], Steven M. Levy[3], and Punam K. Saha[1,2]

[1] Department of Electrical and Computer Engineering,
University of Iowa, Iowa City, IA 52242, USA
{cheng-chen,dakai-jin,xiaoliu-zhang}@uiowa.edu,
pksaha@healthcare.uiowa.edu
[2] Department of Radiology, University of Iowa, Iowa City, IA 52242, USA
[3] Department of Preventive and Community Dentistry,
University of Iowa, Iowa City, IA 52242, USA
steven-levy@uiowa.edu

Abstract. Characterization of trabecular bone (TB) microarchitecture and computational modelling of bone strength are widely used in research and clinical studies related to osteoporosis, which is associated with elevated risk of fractures. Segmentation of TB network from the background marrow space is essential for quantitative assessment of the quality of TB microarchitecture and bone strength, which are key determinants of fracture risk. Clinical CT is rapidly emerging as a viable modality for *in vivo* TB microarchitectural imaging at peripheral sites. Here, we present a new method for TB segmentation using *in vivo* CT imaging of distal tibia. Our method is primarily based on computing the spatial variation in the background marrow intensity as well as the bone-marrow contrast. First, a new anisotropic diffusion algorithm is developed and applied to improve local separability of TB microstructures that uses Hessian matrix to locally guide the diffusion process. Subsequently, a new multi-scale morphological algorithm is developed and applied to determine spatial distribution of bone and marrow intensity values. The accuracy of the method was examined by comparing its performance with multi-user-selected regional thresholding for bone-marrow separation on *in vivo* CT images of ten subjects each containing twenty random regions of interest (ROIs). High sensitivity (0.93), specificity (0.93), and accuracy (0.93) of the new method were observed from experimental results. In addition, the impact of the new method on predicting bone strength was examined in a cadaveric study. Experimental results have shown that the new TB segmentation method significantly improves the ability ($R^2 = 0.82$) of the computed TB thickness measure to predict actual bone strength determined by mechanical testing on TB cores.

© Springer International Publishing AG 2017
C.-S. Chen et al. (Eds.): ACCV 2016 Workshops, Part III, LNCS 10118, pp. 3–15, 2017.
DOI: 10.1007/978-3-319-54526-4_1

1   Introduction

Osteoporosis is associated with an increased risk of fractures. Its incidence increases progressively with age [1]. In the United States, about 8 million women and 2 million men have osteoporosis with medical costs estimated at $22 billion in 2008 [2]. Clinically, osteoporosis is defined by low bone mineral density (BMD). However, BMD only accounts for approximately 60% to 70% of the variability in bone strength [3]. The remaining variability is due to the cumulative and synergistic effects of various factors, including trabecular bone (TB) microstructure [4,5]. Thus, reliably measuring TB microstructure could be of clinical significance, particularly as trabecular bone may be more susceptible to hormonal, pharmacological, and toxic effects.

TB is a complex interconnected network of individual trabecular microstructures. The conventional tool for assessing TB structure is two-dimensional (2D) cross-sectional histomorphometry from bone biopsies [6]. Recent advancements in volumetric bone imaging, such as magnetic resonance (MR) [3,7,8], high resolution peripheral quantitative computed tomography (HR-pQCT) [9,10], and clinical computed tomography (CT) [11–13], allow characterization of bone micro-architecture at peripheral sites without the need for biopsy. Various methods related to digital topology and geometry [14], scale [15,16], and computational mechanics [17–19] have been reported in literature [20]. Saha *et al.* developed a skeletonization [21] based method of digital topological analysis [22–25] for regional plate-rod characterization of individual trabeculae. This basic method has been further modified and applied by other research groups [26–29] for assessment bone strength and fracture-risk. Segmentation of TB from the marrow space is essential for quantitative analysis of morphometric and biomechanical properties of TB microstructure. Irrespective of the methods used for quantitative *in vivo* assessment of TB microstructure, the fidelity of measures largely depends on the accuracy of TB segmentation. Although, current *in vivo* imaging technologies allow characterization of TB microstructure, accurate segmentation of TB network in such images remains challenging due to significant partial voluming effects, noise, and space varying background intensity for marrow voxels.

Clinical CT imaging is rapidly emerging as a frontrunner for imaging bone microstructure at peripheral sites due to its high availability in clinical environments, high spatial resolution, ultra-high speed scanning, ultra-low dose radiation, and large scan-length. A global threshold scheme [30,31] is often used for TB segmentation in CT imaging. However, due to partial voluming, scattering and other CT imaging artifacts, trabecular bone regions with dense and thick trabeculae require higher values of the threshold to preserve marrow holes, while regions with sparse and thin trabeculae demands lower values of the threshold to preserve local connectivity of trabecular network. Therefore, locally adaptive thresholding is needed for accurate and robust segmentation of TB that preserves both marrow holes as well as trabecular network connectivity. Waarsing *et al.* applied local thresholding on high resolution micro-CT imaging [32]. Burghardt *et al.* developed a locally adaptive thresholding approach using a hysteresis-based algorithm on gradient map, and applied the method on high resolution peripheral quantitative CT (HRpQCT) images [33]. In addition to threshold based methods, Scherf and Tilgner proposed the Ray Casting Algorithm, which utilizes local maximum, to segment fossil and cancellous bone in micro-CT images [34]. Tassani *et al.* applied a modified Chan-Vese method to segment TB on micro-CT images [35].

In this paper, we present a new method for TB segmentation from *in vivo* clinical CT imaging of distal tibia by computing spatial variation in the background marrow intensity and the bone-marrow contrast. First, it enhances local TB separability using a new anisotropic diffusion algorithm that uses Hessian matrix to encourage along-structure smoothing, while restricting cross-structure diffusion, thus, arresting edge-blurring. Subsequently, the method determines the spatial variation in the background marrow intensity and the bone-marrow contrast using a new multi-scale morphological algorithm. Essentially, the algorithm is designed to determine local intensity values for TB structures resembling mountain-ridges as well as for marrow regions forming valleys running quasi-parallel to TB ridges. These two steps are described in Sect. 2. The performance of the method is examined by comparing with the results from regional manual thresholding, and the implication of the new method for TB microarchitectural analysis and ability to predict bone strength is studied. Experimental methods and results are presented in Sect. 3. Finally, the conclusions are drawn in Sect. 4.

2   Method

In this section, we describe the Hessian matrix-guided anisotropic diffusion and the multi-scale morphological algorithms in Sects. 2.1 and 2.2, respectively.

2.1   Anisotropic Diffusion

Anisotropic diffusion, first proposed by Perona and Malik [36], can be described using a divergence operator on an intensity flow vector field $\mathbf{F}$ as follows:

$$\frac{\partial I}{\partial t} = div\mathbf{F}, \tag{1}$$

where I is the image at time point t. The intensity flow vector, constructed as $\mathbf{F} = G(|\nabla I|) \cdot \nabla I$, controls the diffusion process that facilitates within-region diffusion, while arresting cross-structure blurring. Here, ∇I represents the intensity gradient, and the diffusion-conductance function G is a monotonically decreasing nonlinear function that leads to generous diffusion within a homogeneous region, while a constricted diffusion across boundaries. Following the recommendation by Perona and Malik, we use

$$G(|\nabla I|) = exp(-\frac{|\nabla I|^2}{\mu^2}). \tag{2}$$

The discrete formulation of the diffusion flow process is expressed as

$$I_{t+1}(p) = I_t(p) + \frac{k}{|N_8(p)| + 1} \sum_{q \in N_8(p)} G(|\nabla I_{p,q}|) \nabla I_{p,q}, \tag{3}$$

where $t \geq 0$ is the iteration number; $I_0 = I$; N_8 is the 8-neighborhood; $\nabla I_{p,q}$ is the intensity gradient between two voxels p, q; and $k \in [0, 1]$ is a constant determining the overall speed of diffusion. In the above equation, the parameter μ serves as the fulcrum in the entire diffusion process selecting between within- and across-region image gradients. Thus, the choice of μ is critical, and a robust locally adaptive strategy is needed to improve the performance of anisotropic diffusion [37]. Inspired by Frangi's vesselness enhancement using Hessian matrix [38], we develop a locally adaptive strategy using the eigen values and eigen vectors of Hessian matrix that will facilitate diffusion along the structure, while constricting diffusion across an edge. The formulation of our locally adaptive diffusion process using Hessian matrix is described in the following.

Let $H(p)$ denote the Hessian at a voxel p; let $\lambda_1(p)$ and $\lambda_2(p)$ be the two eigenvalues of $H(p)$ and $\mathbf{i}_1(p)$ $\mathbf{i}_2(p)$ be corresponding eigenvectors. Let us assume that $|\lambda_1(p)| \leq |\lambda_2(p)|$. In general, when both $|\lambda_1(p)|$ and $|\lambda_2(p)|$ are small, the voxel p belongs to a homogeneous region; in this case, a generous diffusion should be allowed in all directions. When both $|\lambda_1(p)|$ and $|\lambda_2(p)|$ are large, p is an end- or a sharp corner-point, and the diffusion should be restricted in all directions. When $|\lambda_1(p)|$ is small but $|\lambda_2(p)|$ is large, p lies on an edge and a generous diffusion can be allowed along $\mathbf{i}_1(p)$, the eigenvector associated to the smaller eigenvalue, representing the direction along the structure, while the diffusion along $\mathbf{i}_2(p)$ representing the direction across the edge should be prohibited.

Here, we describe the computation of the parameter $\mu_{p,q}$ for the diffusion flow from a voxel p to its neighbor q. Let $\mathbf{i}_{p,q}$ denote the unit vector from p to q. The second-order gradient magnitude $L_{p,q}$ from p to q is derived by projecting the Hessian matrix $H(p)$ along $\mathbf{i}_{p,q}$ as follows:

$$L_{p,q} = |\mathbf{i}_{p,q}^{\mathrm{T}} H(p) \mathbf{i}_{p,q}|. \tag{4}$$

To constrain diffusion in high-gradient regions, we use the the l_1 norm of the Hessian matrix as follows:

$$S_p = |\lambda_1(p)| + |\lambda_2(p)|. \tag{5}$$

Thus, the local diffusion control parameter $\mu_{p,q}$ should account for both directional gradient component $L_{p,q}$ as well as local isotropic gradient component S_p. The parameter $\mu_{p,q}$ is formulated as follows:

$$\mu_{p,q}^2 = \nabla_{\mathrm{avg}} \exp\left(-\frac{L_{p,q} + S_p}{2\lambda_{\mathrm{avg}}}\right), \tag{6}$$

where ∇_{avg} is the average gradient magnitude, and λ_{avg} is the average of $|\lambda_1|$ and $|\lambda_2|$ within the trabecular bone regions. Results of the new Hessian matrix-guided anisotropic diffusion is presented in Fig. 1. As it visually appears, the new

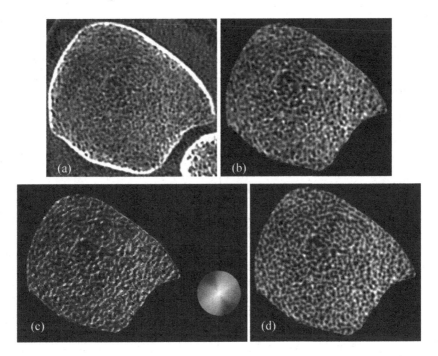

Fig. 1. Results of Hessian matrix-guided anisotropic diffusion. (a) An axial view of an original BMD image and the region of interest (ROI). (b) BMD image within the ROI. (c) Color coded illustration of local flow tensor. Here, hue is the primary flow direction; saturation is the eccentricity of the flow tensor; and intensity is the BMD value. The color disk is shown. (d) Diffusion enhanced BMD image. (Color figure online)

diffusion algorithm has increased the contrast and local separability between trabecular bone microstructures and the marrow spacing. It is encouraging to note that, in bright regions containing thicker trabeculae, the algorithm enhances trabecular separation and preserves small marrow holes. At the same time, in darker regions containing thinner and sparse trabeculae, the algorithm preserves the trabecular structure. Therefore, it may be claimed that the algorithm improves local separability of trabecular structure.

2.2 Grayscale Morphological Reconstruction

The new TB segmentation algorithm models TB microstructures in a CT image as bright structures resembling ridges accompanied with quasi-parallel and relatively darker marrow regions resembling valleys. Our algorithm aims to determine spatial variation of intensity values for marrow valleys and TB ridges. Grayscale morphological algorithms [39,40] are used as the underlying tool to determine local valley and ridge intensity profiles. A multi-scale strategy is incorporated to approximate the valleys and ridges, which are recognized over a large range of scales of morphological operations reducing the subjectivity of selecting a specific scale. Let V and R denote the smooth intensity maps of marrow valleys and TB ridges, respectively, computed using multi-scale morphological algorithms. The segmentation of a TB network can be obtained using a space-varying thresholding scheme that accounts for spatial variation of both the valley intensity V and the trabeculae-marrow contrast $R - V$. The computation of V and R together with the space-varying thresholding algorithms are described in the following.

Let E_s denote a binary structuring element (SE) at a given scale s; the intensity map of the valley V_s is obtained after removing the trabecular structures using morphological opening as follows,

$$V_s^1 = I \circ E_s = (I \ominus E_s) \oplus E_s, \tag{7}$$

where $\circ, \ominus, \oplus$ denote the morphological opening, erosion, dilation operators, respectively. The morphological opening algorithm is repeated for a range of scales of SE. At a relatively smaller scale, the opening algorithm captures small variations in the background intensity, but, suffers the risk of being influenced by thicker trabeculae. At a larger scale, relatively small scale variations in the background intensity are ignored. To overcome these challenges and to reduce subjectivity errors of selecting a specific morphological scale, we used averaged intensity map over a large range of scales. The regional valley intensity V for marrow using the multi-scale strategy is given by

$$V = \sum_s w_s V_s, \tag{8}$$

where w_ss are equal weights such that $\sum_s w_s = 1$. Here, circular SEs with diameters $\{9, 11, \cdots, 21\}$ voxels are used; uniform weights are used for all scales.

The intensity map of the TB ridge R_s at a given scale s is computed using a morphological closing operation as follows:

$$R_s^1 = I \bullet E_s = (I \oplus E_s) \ominus E_s. \tag{9}$$

The regional TB ridge intensity R using the multi-scale strategy is given by

$$R = \sum_s w_s R_s. \tag{10}$$

Finally, the binary segmentation of TB region is obtained by using a space varying thresholding map T that accounts both regional marrow intensity as well as bone-marrow contrast as defined in the following.

$$T = (1 - \alpha)V + \alpha R. \tag{11}$$

The value of α was empirically determined as 0.4, and used for all experiments presented in this paper; See Fig. 3. Results of computing spatial distribution of marrow valley and TB ridge intensity values are presented in Fig. 2. Detection of low-intensity TB microstructures while preserving high-intensity marrow holes is visually apparent in (d) and (e) as compared to the results of global manual thresholding in (f).

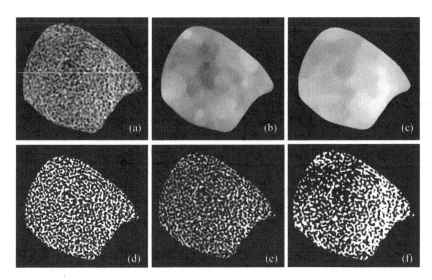

Fig. 2. Illustration of TB segmentation using multi-scale morphological algorithms. (a) Original BMD image. (b) Regional intensity distribution of marrow valleys. (c) Same as (b) but for TB ridges. (d) Binary segmentation mask for TB microstructures. (e) BMD image on the TB mask of (d). (f) Segmented TB mask using manual `global thresholding on the BMD image of (a).

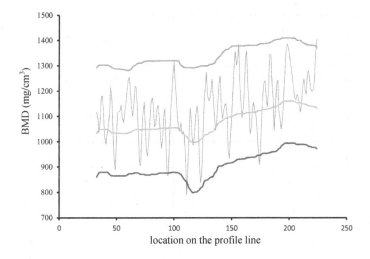

Fig. 3. Illustration of marrow valley, TB ridge, and threshold detection on the intensity profile of the straight line shown in Fig. 2(a). The intensity profile is shown in blue, while the valley, ridge, and threshold profiles are shown in red, green, and gold colors, respectively. (Color figure online)

3   Experiments and Results

The performance of the new TB segmentation method was evaluated in terms of sensitivity, specificity, and accuracy by comparing with the results of regional manual thresholding. To study the implication of the new TB segmentation method on the TB thickness measure, we computed the measures from TB segmentations using the new and other methods and examined their abilities of predicting actual bone strength. These experiments and results are described in the following.

3.1   Segmentation Accuracy

To quantitatively examine the sensitivity, specificity, and accuracy, ten *in vivo* CT images of distal tibia of 10 healthy volunteers (age: 19 to 21 years; 5 males) were used. Images were acquired at 120 kV, 200 effective mAs, 1 s rotation speed, pitch factor: 1.0, and reconstructed at 0.2 mm isotropic voxel. CT HU numbers were converted to BMD using a Gammex calibration phantom. For each image, twenty small region of interests (ROIs) ($30 \times 30 \times 3$ voxels) were randomly selected from the 30% peel ROI at 8% proximal site of the distal tibia covering 8% of the tibial length (see Fig. 1). Thus, we generated a total of 200 small ROIs for the accuracy experiment. As illustrated in Fig. 4, segmentation results on one slice of the TB image and three small ROIs are shown for each segmentation method.

Three mutually blinded readers independently selected the threshold value for TB segmentation in each small ROI, and the true TB segmentation for each ROI was determined using the average threshold value from three readers. For each method, true positive (TP), false negative (FN), true negative (TN) and false positive (FP) were computed. Sensitivity $= \mathrm{TP}/(\mathrm{TP+FN})$, specificity $= \mathrm{TN}/(\mathrm{TN+FP})$, accuracy $= (\mathrm{TP+TN})/(\mathrm{TP+TN+FP+FN})$, and the results are summarized in Table 1.

The proposed method produced high sensitivity (0.93), specificity (0.93), and accuracy (0.93). Otsu's method achieved high sensitivity but significantly low specificity suggesting its bias toward lower threshold values leading to trabecular thickening and small marrow hole filling.

3.2   Bone Strength

To study the implication of the new TB segmentation method on the TB microstructural measures, we computed the thickness of TB network, from TB segmentations using the new and other methods and examined their abilities of predicting actual bone strength. For this purpose, thirteen cadaveric ankle specimens were obtained from 11 body donors (age: 55–91 years). The ankle specimens were removed at the mid-tibia region. Exclusion criteria for this study were evidence of previous fracture of knowledge of bone tumor or bone metastasis. CT images were acquired using the same protocols as described in Sect. 3.1.

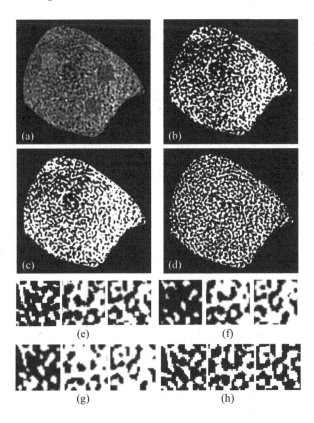

Fig. 4. Comparison of regional TB segmentation using different methods. (a) Original BMD image and three random ROIs. (b, c, d) Segmentation results using global thresholding, Otsu's method, and our method, respectively. (e) Segmentation results using manual thresholding separately selected for each ROI. In the accuracy analysis, this result is considered as the ground truth. (f, g, h) Segmentation results in each ROI using global thresholding, Otsu's method, and our method, respectively.

Table 1. Accuracy results of different methods on ten *in vivo* CT images.

	Sensitivity	Specificity	Accuracy
Global threshold	0.90 ± 0.17	0.85 ± 0.19	0.88 ± 0.07
Otsu's method	0.95 ± 0.09	0.81 ± 0.20	0.85 ± 0.11
Our method	0.93 ± 0.04	0.93 ± 0.06	0.93 ± 0.04

To determine TB strength, mechanical tests were applied on a cylindrical TB core from distal tibia of 8 mm in diameter and 20.9 ± 3.3 mm in length for each subject using an electromechanical materials testing machine. To minimize specimen end effects, strain was measured with a 6 mm gauge length extensometer attached directly to the midsection of the bone. A compressive preload of 10 N

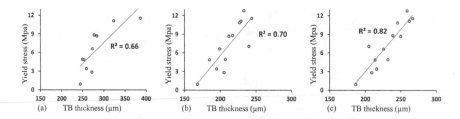

Fig. 5. Ability to predict bone strength using TB thickness measures computed by different methods. Correlation between yield stress and TB thickness computed by Otsu's method (a), manual global thresholding (b), and our method (c).

was applied and strains were set to zero. At a strain rate of $0.005\,\mathrm{s}^{-1}$, each specimen was preconditioned to a low strain with at least ten cycles and then loaded to failure. Yield stress was determined as the intersection of the stress-strain curve and a 0.2% strain offset of the modulus.

The size and location of ROIs for image analysis was chosen as the information recorded during specimen preparation for mechanical tests. First the image was rotated to align the bone axis along the coordinate z-axis using two steps: (1) generation of a cylinder C with its axis lying on the coordinated z-axis and its cross-sectional area equating to the average tibial cross sectional area; and (2) reorientation of the tibial volume to align its axis with C by maximizing the overlap between the tibial volume and the cylinder C. After reorienting the bone image, a ROI cylinder of 8 mm diameter along the coordinate z-axis was generated and its proximal end was manually positioned at the center of the cortical rim using in-plane translation through a graphical user interface. The location of the distal end of the VOI cylinder in the slice direction and its length were determined per the core location and length recorded during specimen preparation; the growth plate was visually located in the CT data of each specimen.

The TB structures were obtained by segmentation using the proposed method, a global thresholding for each subject and Otsu's method. TB thickness was computed using the method described in [41]. The linear correlation between experimental yield stress and computed TB thickness is shown in Fig. 5. As observed from the figure, the TB thickness measure using segmentation with the new method has higher correlation with actual bone strength as compared to those using TB segmentation with other two methods. The prediction error for each method was computed using the linear regression function. The paired t-tests of prediction errors showed that the prediction error of the new method was significantly smaller than the global thresholding ($p = 0.03$) and Otsu's method ($p = 0.01$), respectively.

4   Conclusions

A new TB segmentation algorithm using Hessian-based anisotropic diffusion and multi-scale morphological operations has been developed for *in vivo*

CT imaging. Current implementation of the algorithm on a desktop with a 3.50-GHz Intel(R) Xeon(R) CPU and Linux OS requires approximately 90.5 s for a typical MD-CT image analysis of human TB over a $300 \times 300 \times 300$ voxel array at an isotropic resolution of 150 μm. Experimental results have established high sensitivity, specificity, and accuracy of the new method as compared to manual global thresholding and individual ROI-specific regional thresholding with Otsu's method. The main reason behind the improved performance of the new segmentation method is that the method simultaneously accounts for background variation as well as local contrast. It was observed in experimental results of a cadaveric study that TB segmentation using the new method significantly improves the ability of computed TB thickness measure to predict actual bone strength determined by mechanical testing. These results reinforce the importance of accurate TB segmentation for assessment of TB micro-architecture and prediction of bone strength.

References

1. Riggs, B.L., Khosla, S., Melton, L.J.: Better tools for assessing osteoporosis. J. Clin. Invest. **122**, 4323–4324 (2012)
2. Blume, S.W., Curtis, J.: Medical costs of osteoporosis in the elderly medicare population. Osteoporos. Int. **22**, 1835–1844 (2011)
3. Wehrli, F.W., Saha, P.K., Gomberg, B.R., Song, H.K., Snyder, P.J., Benito, M., Wright, A., Weening, R.: Role of magnetic resonance for assessing structure and function of trabecular bone. Top. Magn. Reson. Imaging **13**, 335–356 (2002)
4. Kleerekoper, M., Villanueva, A., Stanciu, J., Rao, D.S., Parfitt, A.: The role of three-dimensional trabecular microstructure in the pathogenesis of vertebral compression fractures. Calcif. Tissue Int. **37**, 594–597 (1985)
5. Recker, R.R.: Architecture and vertebral fracture. Calcif. Tissue Int. **53**(Suppl. 1), S139–S142 (1993)
6. Parfitt, A.M., Drezner, M.K., Glorieux, F.H., Kanis, J.A., Malluche, H., Meunier, P.J., Ott, S.M., Recker, R.R.: Bone histomorphometry: standardization of nomenclature, symbols, and units: report of the ASBMR histomorphometry nomenclature committee. J. Bone Miner. Res. **2**, 595–610 (1987)
7. Majumdar, S.: Magnetic resonance imaging of trabecular bone structure. Top. Magn. Reson. Imaging **13**, 323–334 (2002)
8. Chang, G., Pakin, S.K., Schweitzer, M.E., Saha, P.K., Regatte, R.R.: Adaptations in trabecular bone microarchitecture in olympic athletes determined by 7T MRI. J. Magn. Reson. Imaging **27**, 1089–1095 (2008)
9. Boutroy, S., Bouxsein, M.L., Munoz, F., Delmas, P.D.: In vivo assessment of trabecular bone microarchitecture by high-resolution peripheral quantitative computed tomography. J. Clin. Endocrinol. Metab. **90**, 6508–6515 (2005)
10. Burghardt, A.J., Pialat, J.B., Kazakia, G.J., Boutroy, S., Engelke, K., Patsch, J.M., Valentinitsch, A., Liu, D., Szabo, E., Bogado, C.E., Zanchetta, M.B., McKay, H.A., Shane, E., Boyd, S.K., Bouxsein, M.L., Chapurlat, R., Khosla, S., Majumdar, S.: Multicenter precision of cortical and trabecular bone quality measures assessed by high-resolution peripheral quantitative computed tomography. J. Bone Miner. Res. **28**, 524–536 (2013)

11. Link, T.M., Vieth, V., Stehling, C., Lotter, A., Beer, A., Newitt, D., Majumdar, S.: High-resolution MRI vs multislice spiral CT: which technique depicts the trabecular bone structure best? Eur. Radiol. **13**, 663–671 (2003)
12. Saha, P.K., Xu, Y., Duan, H., Heiner, A., Liang, G.: Volumetric topological analysis: a novel approach for trabecular bone classification on the continuum between plates and rods. IEEE Trans. Med. Imaging **29**, 1821–1838 (2010)
13. Saha, P.K., Liu, Y., Chen, C., Jin, D., Letuchy, E.M., Xu, Z., Amelon, R.E., Burns, T.L., Torner, J.C., Levy, S.M., et al.: Characterization of trabecular bone plate-rod microarchitecture using multirow detector CT and the tensor scale: algorithms, validation, and applications to pilot human studies. Med. Phys. **42**, 5410–5425 (2015)
14. Saha, P.K., Strand, R., Borgefors, G.: Digital topology and geometry in medical imaging: a survey. IEEE Trans. Med. Imaging **34**, 1940–1964 (2015)
15. Saha, P.K.: Tensor scale: a local morphometric parameter with applications to computer vision and image processing. Comput. Vis. Image Underst. **99**, 384–413 (2005)
16. Hildebrand, T., Rüegsegger, P.: Quantification of bone microarchitecture with the structure model index. Comput. Methods Biomech. Bio Med. Eng. **1**, 15–23 (1997)
17. Rajapakse, C.S., Leonard, M.B., Bhagat, Y.A., Sun, W., Magland, J.F., Wehrli, F.W.: Micro-MR imaging-based computational biomechanics demonstrates reduction in cortical and trabecular bone strength after renal transplantation. Radiology **262**, 912–920 (2012)
18. MacNeil, J.A., Boyd, S.K.: Bone strength at the distal radius can be estimated from high-resolution peripheral quantitative computed tomography and the finite element method. Bone **42**, 1203–1213 (2008)
19. Keaveny, T.M., Morgan, E.F., Niebur, G.L., Yeh, O.C.: Biomechanics of trabecular bone. Annu. Rev. Biomed. Eng. **3**, 307–333 (2001)
20. Link, T.M., Majumdar, S.: Osteoporosis imaging. Radiol. Clin. North Am. **41**, 813–839 (2003)
21. Saha, P.K., Borgefors, G., di Baja, G.S.: A survey on skeletonization algorithms and their applications. Pattern Recogn. Lett. **76**, 3–12 (2016)
22. Saha, P.K., Chaudhuri, B.B.: Detection of 3-D simple points for topology preserving transformations with application to thinning. IEEE Trans. Pattern Anal. Mach. Intell. **16**, 1028–1032 (1994)
23. Saha, P.K., Chaudhuri, B.B.: 3D digital topology under binary transformation with applications. Comput. Vis. Image Underst. **63**, 418–429 (1996)
24. Saha, P.K., Chaudhuri, B.B., Majumder, D.D.: A new shape preserving parallel thinning algorithm for 3D digital images. Pattern Recogn. **30**, 1939–1955 (1997)
25. Saha, P.K., Gomberg, B.R., Wehrli, F.W.: Three-dimensional digital topological characterization of cancellous bone architecture. Int. J. Imaging Syst. Technol. **11**, 81–90 (2000)
26. Stauber, M., Muller, R.: Volumetric spatial decomposition of trabecular bone into rods and plates-a new method for local bone morphometry. Bone **38**, 475–484 (2006)
27. Stauber, M., Rapillard, L., Lenthe, G.H., Zysset, P., Muller, R.: Importance of individual rods and plates in the assessment of bone quality and their contribution to bone stiffness. J. Bone Miner. Res. **21**, 586–595 (2006)
28. Liu, X.S., Sajda, P., Saha, P.K., Wehrli, F.W., Bevill, G., Keaveny, T.M., Guo, X.E.: Complete volumetric decomposition of individual trabecular plates and rods and its morphological correlations with anisotropic elastic moduli in human trabecular bone. J. Bone Miner. Res. **23**, 223–235 (2008)

29. Wehrli, F.W., Gomberg, B.R., Saha, P.K., Song, H.K., Hwang, S.N., Snyder, P.J.: Digital topological analysis of in vivo magnetic resonance microimages of trabecular bone reveals structural implications of osteoporosis. J. Bone Miner. Res. **16**, 1520–1531 (2001)
30. Otsu, N.: A threshold selection methods from grey-level histograms. IEEE Trans. Pattern Anal. Mach. Intell. **9**, 62–66 (1979)
31. Saha, P.K., Udupa, J.: Optimum threshold selection using class uncertainty and region homogeneity. IEEE Trans. Pattern Anal. Mach. Intell. **23**, 689–706 (2001)
32. Waarsing, J.H., Day, J.S., Weinans, H.: An improved segmentation method for in vivo μCT imaging. J. Bone Miner. Res. **19**, 1640–1650 (2004)
33. Burghardt, A.J., Kazakia, G.J., Majumdar, S.: A local adaptive threshold strategy for high resolution peripheral quantitative computed tomography of trabecular bone. Ann. Biomed. Eng. **35**, 1678–1686 (2007)
34. Scherf, H., Tilgner, R.: A new high-resolution computed tomography (CT) segmentation method for trabecular bone architectural analysis. Am. J. Phys. Anthropol. **140**, 39–51 (2009)
35. Tassani, S., Korfiatis, V., Matsopoulos, G.: Influence of segmentation on micro-CT images of trabecular bone. J. Microsc. **256**, 75–81 (2014)
36. Perona, P., Malik, J.: Scale-space and edge detection using anisotropic diffusion. IEEE Trans. Pattern Anal. Mach. Intell. **12**, 629–639 (1990)
37. Tsiotsios, C., Petrou, M.: On the choice of the parameters for anisotropic diffusion in image processing. Pattern Recogn. **46**, 1369–1381 (2013)
38. Frangi, A.F., Niessen, W.J., Vincken, K.L., Viergever, M.A.: Multiscale vessel enhancement filtering. In: Wells, W.M., Colchester, A., Delp, S. (eds.) MICCAI 1998. LNCS, vol. 1496, pp. 130–137. Springer, Heidelberg (1998). doi:10.1007/BFb0056195
39. Serra, J.: Image Analysis and Mathematical Morphology. Academic Press, London (1982)
40. Serra, J.: Image Analysis and Mathematical Morphology. Theoretical Advances, vol. II. Academic Press, New York (1988)
41. Saha, P.K., Wehrli, F.W.: Measurement of trabecular bone thickness in the limited resolution regime of in vivo MRI by fuzzy distance transform. IEEE Trans. Med. Imaging **23**, 53–62 (2004)

Approximation of N-Way Principal Component Analysis for Organ Data

Hayato Itoh[1]([✉]), Atsushi Imiya[2], and Tomoya Sakai[3]

[1] School of Advanced Integration Science, Chiba University, Chiba, Japan
hayato-itoh@graduate.chiba-u.jp
[2] Institute of Management and Information Technologies,
Chiba University, Chiba, Japan
[3] Graduate School of Engineering, Nagasaki University, Nagasaki, Japan

Abstract. We apply multilinear principal component analysis to dimension reduction and classification of human volumetric organ data, which are expressed as multiway array data. For the decomposition of multiway array data, tensor-based principal component analysis extracts multilinear structure of the data. We numerically clarify that low-pass filtering after the multidimensional discrete cosine transform efficiently approximates data dimension reduction procedure based on the tensor principal component analysis.

1 Introduction

For computer assisted diagnosis, inspection and biopsy in precision medicine, abnormality detection based on pattern recognition is a fundamental technique. From cell to human body, medical data used for diagnosis are multiway array data. This paper aims to apply the multiway principal component analysis (mPCA) to volumetric data analysis and processing. The method allows dimension reduction and retrieval of volumetric data preserving both geometric and statistic properties as objects and textures, respectively, of volumetric objects. Furthermore, these PCA-based data dimension reduction for multiway data array allows to extract outline volumetric shapes. Moreover, we develop multiclass classifier based on the multilinearity of volumetric data.

Organs, cells in the organ and microstructure in cells, which are dealt with in biomedical image analysis, are spatial texture. Furthermore, these biological objects possess spatial geometric and topological properties of volumetric structure as three-dimensional objects. Although, local structures of them as volumetric data are computed from geometric and topological properties, texture allows to estimate both local and global statistical properties of these objects. For the data analysis of these volumetric data, the methods simultaneously process geometrical and topological structures and texture properties.

Human bodies and organs in the body are three-dimensional volumetric objects. Since these medical objects are large-size volumetric data and sequence of volumetric data, this paper introduces a pattern recognition method for

© Springer International Publishing AG 2017
C.-S. Chen et al. (Eds.): ACCV 2016 Workshops, Part III, LNCS 10118, pp. 16–31, 2017.
DOI: 10.1007/978-3-319-54526-4_2

dynamics volumetric data using tensor PCA. Toward as abnormality detection of beating-heart sequences, we develop an algorithm to identify individual using a volumetric image from a collection of sequences of a dynamic hearts. Since it is possible to model the beating of a heart in sequence of images as geometrically periodic small deformations to the geometric average-shape of a beating sequence, a recognition method is stable and robust against geometric perturbation of the shape. Mutual subspace method processes stable and robust recognition properties against small geometric perturbations [1]. Therefore, we develop the mutual subspace method for three-way volumetric data.

For numerical computation in pattern recognition, we deal with sampled patterns. In traditional pattern recognition, these sampled patterns are embedded in an appropriate-dimensional Euclidean space as vectors. The other way is to deal with sampled patterns as multiway array data [2–9]. These multiway array data are expressed as tensors to preserve multilinearity of in the original pattern space, since tensors allows to express multiway array data in multilinear forms. Furthermore, for applications of modern pattern recognition techniques such as deep learning [10] and machine learning [11] for big data, we are mathematically and numerically required to evaluate the performance of tensor-based pattern recognition of multilinear data. Importantly, for fast image pattern recognition, a compact representation of these image data is desirable. Therefore, we need dimension-reduction procedure that reduce size of tensor without significant loss of the recognition rate. Tensor expressions fulfill these requirements in applications of pattern recognition of multidimensional array data. By adopting a tensor representation, we can use the spatio-temporal and temporal structure of volumetric data for dimension reduction and classification of the data. Furthermore, the tensor representation of volumetric data does not require the procedure of transforming the data to high-dimensional vectors required for vector-based dimension-reduction and recognition methods.

We develop an approximate closed form for the Tucker-3 tensor PCA, though the Tucker-3 tensor decomposition is achieved by solving variational optimisation iteratively. Our method solves a system of variational optimisation problems derived from the original expression of the Tucker-3 decomposition with the orthogonal constraints for solutions. We also numerically clarify that data reduction by discrete cosine transform (DCT) [12] efficiently approximates dimension reduction based on the tensor PCA, since DCT approximates Karhunen-Loéva (KL) transform [13]. For the validation of these approximation in the recognition of volumetric data, we develop mutual subspace method. Using multilinearity, this method can classify volumetric data without vectorisation of data.

2   Tensor Decomposition

2.1   Tensor Representation for N-way Arrays

We briefly summarise the multilinear projection for N-dimensional arrays from ref. [14]. A Nth-order tensor $\mathcal{X}$ defined in $\mathbb{R}^{I_1 \times I_2 \times \cdots \times I_N}$ is expressed as

$$\mathcal{X} = (x_{i_1, i_2, \ldots, i_N}) \tag{1}$$

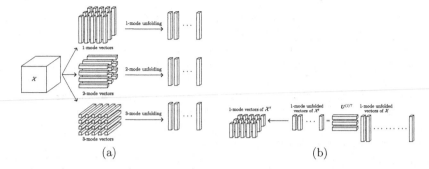

Fig. 1. Unfolding and n-mode projection of a tensor $\mathcal{X}$. (a) Unfoldings for a third-order tensor $\mathcal{X}$. For a tensor $\mathcal{X} \in \mathbb{R}^{4 \times 5 \times 3}$, unfoldings for 1-, 2- and 3-modes give 15 1-mode vectors, 12 2-mode vectors and 20 3-mode vectors, respectively. (b) 1-mode product for $\mathcal{X}$ represented by a linear projection $\mathcal{X}'_{(1)} = U^{(1)\top} \mathcal{X}_{(1)}$. Lower subscription denote 1-mode unfolding of a third-order tensor.

for $x_{i_1, i_2, \ldots, i_N} \in \mathbb{R}$, using N indices i_n. Each subscript n denotes the n-mode of $\mathcal{X}$. For $\mathcal{X}$, the n-mode vectors, $n = 1, 2, \ldots, N$, are defined as the I_n-dimensional vectors obtained from $\mathcal{X}$ by varying this index i_n while fixing all the other indices. The unfolding of $\mathcal{X}$ along the n-mode vectors of $\mathcal{X}$ is defined as

$$\mathcal{X}_{(n)} \in \mathbb{R}^{I_n \times (I_1 \times I_2 \times \ldots I_{n-1} \times I_{n+1} \times \cdots \times I_N)}, \tag{2}$$

where the column vectors of $\mathcal{X}_{(n)}$ are the n-mode vectors of $\mathcal{X}$. Figure 1(a) illustrates unfoldings for a third-order tensor as an example of unfolding of Nth-order tensor. The n-mode product $\mathcal{X} \times_n U$ of a matrix $U \in \mathbb{R}^{J_n \times I_n}$ and a tensor $\mathcal{X}$ is a tensor $\mathcal{G} \in \mathbb{R}^{I_1 \times I_2 \times \cdots \times I_{n-1} \times J_n \times I_{n+1} \times \cdots \times I_N}$, with elements

$$g_{i_1, i_2, \ldots, i_{n-1}, j_n, i_{n+1}, \ldots, i_N} = \sum_{i_n=1}^{I_n} x_{i_1, i_2, \ldots, I_N} u_{j_n, i_n}, \tag{3}$$

by the manner in ref. [15]. A linear projection form of n-mode product in Eq. (3) is given by

$$\mathcal{G}_{(n)} = U \mathcal{X}_{(n)}. \tag{4}$$

Figure 1(b) shows a linear projection form of 1-mode projection for a third-order tensor. For the m- and n-mode product by matrices U and V, we have

$$\mathcal{X} \times_m U \times_n V = \mathcal{X} \times_n V \times_m U \tag{5}$$

since n-mode projections are commutative [15]. We define the inner product of two tensors $\mathcal{X} = (x_{i_1, i_2, \ldots, i_N}), \mathcal{Y} = (y_{i_1, i_2, \ldots, i_N}) \in \mathbb{R}^{I_1 \times I_2 \times \cdots \times I_N}$ by

$$\langle \mathcal{X}, \mathcal{Y} \rangle = \sum_{i_1} \sum_{i_2} \cdots \sum_{i_N} x_{i_1, i_2, \ldots, i_N} y_{i_1, i_2, \ldots, i_N}. \tag{6}$$

Using this inner product, we have the Frobenius norm of a tensor $\mathcal{X}$ by $\|\mathcal{X}\|_F = \sqrt{\langle \mathcal{X}, \mathcal{X} \rangle}$. For the Frobenius norm of a tensor, we have $\|\mathcal{X}\|_F = \|\text{vec }\mathcal{X}\|_2$, where vec and $\|\cdot\|_2$ are the vectorisation operator for a tensor and Euclidean norm for a vector, respectively. For the two tensors $\mathcal{X}_1$ and $\mathcal{X}_2$, we define the distance between them by

$$d(\mathcal{X}_1, \mathcal{X}_2) = \|\mathcal{X}_1 - \mathcal{X}_2\|_F. \tag{7}$$

Although this definition is a tensor-based measure, this distance is equivalent to the Euclidean distance between the vectorised tensors $\mathcal{X}_1$ and $\mathcal{X}_2$.

For a tensor, a multilinear projection maps the input tensor data from one space to another space. We have three basic multilinear projections, that is, the vector-to-vector projection (VVP), tensor-to-vector projection (TVP) and tensor-to-tensor projection (TTP). The VVP is a linear projection from a vector to another vector. To use the VVP for tensors, we need to reshape tensors into vectors before the projection. The TVP, which is also referred to as the rank-one projection [16–18], consists of elementary multilinear projections (EMPs). An EMP projects a tensor to a scalar. Using d EMPs, the TVP obtains a d-dimensional vector projected from a tensor. The TTP projects a tensor to another tensor of the same order. In this paper, we focus on methods of finding the optimal projection for the TTP.

As the tensor $\mathcal{X}$ is in the tensor space $\mathbb{R}^{I_1} \otimes \mathbb{R}^{I_2} \otimes \cdots \otimes \mathbb{R}^{I_N}$, the tensor space can be interpreted as the Kronecker product of N vector spaces $\mathbb{R}^{I_1}, \mathbb{R}^{I_2}, \ldots, \mathbb{R}^{I_N}$. To project $\mathcal{X} \in \mathbb{R}^{I_1} \otimes \mathbb{R}^{I_2} \otimes \cdots \otimes \mathbb{R}^{I_N}$ to another tensor $\mathcal{Y}$ in a lower-dimensional tensor space $\mathbb{R}^{P_1} \otimes \mathbb{R}^{P_2} \otimes \cdots \otimes \mathbb{R}^{P_N}$, where $P_n \leq I_n$ for $n = 1, 2, \ldots, N$, we need N projection matrices $\{\boldsymbol{U}^{(n)} \in \mathbb{R}^{I_n \times P_n}\}_{n=1}^N$. Using the N projection matrices, the TTP is given by

$$\mathcal{Y} = \mathcal{X} \times_1 \boldsymbol{U}^{(1)\top} \times_2 \boldsymbol{U}^{(2)\top} \cdots \times_N \boldsymbol{U}^{(N)\top}. \tag{8}$$

This projection is established in N steps, where at the nth step, each n-mode vector is projected to a P_n-dimensional space by $\boldsymbol{U}^{(n)}$. We call this operation the orthogonal projection of $\mathcal{X}$ to $\mathcal{Y}$.

2.2   Principal Component Analysis for N-way Arrays

A Nth-order tensor $\mathcal{X} \in \mathbb{R}^{I_1 \times I_2 \times \cdots \times I_N}$, which is the array $\boldsymbol{X} \in \mathbb{R}^{I_1 \times I_2 \times \cdots \times I_N}$, is denoted as a set of indices $(i_1, i_2, \ldots, i_N)$. Here we summarise the higher-order singular value decomposition (HOSVD) for Nth-order tensors since N-way principal component is numerically computed by HOSVD. The HOSVD is the Tucker-3 decomposition with orthogonal constraints. For a collection of tensors $\{\mathcal{X}_i\}_{i=1}^M \in \mathbb{R}^{I_1 \times I_2 \times \cdots \times I_N}$ satisfying the zero expectation condition $\mathrm{E}(\mathcal{X}_i) = 0$, we compute the

$$\hat{\mathcal{X}}_i = \mathcal{X}_i \times_1 \boldsymbol{U}^{(1)\top} \times_2 \boldsymbol{U}^{(2)\top} \cdots \times_N \boldsymbol{U}^{(N)\top}, \tag{9}$$

where $\boldsymbol{U}^{(j)} = [\boldsymbol{u}_1^{(j)}, \ldots, \boldsymbol{u}_{I_j}^{(j)}]$, that minimises the criterion

$$J_- = \mathrm{E}\left(\|\mathcal{X}_i - \hat{\mathcal{X}}_i \times_1 \boldsymbol{U}^{(1)} \times_2 \boldsymbol{U}^{(2)} \cdots \times_N \boldsymbol{U}^{(N)}\|_{\mathrm{F}}^2\right) \tag{10}$$

and maximises the criterion

$$J_+ = \mathrm{E}\left(\|\hat{\mathcal{X}}_i\|_{\mathrm{F}}^2\right), \tag{11}$$

with respect to the conditions

$$\boldsymbol{U}^{(j)\top}\boldsymbol{U}^{(j)} = \boldsymbol{I}_j, \tag{12}$$

where $\boldsymbol{I}_j$, $j = 1, 2, \ldots, N$ are the identity matrices in $\mathbb{R}^{I_j \times I_j}$. By fixing $\{\boldsymbol{U}^{(j)}\}_{j=1}^N$ except $\boldsymbol{U}^{(j')}, j' \in \{1, 2, \ldots, N\}$, we have

$$J_j = \mathrm{E}\left(\|\boldsymbol{U}^{(j)\top}\mathcal{X}_{i,(j)}\mathcal{X}_{i,(j)}^\top\boldsymbol{U}^{(j)}\|_{\mathrm{F}}^2\right), \tag{13}$$

where $\mathcal{X}_{i,(j)}$, $j = 1, 2, \ldots, N$, are the j-mode unfolded tensors of $\mathcal{X}_i$.

Eigendecomposition problems are derived by computing the extremals of

$$E_j = J_j + tr((\boldsymbol{I}_j - \boldsymbol{U}^{(j)\top}\boldsymbol{U}^{(j)})\boldsymbol{\Sigma}^{(j)}), \ j = 1, 2, \ldots, N. \tag{14}$$

For matrices $\boldsymbol{M}^{(j)} = \frac{1}{N}\sum_{i=1}^N \mathcal{X}_{i,(j)}\mathcal{X}_{i,(j)}^\top$, $j = 1, 2, \ldots, N$, the optimisation of J_- and J_+ derives the eigenvalue decomposition

$$\boldsymbol{M}^{(j)}\boldsymbol{U}^{(j)} = \boldsymbol{U}^{(j)}\boldsymbol{\Sigma}^{(j)}, \tag{15}$$

where $\boldsymbol{\Sigma}^{(j)} \in \mathbb{R}^{I_j \times I_j}$, $j = 1, 2, \ldots, N$, are diagonal matrices satisfying the relationships $\sigma_k^{(j)} = \sigma_k^{(j')}$, $k \in \{1, 2, \ldots, K\}$ for

$$\boldsymbol{\Sigma}^{(j)} = \mathrm{diag}(\lambda_1^{(j)}, \lambda_2^{(j)} \cdots, \lambda_K^{(j)}, 0 \cdots, 0). \tag{16}$$

The optimisation of each J_j derives the eigendecomposition problems in Eq. (15). However, for the optimisation of $\{J_j\}_{j=1}^N$, there is no closed-form solution to this maximisation problem [19,20]. Algorithm 1 is the iterative procedure of the N-way PCA [2]. This algorithm is one of alternating-least-square (ALS) algorithms for tensors. In Algorithm 1 of K iteration, time complexity is $\mathcal{O}(KI_k^2)$, where $I_j \leq I_k$, $j \neq k$, due to the eigendecomposition problem. For the N-way PCA for volumetric data represented in $\mathbb{R}^{n \times n \times n}$, we then have time complexity $\mathcal{O}(Kn^2)$.

For Algorithm 1, we have the following property.

Property 1. *The N-way PCA without iteration in Algorithm 1 is equivalent to the HOSVD for Nth-order tensors if dimensions of a projected tensor are coincident to ones of each mode of an original tensor.*

For Nth-order tensors, there are $N!$ combinations in selecting the order of modes in a tensor-to-tensor projection for Algorithm 1. For the selection of combinations for Algorithm 1, we have the following property [2].

Algorithm 1. Iterative method for N-way PCA (ALS algorithm)

Input: A set of tensors $\{\mathcal{X}_i\}_{i=1}^M$. Dimension of projected tensors $\{k_j\}_{j=1}^N$
     A maximum number of iteration K. A sufficiently small number η
Output: A set of projection matrices $\{U^{(j)}\}_{j=1}^N$
1: Compute the eigendecomposition of a covariant matrix
  $M^{(j)} = \frac{1}{M}\sum_{i=1}^M \mathcal{X}_{i,(j)}\mathcal{X}_{i,(j)}^\top$, where $\mathcal{X}_{i,(j)}$ is an j-mode unfolded $\mathcal{X}_i$, for $j = 1, 2, \ldots, N$
2: Construct projection matrices by selecting eigenvectors
  corresponding to the k_j largest eigenvalues for $j = 1, 2, \ldots, N$
3: Compute $\Psi_0 = \sum_{i=1}^M \|\mathcal{X}_i \times_1 U^{(1)\top} \times_2 U^{(2)\top} \cdots \times_N U^{(N)\top}\|_F$
4: Iteratively Compute the following procedure
  for $k = 1, 2, \ldots, K$
    for $j = 1, 2, \ldots, N$
      Update $U^{(j)}$ by decomposing matrix $\sum_{i=1}^M W_{i,(j)}^{(-j)} W_{i,(j)}^{(-j)\top}$, where $W_{i,(j)}^{(-j)}$ is an
      unfolding of $\mathcal{X}_i \times_1 U^1 \cdots \times_{j-1} U^{(j-1)} \times_{j+1} U^{(j+1)} \cdots \times_N U^{(N)}$ for a mode j
    end
    Compute $\Psi_k = \sum_{i=1}^N \|\mathcal{X}_i \times_1 U^{(1)\top} \times_2 U^{(2)\top} \cdots \times_N U^{(N)\top}\|_F$
    if $|\Psi_k - \Psi_{k-1}| < \eta$
      break
  end

Property 2. *For Nth-order tensors, the selection of order of modes does not effect to the results of a tensor-to-tensor projection since n-mode projection is cumulative.*

From these two properties, we adopt Algorithm 1 [2] to solve the optimisation of $\{J_j\}_{j=1}^N$. For a set of orthonormal vectors $\{e_k\}_{k=1}^K$, $e_i^\top e_j = \delta_{ij}$, we set orthogonal projection matrices $P^{(j)} = \sum_{k=1}^{k_j} e_k e_k^\top$ for $j = 1, 2, 3$. Using these $\{P^{(j)}\}_{j=1}^N$, the low-rank tensor approximation [20] is achieved by

$$\mathcal{Y} = \mathcal{X} \times_1 (P^{(1)}U^{(1)}) \times_2 (P^{(2)}U^{(2)}) \cdots \times_N (P^{(N)}U^{(N)}), \qquad (17)$$

where $P^{(j)}$ selects k_j bases of orthogonal matrices $U^{(j)}$. The low-rank approximation using Eq. (17) is used for compression in the N-way PCA.

For the HOSVD for Nth-order tensors, we have the following theorem.

Theorem 1. *The HOSVD method is equivalent to the vector PCA method in the compression of Nth-order tensors.*

(Proof). The equation

$$\mathcal{X} \times_1 (P^{(1)}U^{(1)})^\top \times_2 (P^{(2)}U^{(2)})^\top \cdots \times_N (P^{(N)}U^{(N)})^\top = \mathcal{Y} \qquad (18)$$

is equivalent to

$$(P^{(N)}U^{(N)} \otimes \ldots P^{(2)}U^{(2)} \otimes P^{(1)}U^{(1)})^\top \text{vec}\mathcal{X} = \text{vec}\mathcal{Y}. \qquad (19)$$

(Q.E.D.)

This theorem implies that N-dimensional discrete cosine transform (NDDCT) is an acceptable approximation of the HOSVD for Nth-order tensors since this is the analogy of the approximation of the PCA of two-dimensional images by the 2DDCT [13,21].

Furthermore, we have the following theorem.

Theorem 2. *The compression of Nth-order tensors computed by the HOSVD is equivalent to the compression computed by the TPCA.*

(Proof). The projection that selects $K = k_1 k_2 \ldots k_N$ bases of the tensor space spanned by $u_{i_1}^{(1)} \circ u_{i_2}^{(2)} \circ u_{i_3}^{(3)}$, $i_j = 1, 2, \ldots, k_j$ for $j = 1, 2, \ldots, N$, is

$$
\begin{aligned}
(\boldsymbol{P}^{(N)} \boldsymbol{U}^{(N)} \otimes \ldots \boldsymbol{P}^{(2)} \boldsymbol{U}^{(2)} \otimes \boldsymbol{P}^{(1)} \boldsymbol{U}^{(1)}) \\
= (\boldsymbol{P}^{(N)} \otimes \ldots \boldsymbol{P}^{(2)} \otimes \boldsymbol{P}^{(1)})(\boldsymbol{U}^{(N)} \otimes \ldots \boldsymbol{U}^{(2)} \otimes \boldsymbol{U}^{(1)}) = \boldsymbol{PW},
\end{aligned}
\tag{20}
$$

where $\boldsymbol{W}$ and $\boldsymbol{P}$ are an orthogonal matrix and the orthogonal projection matrix, respectively. Therefore, HOSVD is equivalent to TPCA for third-order tensors.

2.3 N-Dimensional Discrete Cosine Transform

For a Nth-order tensor $\mathcal{X} \in \mathbb{R}^{I_1 \times I_2 \times \cdots \times I_N}$, we have DCT matrices

$$
D^{(k)} = ((d_{ij}^{(k)})), \quad d_{ij}^{(k)} = \cos\left(\frac{\pi}{I_k}\left((j-1) + \frac{1}{2}\right)(i-1)\right).
\tag{21}
$$

for mode k, $i = 1, 2, \ldots, I_k$, $j = 1, 2, \ldots, N^{(-k)}$, where $N^{(-k)} = I_1 \times I_2 \times \cdots \times I_{k-1} \times I_{k+1} \cdots \times I_N$. For each mode k, we set projection matrices $\boldsymbol{P}^{(k)} \in \mathbb{R}^{l_k \times I_k}$. These projection matrices select l_k bases of DCT for mode k. Using DCT matrices and projection matrices, we define compression by NDDCT for Nth-order tensors as

$$
\mathcal{Y} = \mathcal{X} \times_1 (\boldsymbol{P}^{(1)} \boldsymbol{D}^{(1)})^\top \times_2 (\boldsymbol{P}^{(2)} \boldsymbol{D}^{(2)})^\top \cdots \times_N (\boldsymbol{P}^{(N)} \boldsymbol{D}^{(N)})^\top.
\tag{22}
$$

This NDDCT is an acceptable approximation for the NDTPCA.

In our application for volumetric data, an $n \times n \times n$ digital array is directly compressed by the NDDCT-II with order $\mathcal{O}(n^3)$. If we apply the fast Fourier transform to the computation of the NDDCT-II, the computational complexity is $\mathcal{O}(n \log n)$.

3   Classification of Tensor Data

Tensor Subspace of Categories. Setting $\{\boldsymbol{U}_k^{(j)}\}_{j=1}^N$ to be orthogonal matrices of a tensor projection for Nth-order tensors, we have a tensor subspace spanned by $\{\boldsymbol{U}_k^{(j)}\}_{j=1}^N$ for kth category. Therefore, we can define a tensor subspace of a category by

$$
\mathcal{C}_k = \{\mathcal{X} \mid \mathcal{X} \times_1 \boldsymbol{U}_k^{(1)\top} \times_2 \boldsymbol{U}_k^{(2)\top} \cdots \times_n \boldsymbol{U}_k^{(N)} = \mathcal{X}\}.
\tag{23}
$$

Since a pattern represented by tensors contains perturbation, we define kth category by

$$\mathcal{C}_k(\delta) = \{\mathcal{X} \mid \|\mathcal{X} \times_1 U^{(1)\top} \times_2 U^{(2)\top} \cdots \times_n U^{(N)} - \mathcal{X}\|_F \ll \delta\}, \quad (24)$$

where a positive constant δ is the bound for a small perturbation to a pattern. Therefore, by defining similarity and dissimilarity between a tensor subspace and query, we can construct tensor-subspace-based classifiers that are robust and stable against small perturbations to patterns.

Tensor Subspace Method. As an extension of the subspace method [22,23] for N-way data, we introduce a new linear tensor subspace method for Nth-order tensors. This method is a N-dimensional version of the 2DTSM [24].

For a Nth-order tensor $\mathcal{X}$, we set $U^{(j)}$, $j = 1, 2, \ldots, N$, to be projection matrices of the tensor-to-tensor projection of $\mathcal{X}$ to $\mathcal{Y}$. For a collection of normalised tensors $\{\mathcal{X}_i\}_{i=1}^M$, such that $\mathcal{X}_i \in \mathbb{R}^{I_1 \times I_2 \times \cdots \times I_N}$, $\|\mathcal{X}_i\|_F = 1$ and $E(\mathcal{X}_i) = 0$, the solutions of

$$\{U^{(j)}\}_{j=1}^N = \arg \max E\left(\|\mathcal{X} \times_1 U^{(1)\top} \times_2 U^{(2)\top} \cdots \times_N U^{(N)\top}\|_F / \|\mathcal{X}_i\|_F\right) \quad (25)$$

with respect to $U^{(j)\top} U^{(j)} = I$ for $j = 1, 2, \ldots, N$ define a multilinear subspace that approximates $\{\mathcal{X}_i\}_{i=1}^M$. Therefore, using projection matrices $\{U_k^{(j)}\}_{j=1}^N$ obtained as the solutions of Eq. (25) for kth category, if a query tensor $\mathcal{G}$ satisfies the condition

$$\arg\left(\max_l \|\mathcal{G} \times_1 U_l^{(1)\top} \times_2 U_l^{(2)\top} \cdots \times_N U_l^{(N)\top}\|_F / \|\mathcal{G}\|_F\right) = \{U_k^{(j)}\}_{j=1}^N, \quad (26)$$

we conclude that $\mathcal{G} \in \mathcal{C}_k$, $k, l = 1, 2, \ldots, N_{\mathcal{C}}$, where $\mathcal{C}_k$ and $N_{\mathcal{C}}$ are the tensor subspace of kth category and the number of categories, respectively.

Tensor Subspace of Queries. We have a collection of query tensors $\{\mathcal{G}_{i'}\}_{i'=1}^{M'}$ normalised by $\mathcal{G}_{i'}/\|\mathcal{G}_{i'}\|_F$. We assume that these queries belong to the same category. Then, using multiway PCA for a collection of queries $\{\mathcal{G}_{i'}\}_{i'=1}^{M'}$, we obtain orthogonal matrices $\{V^{(j)}\}_{j=1}^N$, which are given by Eq. (25) for each mode. The column vectors of orthogonal matrices $\{V^{(j)}\}_{j=1}^N$ are eigenvectors of a set of unfolded tensors for each mode. We note that even if we have only one query $\mathcal{G}$ we can obtain eigenvectors of j-mode by the multiway PCA, since j-mode unfolding gives a set of column vectors of mode-j as shown in Fig. 1. That is, for a query $\mathcal{G}$, we have orthogonal matrices by the decomposition [19]

$$\mathcal{G} = \mathcal{A} \times_1 V^{(1)} \times_2 V^{(2)} \cdots \times_N V^{(N)}. \quad (27)$$

Mutual Tensor Subspace Method. As the extension of mutual subspace method for vector data [1], we define a classifier for two tensor subspaces. For each of N_C categories of tensor data, we set a collection of normalised Nth-order tensors $\{\mathcal{X}_i\}_{i=1}^M$, such that $\mathcal{X}_i \in \mathbb{R}^{I_1 \times I_2 \times \cdots \times I_N}$, $\|\mathcal{X}_i\|_F = 1$ and $E(\mathcal{X}_i) = 0$. For the kth category in N_C categories, we have an orthogonal matrices $\{U_k^{(j)}\}_{j=1}^N$, which satisfy Eq. (25). The orthogonal matrices for kth category span a tensor subspaces $\mathcal{C}_k$ of the category.

We have a collection of query tensors $\{\mathcal{G}_{i'}\}_{i'=1}^{M'}$ normalised by $\mathcal{G}_{i'}/\|\mathcal{G}_{i'}\|_F$. We assume that these queries belong to the same category. Then, using multiway PCA for a collection of queries $\{\mathcal{G}_{i'}\}_{i'=1}^{M'}$, we obtain orthogonal matrices $\{V^{(j)}\}_{j=1}^N$. The obtained eigenvectors of all modes span a tensor subspace $\mathcal{C}_q$ for queries. For the classification of queries, we measure the dissimilarity between a category subspace $\mathcal{C}_k$ and a query subspace $\mathcal{C}_q$. Since $\mathcal{C}_k$ and $\mathcal{C}_q$ represent patterns with perturbations, we can robustly recognise queries against the pattern perturbations by measuring the dissimilarity between two tensor subspace.

Using orthogonal matrices $\{U_k^{(j)}\}_{j=1}^N$ for kth category, we have a projected tensor in a category subspace $\mathcal{C}_k$ by

$$\mathcal{A}_{i'} = \mathcal{G}_{i'} \times_1 U_k^{(1)\top} \times_2 U_k^{(2)\top} \cdots \times_N U_k^{(N)\top}. \tag{28}$$

Furthermore, using orthogonal matrices $\{V^{(j)}\}_{j=1}^N$, we have a projected tensor in a query subspace $\mathcal{C}_q$ by

$$\mathcal{B}_{i'} = \mathcal{G}_{i'} \times_1 V^{(1)\top} \times_2 V^{(2)\top} \cdots \times_N V^{(N)\top}. \tag{29}$$

For a tensor subspaces $\mathcal{C}_k$ and $\mathcal{C}_q$, we define the dissimilarity of subspaces $d(\mathcal{C}_k, \mathcal{C}_q)$ by

$$E\left(\|\mathcal{A}_{i'} \times_1 PU_k^{(1)} \cdots \times_N PU_k^{(N)} - \mathcal{B}_{i'} \times_1 PV^{(1)} \cdots \times_N PV^{(N)}\|_F^2\right), \tag{30}$$

where a projection matrix P selects bases for each mode of tensors. Therefore, using the dissimilarity given by Eq. (30), if queries $\{\mathcal{G}_{i'}\}_{l=1}^{M'}$ satisfy the condition

$$\arg\left(\min_l d(\mathcal{C}_l, \mathcal{C}_q)\right) = \mathcal{C}_k, \tag{31}$$

we conclude that $\{\mathcal{G}_{i'}\}_{i'=1}^{M'} \in \mathcal{C}_k(\delta)$ for $k, l = 1, 2, \ldots, N_C$.

4   Numerical Examples

We comparatively evaluate performance of the TPCA and the 3D-DCT in dimension reduction and classification of volumetric data. For this evaluation, we adopt the following steps.

1. Extract the volumetric data of the left ventricles from cardiac MRI dataset.
2. Reduce the dimension of the extracted data by using four methods.
3. Divide the dimension-reduced data to taring and test data.

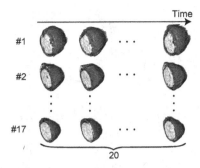

Fig. 2. Illustration of extracted cardiac MRI dataset. These sequences of volumetric data are extracted from cardiac MRI dataset with landmarks of endocardium of left ventricles [25]. As shown in Table 1, we have 17 sequences of volumetric data of left ventricle for 17 patients. Each sequence of volumetric data represents one cardiac beat by 20 frames. Every sequence starts with maximally expanded state. Red and white parts of volume rendering of the data represent muscle and inner space of left ventricles. We set the center of the first sagittal slice of each volume data to the center of the slice. (Color figure online)

4. Construct tensor subspace of each categories by applying the TPCA to the training data.
5. Classify a query (or query subspace) of a category in test data by using tensor-based classifiers.

We extract sequences of volumetric data of left ventricle from cardiac MRI dataset [25], since we need to validate classification of normal organs before abnormality detection. For the extraction, we use the landmarks of endocardium of left ventricle. These landmarks are manually given and provided as the part of the dataset. Table 1 summarises the number and the size of the extracted volumetric data at all phases. Figure 2 illustrates the extracted sequences of volumetric data for 17 patients. Since a beating heart deforms its volumetric shape, we obtain third-order tensors representing shape of heart with geometrical perturbation from a sequence.

For the dimension reduction of volumetric data, we use the TPCA and 3D-DCT. For the practical computation of the TPCA, we use the HOSVD and MPCA. In the MPCA, if we set the number of bases to the size of the original

Table 1. Sizes and number of volumetric data of left ventricles. ♯category represents the number of individuals. ♯data/category represents the number of frames in one sequence of left ventricles. The data size is the original size of the volumetric data. The reduced data size is the size of the volumetric data after reduction. We set $d \in \{8, 16, 32\}$.

	♯category	♯data/category	Data size [voxel]	Reduced data size [voxel]
Volumetric data	17	20	$81 \times 81 \times 63$	$d \times d \times d$

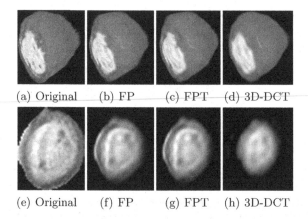

(a) Original   (b) FP   (c) FPT   (d) 3D-DCT

(e) Original   (f) FP   (g) FPT   (h) 3D-DCT

Fig. 3. Shape and inner texture of reconstructed volume data of left ventricle from compressed data. Upper and lower rows show volume rendering and sagittal slice of the volumetric data, respectively. In (a)–(d), red and white parts depict the muscle of heart and inner of heart, respectively, for original and approximation by the FP, the FPT and the 3D-DCT. In these approximation, the data are reduced to the size $16 \times 16 \times 16$. (Color figure online)

tensors in Algorithm 1, we call the method full projection (FP). If we set the number of bases to fewer than the size of the original tensors in Algorithm 1, we call the method the full-projection truncation (FPT). For the dimension reduction by the HOSVD, FP and FPT, we apply these methods to all the extracted volumetric data in all categories. For the evaluation the robustness and stabilizes of methods with respect to the sizes of the data, we set the sizes of the dimension-reduced data to $8 \times 8 \times 8$, $16 \times 16 \times 16$ and $32 \times 32 \times 32$.

Figure 3 illustrates the comparison between original and dimension-reduced data by the three-methods. In Fig. 3(a)–(d), volume rendering of the original and reconstructed volume data are presented. For the data reduced by the FP and FPT, the shapes of volumetric data reconstructed from the compressed data are almost the same in their appearances. The reconstructed data from the data reduced by the 3DDCT is the closest shape to the shape of original volumetric data. In Figs. 3(e)–(h), the differences of appearances between the sagittal slices of reconstructed data and original shape are compared. Compared to the original data shown in Fig. 3(a), the 3D-DCT gives blurred inner texture as shown in Fig. 3(h). As shown in Figs. 3(f) and (g), the dimension reduction by the FP and FPT extract outline shapes of ventricle without inner texture. Figure 4 illustrates reconstructed data from principal components of dimension-reduced volume data. This result show that the principal components of the dimension-reduced volume data are almost the same as shown in Table 2.

In the dimension-reduced data, each sequences consist from 20 frames. We use odd and even frames in dimension reduced data as training and test data, respectively. Applying the FP to training data of each category, we construct

Table 2. Reconstruction error of volumetric data. The reconstruction error is given by distance between tensors of the original and reconstructed volumetric data.

	FP	FPT	3D-DCT
Reconstruction error	11.4×10^3	11.4×10^3	8.23×10^3

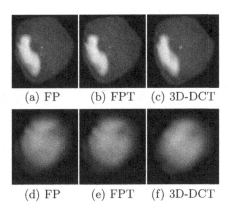

(a) FP   (b) FPT   (c) 3D-DCT

(d) FP   (e) FPT   (f) 3D-DCT

Fig. 4. Extracted principal components of dimension-reduced volume data. For the data dimension reduced by the FP, FPT and 3D-DCT, we apply the FP. Using the extracted principal component, we reconstruct volumetric data. For the extraction, we select the 20 principal eigenvectors of ones of three modes.

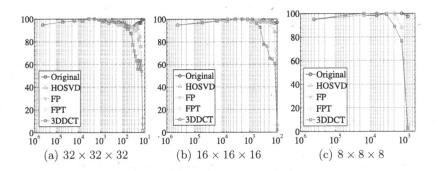

(a) $32 \times 32 \times 32$ (b) $16 \times 16 \times 16$ (c) $8 \times 8 \times 8$

Fig. 5. Recognition rates of the left ventricles for original and compressed tensors. We use tensor subspace method as classifier. The data are reduced to $32 \times 32 \times 32$, $16 \times 16 \times 16$ and $8 \times 8 \times 8$. The HOSVD, FP, FPT and 3D-DCT are used for the reduction. Vertical and horizontal axes represent recognition rate and compression ratio, respectively. For the original size $D = 81 \times 81 \times 63$ and reduced size $K = k \times k' \times k'$, the compression ratio is given by D/K.

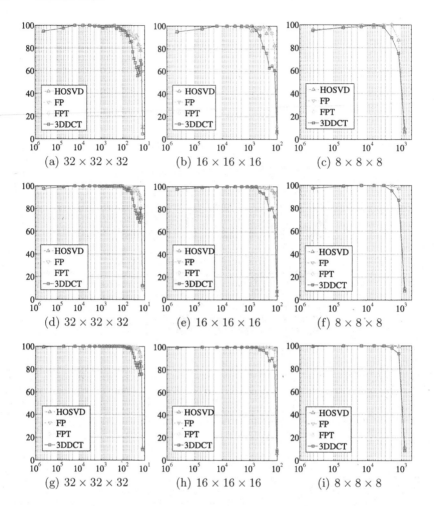

Fig. 6. Recognition rates of left ventricles for compressed tensors. We adopt the reduces sizes of $32 \times 32 \times 32$, $16 \times 16 \times 16$ and $8 \times 8 \times 8$. For compression, we use the HOSVD, FP, FPT and 3D-DCT. In the mutual tensor subspace method, input is a query subspace. The query subspace is spanned by a few queries. To construct a query subspace, we use one, two and three queries. Top, middle and bottom row show recognition rates for the case of one, two and three queries, respectively. Vertical and horizontal axes represent recognition rate and compression ratio, respectively. For the original size $D = 81 \times 81 \times 63$ and reduces size $K = k \times k' \times k'$, the compression ratio is given by D/K.

tensor subspace of 17 categories for the TSM and MTSM. The TSM and MTSM are robust classification methods against geometrical perturbations. Therefore, we use only odd frame for construction of tensor subspaces of categories to evaluate these robustness. If the TSM and MTSM classify a category of even frame, we conclude that these classifiers are robust to the small geometrical change between frames.

The recognition rate is defined as the successful classification ration of individuals in 1000 classifications. We use the TSM and MTSM as classifiers. In the selection of a query for the TSM, we randomly select one of 17 individuals and randomly select one of test data of the individual. From a collection of input data of the heart sequence of a patient, we constructed the subspace of queries for the MTSM. After randomly select an individual, from one to three queries are randomly selected from test data of the selected individual. Applying the FP to the selected queries, we construct a query tensor subspace for the MTSM. Figures 5 and 6 show recognition rate for left ventricles by the TSM and the MTSM, respectively.

In Fig. 5, the profiles of recognition curves for the HOSVD, FP, FPT and 3D-DCT are almost the same in the higher compression ratio than 10^3. Furthermore, for the higher compression ratio than 10^3, the dimension reduced data by four methods derive almost same recognition rates. These recognition rates are the same recognition rate of the tensors of original size. Moreover, the TSM with major five eigenvenctors in each modes processes accurate recognition. Figure 6(a)–(c) show that the MTSM for the query subspace spanned by one query. The results show that the recognition properties are almost the same for data with $8 \times 8 \times 8$, $16 \times 16 \times 16$ and $32 \times 32 \times 32$. The results in Figs. 6(d)–(i) show that the MSM achive more robust recognition against small geometrical perturbations by using a query subspace than the TSM, since a query subspace spanned by a few queries with geometrical perturbations.

These numerical examples show that the 3DDCT accurately approximates performance of the TPCA. Furthermore, recognition of three-way data by the TSM and MTSM is accurate and robust for volumetric data contain geometrical perturbations as temporal deformation.

5   Conclusions

We developed an approximate closed form for the Tucker decomposition for Nth-order tensors. Our method solves a system of variational optimisation problems derived from the original expression of the Tucker decomposition with the orthogonal constraints for solutions. Furthermore, we developed tensor-based multilinear classifiers, the TSM and MTSM, for order-N tensors. In the numerical examples, we evaluated the performance of the dimension reduction by the HOSVD, FP, FPT and 3D-DCT for the recognition of the individual of left ventricles. In the reduction, the FP and FPT extract the outline shapes of left ventricles. The results of the evaluations showed that the dimension reduction by the HOSVD gives the same performance by FP and FPT without iterative computation procedures. The results also numerically clarified that data compression by the DCT efficiently approximates data compression procedure based on the tensor PCA. In the evaluations, the TSM and MTSM accurately recognised individual of left ventricles though cardiac MRI include geometrical perturbations. For the MTSM, using a query subspace spanned by more than one frame, the MTSM achieved more robust and stable recognition to geometrical perturbations than the TSM.

This research was supported by "Multidisciplinary Computational Anatomy and Its Application to Highly Intelligent Diagnosis and Therapy" project funded by a Grant-in-Aid for Scientific Research on Innovative Areas from MEXT, Japan, and by Grants-in-Aid for Scientific Research funded by the Japan Society for the Promotion of Science.

References

1. Maeda, K.: From the subspace methods to the mutual subspace method. In: Cipolla, R., Battiato, S., Farinella, G.M. (eds.) Computer Vision, vol. 285, pp. 135–156. Springer, Heidelberg (2010). doi:10.1007/978-3-642-12848-6_5
2. Lu, H., Plataniotis, K., Venetsanopoulos, A.: MPCA: multilinear principal component analysis of tensor objects. IEEE Trans. Neural Netw. **19**, 18–39 (2008)
3. Jiang, B., Ma, S., Zhang, S.: Tensor principal component analysis via convex optimization. Math. Program. **150**, 423–457 (2014)
4. Lu, H., Plataniotis, K., Venetsanopoulos, A.: Uncorrelated multilinear principal component analysis for unsupervised multilinear subspace learning. IEEE Trans. Neural Netw. **20**, 1820–1836 (2009)
5. Shen, H., Huang, J.Z.: Sparse principal component analysis via regularized low rank matrix approximation. J. Multivar. Anal. **99**, 1015–1034 (2008)
6. Lai, Z., Xu, Y., Chen, Q., Yang, J., Zhang, D.: Multilinear sparse principal component analysis. IEEE Trans. Neural Netw. Learn. Syst. **25**, 1942–1950 (2014)
7. Panagakis, Y., Kotropoulos, C., Arce, G.R.: Non-negative multilinear principal component analysis of auditory temporal modulations for music genre classification. IEEE Trans. Audio Speech Lang. Process. **18**, 576–588 (2010)
8. Vasilescu, M.A.O., Terzopoulos, D.: Multilinear (Tensor) ICA and dimensionality reduction. In: Davies, M.E., James, C.J., Abdallah, S.A., Plumbley, M.D. (eds.) ICA 2007. LNCS, vol. 4666, pp. 818–826. Springer, Heidelberg (2007). doi:10.1007/978-3-540-74494-8_102
9. Bro, R.: PARAFAC. Tutorial and applications. Chemometr. Intell. Lab. Syst. **38**, 149–171 (1997)
10. Dean, J., Corrado, G., Monga, R., Chen, K., Devin, M., Mao, M., Ranzato, M., Senior, A., Tucker, P., Yang, K., Le, Q.V., Ng, A.Y.: Large scale distributed deep networks. In: Proceedings of the Conference on Neural Information Processing Systems, pp. 1232–1240 (2012)
11. Cohen, N., Shashua, A.: Simnets: a generalization of convolutional networks. In: Proceedings NIPS Workshop on Deep Learning (2014)
12. Hamidi, M., Pearl, J.: Comparison of the cosine and fourier transforms of Markov-1 signals. IEEE Trans Acoust. Speech Sig. Process. **24**, 428–429 (1976)
13. Oja, E.: Subspace Methods of Pattern Recognition. Research Studies Press, Brighton (1983)
14. Lu, H., Plataniotis, K., Venetsanopoulos, A.: A survey of multilinear subspace learning for tensor data. Pattern Recogn. **44**, 1540–1551 (2011)
15. Cichoki, A., Zdunek, R., Phan, A.H., Amari, S.: Nonnegative Matrix and Tensor Factorizations. Wiley, Hoboken (2009)

16. Wang, Y., Gong, S.: Tensor discriminant analysis for view-based object recognition. Proc. Int. Conf. Pattern Recogn. **3**, 33–36 (2006)
17. Tao, D., Li, X., Wu, X., Maybank, S.: Elapsed time in human gait recognition: a new approach. Proc. Int. Conf. Acoust. Speech Sig. Process. **2**, II (2006). http://ieeexplore.ieee.org/document/1660308/
18. Hua, G., Viola, P., Drucker, S.: Face recognition using discriminatively trained orthogonal rank one tensor projections. In: Proceedings of the IEEE Conference on Computer Vision and Pattern Recognition (2007)
19. Lathauwer, L., Moor, B., Vandewalle, J.: A multilinear singular value decomposition. SIAM J. Matrix Anal. Appl. **21**, 1253–1278 (2000)
20. Lathauwer, L.D., Moor, B.D., Vandewalle, J.: On the best rank-1 and rank-(r_1, r_2, r_n) approximation of higher-order tensors. SIAM J. Matrix Anal. Appl. **21**, 1324–1342 (2000)
21. Itoh, H., Imiya, A., Sakai, T.: Low-dimensional tensor principle component analysis. Proc. Int. Conf. Comput. Anal. Images Patterns Part I **9256**, 223–235 (2015)
22. Iijima, T.: Theory of pattern recognition. Electron. Commun. Jpn. **1**, 123–134 (1963)
23. Watanabe, S., Pakvasa, N.: Subspace method of pattern recognition. In: Proceedings of the 1st International Joint Conference of Pattern Recognition (1973)
24. Itoh, H., Sakai, T., Kawamoto, K., Imiya, A.: Topology-preserving dimension-reduction methods for image pattern recognition. In: Kämäräinen, J.-K., Koskela, M. (eds.) SCIA 2013. LNCS, vol. 7944, pp. 195–204. Springer, Heidelberg (2013). doi:10.1007/978-3-642-38886-6_19
25. Andreopoulos, A., Tsotsos, J.K.: Efficient and generalizable statistical models of shape and appearance for analysis of cardiac MRI. Med. Image Anal. **12**, 335–357 (2008)

Bayesian Saliency Model for Focal Liver Lesion Enhancement and Detection

Xian-Hua Han[1,2(✉)], Jian Wang[1,2], Yuu Konno[1,2], and Yen-Wei Chen[1,2]

[1] National Institute of Advanced Industrial Science and Technology,
2-3-26, Aomi, Koutou-ku, Tokyo 135-0064, Japan
han-xhua@aist.go.jp
[2] Ritsumeikan University, 1-1-1 Nojihigashi, Kusatsu 524-8577, Japan

Abstract. Focal liver lesion enhancement and detection has an essential role for the computer-aided diagnosis and characterization of lesion regions in CT volume data. This paper proposes a novel focal lesion enhancement strategy by extracting a lesion saliency map, which represents the deviation degree of the uncommon or lesion tissue from the common tissues (liver and vessel) in CT volumes. The saliency map can be constructed by exploring the existing probability of lesion for any voxel. However, due to the large diversity of liver lesions, it is difficult to construct an universal model for all types of lesions. Therefore, this study proposes to construct probability models of the common tissues, which have comparably small variability even for different samples and is relatively easy to obtain the prototype regions even from the under-studying CT volume. In order to robustly and flexibly characterize the common tissues, we explore a Bayesian framework by combining a general model, which is constructed oriented to all CT samples, and an adaptive model, which is constructed specific to the under-studying CT sample, for calculating the existing probability of the common tissues (liver or vessel). Then, the saliency map (the existing probability) of focal lesion can be deduced from that of liver or vessel. The advantages of our proposed strategy mainly include three aspects: (1) it only needs to prepare the prototypes of common tissue such as liver or vessel region, which are easily obtained in any CT liver volume; (2) it proposes to combine the general and adaptive model as Bayesian framework for more robust and flexible characterization of the common tissue; (3) dispensable to remove the other different structure such as vessel in liver volume as a pre-processing step. Experiments validate that the proposed Bayesian-based saliency model for focal liver lesion enhancement can perform much better than the conventional approaches such as EM, EM/MPM based lesion detection and segmentation methods.

1 Introduction

Liver cancer is a severe disease with high-frequency on the world, and is considered as one of the major causes of death in humans. Early diagnosis and

Electronic supplementary material The online version of this chapter (doi:10. 1007/978-3-319-54526-4_3) contains supplementary material, which is available to authorized users.

accurate appraisal of focal lesion in surgical resection are critical for increasing the survival chances of patients. Recent advancements in medical imaging have enabled the acquisition of high-resolution CT volumes, and thus, allows not-bad detection rates for some lesion types. However, due to the small observable variations between healthy tissues and lesion ones, and large diversity of different lesion types that includes cysts, focal nodular hyperplasia (FNH), haemangiomas (HEM), hepatocellular carcinoma (HCC) and metastasis (METS), the detection of liver lesions is challenging even for clinic experts, where an important identification variation (15–25%) for a volume can be appeared dependent on physician's experience. In addition, owing to the increasing large number of images in medical volumes, it is extremely difficult to manually analyze all images, which possibly takes several hours, and then some useful diagnosis information may be overlooked. Therefore, automatic early detection of focal liver lesions for computer assisted diagnosis (CAD) becomes a more and more important research line.

There are several proposed methods for lesion detection and segmentation in liver CT volumes, which can be classified as semi-automatic [1,2] and automatic strategies [3,4]. Mala *et al.* have proposed a neural network (NN) based texture analysis for liver lesion classification, which employed wavelet-based texture for identifying tumor tissue with the trained NN model [4]. However, the NN-based approach requires to construct the prototypes of the lesion and healthy liver tissues for learning the NN-based classification model in training step, which is extremely difficult for lesion tissue due to variation of different types and variable imaging condition between patients. Smeets *et al.* have explored a semi-automatic level set method only for large-size tumor segmentation, which combine a spiral scanning technique with supervised fuzzy pixel classification [2]. In order to reduce ill effect of other healthy tissues, Park *et al.* have proposed to firstly remove the vessels in liver volumes as pre-processing step, and then estimate the distribution parameters of liver and lesion tissues by assuming a bimodal histogram of vessel-deleting volume (including healthy liver and lesions) [4]. Then, an optimal threshold can be determined by a mixture probability density algorithm to segment lesion regions. However, in low-contrast CT volumes, it is difficult to accurately remove vessels from liver volumes. Thus, Masuda *et al.* [5] proposed to adaptively enhance the contrast of CT images firstly, and then, applied a expectation maximization and maximization of the posterior marginal (EM/MPM) algorithm by integrating the spatial information for robust lesion detection and segmentation. However, due to the variation of tumor number amount and size in different CT volumes, the optimal threshold is difficult to being determined even considering the neighborhood (spatial) information. Furthermore, all the previous methods conducted lesion detection only on a single-phase CT data, which led to unacceptable detection results for some types of liver lesions. Recently, multi-phase contrast-enhanced CT has became the primary imaging technique employed by clinicians for detection, diagnosis, and monitoring of liver lesions. Multi-phase CT data mainly includes four phases: un-enhanced (non-contrast) scan, arterial phase scan (30–40 s after contrast injection), portal venous phase scan (70–80 s after injection) and delayed phase scan (3–5 min after injection). There are different visual characteristics of Liver lesions at the different phase data. Clinical experiences have already shown that discrimination of different lesion types can be observed in different phases of CT data. Therefore, it would be more reasonable to detect different types of liver lesion using multi-phase CT data.

In this paper, we propose a novel enhancement strategy of all lesion types by extracting a lesion saliency map, which attempts to explore the existing probability of the uncommon or lesion tissue compared to the liver and vessel ones in CT volumes. As we know that the healthy liver and vessel tissues in any CT volume is generally existed in all slices, and then, is easily obtained as prototypes of common tissues. However, due to large variations of different types of lesions and impossibility of achieving the adaptive lesion tissue regions for a specific patient, the lesion prototype is extremely difficult for extraction. Therefore, this paper explores to only sample the liver and vessel tissues as common prototype, and then calculate a saliency map of uncommon tissues such as lesion region based on the collected prototypes of common tissues. The saliency map of lesion tissue can be obtained by firstly applying a Bayesian framework for calculating the existing probability of liver or vessel and then deduce the deviation degree of the liver lesion. Experiments validate that the proposed Bayesian-based saliency model can greatly enhance lesion regions and suppress other healthy structure. Furthermore, for lesion detection, it can perform much better than the conventional approaches such as EM, EM/MPM based methods.

2   Bayesian Saliency Model for Focal Liver Lesion Enhancement

Visual saliency describes the state by which an object stands out from its surrounding, and provides a mechanism to prioritize the processing of the overloaded visual information. The general computational modeling of visual saliency aims to construct a saliency map that represents the presence of uncertainty in the corresponding scene. Because of the possible wide applications such as image compression, visualization, saliency detection has recently become a hot research topic for general scenes. The reported saliency modeling techniques can be broadly classified into two categories depending on whether learning is involved. The typical non-learning technique is mainly based on the idea of feature integration theory (FIT) [6] such as [7,8], which computes saliency based on the difference of filter response within different color channels at different image scales. Some other works explored the saliency model based on the complexity of image regions that is captured by image variance or image entropy [9]. However, due to the limited available information and the existing noise in the medical data, the current non-learning saliency model is impossible to be applied for lesion saliency detection in liver volume. In the other hand, the learning-based techniques firstly learn a set of saliency features from either the image under study or a pool of natural images, and then, the saliency of an image region is computed based on similarity between the learned features and that of the image region under study. However, for our purpose of lesion enhancement and detection, it is really difficult to construct a universal model for well fitting all types of lesions, and also is naturally impossible to extract lesion regions of the under-studying medical volume for learning saliency features.

Therefore, we propose a novel saliency model for liver lesion enhancement, which does not need to prepare any lesion region for learning. As previously mentioned, we only aim to enhance the uncommon or surprise observation: lesion tissue. The expected common prototypes should include both healthy liver and vessel. In our proposed strategy, we firstly need to prepare some prototypes regions from healthy tissues of liver and vessel, and then extract a set of prototype features M such as with size K*K*L for constructing probability model of health tissues.

Given some observations $\mathbf{M}$ of background information (the healthy liver and vessel tissues in our application), let's assume that it can be captured by the prior probability distribution $P(\mathbf{M})_{\mathbf{M}\in\mathfrak{M}}$ over the hypotheses or Models $\mathbf{M}$ in a model space $\mathfrak{M}$. With the available belief prior distribution, the fundamental effect of a new data observation $\mathbf{D}$ on the observed background model is to change the prior distribution $P(\mathbf{M})_{\mathbf{M}\in\mathfrak{M}}$ into the posterior distribution $P(\mathbf{M}|\mathbf{D})_{\mathbf{M}\in\mathfrak{M}}$ via Bayes theorem as:

$$P(\mathbf{M}|\mathbf{D}) = \frac{P(\mathbf{D}|\mathbf{M})}{P(\mathbf{D})}P(\mathbf{M}) \tag{1}$$

In the above framework, the new observed data $\mathbf{D}$ carries no uncommon information (means possibly abnormal or lesion) if it leave the background observer beliefs unaffected, which means the posterior is identical to the prior; otherwise, $\mathbf{D}$ would include non-common tissues if the posterior distribution due to the observing data $\mathbf{D}$ significantly changes from the prior distribution. Therefore, the uncommon information contained in the data can formally measured by some distances between the posterior and prior distributions, such as the Kullback-Leibler (KL) divergence for uncommon saliency map:

$$S(\mathbf{D},\mathbf{M}) = KL(P(\mathbf{M}|\mathbf{D}), P(\mathbf{M})) = \int_{\mathfrak{M}} P(\mathbf{M}|\mathbf{D}) \log \frac{P(\mathbf{M}|\mathbf{D})}{P(\mathbf{M})} dM \tag{2}$$

taken with respect to the posterior distribution over the model class $\mathfrak{M}$. However, in our lesion-enhancement application, we aim to explore the saliency value of each voxel in the processing CT volumes, and a set of new data $\mathbf{D}$ containing the same tissue: uncommon lesion or common healthy and vessel tissues, are also unavailable for exploring the KL divergence. Therefore, this study investigate probability of any voxel sample $\mathbf{d}$ belonging to the healthy tissues (HT), given the prior distribution of the healthy tissue (HT) observer $\mathbf{M}$ as background model:

$$P(\mathbf{HT}|\mathbf{d}) \approx P(\mathbf{d}|\mathbf{M})P(\mathbf{M}) = P(\mathbf{d}/\mathbf{HT})P(\mathbf{HT}) \tag{3}$$

where $P(\mathbf{d}|\mathbf{M})$ is the likelihood of $\mathbf{d}$ based on the healthy tissue (HT) observer $\mathbf{M}$, $P(\mathbf{M})$ is the prior model of $\mathbf{M}$, and $P(\mathbf{HT}|\mathbf{d})$ is the posterior probability of the feature $\mathbf{d}$ of any voxel belonging to healthy tissue. Then, the saliency map of abnormal tissue can be extracted using the probability to the uncommon tissue (UT) of each CT voxel, which can be approximated as:

$$S(\mathbf{d}) = P(\mathbf{UT}|\mathbf{d}) \approx C - P(\mathbf{HT}|\mathbf{d}) \qquad (4)$$

where C is a constant for transferring the healthy tissue's probability to the uncommon saliency map of any voxel in CT volume. Next section, we explore how to construct the prior and likelihood models for calculating the posterior probability of any voxel.

3   Probability Model Construction for Healthy Tissues

In order to increase availableness for any CT volume and adaptability to a specific CT data, the proposed Bayesian framework explores two healthy tissue models: the prior one leaned from the previously prepared healthy prototypes as the general model, which is universal and unchanged to all CT samples, and the likelihood model constructed using the prototype regions from the understudying CT volume, which is specific to the under-studying sample. Then, the combination of these two models is defined as the Bayesian framework for calculating the posterior probability of any voxel belonging to the healthy tissues. The flowchart of the proposed framework is shown in Fig. 1.

The Prior Model of Healthy Tissues: In order to construct a universal model fitting the healthy tissues form all CT samples, we first prepare several training CT samples for predefining healthy tissue regions (liver and vessel) as shown in the top row in Fig. 1. Then we regard the intensities of all voxels in these regions as samples $\mathbf{x} = [x_1, x_2, \cdots, x_T]$, with $x_i \in \mathbf{R}$ and T being sample number. Since, the variable x represents the voxel intensity from the liver or vessel regions, which generally has a multi-modal histogram, and thus we model

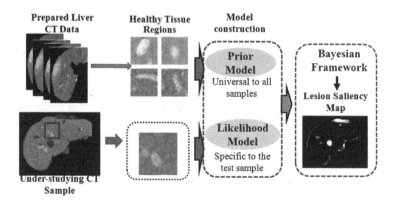

Fig. 1. The flowchart of the proposed saliency model.

them as a Gaussian mixture model (GMM). With the assumed GMM model, we can formulate intensity sample space $\mathbf{x}$ as

$$P(\mathbf{x}|\lambda) = \sum_{k=1}^{K} w_k N(x|\mu_k, \sigma_k) = \sum_{k=1}^{K} w_k \frac{1}{(2\pi)^{\frac{1}{2}}\sigma_k} exp\{-\frac{1}{2\sigma^2}(x - \mu_k)^2\} \quad (5)$$

where λ are the parameters for formulating the probability function, in the Gaussian mixture model with K-components, denoted $\lambda = \{w_k, \mu_k, \sigma_k, k = 1, \ldots, K\}$. w_k, μ_k, σ_k are the mixture weight, mean and standard deviation of Gaussian k, respectively. Since there are only two types of tissues for modeling in our application, we fix the Gaussian component number as 2 with each one representing one type of tissue. Given some samples $\mathbf{x} = [x_1, x_2, \cdots, x_T]$, we can adaptively learn the prior parameters $\lambda = \{w_k, \mu_k, \sigma_k, k = 1, 2\}$ of GMM using Expectation maximization (EM) strategy. Once, the parameters of the prior model are learned from the previously prepared healthy intensity samples, they are fixed and unchanged for calculating the probability of a voxel x_t from any CT sample belonging to healthy tissue as the following:

$$P(\mathbf{HT}) = max\{P(x_t|\mu_1, \sigma_1), P(x_t|\mu_2, \sigma_2)\}. \quad (6)$$

The Likelihood Model of Healthy Tissues: We explore the likelihood model of healthy tissues, which is specific and adaptive to the under-studying CT sample. As we know that there generally are much less observation for lesion regions than healthy liver tissues in most CT volume. Even with randomly sampling from the liver CT data, the most sampled patches as prototypes should also be healthy tissue excluding a few outliers. However, compared to healthy liver tissue, vessel is also a low-frequency appeared observation, and then, there are no or less vessel structures, which is included in the extracted prototype. Thus, we propose to present a randomly selected slice in the under-studying volume to the user, and let the user indicates a region handy, which includes both healthy liver and vessel, for giving the healthy prototypes to construct likelihood model. In the adaptive model, we employ a reshaped vector $\mathbf{m}_i$ of a small patch surrounded the focused voxel as the voxel feature, and extract L prototype features $\mathbf{M}$. Then, we investigates a non-parametric density estimation, which can compute the probability density $P(\mathbf{d}_t|\mathbf{M})$ of sample $\mathbf{d}_t$ (the reshaped vector of a patch) belonging to the background observations (healthy liver and vessel tissues). Given the prototypes $\mathbf{M} = [\mathbf{m}_1, \mathbf{m}_2, \cdots, \mathbf{m}_L]$ of the healthy tissues, the accurate non-parametric approximation of the continuous probability density $P(\mathbf{d}_t|\mathbf{M})$ can be achieved using parzen density estimation as:

$$\hat{P}(\mathbf{d}_t|\mathbf{M}) = \frac{1}{L} \sum_{j}^{L} K(\mathbf{d}_t - \mathbf{m}_j) \quad (7)$$

where $K(\cdot)$ is the Parzen kernel function, which is non-negative and integrate to 1. The Parzen kernel function can generally designed as a Gaussian:

$$K(\mathbf{d}_t - \mathbf{m}_j) = \exp(-\frac{1}{2\sigma^2}||\mathbf{d}_t - \mathbf{m}_j||^2) \quad (8)$$

As the L approaches infinity, and the width σ of the kernel function $K(\cdot)$ reduces, the parzen likelihood $\hat{P}$ converges to the true density $P(\mathbf{d}_t|\mathbf{M})$. In general rule, all the prototypes of the healthy tissues should be used in density estimation of Eq. (7) to obtain high accuracy. However, due to the possible large variation of the prepared prototypes including both healthy liver and vessel, and its computationally complexity requiring distance computation $(\mathbf{d}_t - \mathbf{m}_j)$ for all prototypes $\mathbf{m}_j, (j = 1, 2, \cdots, L)$, it is more feasible to adaptively select the prototypes for each test sample to estimate probability density $\hat{P}(\mathbf{d}_t|\mathbf{M})$. Therefore, we propose a likehood approximation using NN algorithm (LANN).

The LANN Algorithm: Due to the large variation and the long-tail characteristic of healthy tissues (prototypes) distribution, almost all of the samples are rather isolated in the sample space, therefore very far from most prototypes in the observation dataset. Accordingly, all of the terms in the summation of Eq. (7), except for a few, can be negligible (exponentially decreases with distance). Thus we accurately approximate the summation in Eq. (7) using the (few) N largest elements in the sum. These N largest elements correspond to the N nearest neighbors of the sample $\mathbf{d}_t$ within the prototype dataset $\mathbf{m}_1, \mathbf{m}_2, \cdots, \mathbf{m}_L \in \mathbf{M}$:

$$\hat{P}_{NN}(\mathbf{d}_t|\mathbf{HT}) = \hat{P}_{NN}(\mathbf{d}_t|\mathbf{M}) = \frac{1}{N}\sum_{j}^{N} K(\mathbf{d}_t - \mathbf{m}_{NN_j}) \qquad (9)$$

Note that the approximation of Eq. (9) always bonds from below the complete Parzen window estimate of Eq. (7). Finally, the saliency of a lesion voxel t can be estimated as:

$$S(Sal/\mathbf{d}_t, x_t) = C - \frac{1}{N}\sum_{j}^{N} K(\mathbf{d}_t - \mathbf{m}_{NN_j})P(\mathbf{HT}) \qquad (10)$$

After the saliency map of CT input volume is extracted with the proposed strategy, a simple threshold algorithm can be applied to produce the lesion candidates in the CT volume.

3.1   Sampling Strategies for Prototype Extraction

In the proposed strategy, we need to prepare the healthy tissue prototype from under-studying data for constructing adaptive model (likelihood model). We first extract some small regions such as size K*K*L patches as healthy liver tissue prototype by randomly sampling the CT volume under study or achieve the healthy prototypes in a defined tumor-free region by user. The two sampling procedures are shown in Fig. 2(a). As we know that there generally are much less observation for tumor than healthy liver tissues for all most CT volume. Then, the most randomly sampled patches as prototypes should also be structure healthy tissue excluding a few outliers. However, compared to healthy liver tissue, vessel is also a low-frequency appeared observation, and then, there are no or less vessel structures, which is included in the extracted prototype. For our purpose,

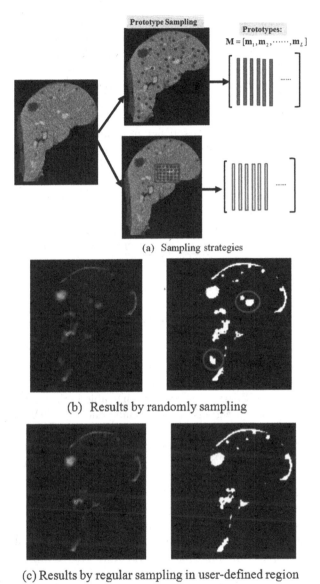

(a) Sampling strategies

(b) Results by randomly sampling

(c) Results by regular sampling in user-defined region

Fig. 2. (a) The sampling strategies for extracting prototypes of healthy tissues. The up row is the randomly sampling strategy, where the randomly defined points would mainly be located in healthy tissues thanking to its high-frequency. However, almost no vessel tissues (healthy tissue) are included in the sampled prototypes, which would result in the vessel tissue as uncommon and enhance it in our proposed strategy. The bottom row strategy ask user to given a region, which includes healthy liver and vessel tissues but no tumor, and then regular sample the points in the user-defined region as prototypes, which surely contain both healthy liver and vessel tissues. The experimental results using different sampling strategies. (b) Results by randomly sampling for prototype extraction of healthy tissues; the left one: saliency image, the right one: detected tumor candidates; (c) Results by regular sampling in a user-defined region; the left one: saliency image, the right one: detected tumor candidates.

we only aim to enhance the uncommon or surprise observation: tumor tissue. The expected common prototypes should include both healthy liver and vessel. Thus, we propose to present a randomly selected slice in the under-studying volume to the user, and let the user indicates a region, which includes both healthy liver and vessel, for sampling the prototypes.

We validate different sampling strategies for prototype extraction. As we mentioned above, randomly sampling the prototypes in the under-studying CT volume is also a feasible way for our saliency model. However, with the randomly sampling, not only the uncommon tumor tissues but also some vessel tissues are enhanced as shown in Fig. 2(b). Therefore, we firstly let user define a region including both healthy liver and vessel tissues, and then, regular sample points for producing prototypes. An experimental results with the extracted prototypes in the user-defined region is shown in Fig. 2(c), which proves enhancement of tumor tissues and suppression of vessel ones. All of the following results are achieved using the extracted prototypes in user-defined region.

4   Experimental Results

We applied our proposed strategy to four sets of single phase CT volumes, and 35 sets of multi-phase CT volumes with NC, ART and PV phases, which includes five types of lesions (CYST, FNH, HEM, HCC and METS). The single phase datasets are from the JAMIT CAD contest in July 2010, and the multi-phase CT volumes are from Zhejiang University, China. The detailed information of the used 4 single phase CT volumes are shown in Table 1. Each single phase data sets includes multiple lesion regions, and especially one of them consists 57 lesion regions, which includes certain size and very small size of ones. Each multi-phase sets generally have one large size of lesion regions, and different types of lesions are existed in different sets. For prototype preparation of healthy tissue in the likelihood model, a 7*7 patch in 2D slice or a 5*5*3 volume patch is obtained for voxel representation. For testing, the small patch (7*7 for 2D, 5*5*3 for 3D) around the focused pixel (voxel) is extracted for calculating likelihood, and the voxel intensity is employed for giving the prior probability, and then the lesion saliency is deduced according to the proposed Bayesian-based saliency model. Because of the achieved similar experimental results in 2D and 3D strategies, we only show the results with 2D strategy, but evaluate the performance of the

Table 1. Detail information of test CT volumes

Data set	Data size			Data spacing (mm)			Number of tumor
	x	y	z	x	y	z	
Data set 1	512	512	201	0.63	0.63	1.00	57
Data set 2	512	512	204	0.78	0.78	1.00	1
Data set 3	512	512	191	0.67	0.67	1.00	3
Data set 4	512	512	271	0.65	0.65	1.00	4

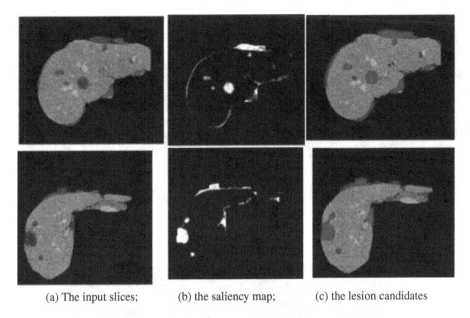

(a) The input slices;   (b) the saliency map;   (c) the lesion candidates

Fig. 3. The experimental results on the single phase dataset. (a) The input slices; (b) the saliency map; (c) the lesion candidates.

detected lesions in 3D volumes after a simple threshold on the enhanced saliency map. Some examples of the detection results using our proposed strategy are shown in supplementary material. We show the saliency map of the single phase data and the detected lesion candidates by thresholding on the saliency map in Fig. 3. It is obvious that the proposed Bayesian-based saliency model can significantly enhance all uncommon (lesion) tissues, and at same time, suppress other healthy tissues. The detection results of different types of lesions are shown in Fig. 4, which manifests that our proposed strategy can successfully detect all types of lesion regions.

We denote lesion as positive, non-lesion (healthy liver, vessel and so on) as negative. If the lesion region is detected out, we consider the result as a true positive (TP), otherwise as false negative (FN). In the other hand, if a non-lesion region is detected as lesion region, we consider the result as false positive (FP). The sensitivity and specificity in ROC curve are defined as the following:

$$Sensitivity = \frac{TP}{TP + FN} \qquad Specificity = 1 - \frac{FP}{TP + FP} \qquad (11)$$

Figure 5 shows the ROC curve of the detection results on one single-phase data with 57 lesion regions, which manifests that all lesion regions can be correctly detected out under the specificity value 0.26, and even with the large specificity value 0.5, almost lesion can be detected with our proposed strategy. In order to compare with the conventional approaches such as EM, EM/MPM in lesion detection [5], we also show the quantitative performance such as true positive (TP), false positive (FT), false negative (FN) for all 4 data sets in

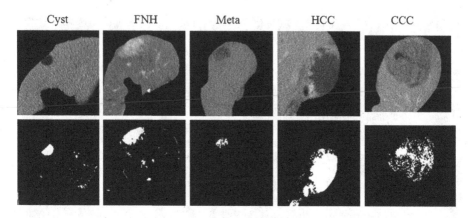

Fig. 4. The detection results of different types of lesions.

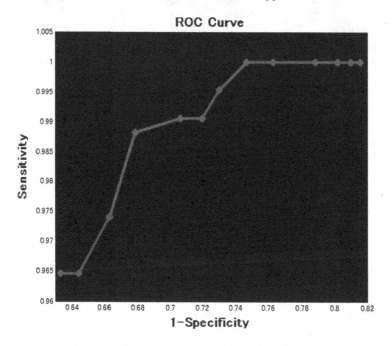

Fig. 5. The ROC curve on the single-phase dataset which includes 57 lesion regions.

Table 2(a). From Table 2(a), it is obvious that our proposed strategy can detect all existing but with much less false positive (FP). The detection results on the multi-phase sets with different lesion types are shown in Table 2(b), and validate that our proposed strategy can descry almost lesion regions of different types by combining multi-phase data and acceptable detection performance even with a single phase for most types of lesion regions. The quantitative evaluations (TP and FP) of 10 multi-phase CT volume (two samples for each lesion type) are show in Fig. 6.

Table 2. The detection lesion results on the single- and multi-phase datasets.

	Method	FP	FN	TP
	EM	624	1	56
Data1	EM /MPM	724	0	57
	Ours	312	0	57
	EM	742	0	1
Data2	EM /MPM	453	0	1
	Ours	25	0	1
	EM	710	0	3
Data3	EM /MPM	421	0	3
	Ours	168	0	3
	EM	602	0	4
Data4	EM /MPM	755	0	4
	Ours	121	0	4

	NC	ART	PV	ALL	Lesion
	TP	TP	TP	TP	Number
Cyst	7	7	4	7	7
FNH	4	5	4	7	7
CCC	4	5	5	6	7
HCC	5	7	6	7	7
Meta	6	6	6	7	7

(a) The compared results on the single-phase data; (b) the detection results on all multi-phase data.

	NC		ART		PV		Lesion Number
	TP	FP	TP	FP	TP	FP	
Cyst1	1	10	1	2	1	17	1
Cyst2	1	4	1	3	1	32	1
FNH1	0	2	1	11	1	6	1
FNH2	0	15	1	17	0	7	1
CCC1	1	9	1	2	0	29	1
CCC2	1	9	0	1	1	11	1
HCC1	1	3	1	5	1	19	1
HCC2	1	16	1	3	1	10	1
Meta1	1	8	0	14	1	4	1
Meta2	1	3	1	9	1	26	1

Fig. 6. The quantitative evaluations (TP and FP) of 10 multi-phase CT volume.

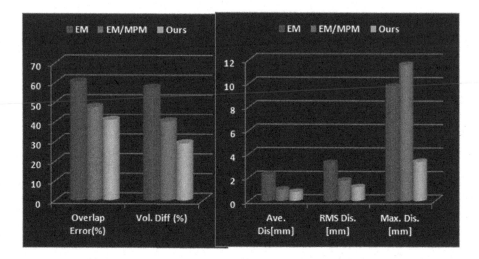

Fig. 7. The compared performance in terms of metric proposed in MICCAI liver segmentation challenge using our proposed, conventional EM, EM/MPM segmentation methods.

Finally, we quantitatively evaluate the performance of our proposed strategy in terms of the metrics used in the MICCAI liver tumor segmentation challenge 2008. The metrics are the volumetric overlap error, absolute relative volume difference (Vol. diff.), average symmetric surface distance (Ave. Dis.), RMS symmetric surface distance (RMS Dis.), and maximum surface distance (Max. Dis.). For ideal lesion extraction, all metrics should be zero. Due to requiring of manually segmenting the annotated tumor in CT volume for comparison, we only show the average metrics of four certain size lesion regions from the four single-phase datasets, respectively. Figure 7 gives the compared results in the metrics given used in the MICCAI liver tumor segmentation challenge 2008, which again proves that our proposed method can achieve much better performance than the conventional methods such as EM, EM/MPM tumor segmentation.

5   Conclusions

This study presented a novel Bayesian-based saliency model for focla liver lesion enhancement. In order to enhance lesion tissue, we firstly construct a saliency map, which represents the uncommon or lesion tissue compared to the liver and vessel ones in CT volumes, and then, detect the lesion candidates using a simple threshold algorithm. By combining multi-phase data, our proposed strategy can accurately detect the lesion regions of different types. Experiments validate that the proposed Bayesian-based saliency model for focal liver lesion enhancement can perform much better than the conventional approaches such as EM, EM/MPM methods.

References

1. Smeets, D., Loeckx, D., Stijnen, B., De Dobbelaer, B., Vandermeulen, D., Suetens, P.: Semiautomatic level set segmentation of liver tumors combining a spiral scanning technique with supervised fuzzy pixel classification. Med. Image Anal. **14**, 13–20 (2010)
2. Hame, Y., Alhonnoro, T., Pollari, M.: Image analysis for liver tumor ablation treatment planning. In: Hands-on Image Processing. Robotiker-Tecnalia (2009)
3. Mala, K., Sadasivam, V., Alagappan, S.: Neural network based texture analysis of liver tumor from computed tomography images. Int. J. Biomed. Sci. **2**, 33–40 (2006)
4. Park, S.-J., Seo, K.-S., Park, J.-A.: Automatic hepatic tumor segmentation using statistical optimal threshold. In: Sunderam, V.S., Albada, G.D., Sloot, P.M.A., Dongarra, J.J. (eds.) ICCS 2005. LNCS, vol. 3514, pp. 934–940. Springer, Heidelberg (2005). doi:10.1007/11428831_116
5. Masuda, Y., Tateyama, T., Xiong, W., Zhou, J.Y., Wakamiya, M., Kanasaki, S., Furukawa, A., Chen, Y.-W.: CT image contrast enhancement and liver tumor detection by using probability density function and EM/MPM algorithm. In: Proceedings of the IEEE International Conference on Image Processing (2011)
6. Treisman, A., Gelade, G.: A feature-integration theory of attention. Cogn. Psychol. **12**, 97–136 (1980)
7. Itti, L., Koch, C.: Computational modeling of visual attention. Nat. Rev. Neurosci. **2**, 194–203 (2001)
8. Gao, D., Vasconcelos, N.: Bottom-up saliency is a discriminant process. In: IEEE 11th International Conference on Computer Vision, pp. 1–6 (2007)
9. Kadir, T., Brady, M.: Saliency, scale and image description. Int. J. Compt. Vis. **45**, 83–105 (2001)

A Novel Iterative Method for Airway Tree Segmentation from CT Imaging Using Multiscale Leakage Detection

Syed Ahmed Nadeem[1(✉)], Dakai Jin[1], Eric A. Hoffman[2],
and Punam K. Saha[1,2]

[1] Department of Electrical and Computer Engineering,
University of Iowa, Iowa City, IA 52242, USA
{syedahmed-nadeem,dakai-jin}@uiowa.edu, pksaha@healthcare.uiowa.edu
[2] Department of Radiology, University of Iowa, Iowa City, IA 52242, USA
eric-hoffman@uiowa.edu

Abstract. Computed tomography (CT)-based metrics of airway phenotypes, wall-thickness, and other morphological features are increasingly
being used in large multi-center lung studies involving many hundreds
or thousands of image datasets. There is an unmet need for a fully reliable, automated algorithm for CT-based segmentation of airways. State-
of-the-art methods require a post-editing step, which is time consuming when several thousands of image data sets need to be reviewed and
edited. In this paper, we present a novel iterative algorithm for CT-based
segmentation of airway trees. Early testing suggests that the method
requires no editing to extract a set of airway segments along a standardized set of bronchial paths extending two generations beyond the segmental airways. It uses simple intensity-based connectivity and new leakage
detection and volume freezing algorithms to iteratively grow an airway
tree. It starts with an initial, automatically determined seed inside the
trachea and a conservative threshold; applies region growing and generates a leakage-corrected segmentation; freezes the segmented volume; and
shifts the threshold toward a more generous value for the next iteration
until a convergence occurs. The method was applied on chest CT scans of
fifteen normal non-smoking subjects. Airway segmentation results were
compared with manually edited results, and branch level accuracy of the
new segmentation method was examined along five standardized segmental airway paths and continuing to two generations beyond the segmental
paths. The method successfully detected all branches up to two generations beyond the five segmental airway paths with no visual leakages.

1 Introduction

There is a growing use of quantitative computed tomography (QCT) to assess the
lung both in terms of parenchymal characteristics as well as characteristics of the
bronchial tree [1–5]. With the labeling of the extracted airway tree [6,7] allowing
for the comparison of spatially matched airway segments across individuals, it
has been demonstrated that new insights into airway phenotypes can emerge.

© Springer International Publishing AG 2017
C.-S. Chen et al. (Eds.): ACCV 2016 Workshops, Part III, LNCS 10118, pp. 46–60, 2017.
DOI: 10.1007/978-3-319-54526-4_4

This is evidenced by the recent observation that, on average, the airway walls of smokers with chronic obstructive pulmonary disease (COPD) actually have thinner rather than thicker airways [8]. There are numerous large multi-center studies incorporating lung imaging as a study component including SPIROMICS [9], COPDGene [10], MESA Lung [11], Severe Asthma Research Project (SARP) [2], CANCOLD [12], and more. To our knowledge, there have been no fully automated methods developed for airway tree segmentation, free of the need for user review to assure that the airway tree has been extracted so as to include a standardized set of bronchial segments. When there is a failure of even a fraction of the segmentations, the airway tree masks must be manually reviewed for all subjects which is extremely cumbersome when evaluating many thousands of image data sets. To simplify the review process and the subsequent use of the many resulting metrics, the radiology center of the SARP has standardized on the airway paths passing through 6 segmental bronchial segments (RB1, RB4, RB10, LB1, LB4, LB10) and continuing two generations beyond these segmental bronchi [1]. This has more recently been reduced to 5 segments as LB4 is often subject to cardiogenic oscillation-derived blurring. In this paper, we present a new airway tree segmentation method which has, in a preliminary evaluation, been shown to reliably extract the bronchial segments along these 5 standardized paths up to two generations beyond the segmental level without the requirement of manual intervention.

Several methods for segmentation of airway trees have been reported in the literature. For example, Sonka *et al.* [13] used a rule-based approach for segmentation of airway trees using the underlying information of its anatomy; others have applied intensity-based classification and fuzzy logic, where voxels undergo a competitive process determining the regions of their belonging [14,15]. Other image processing techniques such as region growing using intensity-based voxel connectivity [16–19], mathematical morphology [20,21], gradient vector flow [22], central-axis analysis [23–26], energy minimization [27], graph-based approaches [28], etc. have been applied for airway segmentation. Also, there are several hybrid methods [29,30] combining multiple approaches listed above. Key performance metrics of an airway segmentation algorithm are primarily related to answering the following questions—(1) on an average, how much manual post-correction time is needed to generate acceptable airway segmentation results, (2) how many airway branches at segmental, children, and grandchildren levels are correctly identified, and (3) performance on lower radiation dose CT scans.

Major challenges with chest CT airway tree segmentation emerge from the following facts. Although airway lumen and wall voxels are expected to receive the values of $-1,000$ HU and -450 HU [31], respectively, due to noise and partial voxel volume effects, reduced contrast appears at several locations on the airway wall creating possible sites for leakages during segmentation (see Fig. 1). Generally, a simple thresholding method [32] fails to work for airway segmentation up to the target level of segmental bronchi and two generations beyond [6]. Also, the CT values of an airway wall along a branch decrease (become more negative) in the proximal-to-distal direction (see Fig. 2) adding further challenges

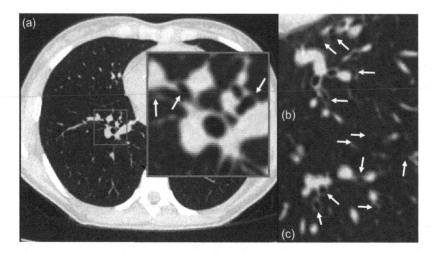

Fig. 1. Illustration of challenges with CT-based segmentation of airway trees. Reduced contrast of airway walls (indicated by white arrows) between airway and lung parenchyma due to noise and partial voxel volume. These low contrast airway walls are possible sites of leakages during segmentation. (Contrast settings: level = −450 HU, window = 1200 HU).

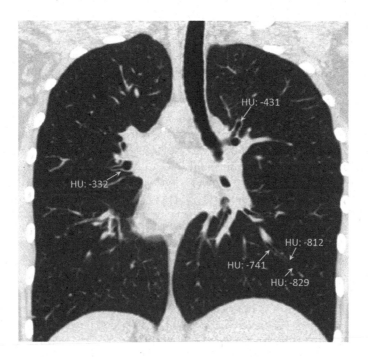

Fig. 2. CT values of airway walls at proximal and distal locations. Airway wall intensity values were reduced from −332 HU at a proximal location to as low as −829 HU at a distal location. (Contrast settings: level = −450 HU, window = 1200 HU).

to a simple threshold-based approach of airway segmentation. Moreover, limited spatial resolution and image reconstruction artifacts, especially using sharp kernels, often cause thin airway walls to appear perforated and discontinuous resulting in leakages or early termination of airway lumen growth [15].

In this paper, we present a novel iterative algorithm for airway segmentation using multi-scale leakage detection and volume freezing techniques. The method is fully automated requiring no manual inputs or post-editing steps. It uses simple intensity-based connectivity and iteratively grows an airway tree starting with an initial seed inside the trachea and a conservative threshold value. During an iteration it executes the following sequential steps—apply region growing and generate a leakage-corrected segmentation; freeze the segmented volume; and shift the threshold toward a more generous value for the next iteration. This iterative strategy of airway tree growing continues until a convergence occurs. Several leakage detection methods have been applied for CT-based airway segmentation [6,33]. Tschirren *et al.* [6] used topological features [34] for leakage detection; specifically, they used the criterion that a leaked volume includes complex topological features, e.g., tunnels. Others have used geometric rule-based approaches for leakage detection for airway tree segmentation [33]. In this paper, we present a new approach of leakage detection using scale analysis along an airway branch and a measure of distance metric based tortuosity [35] to detect spongy leakages. Moreover, the iterative airway segmentation algorithm introduces a unique notion of volume freezing and active seed selection strategy to progressively shift the segmentation strategy from a conservative to a generous thresholding scheme. The algorithm requires no threshold-related parameter. During an iteration, the CT intensity threshold used for region growing is automatically computed that barely causes a leakage in the airway segmentation.

2   Methods and Algorithms

The basic principle of the overall method is described in Sect. 2.1. The major steps in the algorithm, namely, crude segmentation, leakage detection, setup for next iteration and termination criterion are discussed in Sects. 2.2 to 2.4, respectively.

2.1   Basic Principle

The airway segmentation algorithm presented in this paper uses simple intensity-based connectivity together with new methods of leakage detection and volume freezing to iteratively grow an airway tree in a chest CT image starting with an initial seed inside the trachea. The method is fully automated requiring no manual post correction steps. A block diagram of major steps of the algorithm is presented in Fig. 3. Also, schematic illustrations of results after different intermediate steps are shown in Fig. 4.

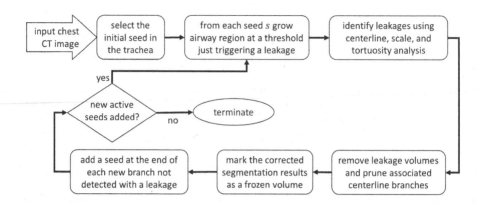

Fig. 3. A block-diagram of major steps of the iterative algorithm of airway segmentation from human chest CT scans presented in this paper.

The initial seed selection inside the trachea is automatically performed using the algorithm by Mori *et al.* [16]. The method applies compactness and size analysis of connected air-space in the upper quartile of axial slices in an acquired chest CT scan to locate the trachea. Finally, the seed is located at the centroid of the trachea. After locating the seed point, the method applies a simple threshold-based connectivity analysis to compute the initial segmentation. The threshold value is automatically selected as the lowest CT intensity triggering an airway segmentation leakage. The results of initial segmentation are illustrated in Fig. 4(a), which includes apparent leakages together with several valid airway branches.

The next and crucial step is to locate and correct for leakages in the initial rough segmentation. This step is based on analysis of scale and a geodesic distance-based measure of tortuosity along centerline paths of the initial airway tree segmentation. The centerline of the segmented region is computed using a centered minimum-cost path approach recently developed by Jin *et al.* [36]. In Fig. 4(b), the centerline is shown using green lines. A scale analysis [37] is performed along the centerline to detect a leakage. Normally the scale, i.e., local diameter of an airway branch follows a quasi-non-increasing pattern along its centerline path in the proximal-to-distal direction. Thus, a leakage in an airway segmentation can be characterized as a rapid inflation in scale values along a centerline path of the segmented region. Although, this scale analysis method successfully detects solid leakages, it often fails to locate a spongy leakage because the scale is small everywhere inside a spongy region. To detect a spongy leakage, we use distance metric tortuosity, defined as the ratio between its geodesic length [38] and the Euclidean distance between its end points. Normally, the Euclidean distance between the end points of the centerline of an airway branch is similar in value to its geodesic length. However, inside a spongy leakage region, a centerline path grows in an uncontrolled manner. Therefore, the geodesic length of a centerline segment inside a spongy region increases rapidly as compared to the Euclidean distance between its end points. This simple strategy successfully

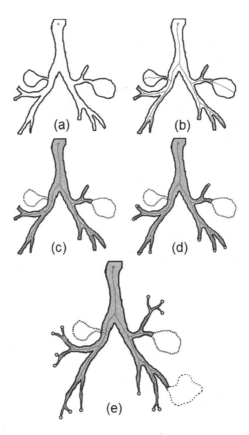

Fig. 4. Schematic description of intermediate results of the new airway segmentation algorithm. (a) Initial segmentation using intensity-based region growing. (b) Centerline Extraction (Green) of (a) and leakage detection and correction using branch pruning. (c) Frozen airway segmentation volume after leakage-correction. In subsequent iterations, connectivity paths are prohibited to enter inside the frozen region. (d) New active seeds (yellow dots) at every new branch not pruned for leakage-correction. (e) The segmentation result after the next iteration. (Color figure online)

detects all complex and spongy leaks in airway segmentation. The results of leakage detection are schematically shown in Fig. 4(b).

After leakage detection, the leakage-correction step works as follows. First, each centerline branch associated with a leakage is pruned, and the leakage-corrected centerline representation of the airway tree is computed. Subsequently, the leakage-corrected airway volume is obtained by dilating individual centerline branches with local scales (Fig. 4(c)). This segmentation volume after the current iteration is considered as a frozen region, and no connectivity paths in subsequent iterations are allowed to enter into it. This step is crucial for developing an iterative-threshold-shift strategy starting with a conservative threshold and progressing toward more generous values. The end point of an airway centerline branch that is not associated with a leak is used as an active seed for further airway growth in following iterations (Fig. 4(d)). Here, it is necessary to ensure that an active seed at the end of a valid branch is not buried inside the frozen volume, which would arrest any further airway segmentation growth. An efficient solution using distance analysis in the dilation process is developed analytically guaranteeing that an active seed is never buried inside the frozen volume. After computing the frozen airway segmentation volume and active seeds, the process enters in to the next iteration. This iterative process continues until convergence, i.e., no new active seed can be detected after an iteration. Individual steps of our airway segmentation algorithm are presented in following sections.

2.2   Initial Rough Segmentation

The algorithm presented here uses simple intensity-based region growing to compute an initial rough segmentation of the airway tree volume. Such an initial segmentation is only a rough estimation, which may include one or more leakages. Two major requisites in this step are—(1) identification of seed points and (2) determination of the intensity threshold value. As discussed in the previous section, only one seed is used in the first iteration, which is automatically located inside the trachea using the algorithm by Mori *et al.* [16]. In a subsequent iteration, active seed(s) located during the previous iteration are used; the method of locating active seeds is described in Sect. 2.4.

During an iteration, airway segmentation volume is independently grown from each seed point. Let s denote a seed point and let t_s denote the intensity value of the seed point. The threshold value for airway region growing from s is automatically selected as the lowest intensity value triggering a segmentation leakage. For efficient computation of the optimum threshold, a binary search process between t_s and $t_s + \Delta_{max}$ is performed. For all experiments presented in this paper, a value of 200 HU is used as Δ_{max}. It should be noted that the final threshold value is independent of the choice of Δ_{max}.

It was noted that, often, the intensity-based region growing algorithm leaves small tunnels and cavities inside the segmented region. These cavities and tunnels are filled using a mathematical morphological closing operation with a structure element of $5 \times 5 \times 5$. This step is imperative to ensure that an initial segmented volume does not contain small holes which would add significant errors in centerline computation, scale analysis, and tortuosity estimation. After filling small tunnels and cavities, the initial rough segmentation is passed to the next module for leakage detection.

2.3   Leakage Detection

The leakage detection method presented in this paper is based on analysis of scale and distance metric tortuosity along airway centerline branches. As described in the previous section, during an iteration, initial rough segmentation of an airway subtree volume is computed from a seed point; let s denote the seed point and

V be the initial subtree volume computed from s. First, a centerline tree is computed for V with s as the root using a robust and efficient centerline detection algorithm recently developed by Jin *et al.* [36]. The centerline detection method uses fuzzy distance transform (FDT) [39] and collision-impact [40] based centered minimum cost paths to locate individual centerline branches. It generates a partially ordered representation of centerline branches. Let $\langle \pi_1, \pi_2, \ldots, \pi_n \rangle$ be the breadth-first traversal of the partially ordered centerline tree, where π_i is a centerline branch, and $\bigcup_{i=1}^{N} \pi_i$ represents the entire centerline tree. Individual centerline branches are examined for leakages in the order of their breadth-first traversal. If a leakage is detected on a centerline branch π_i, the exact leakage point p on the branch is located and it is pruned up to that point. Moreover, the centerline branches descendant to point p are removed from the center tree. Centerline branches are examined for two types of leakages—(1) solid leakages and (2) spongy leakages. The methods for detecting these two specific types of leakages are described in the following.

Solid leakages are characterized by the rapid scale inflation along an airway centerline branch. In general, local scales along an airway tree branch are quasi-non-increasing in the proximal-to-distal direction. Thus, a solid leakage can be detected by locating sudden increase in scale values along an airway centerline branch. This process is illustrated in Fig. 5. The scale at a centerline voxel on a skeletal branch is computed using a star-line approach which determines the length of the shortest object intercept through the centerline voxel [41].

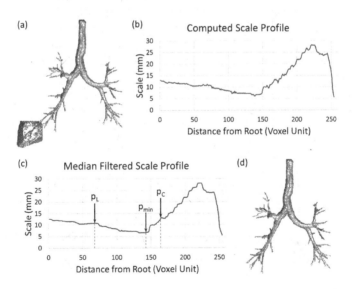

Fig. 5. Detection of a solid leakage using scale analysis along an airway centerline path. (a) Initial rough segmentation of an airway tree with a solid leakage. The centerline path for current checking of leakages is shown in red. (b) Computed scale profile along the centerline path. (c) Leakage detection on the median filtered scale profile. See text for detail. (d) Leakage-corrected airway tree. (Color figure online)

Computed scale profile along a centerline branch of Fig. 5(a) is shown in Fig. 5(b). This computed scale profile is passed through a median filtering (window size = 5) to smooth small noisy dents and protrusions; the scale profile after smoothing is shown in Fig. 5(c). The leakage detection algorithm is applied on the smoothened scale profile, and it works as follows. Let p_c be a voxel on a centerline branch $p_0, p_1, \ldots, p_{c-1}, p_c, p_{c+1}, \ldots$. The ratio between the scale at p_c and that at each of its predecessors $p_i \in p_0, p_1, \ldots, p_{c-1}$ is checked in the proximal-to-distal direction. A leakage on the branch is confirmed on the target branch at or prior to p_c, if a voxel p_l ($l < c$) is found whose scale ratio with p_c exceeds a predefined threshold value of 2. Finally, the exact location of leakage is detected by finding the minimum scale voxel p_{min} on the centerline branch between p_c to p_l (see Fig. 5(c)), and the branch is pruned up to p_{min}. Final results after leakage detection and correction are shown in Fig. 5(d).

After a centerline path is checked for solid leakages, it is further checked for spongy leakages. Such leakage regions contain tunnels and cavities which are not completely filled by the morphological closing operation on initial rough segmentation discussed in Sect. 2.2. Inside these regions, the centerline path propagates in an aimless and uncontrolled manner and the scale-analysis of solid leakage detection fails to detect such spongy leakages due to holes. To detect such leakages, we use distance metric toruosity, defined as the ratio between its geodesic length and the Euclidean distance between its end points. Normally, the Euclidean distance between the end points of the centerline of an airway branch is similar in value to its geodesic length. Due to meandering nature of centerline path in a spongy leakage, the geodesic length of a centerline segment increases rapidly as compared to the Euclidean distance between its end points. By applying a predefined threshold of 2 to the tortuosity of a centerline path, it can be discerned if it is part of a spongy region and pruned. This process is schematically described in Fig. 6.

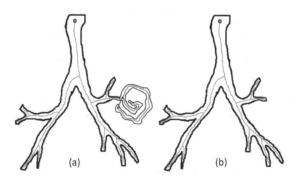

Fig. 6. Detection of a spongy leakage using tortuosity of an airway centerline branch. (a) The tortuosity of each centerline path (between black points) is checked in a breadth-first manner. A centerline branch with a high tortuosity is located (geodesic length: red, euclidean distance: blue). (b) The centerline paths and their descendants are removed during leakage correction. (Color figure online)

2.4   Setup for Next Iteration

During each iteration of the new airway segmentation algorithm, it augments the airway tree, and adds active seeds at the end of each newly added airway branch without a leakage. The leakage detection and correction step, described in the previous section, produces a centerline representation of the newly added leakage-free region of the airway tree. The purpose of the current step is to use the verified centerline to—(1) dilate each centerline to fill the augmented airway tree volume and (2) freeze the newly augmented airway tree volume while preserving the active seed voxel at the end of each augmented centerline branch without a leakage. The purpose of volume freezing is to avoid any new leakage while segmenting forward tree branches at a more generous threshold. Preservation of active seed is essential to ensure new growth for forward airway tree volume.

Using the scale information determined as part of the thickness analysis in the leakage detection step, the volume representation of the airway tree can be reconstructed by dilating along the verified centerline using local scale. The reconstructed airway segmentation volume after the current iteration is dilated by a $3 \times 3 \times 3$ structure element, marked as frozen, and no connectivity paths in subsequent iterations are allowed to enter inside the frozen volume. This step is crucial to design a strategy of iteratively shifting the threshold intensity value starting with a conservative one and then progressing towards more generous threshold values. The dilation step is needed to stop connectivity paths from entering into the narrow region (mostly including partial-volume voxels) between the segmented airway volume and the surrounding airway wall.

The end point of an airway centerline branch without a leakage is used as an active seed for further airway tree growth in the next iteration. Here, it is necessary to ensure that active seeds at the end of valid branches are not buried during the dilation process which would disable any further airway segmentation growth in subsequent iterations. Here, an efficient solution using distance analysis is presented that analytically guarantees that an active seed is never buried during the dilation process. Essentially, the method adds a simple constraint on the dilation process during airway tree volume reconstruction from corrected the centerline tree.

$$V_{airway} = \{p | \exists q \in S_{centerline}\ s.t.\ D(p,q) < scale(q)\ and\ D(p,q) < D(p, p_{end})\}, \tag{1}$$

where p is any voxel; $S_{centerline}$ is the airway centerline branch being reconstructed; $scale(\cdot)$ is the scale function at centerline voxels; and p_{end} is the end voxel of $S_{centerline}$. After reconstructing segmented airway volume, computing and marking the frozen volume, and identifying the active seeds, the process enters in to the next iteration. Finally, the process terminates when no new active seed is detected in an iteration.

3   Experimental and Results

The experiments were aimed to examine the branch-level accuracy of the new airway segmentation method at segmental, sub-segmental, and one level beyond.

The method was applied on chest CT scans of fifteen normal non-smoking subjects (previously acquired under IRB approval; age: 21–48 Yrs, mean: 28.5 Yrs; 7 female) at total lung capacity (TLC: 90% vital capacity) using a volume-controlled breath-hold maneuver. CT scans were acquired on a Siemens Definition Flash 128 (at 120 kV with effective mAs of 200), with images reconstructed at 0.5 mm slice thickness using a standard B35 kernel. Airway segmentation results on a CT image using the new method after different iterations are shown in Fig. 7.

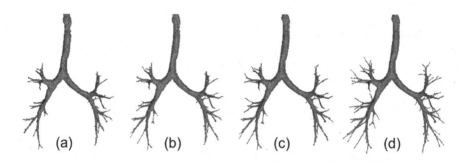

Fig. 7. Results of airway tree segmentation using the new iterative multi-scale leakage detection algorithm after the first (a), second (b), third (c), and the final (d) iterations.

For quantitative experiment, we examined the branch level accuracy of the new segmentation method along five standardized airway paths passing through segmental bronchial segments (RB1, RB4, RB10, LB1, LB10) and continuing two generations beyond these segmental bronchi [1]. An airway tree up to segmental level and two-generations beyond is shown in Fig. 8(a). Branches at segmental level and two-generations beyond along the five anatomic airway paths, used in our experiments, are shown in Fig. 8(b).

To examine the branch-level accuracy of the new method, its segmentation results were compared with the matching airway segmentation results obtained by an expert user's manual editing (both removal of leakage volumes and addition of missing branches) on the airway segmentation results computed using the algorithm [6] provided within the Apollo pulmonary workstation software (VIDA Diagnostics, Coralville, IA). During comparative examination, a blinded expert user compared every branch at the segmental level and two-generations beyond along each of the five anatomic paths on 3-D visual representations of two airway segmentation results—one using the new method and the other being the manually edited one. No leakage was observed in the automatic segmentation results using the new method on any of the fifteen data sets used in this experiment. Numbers of branches observed at segmental, subsegmental, and one-generation beyond levels were 5 ± 0 (mean $\pm$ std.), 10 ± 0, and 15.6 ± 1.5776, respectively compared to the reference method which were 5 ± 0, 10 ± 0, and 15.4 ± 1.3499. The new automatic method successfully detected all branches at the segmental level and two-generations beyond, which were detected in the manually edited

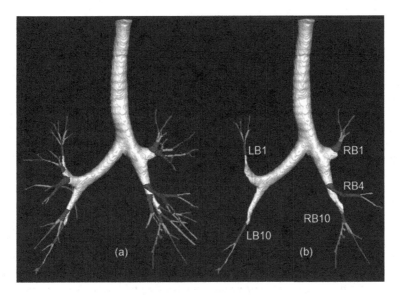

Fig. 8. (a) An airway tree representation including segmental (red), sub-segmental (green), and one level beyond (pink). (b) The five anatomic airway paths used in our experiments. (Color figure online)

results. Moreover, in one data set, the new method detected an additional valid branch at the second generation beyond the segmental level, which was not detected in the manually edited results.

4   Conclusions

An iterative airway segmentation algorithm using new methods of multi-scale leakage detection and volume freezing has been presented, which is suitable for airway analysis in CT-based large cross-sectional and longitudinal studies. The method is fully automated requiring no intensity threshold parameter, manual interaction or post editing enhancing its suitability across different CT scanners and imaging protocols. It was observed in a pilot study that the new method matches the performance of manually edited airway segmentation or excels in terms of branch-level accuracy on chest CT data of fifteen normal non-smoking subjects at total lung capacity at the segmental level and two generations beyond when focusing on a standardized set of paths within the airway tree. Additionally, no leakage was observed in the segmentation results by the new algorithm. Accuracy and reproducibility of the method is currently being examined on larger data sets from two on-going multi-center NIH studies.

Acknowledgements. This work was supported by the NIH grant R01 HL112986.

References

1. Castro, M., Fain, S.B., Hoffman, E.A., Gierada, D.S., Erzurum, S.C., Wenzel, S.: Lung imaging in asthmatic patients: the picture is clearer. J. Allergy Clin. Immunol. **128**, 467–478 (2011)
2. Jarjour, N.N., Erzurum, S.C., Bleecker, E.R., Calhoun, W.J., Castro, M., Comhair, S.A., Chung, K.F., Curran-Everett, D., Dweik, R.A., Fain, S.B.: Severe asthma: lessons learned from the national heart, lung, and blood institute severe asthma research program. Am. J. Respir. Crit. Care Med. **185**, 356–362 (2012)
3. Newell Jr., J.D., Sieren, J., Hoffman, E.A.: Development of quantitative CT lung protocols. J. Thorac. Imaging **28**, 266–271 (2013)
4. Coxson, H.O., Leipsic, J., Parraga, G., Sin, D.D.: Using pulmonary imaging to move chronic obstructive pulmonary disease beyond FEV1. Am. J. Respir. Crit. Care Med. **190**, 135–144 (2014)
5. Hoffman, E.A., Lynch, D.A., Barr, R.G., van Beek, E.J., Parraga, G.: Pulmonary CT and MRI phenotypes that help explain chronic pulmonary obstruction disease pathophysiology and outcomes. J. Magn. Reson. Imaging **43**, 544–557 (2016)
6. Tschirren, J., Hoffman, E.A., McLennan, G., Sonka, M.: Intrathoracic airway trees: segmentation and airway morphology analysis from low-dose CT scans. IEEE Trans. Med. Imaging **24**, 1529–1539 (2005)
7. Tschirren, J., McLennan, G., Palágyi, K., Hoffman, E.A., Sonka, M.: Matching and anatomical labeling of human airway tree. IEEE Trans. Med. Imaging **24**, 1540–1547 (2005)
8. Smith, B.M., Hoffman, E.A., Rabinowitz, D., Bleecker, E., Christenson, S., Couper, D., Donohue, K.M., Han, M.K., Hansel, N.N., Kanner, R.E., Kleerup, E., Rennard, S., Barr, R.G.: Comparison of spatially matched airways reveals thinner airway walls in COPD. The multi-ethnic study of atherosclerosis (MESA) COPD study and the subpopulations and intermediate outcomes in COPD study (SPIROMICS). Thorax **69**, 987–996 (2014)
9. Couper, D., LaVange, L.M., Han, M., Barr, R.G., Bleecker, E., Hoffman, E.A., Kanner, R., Kleerup, E., Martinez, F.J., Woodruff, P.G., et al.: Design of the subpopulations and intermediate outcomes in COPD study (SPIROMICS). Thorax **69**, 492–495 (2014)
10. Regan, E.A., Hokanson, J.E., Murphy, J.R., Make, B., Lynch, D.A., Beaty, T.H., Curran-Everett, D., Silverman, E.K., Crapo, J.D.: Genetic epidemiology of COPD (COPDGene) study design. COPD: J. Chronic Obstr. Pulm. Dis. **7**, 32–43 (2011)
11. Hoffman, E.A., Jiang, R., Baumhauer, H., Brooks, M.A., Carr, J.J., Detrano, R., Reinhardt, J., Rodriguez, J., Stukovsky, K., Wong, N.D.: Reproducibility and validity of lung density measures from cardiac CT scans-the multi-ethnic study of atherosclerosis (MESA) lung study 1. Acad. Radiol. **16**, 689–699 (2009)
12. Tan, W., Sin, D., Bourbeau, J., Hernandez, P., Chapman, K., Cowie, R., FitzGerald, J., Marciniuk, D., Maltais, F., Buist, A.S., et al.: Characteristics of COPD in never-smokers and ever-smokers in the general population: results from the CanCOLD study. Thorax **70**, 822–829 (2015)
13. Sonka, M., Park, W., Hoffman, E.A.: Rule-based detection of intrathoracic airway trees. IEEE Trans. Med. Imaging **15**, 314–326 (1996)
14. Park, W., Hoffman, E.A., Sonka, M.: Segmentation of intrathoracic airway trees: a fuzzy logic approach. IEEE Trans. Med. Imaging **17**, 489–497 (1998)

15. Tan, K.L., Tanaka, T., Nakamura, H., Shirahata, T., Sugiura, H.: Segmentation of airway trees from multislice CT using fuzzy logic. In: 2009 Conference Record of the Forty-Third Asilomar Conference on Signals, Systems and Computers, pp. 1614–1617. IEEE (2009)
16. Mori, K., Hasegawa, J.I., Toriwaki, J.I., Anno, H., Katada, K.: Recognition of bronchus in three-dimensional X-ray CT images with applications to virtualized bronchoscopy system. In: Proceedings of the 13th International Conference Pattern Recognition, vol. 3, pp. 528–532. IEEE (1996)
17. Summers, R.M., Feng, D.H., Holland, S.M., Sneller, M.C., Shelhamer, J.H.: Virtual bronchoscopy: segmentation method for real-time display. Radiology **200**, 857–862 (1996)
18. Law, T.Y., Heng, P.: Automated extraction of bronchus from 3D CT images of lung based on genetic algorithm and 3D region growing. In: Medical Imaging, pp. 906–916. International Society for Optics and Photonics (2000)
19. Schlathoelter, T., Lorenz, C., Carlsen, I.C., Renisch, S., Deschamps, T.: Simultaneous segmentation and tree reconstruction of the airways for virtual bronchoscopy. In: SPIE: Medical Imaging, pp. 103–113. International Society Optics Photonics (2002)
20. Pisupati, C., Wolff, L., Zerhouni, E., Mitzner, W.: Segmentation of 3D pulmonary trees using mathematical morphology. In: Maragos, P., Schafer, R.W., Butt, M.A. (eds.) Mathematical Morphology Applications Image Signal Processing. CIV, vol. 5, pp. 409–416. Springer, Heidelberg (1996). doi:10.1007/978-1-4613-0469-2_48
21. Aykac, D., Hoffman, E.A., McLennan, G., Reinhardt, J.M.: Segmentation and analysis of the human airway tree from three-dimensional X-ray CT images. IEEE Trans. Med. Imaging **22**, 940–950 (2003)
22. Bauer, C., Bischof, H., Beichel, R.: Segmentation of airways based on gradient vector flow. In: International Workshop Pulmonary Image Analysis, Medical Image Computing Computer Assisted Intervention, pp. 191–201. Citeseer (2009)
23. Saha, P.K., Chaudhuri, B.B.: Detection of 3-D simple points for topology preserving transformations with application to thinning. IEEE Trans. Pattern Anal. Mach. Intell. **16**, 1028–1032 (1994)
24. Swift, R.D., Higgins, W.E., Hoffman, E.A., McLennan, G., Reinhardt, J.M.: Automatic axis generation for 3D virtual-bronchoscopic image assessment. In: SPIE: Medical Imaging, pp. 73–84. International Society for Optics and Photonics (1998)
25. Reynisson, P.J., Scali, M., Smistad, E., Hofstad, E.F., Leira, H.O., Lindseth, F., Hernes, T.A.N., Amundsen, T., Sorger, H., Langø, T.: Airway segmentation and centerline extraction from thoracic CT-comparison of a new method to state of the art commercialized methods. PloS One **10**, e0144282 (2015)
26. Saha, P.K., Borgefors, G., di Baja, G.S.: A survey on skeletonization algorithms and their applications. Pattern Recogn. Lett. **76**, 3–12 (2016)
27. Fetita, C., Prêteux, F.: Quantitative 3D CT bronchography. In: Proceedings of the IEEE International Symposium Biomedical Imaging, pp. 221–224. IEEE (2002)
28. Liu, X., Chen, D.Z., Tawhai, M.H., Wu, X., Hoffman, E.A., Sonka, M.: Optimal graph search based segmentation of airway tree double surfaces across bifurcations. IEEE Trans. Med. Imaging **32**, 493–510 (2013)
29. Kiraly, A.P., Higgins, W.E., McLennan, G., Hoffman, E.A., Reinhardt, J.M.: Three-dimensional human airway segmentation methods for clinical virtual bronchoscopy. Acad. Radiol. **9**, 1153–1168 (2002)
30. Xu, Z., Bagci, U., Foster, B., Mansoor, A., Udupa, J.K., Mollura, D.J.: A hybrid method for airway segmentation and automated measurement of bronchial wall thickness on CT. Med. Image Anal. **24**, 1–17 (2015)

31. Coxson, H.O.: Quantitative computed tomography assessment of airway wall dimensions: current status and potential applications for phenotyping chronic obstructive pulmonary disease. Proc. Am. Thorac. Soc. **5**, 940–945 (2008)
32. Saha, P.K., Udupa, J.K.: Optimum image thresholding via class uncertainty and region homogeneity. IEEE Trans. Pattern Anal. Mach. Intell. **23**, 689–706 (2001)
33. Ginneken, B., Baggerman, W., Rikxoort, E.M.: Robust segmentation and anatomical labeling of the airway tree from thoracic CT scans. In: Metaxas, D., Axel, L., Fichtinger, G., Székely, G. (eds.) MICCAI 2008. LNCS, vol. 5241, pp. 219–226. Springer, Heidelberg (2008). doi:10.1007/978-3-540-85988-8_27
34. Saha, P.K., Chaudhuri, B.B.: 3D digital topology under binary transformation with applications. Comput. Vis. Image Underst. **63**, 418–429 (1996)
35. Bullitt, E., Gerig, G., Pizer, S.M., Lin, W., Aylward, S.R.: Measuring tortuosity of the intracerebral vasculature from MRA images. IEEE Trans. Med. Imaging **22**, 1163–1171 (2003)
36. Jin, D., Iyer, K.S., Chen, C., Hoffman, E.A., Saha, P.K.: A robust and efficient curve skeletonization algorithm for tree-like objects using minimum cost paths. Pattern Recogn. Lett. **76**, 32–40 (2016)
37. Saha, P.K., Udupa, J.K., Odhner, D.: Scale-based fuzzy connected image segmentation: theory, algorithms, and validation. Comput. Vis. Image Underst. **77**, 145–174 (2000)
38. Saha, P.K., Strand, R., Borgefors, G.: Digital topology and geometry in medical imaging: a survey. IEEE Trans. Med. Imaging **34**, 1940–1964 (2015)
39. Saha, P.K., Wehrli, F.W., Gomberg, B.R.: Fuzzy distance transform: theory, algorithms, and applications. Comput. Vis. Image Underst. **86**, 171–190 (2002)
40. Jin, D., Chen, C., Saha, P.K.: Filtering non-significant quench points using collision impact in grassfire propagation. In: Murino, V., Puppo, E. (eds.) ICIAP 2015. LNCS, vol. 9279, pp. 432–443. Springer, Heidelberg (2015). doi:10.1007/978-3-319-23231-7_39
41. Liu, Y., Jin, D., Li, C., Janz, K.F., Burns, T.L., Torner, J.C., Levy, S.M., Saha, P.K.: A robust algorithm for thickness computation at low resolution and its application to in vivo trabecular bone CT imaging. IEEE Trans. Biomed. Eng. **61**, 2057–2069 (2014)

Rapid Analytic Optimization of Quadratic ICP Algorithms

Leonid German[1($\boxtimes$)], Jens R. Ziehn[1,2], and Bodo Rosenhahn[1]

[1] Institut für Informationsverarbeitung (TNT),
Leibniz Universität Hannover, Hannover, Germany
`german@tnt.uni-hannover.de`
[2] Fraunhofer IOSB, 76131 Karlsruhe, Germany

Abstract. This paper discusses the efficient optimization of iterative closest points (ICP) algorithms. While many algorithms formulate the optimization problem in terms of quadratic error functionals, the discontinuities introduced by varying changing correspondences usually motivate the optimization by quasi-Newton or Gauss-Newton methods. These disregard the fact that the Hessian matrix in these cases is constant, and can thus be precomputed analytically and inverted a-priori. We demonstrate on the example of Allen et al.'s seminal paper "The space of human body shapes", that all relevant quantities for a full Newton method can be derived easily, and lead to an optimization process that reduces computation time by around 98% while achieving results of almost equal quality (about 1% difference). Along the way, the paper proposes minor improvements to the original problem formulation by Allen et al., aimed at making the results more reproducible.

1 Introduction

ICP (iterative closest point) algorithms are simple, locally optimal algorithms that, given a distance measure, compute correspondences between two sets of points (registration). By establishing this correspondence, it is possible to estimate a rigid or non-rigid transformation, that will align the points. This approach has been successfully applied to morph a known shape towards an unknown shape, for example moving a labeled mesh of a generic face onto a stereo depth surface showing the face of a person, to detect face parts and expressions. Another common application is morphing a labeled mesh of a human body onto a laser scan to detect body parts, assign animations, or estimate pose or body shape. ICP is also used to register 2D (retinal images [1]) as well as 3D (CT, ultrasound [2]) medical image data. Registrations can produce a large image by stitching separate snapshots, reveal connections between images of different modalities and help in diagnosis by exposing change in images taken over time.

[3] gives an overview of various registration methods and categorizes them by different properties, among others by techniques chosen for optimization. Among these, many relevant ICP approaches (such as [4–8]) express the matching error between two shapes (which is to be minimized) exclusively in terms of quadratic

© Springer International Publishing AG 2017
C.-S. Chen et al. (Eds.): ACCV 2016 Workshops, Part III, LNCS 10118, pp. 61–75, 2017.
DOI: 10.1007/978-3-319-54526-4_5

differences, which means that the optimization problem has a constant Hessian matrix. This could be exploited by optimizing via Newton's method, using gradients and the inverted Hessian directly. Instead however, the common optimization approach is to *approximate* the relevant quantities (such as the Hessian) numerically at each iteration, for example through the L-BFGS method. This introduces an extreme computational overhead; whether it is justified has (to the knowledge of the authors) not been studied conclusively. [7] remarks "Because the cost function is evaluated based on the control points in the test scan and their closest counterparts in the deformable model, and the closest counterparts may change due to the adjustment [...], the optimization problem is highly nonlinear", which is true: The problem is a discontinuous piecewise combination of parabolae of identical shape but different offset. It is, however, not evaluated, whether approximation algorithms such as BFGS necessarily perform better in this case. [6] computes correspondences in faces by minimizing an objective function built from squares of distances and squares of scalar products of vectors. A minimum is found via L-BFGS by performing at most 1000 iterations. [9] prefers Newton's method over BFGS for its more analytic and transparent formulation, and notes improvements in solution quality for face registration; however there is no acknowledgment of the fact that the Hessian is constant and need not be recomputed in the process, and no record of any significant improvement in computation speed. A simplification step is considered to produce a minimizable quadratic function. We will show, however, that the original function is already locally quadratic almost everywhere with a constant Hessian, so that the proposed repeated solution of a linear system [9] is unnecessary and can be reduced to a matrix multiplication. This leads to a speed-up that can be motivated by the problem understanding alone.

This paper will demonstrate on the example of [5], that the analytic derivation of gradients and of the constant Hessian can dramatically reduce computational effort in applicable ICP algorithms, while at the same time achieving results of very similar quality, despite the fact that the optimization problem is highly discontinuous. To do so, Sect. 2 will introduce the original problem formulation in [5]; Sect. 3 will discuss the resulting problem structure from an optimizational perspective, and derive the relevant analytic quantities to perform Newton's method. Section 4 compares the proposed analytic solution to several variants of the original BFGS solution, to determine their respective performance in terms of result quality and computation time.

2   Problem Statement

This section introduces the notation used to describe the problem, as well as the formulae that define the optimization goals. As previously stated, the goal is to transform vertices of a mesh M (here considered to be a generic low-poly model of a human body) by a vector of unknown homogeneous transformation matrices $\mathbf{x}$ to fit it to a surface S (here a *finite* high-resolution 3D point cloud from a laser scanner, without a specific topology but including normals).

Definition 1 (Surface, s, S). The *surface* is considered to be a set[1] of points S, whose elements $s \in S$ are assigned the following quantities:

- $\boldsymbol{\psi}_s$, an $\mathbb{R}^4$ vector of homogeneous coordinates that describe where s lies on the surface.
- $\boldsymbol{\nu}_s$, the $\mathbb{R}^3$ normal vector of the surface at s.

Definition 2 (Mesh, m, M). The *mesh* is considered to be a set of points M, whose elements $m \in M$ are assigned the following quantities:

- $\mathbf{p}^m$, an $\mathbb{R}^4$ vector of homogeneous coordinates of m in the input mesh.
- $\mathbf{X}^m$, an $\mathbb{R}^{4 \times 4}$ homogeneous transformation matrix; $\mathbf{X}^m \mathbf{p}^m$ is the position of m in the deformed mesh. These matrices are the optimization variables.
- $\mathcal{N}^m \subset M$, the neighborhood of m in the mesh's topology. It holds that $n \in \mathcal{N}^m \Leftrightarrow m \in \mathcal{N}^n$, but $m \notin \mathcal{N}^m$.
- $\mathbf{n}^m$, the $\mathbb{R}^3$ mesh normal at m *after* the transformation. It must be recomputed several times during an optimization from the current $\mathbf{X}^n \mathbf{p}^n, n \in \mathcal{N}^m$.

Definition 3 (Vector of all unknowns, x, $\mathbf{x}^m$). The concatenation of all $\mathbf{X}^m$ into a vector $\mathbf{x} \in \mathbb{R}^{12|M|}$ is referred to as *vector of all unknowns* (the actual ordering of the indices is of no concern here). $\mathbf{x}^m$ denotes the part of $\mathbf{x}$ containing exactly the elements of $\mathbf{X}^m$.

2.1 Optimization Goal

A given vector $\mathbf{x} \in \mathbb{R}^{12|M|}$ is evaluated by an *energy function* of the form

$$E_{\text{total}}(\mathbf{x}) = \delta \cdot E_{\text{data}}(\mathbf{x}) + \mu \cdot E_{\text{marker}}(\mathbf{x}) + \sigma \cdot E_{\text{smooth}}(\mathbf{x}), \qquad (1)$$

using weights $\delta, \mu, \sigma \in [0, \infty)$, with the goal of finding $\mathbf{x}^* \in \arg\min_{\mathbf{x}} E_{\text{total}}(\mathbf{x})$. The corresponding energy terms will be defined in this section, along with a brief discussion of their relevant properties.

Data Error. The similarity between the transformed mesh M and the surface S is determined via

$$E_{\text{data}}(\mathbf{x}) = \sum_{m \in M} \left\| \mathbf{X}^m \mathbf{p}^m - \boldsymbol{\psi}_{s(m)} \right\|_2^2, \qquad (2)$$

where $s(m)$ picks the *closest compatible surface point* to m out of S.

Definition 4 (Closest Compatible Point, $s(m)$). A point $s(m) \in S$ is called the *closest compatible point of m*, if it is the closest to m among all surface points $S(m)$ whose normal $\boldsymbol{\nu}_s$ is within an angle of $\pi/2$ of the transformed mesh normal $\mathbf{n}^m$. Formally

$$S(m) = \{ s \in S | \boldsymbol{\nu}_s^{\mathsf{T}} \mathbf{n}^m \geqslant 0 \} \quad \text{and} \quad s(m) = \arg\min_{s \in S(m)} \| \mathbf{X}^m \mathbf{p}^m - \boldsymbol{\psi}_s \|_2. \qquad (3)$$

E_{data} thus accumulates the squared distances between all mesh points and their closest partners on the surface.

---

[1] While [5] allows for an infinite $|S|$, all known practical applications (including the ones in [5]) use a finite $|S|$. For this reason, we focus on the finite case, which brings about peculiar effects (discussed in Sect. 3) absent in an infinitely fine S.

Marker Error. To stabilize the optimization by use of limited manual pre-processing, a (usually small) number of *markers* can be specified, which are assigned fixed corresponding surface points (in lieu of the closest compatible point function $s(m)$), for example for the head, hands and feet.

Definition 5 (Markers, M', $\check{s}(m)$). A subset $M' \subset M$ is called *markers*, for which there exists an a-priori function $\check{s} : M' \to S$ which assigns marker points $m' \in M'$ definite correspondence $\check{s}(m')$ in the surface.

The distance of marker points to their corresponding (according to $\check{s}$) surface points is evaluated by the marker error:

$$E_{\text{marker}}(\mathbf{x}) = \sum_{m' \in M'} \left\| \mathbf{X}^{m'} \mathbf{p}^{m'} - \boldsymbol{\psi}_{\check{s}(m')} \right\|_2^2. \tag{4}$$

In the formulation in [5], markers are thus evaluated twice, once in (4) and once in (2), which does not exclude M', but possibly with respect to different points. This formulation is optional; (2) may include or exclude M'.

Smoothness Error. The use of homogeneous matrices for transformation, instead of mere translations, means that transformations such as scaling or rotation can be described as a uniform transformation of several points (such as an arm), as opposed to a diverse variety of individual translations. By asking that the matrices be as uniform as possible over the entire mesh (thus for the first time using the mesh's topology $\mathcal{N}$), it is possible to express that the original mesh shape should be locally preserved as well as possible. This is achieved by stating that the matrices $\mathbf{X}^m$ and $\mathbf{X}^n$ of neighboring points should be as similar as possible. The term given in [5] as a *smoothness error*[2] is

$$E_{\text{smooth}}(\mathbf{x}) = \sum_{m \in M} \sum_{n \in \mathcal{N}^m} \left\| \mathbf{X}^m - \mathbf{X}^n \right\|_{\text{fro}}^2, \tag{5}$$

using the Frobenius norm $\left\| \mathbf{M} \right\|_{\text{fro}}^2 = \sum_{i,j} M_{ij}^2$ as a measure of similarity. Although not the main focus of this paper, (5) as proposed in [5] leaves two potential pitfalls, which may be worth resolving:

- **Unit Invariance.** First, it is risky to sum over all (squared) matrix elements regardless of their unit: While all $X_{i,j}$ with $i,j \in \{1,3\}$ are ratios and thus dimensionless, the translation components $X_{i,4}$ with $i \in \{1,3\}$ represent units of length. Therefore, two identical models in different units (e.g. inches and centimeters) optimized with the same parameters could have different optima. For this reason we propose to include another ratio weight ρ in addition to

---

[2] It should be noted that "smoothness" here refers to the matrix elements, not the mesh: Each matrix element should change smoothly between neighboring points. The resulting mesh need not be smooth in shape at all, but neighboring points should be transformed similarly.

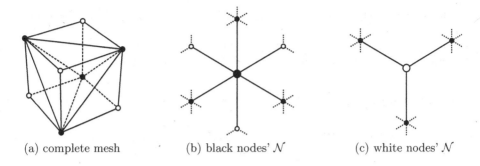

(a) complete mesh   (b) black nodes' $\mathcal{N}$ (c) white nodes' $\mathcal{N}$

Fig. 1. Unevenly distributed neighbor counts on an isotropic 3D model: (a) shows a triangulated cube, which has two types of vertices (shown black and white) with different numbers of neighbors: The black vertices (b) have 6 neighbors each, the white vertices (c) only three. Depending on the smoothness error formulation, black and white vertices behave very differently, even though in the untriangulated cube, all corners are equal.

σ, that turns units into dimensionless values, e.g. $\rho = 2.3/\text{mm}$. Equation (5) thus becomes

$$E_{\text{smooth}} = \sum_{m \in M} \sum_{n \in \mathcal{N}^m} \left\| \underbrace{\begin{bmatrix} 1 & 1 & 1 & \sqrt{\rho} \\ 1 & 1 & 1 & \sqrt{\rho} \\ 1 & 1 & 1 & \sqrt{\rho} \\ \cdot & \cdot & \cdot & \cdot \end{bmatrix}}_{P} \circ (\mathbf{X}^m - \mathbf{X}^n) \right\|_{\text{fro}}^2, \qquad (6)$$

where $(\mathbf{M} \circ \mathbf{N})_{ij} = m_{ij} \cdot n_{ij}$ defines the element-wise (or Hadamard) product. The last row of the weighting matrix is irrelevant because it is identical for all $\mathbf{X}^m$. The square roots represent the fact that all other weights are applied to squared values.

– **Triangulation Invariance.** In the original definition a node with many neighbors will generally be affected more strongly by smoothness than a node with few neighbors, because the errors are summed over the entire neighborhood. This means that the result of the optimization may depend heavily on the topology, but the topology of triangulation does not necessarily reflect the topology of the underlying shape. Figure 1 gives an example: A regular cube is triangulated, but the number of neighbors is very unevenly distributed and varies by a factor of 2. In general this is not avoidable, and thus not a mark of a bad triangulation. For the cube it means that even though a cube in theory has no preferred vertices, its triangulated representation does, and so if this cube is fit to a given surface, its points will have to satisfy very different smoothness requirements and the transformation will likely not be uniform. This can be countered in various ways; the one proposed here is normalizing the smoothness term over the number of neighbors, to obtain (along with the previously introduced weighting of (6))

$$E_{\text{smooth}} = \sum_{m \in M} \frac{1}{|\mathcal{N}^m|} \sum_{n \in \mathcal{N}^m} \left\| \mathbf{P} \circ (\mathbf{X}^m - \mathbf{X}^n) \right\|_{\text{fro}}^2. \qquad (7)$$

In this paper, Eq. (7) was used as a smoothness error, and all following derivations rely on this instead of the original error (5) as given in [5]. The changes necessary to adapt the following considerations to the original formulation are, however, straightforward.

3   Optimization

To (approximately) minimize (1) for $\mathbf{x}^*$, [5] proposes using the limited-memory Broyden–Fletcher–Goldfarb–Shanno (L-BFGS) method, an approximation of Newton's method. We will provide a very brief overview of the key ideas behind both methods in this section, to motivate why for the particular problem structure proposed in [5] (as described in Sect. 2) Newton's method is both simpler and better suited. The reader is referred to [10] for detailed yet accessible accounts of all methods mentioned here.

3.1   Newton's Method and L-BFGS

In optimization, Newton's method takes an objective function $E(x)$ (often assumed to be at least twice continuously differentiable) and an "initial guess" x_i, and improves $x_i \mapsto x_1$ to minimize $E(x_1)$ by only considering local derivatives of E up to second order. The process is usually iterated over $x_i \mapsto x_{i+1}$. It approximates $E(x)$ based on its second-order Taylor expansion

$$E(x_i + s) \approx \tilde{E}_{x_i}(s) = E(x_i) + s \cdot E'(x_i) + s^2 \cdot 1/2 \cdot E''(x_i), \tag{8}$$

where $E' = \mathrm{d}E/\mathrm{d}x$ and $E'' = \mathrm{d}^2 E/(\mathrm{d}x)^2$. The optimum of the parabola $\tilde{E}_{x_i}(s)$ lies at $\hat{s} = -E'(x_i)/E''(x_i)$. If $E'' > 0$, this optimum is a minimum, and the update $x_1 = x_i - \hat{s}$ can be assumed to approximate the (or a) minimum of E. When applied iteratively, the process is repeated until, for example, $E(x_i)$, $|E(x_i) - E(x_{i-1})|$ or $E'(x_i)$ are sufficiently small, or i becomes too large. In n dimensions, Newton's method chooses steps $\hat{s}$ according to

$$\mathbf{H}\hat{s} = -\mathbf{g}, \tag{9}$$

where $\mathbf{H}$ is the local *Hessian* $H_{ij} = \mathrm{d}^2/\mathrm{d}x_i \mathrm{d}x_j E$ and $\mathbf{g} = \nabla E$ is the local *gradient* $g_i = \mathrm{d}E/\mathrm{d}x_i$.

Line Search, Wolfe Conditions and BFGS. Optimization using a *line search* method departs from Newton's method in that it does not compute the step length along with the step direction (as $\hat{s}$), but in a second step and by potentially separate criteria. It is common to use the direction of $\hat{s}$ as determined in (9), and then search along the *line* $\Lambda(\alpha) = E(\mathbf{x} + \alpha \hat{s})$ for a step scale α.

In practice, it is often considered inefficient to attempt to locally or globally minimize even the one-dimensional Λ for α (in particular due to E usually being not well-understood); instead, *Wolfe conditions* are used which provide criteria for so-called *sufficient decrease*. A step that satisfies these conditions is proven to exist, and within these bounds assures progress in minimization. To find the optimal scale α, $\Lambda(\alpha)$ and thereby E usually has to be evaluated several times.

The Broyden–Fletcher–Goldfarb–Shanno (BFGS) method can be applied when the Hessian is not available or too difficult to compute. It approximates the Hessian in this process by starting as a line search along the gradient from a given point (assuming the Hessian e.g. as an identity matrix) and estimating a new Hessian from the *change in gradients* encountered after each iteration. Since the underlying Newton model of quadratic approximation can be used to predict the gradient at the next step, any deviation in the actually computed gradient hints at a defect in the approximated Hessian (either because the approximation was insufficient, or because the Hessian has changed between steps). In this case, the true gradient provides a one-dimensional information to perform a rank-one update on the approximated Hessian. This process is well defined if BFGS is used in combination with *line search* and *Wolfe conditions*.

L-BFGS. In cases where the Hessian is further very large[3], it may be considered inefficient to store and update the full approximation. Therefore, limited-memory BFGS (L-BFGS) stores just a fixed number of previous rank-one updates and expresses the approximated Hessian exclusively in terms of these.

We conclude that the aptness of the three approaches depends on properties of the optimization problem, namely whether the Hessian can and should be computed analytically, and whether function evaluation is so computationally inexpensive that line search and Wolfe conditions should be applied.

Furthermore, the actual performance of an optimization method cannot be predicted exclusively from these metrics; therefore the considerations in the following Sect. 3.2 merely provide a theoretical basis to be validated in Sect. 4.

3.2   Optimizational Properties of the Error Functional

This section will illustrate the properties of E_{total} in terms of optimization, and motivate the choice of Newton's method over L-BFGS. Analytical representations of the gradient and the Hessian will be given. Due to the sum rule, it holds for the gradient that $\nabla E_{\text{total}} = \nabla E_{\text{data}} + \nabla E_{\text{marker}} + \nabla E_{\text{smooth}}$ and respectively for the Hessian that

$$\underbrace{\nabla\nabla^{\mathsf{T}} E_{\text{total}}}_{\mathbf{H}} = \underbrace{\nabla\nabla^{\mathsf{T}} E_{\text{data}}}_{\mathbf{H}_{\text{data}}} + \underbrace{\nabla\nabla^{\mathsf{T}} E_{\text{marker}}}_{\mathbf{H}_{\text{marker}}} + \underbrace{\nabla\nabla^{\mathsf{T}} E_{\text{smooth}}}_{\mathbf{H}_{\text{smooth}}}, \tag{10}$$

so that the gradients and Hessians can be considered for each error term E_{data}, E_{marker} and E_{smooth} independently. It will be shown that the Hessian matrix $\mathbf{H} = \nabla\nabla^{\mathsf{T}} E_{\text{total}}$ is <u>constant</u> over the optimization process for a given application, and depends exclusively on the topology of the input mesh. The reason for this lies in the fact that the energy function E_{total} is piecewise quadratic in the

---

[3] In the present example, there are $\mathcal{O}(|M|^2)$ entries in $\mathbf{H}$, which can be considerable.

variables x_{ij}, and any quadratic function has a constant $\mathbf{H}$. There are, however, discontinuities in $\mathbf{g}$ that propagate into $\mathbf{H}$, when the association between a mesh point and its closest surface point changes. In these places $\mathbf{H}$ is not defined.[4]

Data Error. The term E_{data} of (2) is not everywhere continuously differentiable due to the fact that $\psi_{s(m)}$ is not continuous. The associated closest points may jump at infinitesimally small changes in $\mathbf{X}^m$. It is, however, continuously differentiable *almost* everywhere, since the space of parameters where a mesh point switches the closest surface point has measure zero. Furthermore the term is everywhere *continuous* as even if $\psi_{s(m)}$ jumps from ψ_a to ψ_b when $\mathbf{X}^m$ is changed infinitesimally, the jump occurs just when both ψ_a and ψ_b have the same distance to $\mathbf{X}^m\mathbf{p}^m$, so the limits from both sides match.

Also, using a homogeneous matrix with 12 real parameters instead of an $\mathbb{R}^3$ translation vector to transform individual points means that E_{data} has a huge overhead of redundant parameters. Therefore there is no single minimum but a space of minima with $|M|\cdot 9$ parameters. In general there are $|S|^{|M|}$ disconnected minima, since any point in M can be located directly at any surface point.

Data Gradient. From Eq. (2) it follows that the derivative of E_{data} by the m-th unknown transformation matrix $\mathbf{X}^m$ is:

$$\frac{dE_{\text{data}}}{d\mathbf{X}^m} = \frac{d}{d\mathbf{X}^m}\sum_{\bar{m}\in M}\left\|\mathbf{X}^{\bar{m}}\mathbf{p}^{\bar{m}} - \psi_{s(\bar{m})}\right\|_2^2 \tag{11}$$

$$= \frac{d}{d\mathbf{X}^m}\left\|\mathbf{X}^m\mathbf{p}^m - \psi_{s(m)}\right\|_2^2 = 2\left(\mathbf{X}^m\mathbf{p}^m - \psi_{s(m)}\right)\mathbf{p}^{m\mathsf{T}}, \tag{12}$$

where step (11)–(12) makes use of the fact that each matrix is independent of all others in the sum.[5] As previously discussed, the gradient is not defined where $s(m)$ is not uniquely defined, i.e. when two different surface points are equally close to $\mathbf{X}^m\mathbf{p}^m$. These discontinuities lead to a non-convex problem.

Data Hessian. The derivation of Eq. (12) by another matrix $\mathbf{X}^o$ (because "n" is still reserved for neighbors in the smoothness error) gives

$$\frac{d^2E_{\text{data}}}{d\mathbf{X}^o d\mathbf{X}^m} = \frac{d}{d\mathbf{X}^o}2\left(\mathbf{X}^m\mathbf{p}^m - \psi_{s(m)}\right)\mathbf{p}^{m\mathsf{T}} = \begin{cases} 2\,\mathbf{p}^m\mathbf{p}^{m\mathsf{T}} & \text{for } m = o \\ 0 & \text{else,} \end{cases} \tag{13}$$

which is, as can be seen, independent of all variable properties, and exclusively depends on the untransformed coordinates of the initial mesh. Nevertheless the Hessian is not formally defined at those $\mathbf{x}$ where the closest point associations

---

[4] In the theoretical limit of a continuous surface, this happens almost permanently. In this case, $\mathbf{g}$ is almost always continuous and $\mathbf{H}$ cannot be given as shown here (i.e. in terms of a polygon topology). In general, $\mathbf{H}$ is not constant then. As initially stated, this case is not considered.

[5] The aforementioned overhead in parameters (Sect. 2.1) is mirrored here in the fact that the gradient is of the form $\mathbf{u}\mathbf{v}^\mathsf{T}$, which means it has rank 1 (dimension 3), corresponding to the expected 3° of freedom out of 12 parameters.

$s(m)$ change, because the gradient is not defined there—in spite of the fact that these places constitute a *removable* singularity on behalf of the Hessian, due to the fact that the Hessian is the same everywhere.

Marker Error. The term E_{marker} of (4) is continuous, convex (but not *strictly* convex), quadratic and has a global minimum value of zero, attained when all $\mathbf{X}^m$ (where $m \in M'$) place their corresponding $\mathbf{p}^m$ exactly at their markers $\psi_{\check{s}(m)}$. Again there is no single minimum, once due to the homogeneous coordinates as for the data error, but additionally because once all points in M' are set, all other points $M \setminus M'$ can be assigned arbitrary matrices. This leads to a single connected space of minima with $|M \setminus M'| \cdot 12 + |M'| \cdot 9$ parameters.

Marker Gradient and Hessian. The marker gradient and Hessian are identical to the data gradient and Hessian, only that the corresponding surface points are fixed. This also includes that the marker gradient and Hessian are defined everywhere without exception.

Smoothness Error. The term E_{smooth} of (5) or (7) is continuous and has a global minimum value of zero, attained when all $\mathbf{X}^m$ are identical. It is no single minimum either: $\mathbf{X}^1 = \cdots = \mathbf{X}^{|M|}$ can be arbitrary. Thus the space of minima has 12 free parameters. This is, by the way, independent of which error metric (of the three alternatives given in Sect. 2) is used.

Smoothness Gradient. The derivation of Eq. (7) gives

$$\frac{\mathrm{d}E_{\text{smooth}}}{\mathrm{d}\mathbf{X}^m} = \frac{\mathrm{d}}{\mathrm{d}\mathbf{X}^m} \sum_{\bar{m} \in M} \frac{1}{|\mathcal{N}^{\bar{m}}|} \sum_{\bar{n} \in \mathcal{N}^{\bar{m}}} \|\mathbf{P} \circ (\mathbf{X}^{\bar{m}} - \mathbf{X}^{\bar{n}})\|_{\text{fro}}^2 \qquad (14)$$

$$= \mathbf{P}^2 \circ \left(\frac{2}{|\mathcal{N}^m|} \sum_{\bar{n} \in \mathcal{N}^m} (\mathbf{X}^m - \mathbf{X}^{\bar{n}}) + \sum_{\bar{n} \in \mathcal{N}^m} \frac{2}{|\mathcal{N}^{\bar{n}}|} (\mathbf{X}^{\bar{n}} - \mathbf{X}^m) \right), \qquad (15)$$

where the first addend of Eq. (15) represents the case where $\mathbf{X}^m$ is "here" (in the role of m in Eq. (7)), the second addend is where $\mathbf{X}^m$ is one of the neighbors (in the role of n in Eq. (7)), and $\mathbf{P}^2 = \mathbf{P} \circ \mathbf{P}$, not $\mathbf{PP}$.

Smoothness Hessian. The smoothness Hessian is the derivation of Eq. (15) by another matrix $\mathbf{X}^o$

$$\frac{\mathrm{d}^2 E_{\text{smooth}}}{\mathrm{d}\mathbf{X}^o \mathrm{d}\mathbf{X}^m} = \frac{\mathrm{d}}{\mathrm{d}\mathbf{X}^o} \mathbf{P}^2 \circ \left(\frac{2}{|\mathcal{N}^m|} \sum_{\bar{n} \in \mathcal{N}^m} (\mathbf{X}^m - \mathbf{X}^{\bar{n}}) + \sum_{\bar{n} \in \mathcal{N}^m} \frac{2}{|\mathcal{N}^{\bar{n}}|} (\mathbf{X}^{\bar{n}} - \mathbf{X}^m) \right), \qquad (16)$$

which can be simplified as all X_{ij}^m in one $\mathbf{X}^m$ share the same smoothness Hessian entries. These entries can be denoted by a scale factor η^{om}, which depends only on the mesh points o and m (namely their topology), but not indices i and j. It can be defined via the formula

$$P_{ij}^2 \cdot \eta^{om} = \frac{\mathrm{d}^2}{\mathrm{d}X_{ij}^o \mathrm{d}X_{ij}^m} E_{\text{smooth}}, \qquad (17)$$

where P_{ij}^2 is either 1 or ρ (cf. Eq. (7)). If, as above, this is factored out, η^{om} is fully independent of ij.

The above equation only differentiates twice by the index ij. It must also be considered that the gradient ij could be differentiated by different indices kl. However in this case it holds that

$$\frac{d^2}{dX_{ij}^o dX_{kl}^m} E_{\text{smooth}} = 0 \quad \text{for ij} \neq kl, \tag{18}$$

because in E_{smooth} each matrix entry ij only depends on the entries ij of its neighbors, not on any other indices kl. Also, if o and m are not identical or neighbors, η^{om} will always be 0, because distinct, non-neighboring matrices do not depend on each other in terms of smoothness.

Using these considerations, Eq. (16) can be solved for η^{om} to obtain

$$\eta^{om} = \begin{cases} 2 - \sum_{n \in \mathcal{N}^m} \frac{2}{|\mathcal{N}^n|} & \text{if } o = m \\ \frac{2}{|\mathcal{N}^o|} - \frac{2}{|\mathcal{N}^m|} & \text{if } o \in \mathcal{N}^m \\ 0 & \text{else,} \end{cases} \tag{19}$$

which in turn can be used to set up the smoothness Hessian (using $\mathbf{D}$, which is a $\mathbb{R}^{12 \times 12}$ diagonal matrix whose diagonal elements consist of the $P_{ij} \in \{1, \rho\}$ in the order fitting how $\mathbf{x}^m$ is obtained from $\mathbf{X}^m$):

$$\mathbf{H}_{\text{smooth}} = \begin{bmatrix} \eta^{11}\mathbf{D} & \cdots & \eta^{1n}\mathbf{D} \\ \vdots & \ddots & \vdots \\ \eta^{n1}\mathbf{D} & \cdots & \eta^{nn}\mathbf{D} \end{bmatrix}. \tag{20}$$

Again it can be found that, as with the data Hessian, the smoothness Hessian does not depend on any variables, just on the topology of the input mesh. Taken together this proves the expected result that the complete $\mathbf{H}$ be independent of $\mathbf{x}$, and thus constant throughout the process, because E_{total} is piecewise quadratic.

Combined Error Term. The combined error term E_{total} is the sum of the previously described individual errors (as given in Eq. (1)). Unless some weights are set to zero, it is influenced by all previously described properties. The first important property is that while none of the previous terms defined exact local minima (each had a whole space of minima due to an overhead of parameters) the total error in general requires the full set of parameters, and thus local minima are points, not spaces. The smoothness error makes use of the full homogeneous matrices, and the data and marker errors assure that smoothness is related to an optimal data fit, and thus prefers some transformation matrices over others. Furthermore due to the fact that E_{data} (2) is not continuously differentiable (due to reassignment of closest surface points), while both other errors are, the sum of them must have a discontinuous derivative as well.

3.3   Summary of Optimization Considerations

The previous sections have shown not only that for a given problem the analytic prerequisites for Newton's method, the gradient and the Hessian, can be computed analytically—but that they can be computed easily and efficiently, particularly due to the constant Hessian which need not be recomputed during iterations. The repeated numerical approximation of these values via BFGS is not necessary: The simple problem formulation makes it ideally suited for Newton's method using analytic gradients and Hessians.

Furthermore, it was seen that an evaluation of E requires a considerable number of processing steps for large M and S, due to the effect of normals and point associations. Therefore, line search must be regarded sceptically, since the assumption that function evaluations are less costly than the computation of an exact Newton step is not necessarily satisfied.

However, as initially stated, L-BFGS could still outperform the results of Newton's method, depending on the actual shape of M, S and thus E. To shed light on this, the following Sect. 4 will compare both approaches based on their practical performance on realistic data.

4   Practical Application

To evaluate the practical performance of the proposed analytic optimization, we compare it to the original algorithm in [5] on realistic data. We match three human body meshes of equal topology but different pose, $|M_1| = |M_2| = |M_3| = 1002$ to three laser-scanned human body surfaces, $|S_1| = 347\,644$, $|S_2| = 361\,026$, $|S_3| = 367\,093$. This provides 9 combinations, for each of which we compare Newton's method and different variants of L-BFGS.

[5] proposes a two-pass approach: First minimizing the error using only markers and smoothness, with weights $\delta = 0, \sigma = 1, \mu = 10$, then including the data term with $\delta = 1, \sigma = 1, \mu = 10$.[6] This is meant to place the mesh in a roughly correct position before transforming separate mesh points. We test this method with BFGS, and additionally the "one-pass" approach of using data error right from the start. For BFGS, testing both approaches is valid because (as seen in Fig. 3), the one-pass approach takes only about 2/3 of the computation time, but provides results with about 200% the error. Therefore, quality and speed could be traded off. For Newton's method, omitting the marker-only pass is unnecessary: As for $\delta = 0$ point associations do not change, Newton's method converges after only one iteration, so that the additional effort is negligible.

Additionally, in Newton's method, knowing the Hessian to be constant provides another choice: It is possible to invert the Hessian completely after computing it, and using it to solve (9) via $\hat{\mathbf{s}} = -\mathbf{H}^{-1}\mathbf{g}$, or applying Cholesky decomposition to separate it into two triangular matrices $\mathbf{H} = \mathbf{CC}^{\mathsf{T}}$, and using substitution in each iteration steps to solve (9) for $\hat{\mathbf{s}}$. The former option, which we will denote Newton-INV, features a slower inversion step, but a faster iteration step only involving matrix multiplication; the latter option, denoted Newton-CD, features

---

[6] It should be noted at this point that here the issues discussed in Sect. 2 become apparent, where a lack of specified units will lead to different optimization goals for models in meters, feet or inches, for example. In this case, the unit is considered to be meters.

a faster initial decomposition step, but requires slower substitution operations at each iteration. In general it can be assumed that Newton-INV is faster in cases where many iterations are required, such that $\mathbf{H}^{-1}$ can be reused many times, while Newton-CD is faster for fewer iterations.

Figure 2 shows the comparison between Newton-CD and Newton-INV. In all evaluated cases, Newton-INV is faster, suggesting that already at 13 iterations

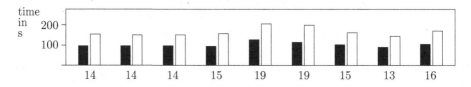

Fig. 2. Comparison of computation times for Newton-INV (black) and Newton-CD (white) on the 9 examples; the required iteration steps until convergence are given below the bars. In all cases, Newton-INV outperforms Newton-CD, at an average 62% of the latter's computation time.

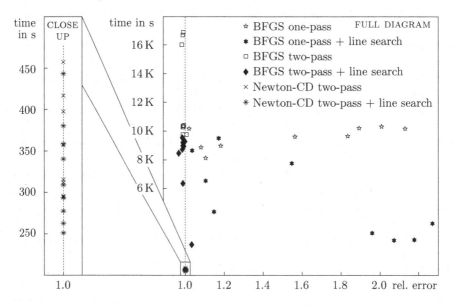

Fig. 3. Performance of Newton's method with Cholesky decomposition and on-the-fly substitution, and different variants of BFGS, computation time vs. result optimality (in error points normalized per mesh/shape combination such that the result of Newton's method is exactly 1. It can be seen that the two-pass BFGS methods usually achieve slightly better results, however at approximately 31× the computation time. One-pass method can match neither the speed nor the quality of Newton's method. Line search almost always reduces computation time considerably. A further acceleration with respect to the Cholesky decomposition, the full a-priori inversion of the Hessian, is given in Fig. 2.

the effort of once inverting **H** outweighs the effort of repeatedly substituting via the Cholesky-decomposed **H** (although implementation details may affect this). In all cases, the same minimum was found in the same number of steps by Newton-CD and Newton-INV, which appears obvious analytically, but indicates that numerical stability is not an issue when choosing between the two approaches.

Figure 3 shows the evaluation of L-BFGS with/without line search, and with one-pass and two-pass, while for Newton's method we only depict the slower Newton-CD and only differentiate between with and without line search. It can be seen that even Newton-CD vastly outperforms the L-BFGS methods in terms of computational effort, which are on average 31 times slower than the Newton-CD approach (and 50 times slower than the Newton-INV approach). The result quality varies; one-pass approaches perform poorly, while the two-pass approaches suggested in [5] usually slightly outperform Newton's method

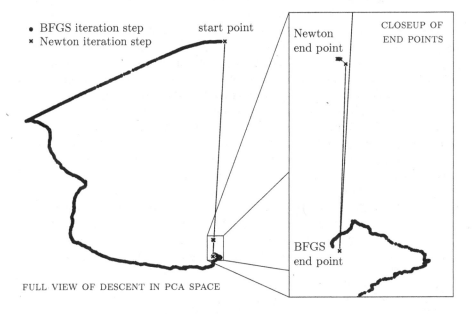

Fig. 4. To visually compare the paths that L-BFGS and Newton's method take, the respective paths of a given transformation optimization were reduced from $\mathbb{R}^{12 \cdot |M|} = \mathbb{R}^{12\,024}$ to $\mathbb{R}^2$. It can be seen that the original L-BFGS approach takes a considerable detour involving many iteration steps, before even reaching the proximity of the first Newton step. L-BFGS terminates approximately there for lack of progress, while Newton's method descends several steps further after this, reaching a lower minimum after just 7 steps.

by an average of 1.06%[7]. Figure 4 visualizes the fundamentally different descent approaches by showing the paths taken by L-BFGS and Newton's method for one of the examples in a projection onto an abstract 2-dimensional space, achieved by computing a PCA on the paths and plotting the two most significant axes. It can be seen that L-BFGS requires many iterations before reaching the vicinity of the optimum; Newton's method on the other hand reaches this region after only one step, but settles in a different local optimum.

5   Conclusion and Outlook

In this paper, we have proposed an analytic solution to ICP algorithms with quadratic error terms, and demonstrated the approach on the example of [5]. It was shown that in such cases, the Hessian matrix is constant and can easily be computed analytically, instead of numerically approximating it during each optimization step, for example via L-BFGS, as is commonly done in ICP methods.

In the given example, an optimization descent with Newton's method using precomputed Hessians was able to reduce computation times on average to (not *by*) 2% of the computation times of BFGS when applied to the same problem. The quality of the optimization result of BFGS was on average slightly better, by around 1.06%, depending on the choice of parameters. Still it can be concluded that Newton's method can achieve very similar result qualities at a fraction of the computation time, rendering quadratic ICP approaches fit for real-time settings or to run on simpler hardware.

On the example of [5], all necessary parameters to adapt the original algorithm to the proposed analytic version were derived comprehensibly and can be readily implemented. Optimizational properties of the problem formulation were analyzed in detail to provide an understanding of the original and the newly proposed optimization process, and facilitate potential further adjustments.

In addition to the improvements in solution efficiency, we have proposed minor improvements to the original problem formulation of Allen et al. that make the formulation more mathematically sound, and allow to reuse weight parameters for models with different mesh triangulation or different units of measurement, without affecting the optimization goal. The original algorithm implicitly required to redetermine weights for different mesh triangulations, and for a different unit system.

---

[7] The approximative algorithm outperforming the exact one in terms of result quality may seem counter-intuitive, but is immanent to the problem structure: Newton's method is exact at finding the closest local optimum where point associations do not change. L-BFGS instead is more likely to miss this "direct" minimum. As its approximation combines information across different point associations, it can thereby "learn" the overall shape of the surface, which is, for high-density surface points, more accurate than the purely local quadratic view. In turn, for sparse surface models, Newtons method is more accurate, as it better captures the dominant structure.

The vast reduction of computation times in quadratic ICPs opens up new optimizational possibilities. As was noted in Sect. 4, Newton's method usually achieves a local minimum that is analytically exact, but slightly greater than the minimum found by the L-BFGS approach. However, as Newton's method required only 1/50 of the computation time, there is plenty of time to search the vicinity and match the result quality of BFGS, and still outperform it in terms of computation time.

References

1. Stewart, C.V., Tsai, C.L., Roysam, B.: The dual-bootstrap iterative closest point algorithm with application to retinal image registration. IEEE Trans. Med. Imaging **22**, 1379–1394 (2003)
2. Almhdie, A., Lger, C., Deriche, M., Lde, R.: 3D registration using a new implementation of the ICP algorithm based on a comprehensive lookup matrix: application to medical imaging. Pattern Recogn. Lett. **28**, 1523–1533 (2007)
3. Tam, G.K.L., Cheng, Z.Q., Lai, Y.K., Langbein, F.C., Liu, Y., Marshall, D., Martin, R.R., Sun, X.F., Rosin, P.L.: Registration of 3D point clouds and meshes: a survey from rigid to nonrigid. IEEE Trans. Vis. Comput. Graph. **19**, 1199–1217 (2013)
4. Allen, B., Curless, B., Popović, Z.: Articulated body deformation from range scan data. ACM Trans. Graph. **21**, 612–619 (2002)
5. Allen, B., Curless, B., Popović, Z.: The space of human body shapes: reconstruction and parameterization from range scans. ACM Trans. Graph. (TOG) **22**, 587–594 (2003)
6. Salazar, A., Wuhrer, S., Shu, C., Prieto, F.: Fully automatic expression-invariant face correspondence. Mach. Vis. Appl. **25**, 859–879 (2014)
7. Lu, X., Jain, A.: Deformation modeling for robust 3D face matching. IEEE Trans. Pattern Anal. Mach. Intell. **30**, 1346–1357 (2008)
8. Sumner, R.W., Popović, J.: Deformation transfer for triangle meshes. ACM Trans. Graph. **23**, 399–405 (2004)
9. Amberg, B., Romdhani, S., Vetter, T.: Optimal step nonrigid ICP algorithms for surface registration. In: 2007 IEEE Conference on Computer Vision and Pattern Recognition, pp. 1–8 (2007)
10. Nocedal, J., Wright, S.J.: Numerical Optimization, 2nd edn. Springer, Berlin (2006)

Segmentation of Partially Overlapping Convex Objects Using Branch and Bound Algorithm

Sahar Zafari[1]([⊠]), Tuomas Eerola[1], Jouni Sampo[2],
Heikki Kälviäinen[1], and Heikki Haario[2]

[1] Machine Vision and Pattern Recognition Laboratory,
School of Engineering Science, Lappeenranta University of Technology,
Lappeenranta, Finland
{Sahar.Zafari,Tuomas.Eerola,Heikki.Kalviainen}@lut.fi
[2] Mathematics Laboratory, School of Engineering Science,
Lappeenranta University of Technology, Lappeenranta, Finland
{Jouni.Sampo,Heikki.Haario}@lut.fi

Abstract. This paper presents a novel method for the segmentation of partially overlapping convex shape objects in silhouette images. The proposed method involves two main steps: contour evidence extraction and contour estimation. Contour evidence extraction starts by recovering contour segments from a binarized image using concave contour point detection. The contour segments which belong to the same objects are grouped by utilizing a criterion defining the convexity, symmetry and ellipticity of the resulting object. The grouping is formulated as a combinatorial optimization problem and solved using the well-known branch and bound algorithm. Finally, the contour estimation is implemented through a non-linear ellipse fitting problem in which partially observed objects are modeled in the form of ellipse-shape objects. The experiments on a dataset of consisting of nanoparticles demonstrate that the proposed method outperforms four current state-of-art approaches in overlapping convex objects segmentation. The method relies only on edge information and can be applied to any segmentation problems where the objects are partially overlapping and have an approximately convex shape.

1 Introduction

Segmentation of overlapping objects aims to address the issue of representation of multiple objects with partial views. Overlapping or occluded objects occur in various applications, such as morphology analysis of molecular or cellular objects in biomedical and industrial imagery where quantitative analysis of individual objects by their size and shape is desired [1–3]. In many such applications, the objects can often be assumed to have approximately elliptical shape. For example, the most commonly measured properties of nanoparticles are their length and width, which can correspond to the major and minor axis of an ellipse fitted over the particle contour [4].

Even with rather strong shape priors, segmentation of overlapping objects remains a challenging task. Deficient information from the objects with occluded

C.-S. Chen et al. (Eds.): ACCV 2016 Workshops, Part III, LNCS 10118, pp. 76–90, 2017.
DOI: 10.1007/978-3-319-54526-4_6

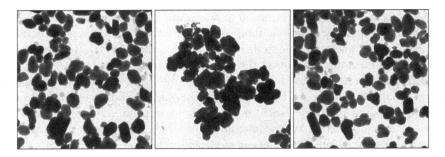

Fig. 1. Overlapping nanoparticles.

or overlapping parts introduces considerable complexity into the segmentation process. For example, in the context of contour estimation, the contours of objects intersecting with each other do not usually contain enough visible geometrical evidence, which can make contour estimation problematic and challenging. Frequently, the segmentation method has to rely purely on edges between the background and the foreground, which makes the processed image essentially a silhouette image (see Fig. 1). Furthermore, the task involves simultaneous segmentation of multiple objects. A large number of objects in the image causes a large number of variations in pose, size and shape of the objects, and leads to a more complex segmentation problem.

Several approaches have been proposed for the segmentation of overlapping objects in various applications. The watershed transform is one of the commonly used approaches in overlapping cell segmentation [5–7]. However, methods based on the watershed transform suffer from a poor or inadequate initialization and may experience difficulties with segmentation of highly overlapped objects in which a strong gradient is not present.

Other commonly used approach is to resolve the segmentation of overlapping objects using active contours [8,9]. The efficiency of the active contour based methods depends highly on the accuracy of the model initialization. Active contour based methods are also computationally heavy when the amount of objects is large.

Morphological operations have been also used for overlapping object segmentation. In [1], an automated morphology analysis coupled with a statistical model to inference and segment partially overlapping nanoparticles. The method is prone to under-segmentation with highly overlapped objects.

In [2], the problem of overlapping objects segmentation was approached using concave points extraction and ellipse fitting. Although this approach is efficient for regular shaped objects, such as bubbles, objects with a shape that deviates from elliptical shape, such as nanoparticles, cause problems for the method. In [10,11], the problem of overlapping objects with non-elliptical shapes were addressed by proposing a method combining various heuristics with the concave point extraction.

In [12], a method for segmentation of overlapping objects with close to elliptical shape was proposed. The method consists of three steps: seed point extraction using radial symmetry, contour evidence extraction via edge to seed point matching, and contour estimation through ellipse fitting.

In this paper, a novel method for the segmentation of partially overlapping convex objects is proposed. The proposed method utilizes both generic and specific object shape properties such as convexity, symmetry and ellipticity. The proposed method follows two sequential steps of contour evidence extraction and contour estimation. The contour evidence extraction step consists of two sub-steps of contour segmentation and segment grouping. In the contour segmentation, contours are divided into separate contour segments. In the segment grouping, contour evidences are built by joining the contour segments that belong to the same object. Once the contour evidence is obtained, contour estimation is performed using numerically stable direct ellipse fitting.

The key contribution of this work is to introduce a novel framework for segment grouping based on the branch and bound (BB) optimization algorithm [13–16]. This is done by optimizing a cost function that combines different object properties including, but not limited to, convexity, symmetry and ellipticity. Another contribution is the integration of the proposed segment grouping method into the segmentation of overlapping objects, enabling improvements compared to four existing methods with higher detection rate and segmentation accuracy.

2   Overlapping Object Segmentation

Figure 2 summarizes the proposed method. Given a gray-scale image as an input, the segmentation process starts with pre-processing to build an image silhouette and the corresponding edge map. The binarization of the image is obtained by background suppression based on the Otsu's method [17] along with morphological opening to smooth the object boundaries. For computational reasons, the connected component are extracted and further analysis is performed for each connected component separately. The edge map is constructed using the Canny edge detector [18]. In the contour evidence extraction steps, edge points that belonged to each object are grouped using concave points and proposed branch and bound algorithm. Once the contour evidence has been obtained, contour estimation is carried out to infer the missing parts of the overlapping objects.

2.1   Contour Segmentation

The first step of the proposed method is to extract the contour evidence containing the visible parts of the objects boundaries that can be used to inference the occluded parts of overlapped objects. The contour evidence extraction involves two separate tasks: contour segmentation and segment grouping.

For contour segmentation, first the image edge are extracted by the Canny edge detector, and then the concave points are obtained through the detection

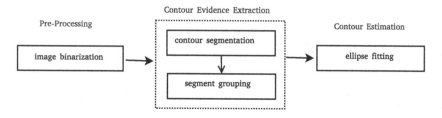

Fig. 2. Proposed method.

of corner points followed by the concavity test. The corner points are detected using the modified curvature scale space (CSS) method based on curvature analysis [19]. The output of the corner detector includes the points with the maximum curvature lying on both concave and convex regions of object contours. Since being only interested in the concave points joining the contours of overlapping objects, the detected corner points are examined if they lie on concave regions.

Let p_i be a detected corner point and p_{i-k} and p_{i+k} its two kth adjacent contour points. The corner point p_i is qualified as concave if the line connecting p_{i-k} to p_{i+k} does not reside inside the object. The obtained concave points are used to split the contours into contour segments. Figure 3 shows an example of concave point extraction and contour segmentation.

A partial overlap between two or more convex shape objects leads to a concave shape with concave edge points that correspond to the intersections of the object boundaries. It is a common practice to utilize these concave points to segment the contour of overlapping objects. Different methods such as polygonal approximation [2,10], curvature [20], and angle [10,21] have been applied to determine the location of concave points in the image.

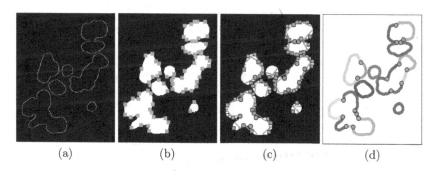

(a)   (b)   (c)   (d)

Fig. 3. Contour segmentation: (a) Edge map; (b) Corner detection by [19]; (c) Concavity test to extract concave corners (green circle) and removed convex corners (pink square); (d) Contour segmentation by concave points (the colors are used only for illustrative purpose to visualize the segmented contour by concave points). (Color figure online)

2.2   Segment Grouping

Due to the overlap between the objects and the irregularities in the object shapes, a single object may produce multiple contour segments. Segment grouping is needed to merge all the contour segments belonging to the same object. In this section, we lay out the mathematical definition of the contour segment grouping problem in terms of a grouping criterion. We formulate the grouping task as a combinatorial optimization problem and solve it using the branch and bound algorithm. The efficiency of this algorithm depends upon the definition of the grouping criterion. We will later study the grouping criterion in more details.

Problem Formulation: Let $S = \{S_1, S_2, ..., S_N\}$ be an ordered set of N contour segments in a connected component of an image. The aim is to group the contour segments into M subsets such that the contour segments that belong to individual objects are grouped together and $M \leq N$. Let ω_i be the group membership indicator giving the group index to which the contour segment S_i belongs to. Denote Ω be the ordered set of all membership indicators: $\{\omega_1, \omega_2, ..., \omega_N\}$.

The grouping criterion is given by a scalar function $J(\cdot)$ of Ω which maps a possible grouping of the given contour segments onto the set of real numbers $\mathbb{R}$. That is, J is the cost of grouping that ideally measures how the grouping Ω resembles the true contour segments of the objects. The grouping problem for the given set of S is to find the optimal membership set Ω^* such that the grouping criterion (cost of grouping) is minimized:

$$\Omega^* = \underset{\Omega}{\text{argmin}} \ J(\Omega; S) \tag{1}$$

Branch and Bound Algorithm: The contour segment grouping problem can be considered as a combinatorial optimization problem. As the number of groups (objects) is not known, there are at most $1 + \frac{2^N}{2!} + ... + \frac{M^N}{M!} ... + \frac{N^N}{N!}$ distinct candidate solutions. The optimal solution can be obtained by evaluating the grouping criterion $J(\cdot)$ for every candidate grouping and by selecting the solution with minimum grouping criterion. However, this optimization problem is NP hard where even with modest values of M and N, the explicit enumeration is impossible due to the exponentially increasing number of candidate solutions. The BB algorithm is known as an efficient approach for NP-hard combinatorial optimization problems [13–15].

BB is efficient for our optimization problem since it avoids exhaustive enumeration using the value of the current optimal solution and defining bounds for the function to be optimized.

The BB algorithm evaluates the partial solutions for the grouping problem as $\Omega_k = \{\omega_1, \omega_2, ... \omega_k\}$, $1 < k < N$. Let us denote $b_1, b_2, ..., b_N$ as a sequence of lower bounds satisfying

$$b_k = b_k(\omega_1, \omega_2, ..., \omega_k) \leq J(\Omega; S). \tag{2}$$

Let Ω_i and Ω_j be two sets of partial solutions for $i < j$. If the lower bound on Ω_i is greater than the lower bound on Ω_j, i.e. $b_i > b_j$, b_j is replaced with b_i. This implies that:

$$b_1 \leq b_2 \leq \ldots \leq b_N = J(\Omega; S). \tag{3}$$

Let us denote B the upper bound on the optimal J, i.e.,

$$J(\Omega^*; S) \leq B. \tag{4}$$

If the lower bound b_k of a given partial solution Ω_k is greater than the upper bound of optimal solution B, $b_k(\Omega_k) \geq B$, then

1. Ω_k is suboptimal and explicitly rejected, since $J(\Omega; S) \geq b_k > B > J(\Omega^*; S)$, and
2. all grouping of the form $\{\omega_1, \omega_2...., \omega_k, \omega_{k+1}, \ldots, \omega_N\}$ are considered to be suboptimal and implicitly rejected since b_k is unchanged.

Hence, given a single candidate solution to the optimization problem and a sequence of lower and upper bounds as defined in Eqs. 2 and 4, one can eliminate the candidate solutions to the grouping problem that are evaluated as suboptimal. To be more illustrative, we demonstrate the branch and bound algorithm in the contour segment grouping using an example shown in Fig. 4. Given the set of extracted contour segments $S = \{S_1, S_2, S_3, S_4\}$, the BB algorithm systematically searches all the candidate groupings and attempts to find the optimum. It is convenient to represent the whole grouping process of the BB algorithm through a search tree as shown in Fig. 5. The root of such search tree represents the initial state where each contour segment is considered as an individual group, that is S_1, S_2, S_3, and S_4 as four distinct groups. The descenders of the root represent all the other candidate grouping obtained by merging of S. By this means, every node at the level m of the search tree represents a distinct grouping of m segments.

The nodes in search tree are explored by the depth-first search such that there are at most four optimal groupings. With a naive strategy to expand the search tree, a large number of redundant nodes is generated. To increase the efficiency of the search, this should be avoided. For example, the grouping $\{S_2, S_1\}$ is not created as it has been already generated in the previous branch as $\{S_1, S_2\}$. The expansion of the search tree is controlled by the grouping criterion. The optimal grouping is the node with the minimum cost. The upper bound for the optimal grouping is set as soon as the first node is generated (see Fig. 6). Provided that the node already takes part in an optimal grouping, the upper bound is set by the cost of the optimal grouping. Otherwise, the upper bound is set by the cost of the node itself. The initial upper bound in S_1, S_2, S_4 are set to the their costs as the nodes have not been appeared in any optimal grouping so far. The initial upper bound in S_3 is set to $J(S_1 S_3)$ as it is already part of the optimal solution $\{S_1, S_3\}$.

The BB terminates search earlier if the current solution is recognized as suboptimal, i.e., the cost of current solution is greater than the upper bound for optimal solution $J(\Omega; S) > B$. The upper bound B is updated when a solution is found which cost is less than the current value of B.

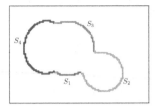

Fig. 4. An example of the contour segmentation result.

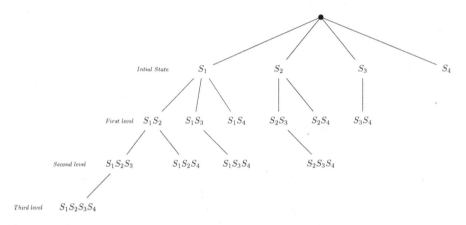

Fig. 5. The corresponding non-redundant search tree for the coutour segments presented in Fig. 4.

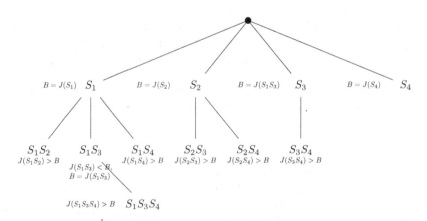

Fig. 6. An example of the termination criterion in the search tree (red color represents the suboptimal solutions and the green color represents the optimal solutions). (Color figure online)

Grouping Criterion: The BB algorithm can be applied with any grouping criterion J. However, the selection of the grouping criterion has significant effect on the overall method performance. Our proposal for the grouping criterion is a hybrid cost function consisting of two parts: (1) *generic part* ($J_{\text{concavity}}$) that encapsulate the general convexity properties of the objects and (2) *specific part* that encapsulates the properties of objects that are exclusive to a certain application, e.g., symmetry (J_{symmetry}) and ellipticity ($J_{\text{ellipticity}}$). The proposed cost function is defined as follows:

$$J = \underbrace{\alpha J_{\text{concavity}}}_{\text{Generic}} + \underbrace{\beta J_{\text{ellipticity}} + \gamma J_{\text{symmetry}}}_{\text{Specific}} \tag{5}$$

where α, β, and γ are the weights for each term respectively.

The generic part encourages the convexity assumption of the objects in order to penalize the grouping of the contour segments belonging to different objects. This is achieved by incorporating a quantitative concavity measure. Given two contour segments S_i and S_j the generic part of the cost function is defined as follows:

$$J_{concavity} = \begin{cases} 1, & \text{if there is concave point between } S_i \text{ and } S_j \\ (1 - \frac{A_{S_i \cup S_j}}{A_{\text{ch}, S_i \cup S_j}}) & \text{otherwise,} \end{cases} \tag{6}$$

where $A_{S_i \cup S_j}$ is the area of a region bounded by S_i, S_j, and the lines connecting the endpoint of S_j to the starting point of S_i and the end point of S_i to the starting point of S_j. $A_{\text{ch}, S_i \cup S_j}$ is the corresponding area of the convex hull bounded by S_i and S_j.

The specific part is adapted to consider the application criteria and certain object properties. Considering the object under examination, several functions can be utilized. Here, the specific term is formed by the ellipticity and symmetry cost functions.

The ellipticity term is defined by measuring the discrepancy between the fitted ellipse and the contour segments [2]. Given the contour segment S_i consisting of n points, $S_i = \{p_k(x_k, y_k)\}_{k=1}^{n}$, and the corresponding fitted ellipse points, $S_{f,i} = \{p_{f,k}(x_{f,k}, y_{f,k})\}_{k=1}^{n}$, the ellipticity term is defined as follows:

$$J_{\text{ellipticity}} = \frac{1}{n} \sum_{k=1}^{n} \sqrt{(x_k - x_{f,k})^2 + (y_k - y_{f,k})^2}. \tag{7}$$

The symmetry term penalize for the non-symmetry of the resulting object. Let O_i and O_j be the center of symmetry of the contour segments S_i and S_j obtained by aggregating the normal vector of the contour segments. The procedure is similar to fast radial symmetry transform [22], but the gradient vectors are replaced by contour segments normal vector. We call this transform as the normal symmetry transform (NST).

In NST every contour segment point gives a vote for the plausible radial symmetry at some specific distance from that point. Given the distance value r of the predefined range $[R_{min}\ R_{max}]$, for every contour segment point (x, y), NST determines the negatively affected pixels p_{-e} and sequentially constructs the orientation projection image $\boldsymbol{O}_r$:

$$p_{-e}(x, y) = (x, y) - round\left(\frac{\boldsymbol{n}(x, y)}{\|\boldsymbol{n}(x, y)\|} \times r\right),\tag{8}$$

$$\boldsymbol{O}_r(p_{-e}(x, y)) = \boldsymbol{O}_r(p_{-e}(x, y)) - 1,\tag{9}$$

In order to construct the orientation image, the radial symmetry contribution $\boldsymbol{S}_r$ for the radius $m \in [R_{min}, R_{max}]$ is formulated as

$$\boldsymbol{S}_r(x, y) = \left(\frac{|\tilde{\boldsymbol{O}}_r(x, y)|}{k_r}\right)^{\alpha},\tag{10}$$

where α and k_r are the radial strictness and the scaling factor that normalizes $\boldsymbol{O}_r$ across different radii, respectively. $\tilde{\boldsymbol{O}}_r$ is defined as

$$\tilde{\boldsymbol{O}}_r(x, y) = \begin{cases} \boldsymbol{O}_r(x, y), & \text{if } \boldsymbol{O}_r(x, y) < k_r. \\ k_r, & \text{otherwise.} \end{cases}\tag{11}$$

The full NST transform $\boldsymbol{S}$ by which the interest symmetric regions are defined is given by the average of the symmetry contributions over all the radii $r \in [R_{min}, R_{max}]$ considered:

$$\boldsymbol{S} = \frac{1}{|N|} \sum_{r \in [R_{min}, R_{max}]} \boldsymbol{S}_r.\tag{12}$$

The center of symmetry O_i and O_j of the contour segments are estimated as the average locations of the detected symmetric regions in $\boldsymbol{S}$. The symmetry term J_{symmetry} is defined as the Euclidean distance between O_i and O_j. The distance is normalized to $[0...1]$ according to maximum diameter of the object.

2.3   Contour Estimation

The last step of proposed method is the contour estimation, where, by means of the visual information produced from the previous step, the missing parts of the object contours are estimated. Ellipse fitting is a very common approach in overlapping object segmentation, especially in the medical and industrial applications.

The most efficient recent ellipse fitting methods based on shape boundary points are generally addressed through the classic least square fitting problem. In this work, the contour estimation is addressed through a stable direct least square fitting method [23] where the partially observed objects are modeled in the form of ellipse-shape objects. Figure 7 shows an example of contour estimation applied to contour evidences.

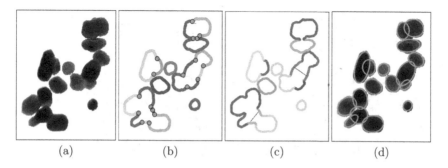

(a)	(b)	(c)	(d)

Fig. 7. Contour estimation: (a) Original image; (b) Contour segmentation; (c) Segment grouping (the thin gray lines are added to illustrate the grouping of non-adjacent segments); (d) Contour estimation.

3   Experiments

3.1   Data

The experiments were carried out using a dataset consisting of nanoparticles images captured by transmission electron microscopy. In total, the dataset contains 11 images of 4008×2672 pixels. Around 200 particles were marked manually in each image by an expert. The annotations consist of manually drawn contours of the objects. Since not all the objects are marked, a pre-processing step was applied to eliminate the unmarked objects from the images. It should be noted that the images consist of dark objects on a white background and, therefore, pixels outside the marked objects could be colored white without making the images considerably easier to analyze.

3.2   Results

To evaluate the method performance and to compare the methods, two specific performance measures, True Positive Rate (TPR) and Positive Predictive Value (PPV), were used:

$$TPR = \frac{TP}{TP + FN} \tag{13}$$

$$PPV = \frac{TP}{TP + FP} \tag{14}$$

where True Positive (TP) is the number of correctly segmented objects, False Positive (FP) is the number of incorrectly segmented objects, and False Negative (FN) is the number of missed objects.

To decide whether the segmentation result was correct or incorrect, Jaccard Similarity coefficient (JSC) [24] was used. Given a binary map of the segmented object O_s and the ground truth particle O_g, JSC is computed as

$$JSC = \frac{O_s \cap Og}{O_s \cup Og}. \tag{15}$$

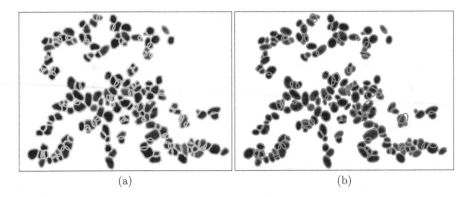

(a)   (b)

Fig. 8. An example of the proposed method segmentation result on nanoparticles dataset: (a) Ground truth; (b) Proposed method.

The threshold value for the ratio of overlap (JSC threshold) in the JSC was set to 0.5. The average JSC (AJSC) value was used as the third measure to evaluate the segmentation performance.

Considering the smallest and largest object axes, the method parameters R_{min} and R_{max} were set to 10 and 20 respectively. The parameter k was experimentally set to 10. The weighting parameters α, β, and γ were set to 0.4, 0.1 and 0.5 based on the prior knowledge about the shape of objects in the image. Figure 8 shows an example of segmentation result.

The performance of the proposed segmentation method was compared to four existing state-of-the-art methods, Concave point detection and Contour evidence extraction (CC) [11], a method based on Seed point extraction, Contour evidence extraction and Contour estimation (SCC) [12], Nanoparticles Segmentation (NPA) [1], and Concave-point Extraction and Contour Segmentation (CECS) [2]. These methods are particularly chosen as previously applied for segmentation of overlapping convex and elliptical shape objects. The implementations made by the corresponding authors were used for NPA, CC, and SCC [25]. CECS was implemented by ourselves based on [2]. The source code of proposed method written in Matlab can be downloaded from [25].

Examples of typical segmentation results are presented in Fig. 9. SCC and NPA suffers from under-segmentation while CC and CECS tend to over-segment the objects. The proposed method neither under- nor over-segments the objects.

The corresponding performance statistics of the competing methods applied to the dataset are shown in Table 1. As it can be seen, the proposed method outperforms the other four with respect to the TPR and JSC and achieves a comparable performance with NPA in terms of PPV. The high JSC value of the proposed method indicates its superiority with respect to the resolved overlap ratio.

Figures 10 shows the effect of the Jaccard similarity threshold on the TPR, PPV, and AJSC scores with the proposed and competing segmentation methods. As expected, the segmentation performance of all methods degrades when the

Table 1. Comparison of the performance of the proposed method for the nanoparticles dataset.

Methods	TPR [%]	PPV [%]	AJSC [%]
Proposed	**86**	88	**77**
CC	85	84	73
SCC	79	89	72
NPA	62	**90**	58
CECS	66	73	53

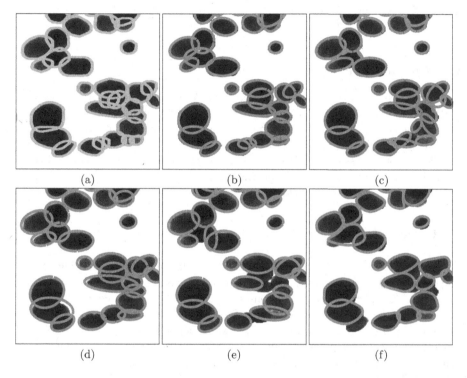

Fig. 9. Examples of segmentation results on the slice of nanoparticles dataset: (a) Ground truth; (b) Proposed method; (c) CC; (d) SCC; (e) CECS; (f) NPA.

JSC threshold is increased. However, the threshold value has only minor effect on the ranking order of the methods and the proposed segmentation method outperforms the other methods with all JSC threshold values below 0.9. With the threshold value of 0.9 NPA is slightly better due to the fact that the objects are not perfect ellipses and the ellipse fitting cannot estimate the contours with a such accuracy.

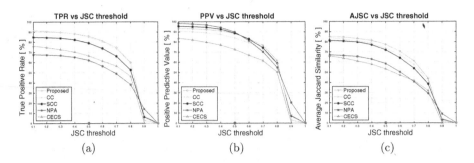

Fig. 10. Performance of the segmentation methods with different values of the JSC threshold in the nanoparticles dataset: (a) TPR; (b) PPV; (c) AJSC.

3.3 Computation Time

The proposed method was implemented in MATLAB, using a PC with a 3.20 GHz CPU and 8 GB of RAM. With the selected combination of parameters the computational time was 170 s per image, while CC demanded 21 s, CECS 77 s, SCC 150 s, NPA demanded 200 s. The computational time breakdown was as follows: contour segmentation 2%, segment grouping 96%, and ellipse fitting 2%. However, it should be noted that the method performance was not optimized and the computation time could be improved.

4   Conclusions

This paper presented a method to segment multiple partially overlapping convex shaped objects in silhouette images using concave points and branch and bound algorithm. The proposed method consisted of two main steps: contour evidence extraction to detected the visible part of each object and contour estimation to estimate the final objects contours. The experiments showed that the proposed method achieved high detection and segmentation accuracies and outperformed four competing methods on a dataset of nanoparticles images. The proposed method relies only on edge information and can be applied also to other segmentation problems where the objects are partially overlapping and have an approximately convex shape.

References

1. Park, C., Huang, J.Z., Ji, J.X., Ding, Y.: Segmentation, inference and classification of partially overlapping nanoparticles. IEEE Trans. Pattern Anal. Mach. Intell. **35**, 669–681 (2013)
2. Zhang, W.H., Jiang, X., Liu, Y.M.: A method for recognizing overlapping elliptical bubbles in bubble image. Pattern Recogn. Lett. **33**, 1543–1548 (2012)

3. Kothari, S., Chaudry, Q., Wang, M.: Automated cell counting and cluster segmentation using concavity detection and ellipse fitting techniques. In: IEEE International Symposium on Biomedical Imaging, pp. 795–798 (2009)
4. Fisker, R., Carstensen, J., Hansen, M., Bødker, F., Mørup, S.: Estimation of nanoparticle size distributions by image analysis. J. Nanopart. Res. **2**, 267–277 (2000)
5. Shu, J., Fu, H., Qiu, G., Kaye, P., Ilyas, M.: Segmenting overlapping cell nuclei in digital histopathology images. In: 35th International Conference on Medicine and Biology Society (EMBC), pp. 5445–5448 (2013)
6. Cheng, J., Rajapakse, J.: Segmentation of clustered nuclei with shape markers and marking function. IEEE Trans. Biomed. Eng. **56**, 741–748 (2009)
7. Jung, C., Kim, C.: Segmenting clustered nuclei using h-minima transform-based marker extraction and contour parameterization. IEEE Trans. Biomed. Eng. **57**, 2600–2604 (2010)
8. Zhang, Q., Pless, R.: Segmenting multiple familiar objects under mutual occlusion. In: IEEE International Conference on Image Processing (ICIP), pp. 197–200 (2006)
9. Ali, S., Madabhushi, A.: An integrated region-, boundary-, shape-based active contour for multiple object overlap resolution in histological imagery. IEEE Trans. Med. Imaging **31**, 1448–1460 (2012)
10. Bai, X., Sun, C., Zhou, F.: Splitting touching cells based on concave points and ellipse fitting. Pattern Recogn. **42**, 2434–2446 (2009)
11. Zafari, S., Eerola, T., Sampo, J., Kälviäinen, H., Haario, H.: Segmentation of partially overlapping nanoparticles using concave points. In: Bebis, G., et al. (eds.) ISVC 2015. LNCS, vol. 9474, pp. 187–197. Springer, Heidelberg (2015)
12. Zafari, S., Eerola, T., Sampo, J., Kälviäinen, H., Haario, H.: Segmentation of overlapping elliptical objects in silhouette images. IEEE Trans. Image Process. **24**, 5942–5952 (2015)
13. Doig, A.G., Land, A.H.: An automatic method for solving discrete programming problems. Econometrica **28**, 497–520 (1960)
14. Principles, E., Clausen, J.: Branch and bound algorithms (2003)
15. Koontz, W.L.G., Narendra, P.M., Fukunaga, K.: A branch and bound clustering algorithm. IEEE Trans. Comput. **24**, 908–915 (1975)
16. Lempitsky, V., Blake, A., Rother, C.: Image Segmentation by Branch-and-Mincut. Springer, Heidelberg (2008)
17. Otsu, N.: A threshold selection method from gray-level histograms. Automatica **11**, 23–27 (1975)
18. Canny, J.: A computational approach to edge detection. IEEE Trans. Pattern Anal. Mach. Intell. **8**, 679–698 (1986)
19. He, X., Yung, N.: Curvature scale space corner detector with adaptive threshold and dynamic region of support. In: Proceedings of the 17th International Conference on Pattern Recognition, pp. 791–794 (2004)
20. Wu, X., Kemeny, J.: A segmentation method for multi-connected particle delineation. In: IEEE Workshop on Applications of Computer Vision, pp. 240–247 (1992)
21. Wang, W.: Binary image segmentation of aggregates based on polygonal approximation and classification of concavities. Pattern Recogn. **31**, 1503–1524 (1998)
22. Loy, G., Zelinsky, A.: Fast radial symmetry for detecting points of interest. IEEE Trans. Pattern Anal. Mach. Intell. **25**, 959–973 (2003)

23. Fitzgibbon, A., Pilu, M., Fisher, R.B.: Direct least square fitting of ellipses. IEEE Trans. Pattern Anal. Mach. Intell. **21**, 476–480 (1999)
24. Choi, S.S., Cha, S.H., Tappert, C.C.: A survey of binary similarity and distance measures. J. Syst. Cybern. Inform. **8**, 43–48 (2010)
25. Zafari, S., Eerola, T., Sampo, J., Kälviäinen, H., Haario, H.: Segmentation of overlapping objects (2016). http://www2.it.lut.fi/project/comphi1/index.shtml. Accessed Aug 2016

Classification of Lung Nodule Malignancy Risk on Computed Tomography Images Using Convolutional Neural Network: A Comparison Between 2D and 3D Strategies

Xingjian Yan[1,3], Jianing Pang[1,2], Hang Qi[1,3], Yixin Zhu[1,3], Chunxue Bai[4], Xin Geng[5], Mina Liu[6], Demetri Terzopoulos[1,3], and Xiaowei Ding[1,3(✉)]

[1] VoxelCloud Inc., Los Angeles, CA 90012, USA
xding@voxelcloud.io
[2] Cedars-Sinai Medical Center, Los Angeles, CA 90048, USA
[3] University of California, Los Angeles, Los Angeles, CA 90095, USA
[4] Shanghai Zhongshan Hospital, Shanghai, China
[5] Department of Cardiothoracic Surgery,
Huashan Hospital of Fudan University, Shanghai, China
[6] Shanghai Chest Hospital of Shanghai Jiao Tong University, Shanghai, China

Abstract. Computed tomography (CT) is the preferred method for non-invasive lung cancer screening. Early detection of potentially malignant lung nodules will greatly improve patient outcome, where an effective computer-aided diagnosis (CAD) system may play an important role. Two-dimensional convolutional neural network (CNN) based CAD methods have been proposed and well-studied to extract hierarchical and discriminative features for classifying lung nodules. It is often questioned if the transition to 3D will be a key to major step forward in performance. In this paper, we propose a novel 3D CNN on the 1018-patient Lung Image Database Consortium collection (LIDC-IDRI). To the best of our knowledge, this is the first time to directly compare three different strategies: slice-level 2D CNN, nodule-level 2D CNN and nodule-level 3D CNN. Using comparable network architectures, we achieved nodule malignancy risk classification accuracies of 86.7%, 87.3% and 87.4% against the personal opinion of four radiologists, respectively. In the experiments, our results and analyses demonstrates that the nodule-level 2D CNN can better capture the z-direction features of lung nodule than a slice-level 2D approach, whereas nodule-level 3D CNN can further integrate nodule-level features as well as context features from all three directions in a 3D patch in a limited extent, resulting in a slightly better performance than the other two strategies.

1 Introduction

Lung cancer is the deadliest type of cancer worldwide. It is estimated that lung cancer caused 158, 040 deaths in the US in 2015, which accounts for nearly 40% of all cancer deaths in the country [1]. Worldwide, it caused 1.69 million deaths

© Springer International Publishing AG 2017
C.-S. Chen et al. (Eds.): ACCV 2016 Workshops, Part III, LNCS 10118, pp. 91–101, 2017.
DOI: 10.1007/978-3-319-54526-4_7

in 2012, nearly 20% of the total cancer deaths in the world [2]. The prognosis of lung cancer depends critically on the stage at which it is diagnosed. The five-year survival rate of early-stage disease is over 50%, whereas that of advanced stage is less than 5%. Currently, more than half of the diagnosed lung cancer cases are in advanced stage [1]. Therefore, effective screening of lung cancer is crucial for detecting the disease in an early and more treatable stage, consequently improving patient survival rates.

Computed tomography (CT) is the current preferred method for non-invasive lung cancer screening due to its high sensitivity. The National Lung Screening Trial, which enrolled more than 50,000 high-risk subjects, demonstrated that low dose CT screening reduced lung cancer mortality by more than 20% compared with chest radiography screening [3]. Despite its promise, the current lung cancer screening method by CT bears several limitations:

1. Interpretation of the CT images requires analyzing hundreds of images at a time, considerably increasing the workload of radiologists;
2. Significant interobserver and intraobserver variations make the screening result subjective and less reliable [4];
3. The false positive rate remains high, limiting the utility of CT as a early screening modality [5].

In this work, we implemented a self-contained artificial neural network for classifying malignant and benign lung nodules, and tested three strategies for feeding the 3D nodule volume into the network: independent 2D slices with nodule-level voting, simultaneous multi-slice input, and full 3D volumetric input. All nodules were extracted from the annotated images in the publicly available Lung Image Database Consortium (LIDC) [6].

1.1 Related Work

Considerable efforts have been devoted to developing efficient, observer-independent computer aided diagnosis (CAD) methods for differentiating malignant from benign nodules [7–14]. The general scheme usually includes first designing and extracting features (e.g. geometry, texture, opacity, etc.) from image patches, and then training a classifier (e.g. linear discriminant analysis, support vector machine, artificial neural network, etc.) to categorize nodules. The performance is usually evaluated by the receiver operating characteristics (ROC).

In contrast to handcrafted features, deep learning methods learn a representation of data via training end-to-end and are capable of automatically extracting features specific to the learning task at hand. The image data is fed directly into a multi-layer convolutional neural network (CNN) that includes convolutional, pooling, and fully connected layers. Leveraging the availability of large training datasets, advances in training algorithms, and increasingly accessible computational power, such methods have achieved impressive performance in various tasks in computer vision (e.g. [15,16]). Several deep learning based approaches for CT lung nodule classification have also been proposed, with differences in network configuration, nodule extraction strategies, and whether the network is self-contained or requires a separate classifier (i.e. network is used for feature extraction only) [17–19]. The major limitation of these works is the use of individual 2D patches as the network input and the subsequent patch-level classifications. Certain information from the nodule-level, e.g. texture and context features in the z-direction, are ignored, which is a sub-optimal setup as the nodules are intrinsically 3D objects.

1.2 Contributions

Our work made the following three contributions:

1. We proposed a 3D convolutional neural network method for lung nodule malignancy risk classification.
2. We compared three neural network input strategies: 2D slice level CNN, 2D nodule level CNN, and 3D nodule level CNN.
3. Using 3D CNN approach, we achieved the best classification accuracy reported (87.4%) on LIDC-IDRI dataset.

2 Data and Method

We implemented three convolutional neural network (CNN) input strategies (slice-level 2D CNN, nodule-level 2D CNN, and 3D CNN) and evaluated their performance on lung nodule malignancy risk classification. All nodules were extracted from the annotated images in the publicly available Lung Image Database Consortium (LIDC) [6] and the reference standard of nodule malignancy risk were obtained using a reader opinion voting procedure described below.

2.1 Data

Both locations and malignancy risk scores of the lung nodules in each CT scan are annotated in 1018 CT images from the LIDC database. The scores range from 1 (not suspicious to be malignant) to 5 (highly suspicious to be malignant) and are given by panels of up to four radiologists. To extract a single score from the panels' readings, majority voting was conducted to account for the subjectivity of each expert's experience. a score of less than 3 was considered one vote for benign, and one above 3 was one vote for malignant, while a score of 3 was discarded. The final classification of a nodule was then determined by the majority vote. If there were equal votes for both classes, the nodule would be removed from the study. A similar voting approach was also used in [17]. Note that the binary label for each nodule was based on the subjective opinions of the readers, therefore was not equivalent to a biopsy or outcome proved ground truth. In the 1882 nodules used in our experiment, the number of nodules voted benign was roughly twice that of nodules voted as malignant. Simple data augmentations were performed (e.g. rotation) to increase the size of the training set and also balance the number of benign and malignant nodules.

2.2   Slice-Level 2D Convolutional Neural Network

A typical self-contained CNN image classifier is a function that takes an image as an input and outputs a $c \times 1$ vector, where c is the number of classes and the i-th element of the vector is the probability of the input image belonging to the i-th class.

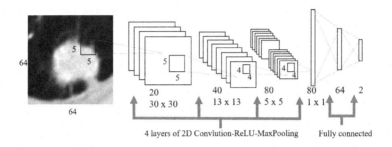

Fig. 1. The network architecture of slice-level 2D CNN.

The configuration of our slice-level 2D CNN is illustrated in Fig. 1. The images are first fed into four sets of 2D convolution-ReLU-pooling layers with 20, 40, 80, and 80 filters of size 5×5, 5×5, 4×4, and 4×4 respectively. The activation function is set as rectified linear unit (ReLU) [20]. The output feature map from each convolution filter all feed into max pooling layers with kernel size 2×2. To maximize the complexity of the model on such input with small size in each direction (64×64), the kernel of the three consecutive convolution-activation-maxpooling layers is set such that the dimension of the last output feature map is 1×1. Finally, we compose a fully connected layer with 64 neurons (with batch-normalization and 50% dropout [21]) and a softmax layer with two outputs. The entire network, as well as for the following two networks, is trained from scratch with randomly initialized weights. To accommodate the large dataset size, the optimization method was chosen to be stochastic gradient descent with 100 noduels per batch (same of the following two models), following the heuristic proposed in [16] to control the learning rate.

The nodule patches are automatically extracted using the labeled center-of-mass, contour, and diameter information. As shown in Fig. 2, the distribution of lung nodule sizes varies from 1.5 to 35 mm, with a mean of 8 mm and standard deviation of 6 mm. Since the neural network requires a fixed size of input patches, for each nodule, we cropped out 5 patches of of size 64×64 pixels in the $x - y$ plane (approximately 40 mm field of view), with the middle patch's center being the center of mass of the nodule. During testing, we choose the majority vote of the output from the network of the five patches to be the final result.

2.3   Nodule-Level 2D Convolutional Neural Network

The slice-level strategy treats all 2D slices as independent examples. Essentially, it "shuffles" nodules in the training examples and discards all information

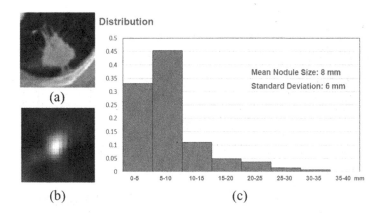

Fig. 2. (a) and (b) are examples of axial slices of two different lung nodules with size of 20 mm and 3 mm. (c) shows the distribution of the nodules' size in the x direction. The horizontal axis is the size range, and the vertical axis is the probability of the total nodules in different size intervals.

along z direction, including texture features and correlations between slices that belonged to the same nodule. To address these shortcomings, we proposed a modified configuration that classified the nodules using multiple 2D slices simultaneously.

The simultaneous multi-slice network took a 3D patch and internally interpreted it as an image with multiple channels. The network has exactly the same architecture as that of the Slice-level 2D CNN as shown in Fig. 1.

The pipeline takes a 3D patch with size $64 \times 64 \times 5$, obtained the same way as in the Slice-level case, and consider it as one image with five channels. The nodule level 2D CNN architecture allows the network to be trained and tested on a nodule by nodule basis and eliminates the need for voting in the slice-level classification approach described above.

2.4   3D Convolutional Neural Network

Given the two configurations above, the intuitive next step is to extract features from all three dimensions at the same time, i.e. building a 3D CNN models that takes the advantages of all 3D information provided by the images. Our implementation is based on [22]. Each convolutional layer contains a number of 3D filters, where each has a size of $w \times h \times z$, where w, h, and z are the width, height, and depth, respectively. Such 3D filters are more powerful than 2D filters [22,23] in the sense that they not only capture spatial relationships between different axial slices, but also are capable to detect the volumetric differences between nodules.

An illustrated in Fig. 3, the 3D CNN network has a similar structure to nodule-level 2D CNN. The network has four sets of 3D convolution-ReLU-pooling layers, followed by two fully connected layers with 50% dropout. The

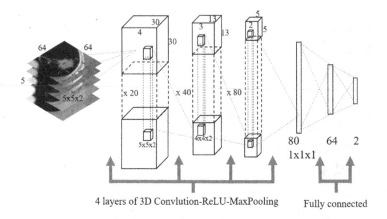

Fig. 3. Illustration of network architecture for 3D CNN.

last layer of the network is a softmax fully connected layer. Each convolutional layer consists of 20, 40, 80, and 80 filters and kernels of size $5 \times 5 \times 2$, $5 \times 5 \times 2$, $4 \times 4 \times 2$, and $4 \times 4 \times 2$, respectively. 2×2 max-pooling in the x and y dimension is applied in the pooling layers. The two fully connected layers have 64 and 2 nodes, respectively (with batch normalization as well as 50% dropout). The the convolution kernels of each layer of the 3D network has one times more weights than the above 2D models. With increased complexity, the 3D network should theoretically produce at least as good result as the 2D networks. Stochastic gradient descent was uses to train the 3D network. For comparison, the same $64 \times 64 \times 5$ patch is used to train the 3D CNN.

3   Experiment Results

3.1   Implementation Details

First, 300 nodules were randomly selected as the testing set with balanced number of begin and malignant nodules, while the remaining over 1500 nodules were selected as training set. Note that the number of benign nodules in the original train set was almost double that of malignant nodules. Thus, for the purpose of balancing the dataset, we doubled the number of malignant nodules in the training set by adding a copy of them with small random translation. Then, we augmented the training set by rotating the nodules 90° four times along the z axis with respect to the center of each patch and flipping to generate over 25000 nodules. Such a setting helped capture a range of translation and rotation invariant features. We split the training set equally into 5 sets and used them to perform a 5-fold cross validation for evaluating classification performance of the three types of CNN we have trained. In each fold, there were 5000 nodules and the number of benign and malignant nodules were very close due to augmentation and shuffling as described above. Each fold of cross validation of the three models was trained for 20 epochs, and the loss generally converges after 10 epochs. The CNN implementation used in this work was the deep learning toolkit Torch [24].

3.2 Results

The three models each with five set of weights produced from cross validation is tested with the testing set of nodules. The five individual accuracies for each network in the 5-fold cross-validation phase are very close (within 3%). The averaged classification accuracy on the testing set achieved by the Slice-level 2D CNN, Nodule-level 2D CNN, and 3D CNN are 86.7%, 87.3% and 87.4% respectively. Table 1 shows the performance metrics, including accuracy, sensitivity, and specificity, averaged over five outcomes of cross validation.

We picked weights of the three models that performed best in testing and drew their testing results as ROC curve as shown in Fig. 4. The area under the ROC curves (ROC AUC) are listed in Table 1. The overall performance of three classifiers suggested that our method can achieve promising results. The 3D approach slightly outperformed the other two in global accuracy and Sensitivity. The advantages of 3D approach in classification performance is limited. It can be ascribed to a factor that although 3D convolutional neural networks can produce useful dimensional reduction without losing the information from the third dimension that is very helpful for lung nodule classification tasks, the low spacial resolution of CT images in z direction (compared to the resolution in x-y directions) limited this capability of 3D CNN.

Table 1. Accuracy measure of three different models

Models	Accuracy	Sensitivity	Specificity	ROC AUC
2D CNN slice-level	86.7%	78.6%	91.2%	0.926 ± 0.014
2D CNN nodule-level	87.3%	88.5%	86.0%	0.937 ± 0.014
3D CNN	87.4%	89.4%	85.2%	0.947 ± 0.014

4   Discussion

Several previous works also utilized the LIDC dataset and proposed CNN based methods for lung nodule malignancy risk classification. Shen et al. [17] proposed to use a multi-scale CNN for feature extraction from center 2D nodule patches, and support vector machine or random forest for classification. The highest achieved accuracy was 86.8%. Kumar et al. [18] proposed to use a five-layer autoencoder to extract features from 2D patches and a decision tree for classification. The mean achieved accuracy was 75.01%. Hua et al. [19] proposed to use a deep belief network (DBN) or CNN for both feature extraction and classification from independent 2D patches. The achieved sensitivity and specificity for the DBN/CNN approach were 73.4/73.3% and 82.2/78.7%, respectively.

Receiver Operating Characteristic Comparison of Three Models

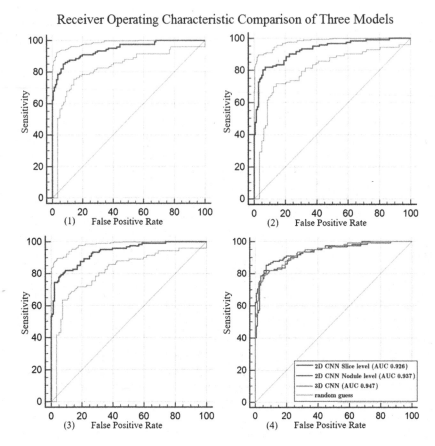

Fig. 4. Receiver operating characteristic comparison of three different models. Plot (1), (2), and (3) are ROC responses of 3D CNN, 2D CNN Nodule Level, and 2D CNN Slice Level respectively. The dashed lines indicate the 95% confidence interval. Plot (4) is a comparison of the ROC responses.

Considering these previous attempts, the major contributions of our work included: (1) implementation of a voting procedure to integrate classification results from multiple 2D patches to provide a nodule-level prediction; (2) implementation of a 3D input strategy to utilize the additional information from the additional dimension; and (3) direct comparison between the 3D approach and 2D approaches for lung nodule malignancy risk classification. Intuitively, lung nodules are 3D objects, therefore it is suboptimal to only extract 2D features from independent slices since all texture and context features in the third dimension are ignored. By constructing the CNN to take 3D volumes directly, it is now possible to extract the object-level features and improve the diagnostic performance over the 2D approaches. For example, for a 2D model, it is not possible to classify a patch that does not intersect or just intersect by a very small area with the nodule itself, while for a 3D implementation such "empty" patches may be included for extracting context features in the third dimension. Although "empty" patches can be avoided on a research dataset where nodules are carefully segmented and boundaries are perfectly defined, in most real world situations only weak labels, e.g. a rough nodule bounding box, are available, and a full 3D model may be superior than the 2D approaches in handling this situation.

Future efforts are warranted to address the limitations of the presented work to further improve and validate the performance. First, our networks were trained and evaluated using the malignancy risk scores given by a panel of radiologists. Such reference standard is subjective and depends on the reader's training and experience. Therefore, it is desirable to use a more objective reference standard such as a biopsy-based malignancy rating. A direct comparison between the performance of the proposed method and a human reader will also become possible. Second, a network with larger feature maps and more layers may further improve the classification performance, as suggested by previous experiences in the general image classification tasks. A larger training set will also be beneficial. Lastly, the locations and sizes of the nodules are given in the LIDC dataset, which is not necessarily provided in a real-world scenario. Hence, it is desirable to integrate a nodule detection module to the current work flow in order to provide a complete solution towards clinical feasibility.

5 Conclusion

In this work, we implemented a 3D CNN based method for lung nodule malignancy classification, trained and tested using the LIDC images. We compared three strategies: 2D slice-level, 2D nodule-level, and full 3D nodule-level. Using the malignancy risk score from a panel of readers as the reference labels, the accuracies of the three input methods were 86.7%, 87.3% and 87.4%, respectively. With comparable network architectures, it was found that incorporating 3D context can only slightly improve the risk prediction performance of the CNN based classifier. But 3D CNN model may be superior than the 2D approaches in handling the situation that only weakly labeled or rough nodule region is available.

References

1. Siegel, R.L., Miller, K.D., Jemal, A.: Cancer statistics, 2016. CA Cancer J. Clin. **66**, 7–30 (2016)
2. Stewart, B., Wild, C.P., et al.: World cancer report 2014. World (2015)
3. Team, N., et al.: Reduced lung-cancer mortality with low-dose computed tomographic screening. N. Engl. J. Med. **365**, 395 (2011)
4. Erasmus, J.J., Gladish, G.W., Broemeling, L., Sabloff, B.S., Truong, M.T., Herbst, R.S., Munden, R.F.: Interobserver and intraobserver variability in measurement of non-small-cell carcinoma lung lesions: implications for assessment of tumor response. J. Clin. Oncol. **21**, 2574–2582 (2003)

5. Swensen, S.J., Jett, J.R., Hartman, T.E., Midthun, D.E., Mandrekar, S.J., Hillman, S.L., Sykes, A.M., Aughenbaugh, G.L., Bungum, A.O., Allen, K.L.: CT screening for lung cancer: five-year prospective experience 1. Radiology **235**, 259–265 (2005)
6. Armato, S.G., McLennan, G., Bidaut, L., McNitt-Gray, M.F., Meyer, C.R., Reeves, A.P., Zhao, B., Aberle, D.R., Henschke, C.I., Hoffman, E.A., et al.: The lung image database consortium (LIDC) and image database resource initiative (IDRI): a completed reference database of lung nodules on ct scans. Med. Phys. **38**, 915–931 (2011)
7. Rubin, G.D., Lyo, J.K., Paik, D.S., Sherbondy, A.J., Chow, L.C., Leung, A.N., Mindelzun, R., Schraedley-Desmond, P.K., Zinck, S.E., Naidich, D.P., et al.: Pulmonary nodules on multi-detector row CT scans: performance comparison of radiologists and computer-aided detection 1. Radiology **234**, 274–283 (2005)
8. Furuya, K., Murayama, S., Soeda, H., Murakami, J., Ichinose, Y., Yauuchi, H., Katsuda, Y., Koga, M., Masuda, K.: New classification of small pulmonary nodules by margin characteristics on highresolution CT. Acta Radiol. **40**, 496–504 (1999)
9. Gurney, J.W., Swensen, S.J.: Solitary pulmonary nodules: determining the likelihood of malignancy with neural network analysis. Radiology **196**, 823–829 (1995)
10. Kawata, Y., Niki, N., Ohmatsu, H., Kusumoto, M., Kakinuma, R., Mori, K., Nishiyama, H., Eguchi, K., Kaneko, M., Moriyama, N.: Computerized analysis of 3-d pulmonary nodule images in surrounding and internal structure feature spaces. In: Proceedings of 2001 International Conference on Image Processing, vol. 2, pp. 889–892. IEEE (2001)
11. Kido, S., Kuriyama, K., Higashiyama, M., Kasugai, T., Kuroda, C.: Fractal analysis of internal and peripheral textures of small peripheral bronchogenic carcinomas in thin-section computed tomography: comparison of bronchioloalveolar cell carcinomas with nonbronchioloalveolar cell carcinomas. J. Comput. Assist. Tomogr. **27**, 56–61 (2003)
12. Shiraishi, J., Abe, H., Engelmann, R., Aoyama, M., MacMahon, H., Doi, K.: Computer-aided diagnosis to distinguish benign from malignant solitary pulmonary nodules on radiographs: ROC analysis of radiologists' performance - initial experience 1. Radiology **227**, 469–474 (2003)
13. Armato, S.G., Altman, M.B., Wilkie, J., Sone, S., Li, F., Doi, K., Roy, A.S.: Automated lung nodule classification following automated nodule detection on CT: a serial approach. Med. Phys. **30**, 1188–1197 (2003)
14. Mori, K., Niki, N., Kondo, T., Kamiyama, Y., Kodama, T., Kawada, Y., Moriyama, N.: Development of a novel computer-aided diagnosis system for automatic discrimination of malignant from benign solitary pulmonary nodules on thin-section dynamic computed tomography. J. Comput. Assist. Tomogr. **29**, 215–222 (2005)
15. Yang, J., Yu, K., Gong, Y., Huang, T.: Linear spatial pyramid matching using sparse coding for image classification. In: IEEE Conference on Computer Vision and Pattern Recognition 2009, CVPR 2009, pp. 1794–1801. IEEE (2009)
16. Krizhevsky, A., Sutskever, I., Hinton, G.E.: Imagenet classification with deep convolutional neural networks. In: Advances in Neural Information Processing Systems, pp. 1097–1105 (2012)
17. Shen, W., Zhou, M., Yang, F., Yang, C., Tian, J.: Multi-scale convolutional neural networks for lung nodule classification. In: Ourselin, S., Alexander, D.C., Westin, C.-F., Cardoso, M.J. (eds.) IPMI 2015. LNCS, vol. 9123, pp. 588–599. Springer, Heidelberg (2015). doi:10.1007/978-3-319-19992-4_46
18. Kumar, D., Wong, A., Clausi, D.A.: Lung nodule classification using deep features in CT images. In: 2015 12th Conference on Computer and Robot Vision (CRV), pp. 133–138. IEEE (2015)

19. Hua, K.L., Hsu, C.H., Hidayati, S.C., Cheng, W.H., Chen, Y.J.: Computer-aided classification of lung nodules on computed tomography images via deep learning technique. Onco Target Ther. **8**, 2015–2022 (2015)
20. Nair, V., Hinton, G.E.: Rectified linear units improve restricted Boltzmann machines. In: Proceedings of the 27th International Conference on Machine Learning (ICML 2010), pp. 807–814 (2010)
21. Srivastava, N., Hinton, G., Krizhevsky, A., Sutskever, I., Salakhutdinov, R.: Dropout: a simple way to prevent neural networks from overfitting. J. Mach. Learn. Res. **15**, 1929–1958 (2014)
22. Tran, D., Bourdev, L.D., Fergus, R., Torresani, L., Paluri, M.: C3D: generic features for video analysis. CoRR, abs/1412.0767 **2** 7 (2014)
23. Ji, S., Xu, W., Yang, M., Yu, K.: 3D convolutional neural networks for human action recognition. IEEE Trans. Pattern Anal. Mach. Intell. **35**, 221–231 (2013)
24. Collobert, R., Kavukcuoglu, K., Farabet, C.: Torch7: a matlab-like environment for machine learning (2011)

A Hybrid Convolutional Neural Network for Plankton Classification

Jialun Dai, Zhibin Yu, Haiyong Zheng$^{(\boxtimes)}$, Bing Zheng, and Nan Wang

College of Information Science and Engineering,
Ocean University of China, Shandong 266100, China
zhenghaiyong@ouc.edu.cn

Abstract. Plankton are fundamental and essential to marine ecosystem, and its survey is significant for sustainable development and ecosystem balance of oceans. The large amount of plankton species and complex relationship among different classes bring difficulty for us to design an automatic plankton classification system. Thus, we develop our model based on convolutional neural network and aim to overcome these shortages. We consider two different ways to extract global and local features to describe shape and texture information of plankton. Furthermore, we design a pyramid fully connected structure to merge different inner products from each sub networks. The experimental results prove our model can take advantage of multiple features and performs better than original convolutional neural network.

1 Introduction

Plankton are a diverse group of organisms that live in the water column of large bodies of water and that cannot swim against a current, which provide a crucial source of food to many large aquatic organisms, such as fish and whales [1]. Diversity and abundance of plankton vary with the currents, geography of seas and ocean basins, and atmospheric conditions. Plankton composition also changes with the seasons, climate, and pollution [2]. So the plankton survey, including species composition, abundance distribution as well as their spatial and temporal changes, has a very important scientific and practical significance for our marine ecosystem, environmental monitoring and marine fishery.

Traditional plankton survey by net and water sampling is hard to meet the demand of *in situ* long-time continuous observation and large-scale fast real-time analysis, especially more and more *in situ* plankton imaging systems are developed and deployed currently [3], such as VPR (Video Plankton Record) [4] and SPC (Scripps Plankton Camera) [5]. Scientists are increasingly using imaging-based technologies to study these creatures in their natural habit, and images from such systems provide an unique opportunity to model and understand plankton ecosystems. Therefore, more and more image datasets for plankton classification are published recently [5,6].

Due to the increasingly plankton image datasets, the corresponding plankton analysis attracts more and more attention [7–9]. Based on the *in situ* images

© Springer International Publishing AG 2017
C.-S. Chen et al. (Eds.): ACCV 2016 Workshops, Part III, LNCS 10118, pp. 102–114, 2017.
DOI: 10.1007/978-3-319-54526-4_8

captured by VPR [4] and SIPPER (Shadowed Image Particle Profiling Evaluation Recorder) [10], Xiaoou Tang *et. al.* combined the shape and texture features and studied the plankton classification (at most 7 classes) via neural networks (NN), principle component analysis (PCA), support vector machine (SVM), etc. [11,12]. By using ZooScan integrated system [13], Gaby Gorsky *et. al.* acquired the zooplankton images and extracted more than sixty properties as features to train six different classifiers (NN, SVM, and Random Forest) for predicting abundance of 20 categories [14]. Besides, Heidi Sosik and Robert Olson [15] studied the automated taxonomic classification of phytoplankton in 22 categories sampled with imaging-in-flow cytometry. Also, ADIAC (Automatic Diatom Identification And Classification) [16] and DiCANN (Dinoflagellate Categorisation by Artificial Neural Network) [17] focused the classification on Diatom and Dinoflagellate respectively, while also used the basic shape and texture feature descriptors and traditional classifiers such as NN, SVM, RF (Random Forest), and DT (Decision Tree), on classification of less than 6 categories.

As we see, although plankton image classification has been addressed for more than two decades, this issue is still very challenging because:

1. both the species and the morphologies are huge;
2. both the intra-class variance and the inter-class similarity are large;

which make this problem a very hard fine-grained visual recognition [18], so that the classical classifiers (NN, SVM, RF, DT, etc.) with traditional hand-crafted features (SIFT, HOG, LBP, etc.) cannot be work well due to the overfitting [19].

Benefit from the big data and high performance computing, deep learning has been proved more and more workable in many fields such as image classification, speech recognition, and bioinformatics [20].

Deep learning methods, which bring a possible solution for big data analysis and image classification problems, can be used to overcome the shortage of the current plankton research. Convolutional neural networks (CNNs) belong to a typical deep neural network. With the deeper and larger of networks, the CNNs can be more capable to solve the problem of large-scale and difficult image recognition. And CNNs have achieved remarkable performance in image and video recognition [21–23].

The design of convolutional neural networks follows the discovery of visual mechanisms in living organisms. CNNs are originated from the neocognitron. The necognitron is introduced in 1980 by Fukushima [24]. It can be seen as a predecessor to convolutional networks. The neocognitron differs from convolutional networks because it does not force units located at several positions to have the same trainable weights. In 1989, LeCun integrated some constraints in backpropagation network to enhance the ability of learning networks to improve the final predictions [25]. And it is shown to outperform all other techniques in 1998 [26]. CNNs are further improved by Krizhevsky and his colleagues in 2012 [21]. They develop AlexNet and become the champion of ImageNet Large Scale Visual Recognition Challenge in 2012. With the powerful learning capacity, convolutional neural networks also can be applied to various tasks for resolving different problems.

For convolutional neural network, it can achieve remarkable performance in image classification under the conditions that it requires many images, hardware and good-designed network structure. Feature designed methods always can obtain good results in some special occasions, and can not apply to more classes or images further. However, based on data and learning capacity, convolutional neural network can explore more abstract and high-level information compared with feature designed methods, which means that it can find more useful patterns to obtain better results in image classification.

Unlike the common features used to classify faces, cars or other objects in daily life, plankton have their own special features such like setae and texture. In this paper, we aim to use different feature descriptors to extract those special plankton features. We not only use original images but also consider 2 traditional feature extraction methods to capture local and global plankton features for classification tasks based convolutional neural network. We believe that the global feature images can represent plankton appearance information and the local feature images describe the plankton texture. In order to take advantage of 3 different features, we designe a pyramid structure in fully connected layer to emerge the information from three channels.

As we mentioned before, inter-class similarity and intra-class variance make plankton classification difficult. Inter-class similarity means that plankton from different classes may be similar and intra-class variance means that plankton from the same class may be various. Humans always classify different objects from shape at first. However, this way does not work well in plankton classification. For different plankton class, the shape of different plankton may be similar. And under this condition, we can distinguish plankton texture, which is the inner-feature in plankton. Texture are various for different plankton class. For the same plankton class, texture are always similar though the shape of plankton from the same class may be different. So shape and texture are very important in plankton classification. And we can transform original images to global feature images and local feature images to represent shape and texture, and fuse these features in convolutional neural networks to improve performance in plankton classification.

Our main contribution is not only a hybrid convolutional neural network model, but a way to analyze plankton images. In biological point of view, both shape and texture are important to classify planktons. And we try to take advantage of this point to design the network model. Experimental results show that our model can achieve better result compared with a single CNN model.

The structure of our paper is organized as follows: some related methods about plankton classification are introduced in Sect. 2. The proposed approach is illustrated in Sect. 3, where we give a detail description about the network composition. In Sect. 4, some experimental results and analysis are discussed in this part. The final conclusion is given in Sect. 5.

2   Research Background

2.1   Convolutional Neural Network

A typical convolutional neural network (CNN) consists of 3 different neuron layers: convolutional layers, pooling layers and fully connected layers.

Convolutional Layers. As a typical deep neural network, each neuron of CNN in the convolutional layer is formed using the input from a local receptive field in the preceding layer and the learned kernels (weights). Neurons within the same feature map share the same kernels but are obtained using different input receptive fields. The kernels used from different feature maps in the same layer are different. Subsequently, an activation function is used. This can be represented as:

$$y_{ij}^{kl} = f((W^k * x)_{ij} + b_j^{kl}) \tag{1}$$

x is the input; i and j is the central position of convolutional computation between kernel and input; y_{ij}^{kl} is the output value of the kth feature map from the lth layer; W^k is the kernel weights; b_j^{kl} denotes the bias of this feature map and $f(x)$ is the activation function.

Pooling Layers. The pooling layer is a form of non-linear down-sampling. Its function is to progressively reduce the spatial size of the representation to reduce the amount of parameters and computation in the network, and hence to also control overfitting. Besides, it provides a form of translation invariance. The pooling layer only changes the size of the input maps while not altering the number of input maps. Averaging and the maximum are the most popular ways to implement the pooling operations. And maxing pooling is the most common implement which is also used in our experiment.

$$y_j^l = f(\beta_j^l \cdot down(y_j^{l-1}) + b_j^l) \tag{2}$$

$beta$ is a constant; $down(\cdot)$ expresses a sub-sampling function, which works as max-pooling or average pooling.

Fully Connected Layer. Neurons in a fully connected layer have full connections to all activations in the previous layer, as seen in regular neural networks. Their activations can hence be computed with a matrix multiplication followed by a bias offset.

$$y_j^l = f(x_j^l) = f(\sum_{i=1}^{N} y_i^{l-1} * w_{ij} + b_j^l) \tag{3}$$

y_j^l is the output value of the jth neuron from l layer; x_j^l is the presynaptic value; N is the total number of input neurons of the previous layer; b_j^l denotes the bias in l layer and $f(x)$ is the activation function.

3   Proposed Method

In order to enhance the plankton image feature extraction problem, we considered 2 different ways to extract plankton features along with original images. Some plankton have significant differences on the shape of cell wall with similar texture of nucleus and cytoplasm, while some of them are similar in the cell wall shape with different internal texture. Our architecture uses multi-sources data that are composed of original images, global feature images and local feature images as the input at the beginning of our architecture for extracting more abstract and representative features. In the first step we extract global and local features from original plankton images for pre-processing. And then we build three sub networks based on AlexNet to load different plankton features respectively. Each sub network has its own input but share the output error (Fig. 3).

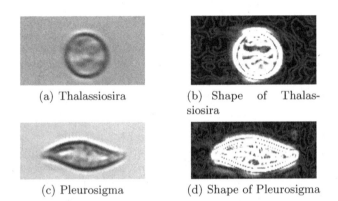

(a) Thalassiosira   (b) Shape of Thalassiosira

(c) Pleurosigma   (d) Shape of Pleurosigma

Fig. 1. The two images on the left are similar in texture. It's difficult to distinguish them just depending on texture information. And the internal texture may not work in some plankton classes. And two images on the right are processed into global feature images, which represents the shape information of plankton.

3.1   Feature Extraction

Global Feature. Setae and shape, which can be considered as a kind of global feature, are always used for plankton classification [27]. We develop a method to extract the global features of plankton. We wish this kind of features can describe the appearance of plankton and omit the internal texture.

At first, bilateral filter [28] is used to remove some noise of plankton images and smooth the plankton shape. Then, scharr operator [29], which is optimized by Sobel operator, can mark plankton appearance information accurately. In the end, in order to distinguish global features (including shape and setae information), we increase the contrast value of image to make shape information outstanding and omit the information inside the plankton. And the image after contrast enhancement can be seen to represent the global feature.

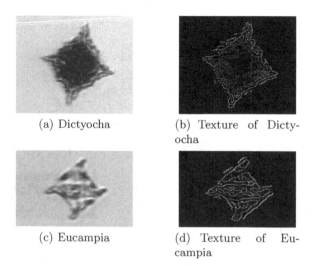

(a) Dictyocha

(b) Texture of Dictyocha

(c) Eucampia

(d) Texture of Eucampia

Fig. 2. The two images on the left are similar in shape. It is difficult to classify these two images correctly just based on shape information. And the internal texture can provide valuable information to distinguish them. And two images on the right are detected by canny edge detector, which describe the important internal information of plankton. The extra information can be used to improve accuracy.

Local Feature. The texture of plankton is also a useful feature for classification. Some plankton, such as Fig. 2(a) and (c) which are described in Fig. 2, have similar shape but different transparency. In this case, the internal texture provides a possible solution to distinguish them. Canny edge detector is a multistage algorithm to detect a wide range of edges in images [30], which extracts useful texture information from different vision objects and dramatically reduces the amount of data to be processed. It can generate a series of edges by the intensity gradients of the image and tracking edge by hysteresis. Here, we use canny edge detector to extract the internal texture (including nucleus and cytoplasm) of plankton.

3.2 Network Structure

The architecture we propose is shown in Fig. 3. Our network structure is composed of three sub Alex networks: network A for training global feature images; network B for training original images and network C for training local feature images. Global and local images for network A and C are preprocessed. Although this model has three different inputs, these three sub networks share the same label on backpropagation. We used a pyramid fully connected structure after convolutional and pooling layers. Because the images with global and local features have the same size of original images, the size of convolutional and pooling layers in three sub networks are the same. Considering the large gap between global and local features, we combine the inner products of two sub

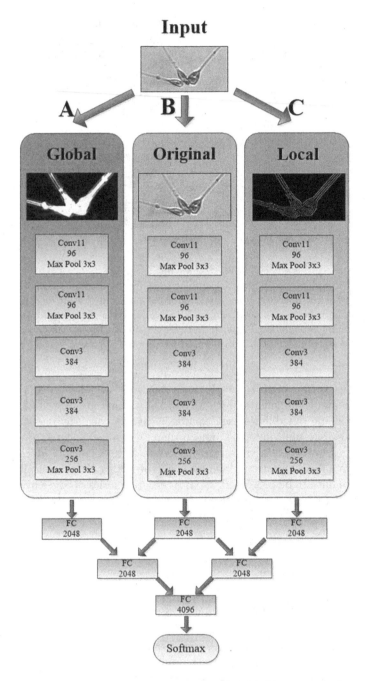

Fig. 3. Proposed architecture for plankton classification. The network A operates on global feature images. The network B is trained on original plankton images, which hold all information of plankton. And the network C operates on local feature images. The hybrid model can explore more detailed and dimensional features based on these features fusion to improve accuracy.

networks respectively (original and local; original and global) and then merge them together. We use the inner products from original images as a bridge to connect global and local features. Please note that there is no direct connection between network A and C before the 8th layer. Unlike the forward equations mentioned in Sect. 2, the forward process from the 6th to 7th fully connected layers can be concluded as the following equations:

$$y_{ab7m} = f\left(\sum_{i}^{2048} w_{71im}y_{a6i} + \sum_{j}^{2048} w_{72jm}y_{b6j} + b_{ab7}\right) \tag{4}$$

$$y_{bc7m} = f\left(\sum_{j}^{2048} w_{73jn}y_{a6j} + \sum_{k}^{2048} w_{74kn}y_{b6k} + b_{ab7}\right) \tag{5}$$

while y_{ab7m} denotes the output of the mth neuron on the left 7th fully connected layer; y_{bc7n} denotes the output of the mth neuron on the right 7th fully connected layer; w_{71im} is the weight between the ith neuron on the 6th and the mth neuron on the 7th layer; y_{a6i} is the ith neuron output of the 6th layer in network A; y_{b6i} is the jth neuron of the 6th layer in network B; y_{b6k} is the kth neuron of the 6th layer in network C and b_{ab7} is the left 7th layer bias. Similarly, we can deduce the remaining forward path to obtain y_{abc8} which is shown in Fig. 4.

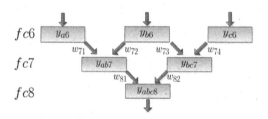

Fig. 4. Detail of pyramid structure in the fully connected layers

The back propagation rule should also follow the proposed pyramid structure. Specially, the error derivative of the 6th layer in network B is calculated as:

$$\delta_{b6j} = f'(x_{b6j})(\delta_{ab7m}w_{72} + \delta_{bc7n}w_{73}) \tag{6}$$

where δ_{b6j} is the error of the jth neuron in the 6th layer of network B and $f'(x)$ is the derivative of the activation function. Please note that only network B will receive the full back propagated error from the last layer. Network A and C only receive a part of error from the 8th layer.

4   Experimental Analysis

4.1   Dataset

The Imaging FlowCytobot (IFCB) at Woods Hole Oceanographic Institution is a system that has been continuously imaging plankton since 2006. They provide WHOI-Plankton: a large scale, fine-grained visual recognition dataset for plankton classification, which comprises over 3.4 million labeled images across 70 classes [6]. Actually, we can obtain over 3 million images across 103 classes from their website. But in such 103 classes dataset, the plankton number of each class is quite unbalanced that one class is over 2 million, some classes are also over 100 thousand and many classes are below 500. And the accuracy on such big 103 classes plankton dataset is unreasonable. Therefore, we collect 30 plankton classes randomly whose images are over 1000 and sample each class with 1000 images for fairly comparison. Here the dataset we use to train and test our architecture contains 30000 images totally in 30 classes. Plankton images are divided into two parts: training set and test set. And training set and test set are sampled randomly. The images for training and test are set as 4:1. And the images in the dataset are shown in Fig. 5. Each sub image in Fig. 5 denotes a kind of plankton. The detail of each class is described in Table 1. And the number of Table 1 is arranged as row-major order in Fig. 5.

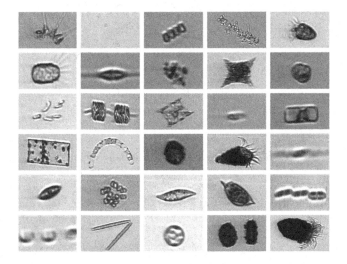

Fig. 5. This image shows that 30 images of 30 classes in dataset respectively. And the size of each images are different, we resize the image and arrange these images for convenience.

Table 1. The table demonstrate that the exact name of plankton classes in dataset. And the number order in table is follow by row-major order in above picture.

Number	Class name	Number	Class name
1	Asterionellopsis	16	Guinardia_flaccida
2	Bad	17	Guinardia_striata
3	Chaetoceros	18	Heterocapsa_triquetra
4	Chaetoceros_flagellate	19	Laboea_strobila
5	Ciliate_mix	20	Leptocylindrus
6	Corethron	21	Pennate
7	Cylindrotheca	22	Phaeocystis
8	Detritus	23	Pleurosigma
9	Dictyocha	24	Prorocentrum
10	Dino30	25	Pseudonitzschia
11	Dinobryon	26	Skeletonema
12	Ditylum	27	Thalassionema
13	Eucampia	28	Thalassiosira
14	Flagellate_sp3	29	Thalassiosira_dirty
15	Guinardia_delicatula	30	Tintinnid

4.2 Experimental Result

To start with, we train AlexNet [21] on the 30 classes dataset, the accuracy of which is set as the baseline. the following experimental results in this part are all based on the baseline.

All these results from single network to multi networks are presented in Table 2. Except training AlexNet on original images, we also have measured the performance of AlexNet on global feature images and local feature images. We can see that the two single network underperform in accuracy comparing with the result of network trained on original images. It is possible that a network only with the global feature images or local feature images may performs poorer than the network with original images.

We also compare the results of three possible architecture with only two sub networks. The two sub networks are merged in the 8th layer. We find that the accuracy is improved a little when we consider both original images and local feature images. We get the best result when we combine all 3 features using the architecture described in Fig. 3. Compared with the baseline, the hybrid architecture has improved more than 1% accuracy.

To prove that our architecture can be extensive to other network structures, we also evaluate it with GoogLeNet [31] based on 30 classes dataset. And the experimental results are shown in Table 3. We have measured the performance of GoogLeNet on original images, global feature images and local feature images

Table 2. The experimental results of AlexNet in 30 plankton classes classification.

Models	Accuracy
AlexNet trained on original images (Baseline)	94.75%
AlexNet trained on global features	94.06%
AlexNet trained on local features	93.09%
Two AlexNets trained on original images and global features	95.32%
Two AlexNets trained on original images and local features	94.50%
Two AlexNets trained on global features and local features	93.33%
Three networks trained on original images, global features and local features	95.83%

Table 3. The experimental results of GoogLeNet in 30 plankton classes classification.

Models	Accuracy
GoogLeNet trained on original images (Baseline)	95.2%
GoogLeNet trained on global features	93.4%
GoogLeNet trained on local features	93.2%
Three GoogLeNets trained on original images, global features and local features	96.3%

respectively. And compared to single GoogLeNet, our hybrid architecture of GoogLeNet also outperforms about 1%.

Our experiments is based on the publicly available deep learning toolbox: Caffe [32]. Four NVIDIA GTX 980Ti 6GB GPUs are used here to implement all above experiments. And we trained our models using stochastic gradient descent with momentum of 0.9 and weight decay of 0.0005. The initialization of the weights in our architecture is set as AlexNet in ImageNet classification [21]. We initialize and train the neuron biases in the 2th, 4th and 5th convolutional layers. And the neuron biases in the remaining layers are set with constant 0. When training AlexNet on original images, we do some image processing by subtracting the mean activity over the training set form each pixel, random mirroring of images and cropping the images into 227×227 randomly. We greedy search the parameters of each architecture to get the best performance.

5   Conclusion

In this paper, we propose a hybrid convolutional neural network to classify 30 classes plankton image set automatically and effectively. We define two different feature extraction operator to calculate the global and local features for plankton classification. We design a pyramid structure to combine original feature, global feature and local feature. Experimental results show that our architecture can extract and classify plankton images more effectively.

The distribution of plankton is highly inhomogeneous. How to classify the plankton effectively on an unbalanced data set is still a challenge. We leave this part in our future work.

Acknowledgement. This work was supported by the National Natural Science Foundation of China under Grant Nos. 61271406, 61301240, and the Fundamental Research Funds for the Central Universities under Grant No. 201562023.

References

1. Lalli, C., Parsons, T.R.: Biological Oceanography: An Introduction. Butterworth-Heinemann, New York (1997)
2. Sardet, C.: Plankton: Wonders of the Drifting World. University of Chicago Press, Chicago (2015)
3. Reynaud, E.G. (ed.): Imaging Marine Life: Macrophotography and Microscopy Approaches for Marine Biology. Wiley, Hoboken (2013)
4. Davis, C.S., Thwaites, F.T., Gallager, S.M., Hu, Q.: A three-axis fast-tow digital video plankton recorder for rapid surveys of plankton taxa and hydrography. Limnolo.Ocean. Methods **3**, 59–74 (2005)
5. Jaffe, J.S., Roberts, P.L.D., Ratelle, D., Laxton, B., Orenstein, E., Carter, M., Hilbern, M.: Scripps plankton camera system (2015)
6. Orenstein, E.C., Beijbom, O., Peacock, E.E., Sosik, H.M.: Whoi-plankton-a large scale fine grained visual recognition benchmark dataset for plankton classification. arXiv preprint arXiv:1510.00745 (2015)
7. Benfield, M.C., Grosjean, P., Culverhouse, P.F., Irigoien, X., Sieracki, M.E., Lopez-Urrutia, A., Dam, H.G., Hu, Q., Davis, C.S., Hansen, A., Pilskaln, C.H., Riseman, E.M., Schultz, H., Utgoff, P.E., Gorsky, G.: RAPID: research on automated plankton identification. Oceanography **20**, 172–187 (2007)
8. MacLeod, N., Benfield, M., Culverhouse, P.: Time to automate identification. Nature **467**, 154–155 (2010)
9. Erickson, J.S., Hashemi, N., Sullivan, J.M., Weidemann, A.D., Ligler, F.S.: In situ phytoplankton analysis: theres plenty of room at the bottom. Anal. Chem. **84**, 839–850 (2011)
10. Samson, S., Hopkins, T., Remsen, A., Langebrake, L., Sutton, T., Patten, J.: A system for high-resolution zooplankton imaging. IEEE J. Ocean. Eng. **26**, 671–676 (2001)
11. Tang, X., Stewart, W.K., Vincent, L., Huang, H., Marra, M., Gallager, S.M., Davis, C.S.: Automatic plankton image recognition. Artif. Intell. Biol. Agric. **12**, 177–199 (1998)
12. Tang, X., Lin, F., Samson, S., Remsen, A.: Binary plankton image classification. IEEE J. Ocean.Eng. **31**, 728–735 (2006)
13. Grosjean, P., Picheral, M., Warembourg, C., Gorsky, G.: Enumeration, measurement, and identification of net zooplankton samples using the zooscan digital imaging system. ICES J. Mar. Sci. J. Conseil **61**, 518–525 (2004)
14. Gorsky, G., Ohman, M.D., Picheral, M., Gasparini, S., Stemmann, L., Romagnan, J.B., Cawood, A., Pesant, S., García-Comas, C., Prejger, F.: Digital zooplankton image analysis using the zooscan integrated system. J. Plankton Res. **32**, 285–303 (2010)

15. Sosik, H.M., Olson, R.J.: Automated taxonomic classification of phytoplankton sampled with imaging-in-flow cytometry. Limnol. Oceanogr. Methods **5**, 204–216 (2007)
16. Buf, H., Bayer, M.M.: Automatic Diatom Identification. World Scientific, Singapore (2002)
17. Culverhouse, P.F., Williams, R., Reguera, B., Herry, V., González-Gil, S.: Do experts make mistakes? A comparison of human and machine identification of dinoflagellates. Mar. Ecol. Prog. Ser. **247**, 17–25 (2003)
18. Khosla, A., Jayadevaprakash, N., Yao, B., Li, F.F.: Novel dataset for fine-grained image categorization: stanford dogs. In: Proceedings of CVPR Workshop on Fine-Grained Visual Categorization (FGVC) (2011)
19. Srivastava, N., Hinton, G., Krizhevsky, A., Sutskever, I., Salakhutdinov, R.: Dropout: a simple way to prevent neural networks from overfitting. J. Mach. Learn. Res. **15**, 1929–1958 (2014)
20. LeCun, Y., Bengio, Y., Hinton, G.: Deep learning. Nature **521**, 436–444 (2015)
21. Krizhevsky, A., Sutskever, I., Hinton, G.E.: Imagenet classification with deep convolutional neural networks. In: Advances in neural information processing systems, pp. 1097–1105 (2012)
22. Zeiler, M.D., Fergus, R.: Visualizing and understanding convolutional networks. In: Fleet, D., Pajdla, T., Schiele, B., Tuytelaars, T. (eds.) ECCV 2014. LNCS, vol. 8689, pp. 818–833. Springer, Heidelberg (2014). doi:10.1007/978-3-319-10590-1_53
23. Simonyan, K., Zisserman, A.: Very deep convolutional networks for large-scale image recognition. arXiv preprint arXiv:1409.1556 (2014)
24. Fukushima, K.: Neocognitron: a self-organizing neural network model for a mechanism of pattern recognition unaffected by shift in position. Biol. Cybern. **36**, 193–202 (1980)
25. LeCun, Y., Boser, B., Denker, J.S., Henderson, D., Howard, R.E., Hubbard, W., Jackel, L.D.: Backpropagation applied to handwritten zip code recognition. Neural Comput. **1**, 541–551 (1989)
26. LeCun, Y., Bottou, L., Bengio, Y., Haffner, P.: Gradient-based learning applied to document recognition. Proc. IEEE **86**, 2278–2324 (1998)
27. Zheng, H., Zhao, H., Sun, X., Gao, H., Ji, G.: Automatic setae segmentation from chaetoceros microscopic images. Microsc. Res. Tech. **77**, 684–690 (2014)
28. Tomasi, C., Manduchi, R.: Bilateral filtering for gray and color images. In: 1998 Sixth International Conference on Computer Vision, pp. 839–846. IEEE (1998)
29. Scharr, H.: Optimal operators in digital image processing. Ph.D. thesis (2000)
30. Canny, J.: A computational approach to edge detection. IEEE Trans. Pattern Anal. Mach. Intell. **8**, 679–698 (1986)
31. Szegedy, C., Liu, W., Jia, Y., Sermanet, P., Reed, S., Anguelov, D., Erhan, D., Vanhoucke, V., Rabinovich, A.: Going deeper with convolutions. In: Proceedings of the IEEE Conference on Computer Vision and Pattern Recognition, pp. 1–9 (2015)
32. Jia, Y., Shelhamer, E., Donahue, J., Karayev, S., Long, J., Girshick, R., Guadarrama, S., Darrell, T.: Caffe: Convolutional architecture for fast feature embedding. arXiv preprint arXiv:1408.5093 (2014)

International Workshop on Driver Drowsiness Detection from Video

Driver Drowsiness Detection via a Hierarchical Temporal Deep Belief Network

Ching-Hua Weng, Ying-Hsiu Lai, and Shang-Hong Lai[✉]

Department of Computer Science, National Tsing Hua University, Hsinchu, Taiwan
lai@cs.nthu.edu.tw

Abstract. Drowsy driver alert systems have been developed to minimize and prevent car accidents. Existing vision-based systems are usually restricted to using visual cues, depend on tedious parameter tuning, or cannot work under general conditions. One additional crucial issue is the lack of public datasets that can be used to evaluate the performance of different methods. In this paper, we introduce a novel hierarchical temporal Deep Belief Network (HTDBN) method for drowsy detection. Our scheme first extracts high-level facial and head feature representations and then use them to recognize drowsiness-related symptoms. Two continuous-hidden Markov models are constructed on top of the DBNs. These are used to model and capture the interactive relations between eyes, mouth and head motions. We also collect a large comprehensive dataset containing various ethnicities, genders, lighting conditions and driving scenarios in pursuit of wide variations of driver videos. Experimental results demonstrate the feasibility of the proposed HTDBN framework in detecting drowsiness based on different visual cues.

1 Introduction

Recent reports have suggested that drowsy driving is one of the main factors in fatal motor vehicle crashes each year [1–3]. In 2014, the National Sleep Foundation (NSF) pledged an initiative that seeks to raise public awareness on drowsy driving and asked legislators to have law enforcement, regulations and recommendations on drowsy driving and distraction prevention [4]. Therefore, developing active monitoring systems that help drivers avoid accidents in a timely manner is of utmost importance [5,6].

In recent drowsy driver detection systems, most of the work focus on using limited visual cues (often just one) [7]. However, human drowsiness is a complicated mechanism. If various cues are combined dynamically [8], results can be improved. Furthermore, drowsiness has an accumulative property and the decision cannot usually be made in a short period of time, *i.e.* drowsy status at a previous time point is a factor for the drowsy status at the current time point and the duration depends on the behavior of individuals. For example, frequent yawning is an important behavioral feature, but it does not always occur before the driver goes into a drowsy state. It should be used as a preemptive measure

© Springer International Publishing AG 2017
C.-S. Chen et al. (Eds.): ACCV 2016 Workshops, Part III, LNCS 10118, pp. 117–133, 2017.
DOI: 10.1007/978-3-319-54526-4_9

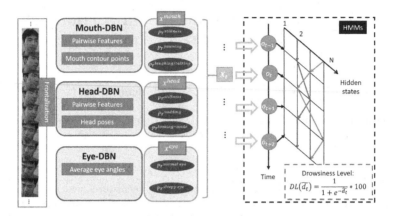

Fig. 1. A Hierarchical Temporal Deep Belief Network (HTDBN) for drowsy driver detection: inputs from the face or head after frontalization and pairwise feature extraction are first fed to deep belief networks to extract high level features and output each motion probabilities to form an observable vector X_t at each time stamp t. The deep neural nets are first pre-trained and then fine-tuned by the target drowsiness-related class. Then X_t is regarded as observation vector o_t for HMM. Two HMMs are learned on top of the deep neural nets and employed to analyze the likelihoods of the observable vector sequence within a fixed length duration. Finally, the driver's drowsiness level is predicted by inverse logit transform.

and memorized until other symptoms are captured. Otherwise, the probability of drowsiness will drop as nothing is detected for some period of time.

To resolve these issues, we introduce a novel and unified Hierarchical Temporal Deep Belief Network (HTDBN) for detecting drowsiness. The overall architecture is shown in Fig. 1. Through the proposed HTDBN framework, a set of high-level facial landmark features can first be extracted and used to learn representations for classification of several drowsiness-related symptoms. We demonstrate that drowsiness-related symptoms can be well classified using facial landmark points and head posture and the results are further composed to form observation vector sequences. For modeling temporal information, we resort to a continuous-Hidden Markov Model (HMM) which can be extended to long-term temporal information. We train two sets of parameters, drowsiness-HMM and non-drowsiness-HMM, covering many possible kinds of driving scenarios using the Baum-Welch algorithm. Finally, by using the forward-backward algorithm, the maximum likelihoods can be calculated and their differences are accumulated over a predetermined period of time.

Despite the importance of research in a practical drowsy driver detection system, most research have used relatively limited datasets. The generalization of different approaches to drowsy driver detection analysis remains unknown. In the absence of performance evaluation on a common public dataset, the comparative strength and weakness of different approaches is difficult to determine. In the field of facial expression and action recognition, comparative performance evaluations have proven valuable [9,10], and similar benefits should be gained in the field of driver drowsiness detection. Therefore, we also describe in this paper a dataset that we specifically designed and collected for drowsy driver detection. This dataset will be soon made publicly available to researchers in the field. The dataset contains a wide variety of human subjects with various races, ethnicities and genders. The data was also collected at different situations like wearing glasses, sunglasses and various lighting conditions. We expect this dataset to be a representative test-bed for drowsy driver detection approaches.

In summary, we make the following contributions in this paper: (1) To the best of our knowledge, we are the first to combine DBN with HMM for drowsy driver detection. (2) The proposed framework captures the temporal information as well as the interactive relation among eyes, mouth and head. (3) We provide a dataset that contains drowsiness and non-drowsiness videos captured under various kinds of driving circumstances.

2   Related Work

Drowsiness-related symptom measurement methods can be generally grouped into two categories [5]: Physiological and Physical. Physiological methods offer an objective and precise way to measure sleepiness. They are based upon the fact that physiological signals start to change in earlier stages of drowsiness [11]. Despite of their reliability, the intrusive nature of measuring physiological signals remains an issue that makes them unacceptable for real-world applications. Physical methods are based upon non-invasive observation of a driver's external state. A typical focus is on facial expressions that might express some characteristics, such as eyelid movement, head movement, gaze, and facial expression [12]. The research in this area can be classified into four groups [13]:

Threshold-Based Approach. The simplest method to predict a driver's drowsiness level is to set a threshold on extracted drowsiness-related symptoms. In the system presented in [14], the percentage of eyelid closure (PERCLOS) in a time window has shown to provide meaningful message of drowsiness. Teyeb *et al.* [7] further showed that when the head inclination angle exceeds a certain value and duration, the level of alertness of the driver is lowered. In [15], yawning is detected based on the rate of change of the mouth contour and is determined as the only sign of drowsiness. This approach may encounter false-alarms when the required visual cues cannot be distinguished from the similar motions, e.g. talking or laughing.

Knowledge-Based Approach. In the knowledge-based approaches, decision of driver's drowsiness is made based on knowledge of an expert. In this approach, knowledge usually appears to be evaluated according to *if-then* rules. Rezaei and Klette [16] implemented a fuzzy control fusion system to prevent road crush. In [17], a Finite State Machine (FSM) was used for hypo-vigilance detection. However, it is difficult to provide an accurate definition of driver drowsiness with some rules, since rules defined in the system do not provide sufficient expressive power to accommodate the large variations and uncertainties in driver videos.

Probability Theory Based Approach. Ji *et al.* [8], proposed a Dynamic Bayesian Network (DBN) system to determine the level of driver's drowsiness. In [12], they included frequent yawning, nodding, gaze distribution and eyelid movement as observation nodes and many other subjective factors such as sleep quality, sleeping time and driving environment as contextual nodes in DBN. Although DBN has the ability to represent the spatio-temporal characteristics for determining drowsiness, it also has a large computational complexity. In addition, the statistical analysis of large-scale subjective training data is difficult to obtain.

Statistical Approach. Support Vector Machine (SVM) and Neural network (NN) are the main methods in statistical pattern recognition. Jin *et al.* [7] utilized SVM and Eskandarian and Sayed [18] applied NN on distinguishing driver's drowsiness both based on the combination of driver behavioral measures and driving performance measures. Neither SVM nor NN was able to model temporal information. Thus, it is unrealistic to port such methods into real-world applications.

In summary, most of the approaches have drawbacks due to impractical reasons mentioned above or do not provide sufficient discrimination to capture the uncertainties. Moreover, most of the existing methods do not evaluate the robustness of their system against subjects from different ethnicities, races, genders, various illumination conditions and partial occlusion (e.g. glasses, sun-glasses and facial hair). In contrast, our work aims to systematically integrate spatio-temporal information using pure visual cues representing the driver's behaviors. Our algorithm is evaluated on a driver video dataset we collected while covering a wide variety of scenerios. All the knowledge needed in the detection system is learned from the training data itself without subjective and sophisticated parameter tuning.

3   Proposed HTDBN Algorithm

3.1   System Overview

The proposed model is inspired by the framework successfully applied to the speech and gesture recognition [13,19]. But instead of limiting the learning ability to Restricted Boltzmann Machine (RBM), the proposed HTDBN based method utilizes the nice property of DBN [20], i.e. modeling high-order dependencies.

As shown in Fig. 1, the learned DBNs are used to extract drowsiness-related symptoms after frontalization along with pairwise feature extractions. On top of the DBNs, two continuous-HMMs are adopted for modeling higher level temporal relationship among drowsiness and non-drowsiness using the observation vectors obtained from the probabilities of mouth-, eye- and head-motions. To evaluate driver's drowsiness level, the accumulated differences of both HMM maximum likelihoods collected from previous time stamps to current time stamp is passed to the inverse logit transform.

3.2 Frontalization

Specifically with faces, the success of the learned network in capturing facial appearance is highly dependent on a rapid 3D normalization step. Faces captured by the camera are considered unconstrained (due to the non-planarity of the face) and non-rigid expressions. Rezaei and Klette [16] focused on head-pose estimation, but ignored the importance of pose normalization on facial cues. Similar to the recent literature [21], our alignment is based on using facial landmark point detectors to direct the normalization process but is simplified by discarding pixel-wise transformation. In our framework, the recent algorithm of [22], based on the supervised descent concept, is to detect and track 49 facial landmarks in a video sequence.

We start our alignment process by extracting 9 2D facial landmark points as temporary anchor points. The points are corners of both eyes and mouth, center of the brows and tip of the nose as illustrated in Fig. 2a. They are used to approximate the pose matrix P given by

$$P = \begin{bmatrix} R_{3\times3} & \mathbf{t}_{3\times1} \\ \mathbf{0}_{1\times3} & 1 \end{bmatrix} \tag{1}$$

by applying POSIT algorithm [23] in which R is a rotation matrix, $\mathbf{t}$ is a translation vector. After obtaining these parameters, the reference 3D face model is then rotated and translated to align with the 2D facial image plane. Four affine transformations are respectively applied on each eye, nose and mouth regions to warp 2D anchor points to the $\mathbf{xy}$-image plane of the 3D reference face (Fig. 2b).

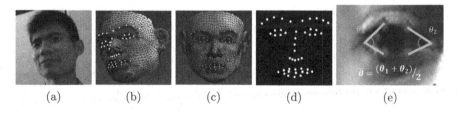

(a)   (b)   (c)   (d)   (e)

Fig. 2. (a) The detected face with 9 initial facial landmark anchor points. (b) Four affine transformations are respectively applied on each eye, nose and mouth to warp 2D anchors points to the $\mathbf{xy}$-image plane of the 3D reference model. Red dots are the original points on the 2D detected face, yellow dots are the results after warping. (c) Frontalization of the 3D shape model along with the aligned points. (d) Project the 3D frontalized points to 2D image plane. (e) The points used and the equation to calculate the average eye angle $\bar{\theta}$ for feature extraction. (Color figure online)

This generates a 3D-aligned version of the 2D facial contour and the corresponding depth information of each facial point can be estimated [24].

Since texture of the face is not considered in our framework, the fitted P can be directly used without concerning about the corruption between pixel-wise warping. Finally, the frontalization is achieved by using transpose of the rotation matrix $R^\top$ on the 3D-aligned facial points, as illustrated in Fig. 2d.

3.3 Pairwise Feature Extraction

Once the system is activated, the system will collect frontalized facial landmark points to form the pairwise features and then the features are passed to mouth-DBN, head-DBN or eye-DBN. The 2D coordinates of facial landmark points of current frame-c are given as: $F_c = \{f_1^c, f_2^c, \ldots, f_\mathbb{N}^c\}$, where $\mathbb{N}$ is the number of points used. We deploy 2D pairwise differences of points for input in the first layer of DBN. The idea comes from the usage of joints in action recognition [25], because we believe that facial landmark points are similar to 3D joints in a way that they can both use position differences to characterize motion information; however, we introduce innovations on some expects. The pairwise differences of points capture posture features, motion features, and max-pool motion features by directed concatenation: $\mathcal{A} = [f_{cc}, f_{cp}, f_{cd}]$ in which:

$$
\begin{aligned}
f_{cc} &= \{f_i^c - f_j^c | i, j = 1, 2, \ldots, \mathbb{N}; i \neq j\} \\
f_{cp} &= \{f_i^c - f_i^p | f_i^c \in F_c; f_i^p \in F_p\} \\
f_{cd} &= \{f_i^c - f_{i,w}^d | f_i^c \in F_c; f_{i,w}^d \in F_d; w = 10, 20, \ldots, 60\}
\end{aligned}
\tag{2}
$$

where f^p denotes the landmark points extracted from preceding frame-p and f^d denotes the landmark points in the frame-d whose sum of differences of points to current frame is maximum in a window size w. Since our system cannot discover where the initial frame of motion is, to characterize more motion information on continuous frames, we replace the offset features f_{ci} in [25] with max-pool motion features f_{cd}. The illustration is demonstrated in Fig. 3.

Specifically, we consider that a motion can be captured by computing the differences of points between current frame and dozens of proceeding frames (previous 2, 3 seconds). However, the more the number of proceeding frames be used for features, the more dimension of features would be, which makes the following DBNs on drowsiness-related symptoms cost too much time on computation. Therefore, we design an innovative way just like max-pool layer in convolutional neural network [26] to discard less important information in time order, *i.e.* to find frames which has maximum sum of differences on certain window size, and then calculate their differences of points as features.

Dimension of $\mathcal{A}$ results in $N_\mathcal{A} = (\mathbb{N} \times (\mathbb{N} - 1)/2 + \mathbb{N} + \mathbb{N} \times 6) \times 2$. Ten facial landmarks ($\mathbb{N} = 10$) at both mouth corners, the middle of uppers and lower lips are considered for yawning and talking/laughing detection. All the points are normalized by the scheme mentioned in Sect. 3.2 before extraction. As for head features, they are extracted in the same fashion as facial landmark points whilst

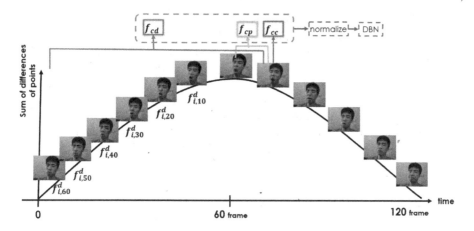

Fig. 3. Raw feature formation: three feature channels f_{cc}, f_{cp}, f_{cd} for capturing information of posture, motion, max-pool motion.

the 2D positional points are replaced by yaw, pitch and roll angles decomposed from the rotation matrix R.

Nevertheless, as for eye features, we consider the average eye angles $\overline{\theta}_{left}, \overline{\theta}_{right}$ at both eyes (Fig. 2e) in previous 10 seconds as features, because eyes have small variation that only can be identified to 3 states, opened, half-closed and closed eyes, which may make extracting eye pairwise features be less meaningful as mouth or head motions.

3.4   Learning the Higher Level Representation

RBMs were originally developed using binary stochastic units for both the visible and hidden units, but logistic units are a very poor representation for real-world data such as pixel intensities in natural images [27]. One solution is to replace the binary visible units by linear units with independent Gaussian noise [28], so called Gaussian RBM (GRBM). The original energy function can be replaced by:

$$E(\mathbf{v}, \mathbf{h}) = \sum_{i \in vis} \frac{(v_i - a_i)^2}{2\sigma_i^2} - \sum_{j \in hid} b_j h_j - \sum_{i,j} \frac{v_i}{\sigma_i} h_j w_{ij} \tag{3}$$

where w_{ij} denotes the matrix of connection between visible unit i and hidden unit j with their bias terms a_i and b_j and σ_i is the standard deviation of the Gaussian noise for visible unit i. Because our pairwise features are continuous features, we use the above GRBM in the same fashion of [10] to model the energy term of the first visible layer. It is possible to learn the variance of the noise for each visible unit but it is much easier to normalize the data (mean subtraction and standard deviation division) to have zero mean and unit variance (e.g. $\sigma_i^2 = 1$) in the preprocessing phase.

Specifically, the DBN in our system consists of one visual layer (*i.e.* the lowest layer in GRBM) with continuous pairwise features and four hidden layers to learn a hierarchical feature representation given training data extracted from mass pool of facial landmark points and head posture data. Given the nodes at the third hidden layer and the classification labels (*i.e.* stillness, yawning or laughing/talking; stillness, nodding or looking aside; normal eye or sleepy eye), the output of the DBN (*i.e.* last hidden layer) can classify by comparing the values of these nodes. The DBNs in HTDBN are activated at all times to detect mouth-motion, head-motion and eye-motion. For mouth-DBN, the model aims to discriminate yawning from mouth stillness and laughing/talking, and for head-DBN, it is used to classified head stillness, nodding and head looking aside. In addition, eye-DBN can differentiate sleepy eye from normal eye.

3.5   Continuous-HMM for Drowsy Driver Detection

A hidden Markov model describes the statistical behavior of a process in time. At each time step t, we have one 8-dimensional feature vector X_t taking values of each motion class probabilities obtained from DBNs, *i.e.* $X_t = \{x^{eye}, x^{mouth}, x^{head}\}$, where $x^{eye} = \{p^{normal}, p^{sleepy}\}$, $x^{mouth} = \{p^{stillness}, p^{yawning}, p^{laughing/talking}\}$, $x^{head} = \{p^{stillness}, p^{nodding}, p^{looking-aside}\}$. The intuition behind this is that a motion in drowsiness-related symptoms consists of a sequence of state transition, e.g. yawning is a long-term motion of mouth that has its fullest open state in between stillness at onset and offset. Therefore, the variation of motion probabilities in a sequence of feature vectors can describe a motion or a transition from different motions.

Here, X_t not only serves as an observation vector O_t for HMM but also describes the relation between eyes, mouth and head. Existing work tend to neglect such relations when detecting drowsiness-related symptoms, e.g. eyes might be closed when yawning, but only mouth features are considered.

Assume a HMM has J (unobserved) states $\{s_1, s_2, \ldots, s_J\}$ and K observation vectors $\{o_1, o_2, \ldots, o_K\}$. At time t, HMM occupies a state s_i and may undergo a state transition from the $s_i = i$ to a state $s_{i+1} = j$ at time $t+1$ with the state transition probability $a_{ij} = P(s_{t+1} = j|s_t = i)$. Associated with each state is a set of observation vectors o_t with their respective observation probability densities, Gaussian M-component mixture densities, $b_i(o) = \sum_{k=1}^{M} c_{ik} \mathcal{N}[o, \mu_{ik}, \mathbb{U}_{ik}]$, where c_{jk} is the mixture weight, $\mathcal{N}$ is the normal density and μ_{ik} and $\mathbb{U}_{ik}$ are the mean vector and co-variance matrix associated with state i, mixture k. Starting from an initial state $s_1 = i$ with probability $\pi = P(s_1 = i)$, the process undergoes a sequence of state transitions over a time duration T and generates an observation vector sequence $O = \{o_1, o_2, \ldots, o_T\}$ with a certain probability. In a sense, the HMM is specified by a triplet $\lambda = (\Pi, A, B)$, where $A = \{a_{ij}\}$ denotes transition probabilities, $B = \{b_j(o)\}$ denotes observation symbol probabilities, and $\Pi = \{\pi_i\}$.

For each HMM, the model parameters can be trained from the selected training vector sequences by applying Baum-Welch algorithm [29]. Here we assume our HHM to be an ergodic model. A model is ergodic means its transition matrix is fully connected, which means all transitions have non-zero probabilities.

With the above definition for a model λ, a given observation symbol sequence O may be generated from one or more state sequences $S = \{s_1, s_2, \ldots, s_T\}$. The probability requires summation over all possible state sequences and an efficient way to do so is a forward-backward algorithm [30]:

$$
\begin{aligned}
P(O|\lambda) &= \sum_{all\,S} P(O|S, \lambda)P(S|\lambda) \\
&= \sum_{all\,S} \pi_{s_1} b_{s_1}(o_1) \prod_{t=2}^{T} a_{s_{t-1}s_t} b_{s_t}(o_t).
\end{aligned}
\tag{4}
$$

Given a video sequence with length T, we obtain the likelihood difference $d = \mathrm{P}(O|\lambda^{drowsy}) - \mathrm{P}(O|\lambda^{nondrowsy})$ from every 300 frames. We then accumulate the likelihood differences within a specific among of time τ to obtain $\tilde{d}_t$. Finally, the drowsiness level (DL) is determined using the inverse logit transform (Eq. 5). While drowsiness level is more than 50%, then the drowsiness detector would consider the current state as drowsiness.

$$
DL(\tilde{d}_t) = \frac{1}{1 + e^{-\tilde{d}_t}} * 100\%
\tag{5}
$$

4   Dataset Acquisition

Most of the previous works on drowsy driver detection attempted to recognize a small set of cases for driver drowsiness detection. Although [31] provided a freely-available dataset for yawning detection, it is still insufficient for a comprehensive drowsy driver study based on pure visual cues. Drowsiness detection from yawning alone is too restricted for use in practice. It should be combined with some additional indicators of drowsiness. Therefore, we collect a large video dataset for performance evaluation of drowsy driver detection methods.

Camera Setting. To cope with the night time or poor lighting problem, we used the active infrared (IR) illumination and acquire IR videos in the dataset collection. To ensure a realistic setup, all the videos were captured by D-Link DCS-932L, a stand-alone surveillance digital camera with the resolution set at 640×480 pixels. The built-in infrared LEDs allow us to view in any light condition from daytime to nighttime. The advantage of activating IR illuminators also in daytime is because it captures occluded eyes with people wearing sunglasses better than using RBG camera (Fig. 4b). However, for the sake of completeness, 24-bit true color (RGB) Logitech C310 HD at 30 frames per second webcam was also set up simultaneously to record the data at 720p in daytime only (Fig. 4c).

Environment Setting. In the collection of our dataset, two rounds of video recordings were performed for each subject. The first round was recorded during the daytime and the second one was performed in the night. In order to

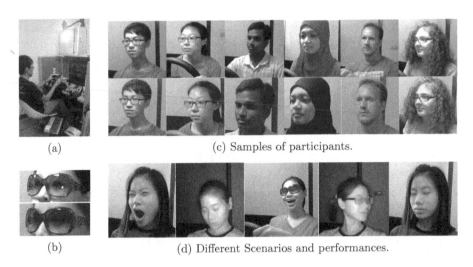

(a)

(c) Samples of participants.

(b)

(d) Different Scenarios and performances.

Fig. 4. (a) Participants were equipped with a fixed but tunable chair and a simulated driving wheel with pedals; they are instructed to perform a series of facial displays shown in Fig. 4d. A standalone IR camera and RBG webcam were placed at the left hand side of the driver while the ambient light was augmented with high-intensity lamp to simulate the condition of sunny day. (b) IR images can capture better occluded eyes for subjects wearing sunglasses. (c) Diversity of skin colors, genders and ethnicities among the participants in the dataset collection. IR and RGB videos were taken simultaneously. (d) Situations like yawning in *BareFace*, nodding when falling asleep in *Night-BareFace*, laughing in *Sunglasses*, looking aside in *Night-Glasses*, blink slowly in *BareFace* and *etc.* are considered and separately recorded.

simulate the condition of sunny day, for approximately one third of subjects, ambient room lighting augmented by a high-intensity lamp was used (Fig. 4a) for the daytime recording. Illuminance was measured by a light meter to ensure environment was well-established.

Our camera was placed on the top left hand side of the subject to emulate the position in the A-Pillar, a common used location in cars. In contrast, most datasets and the corresponding algorithms were based upon fully frontal face views, which is impractical to set up in a real cars since the camera would block the driver's view and the dashboard.

Participants. To make sure the algorithm works for various skin race and genders, 36 adults, aged from 18 to 40 years old with various ethnicities and diverse skin colors (32.5% of black or brown, 32.5% of white, and 35% of yellow) and genders (50% of female), participated in the video collection. The subjects with different hairstyles and clothing were recorded with and without glasses/sunglasses to simulate a wide variety of driving scenarios. Figure 4c and d show some samples of the participants under different conditions and driving scenarios.

Driver Videos. Subjects were recorded when they sit on a chair and play a plain driving game with simulated driving wheel and pedals; meanwhile, they were instructed by an experimenter to perform a series of 8 actions under 5 kinds of scenarios: *BareFace, Glasses, Sunglasses, Night-BareFace* and *Night-Glasses.* The sequences recorded from each subject can be regarded as two branches: drowsiness and non-drowsiness. For drowsiness-related sequences, yawning, slow blink rate (high PERCLOS) and falling asleep (high PERCLOS followed by frequent nodding) were taken about 1 min long, and the combination of drowsiness-related symptoms sequences (yawning, high PERCLOS, frequent nodding) were recorded about 1.5 min. On the other hand, sequences of normal driving (low PERCLOS), shocked face and burst out laughing in about 1 min and the combination of non-drowsiness actions (talking, laughing, looking at both sides) recorded about 1.5 min are represented as the non-drowsiness data. Some examples can be seen in Fig. 4d. Overall, *360 videos* were taken to complete the dataset.

Moreover, to simulate more practical driving situations, 18 subjects from the proposed dataset are randomly selected yet kept the balance of various gender, skin races. Their sequences are edited and combined into a 2–10 min mixing video for each subjects under 5 kinds of scenarios which contains various situations with different number of transitions from non-drowsiness state to drowsiness state, or drowsiness state to non-drowsiness states. Overall, there are 90 mixing videos be added to the dataset for evaluation.

5   Experimental Results

5.1   Experimental Setup

We evaluate the proposed *HTDBN* framework by using the provided dataset mentioned in Sect. 4. We first train the DBNs with a four-hidden-layer structure, where the numbers of the nodes in all the layers are $[N_A, 1000, 1000, 500, N_y]$ from the lowest layer to the highest one, respectively. The numbers of nodes in the visible layer are $N_A = 230$, 45, and 200 for mouth-, head-, and eye-DBN, respectively. The number of outputs N_y for each DBNs is equivalent to the number of classes, *i.e.* there are 3 outputs for mouth- and head-DBN, 2 outputs for eye-DBN.

In our experiments, the dataset is divided into two parts: training, testing dataset. The subjects that have edited mixing videos are for testing, and the sequences from the other subjects are for training. In drowsiness-related symptoms detection, all the sequences from each subject in the training dataset are taken to train DBNs. For each of the sequence, the mouth-motion, head-motion, and eye-motion class probabilities are extracted at each frame and the observable vectors can be obtained. In drowsy driver detection, to maintain accurate classification capability, the selection of training observable vectors sequences is very important. This is since the chosen sequences can be used to adjust the model parameters that can also be used to recognize other sequences of observable symbols, e.g. a drowsy observable vector sequence should get higher probability of re-generation from drowsiness HMM than non-drowsiness HMM. The data we captured for each person consists of long videos. To perform our training, two kinds of videos, the combination of drowsiness-related symptoms videos and the combination of non-drowsiness-related actions videos, are used for training data. We randomly subdivide each of the videos into various shorter overlapping and non-overlapping videos with fixed length 300 frames.

5.2 Drowsy Driver Detection Performance Evaluation

We present the performance of the proposed HTDBN for driver's drowsiness detection using the collected dataset. We use not only accuracy but also F_1-score to evaluate the performance of the proposed detection algorithm since it is a relatively fair measure for unbalanced data:

$$F_1\text{-score} = 2 * \frac{prec(\Delta) * rec(\Delta)}{prec(\Delta) + rec(\Delta)} \tag{6}$$

Δ is the length of the sequence of its likelihoods decoded from drowsiness-HMM and non-drowsiness-HMM. To ensure the stability and to fully scrutinize the strength of the proposed framework, we conduct an experiment to show the performance of the fine-tuned system on *BareFace*, *Glasses*, *Sunglasses*, *Night-BareFace* and *Night-Glasses* separate scenarios. The average scores of different scenarios for drowsiness and non-drowsiness detection are shown in Table 1.

Table 1. The 1st experiment: performance of HTDBN on *separate scenarios* in the drowsy driver detection dataset

Scenario	Drowsiness F_1-score	Non-drowsiness F_1-score	Accuracy
BareFace	92.17%	92.64%	92.42%
Glasses	88.17%	85.04%	86.79%
Sunglasses	74.17%	78.59%	76.58%
Night-BareFace	92.60%	90.97%	91.87%
Night-Glasses	77.74%	73.73%	75.90%
Overall	**85.39%**	**84.19%**	**84.82%**

As shown in the table, the first three scenarios were evaluated in daytime. It can be seen that for the cases with no occlusion on face (*BareFace*) and enough ambient lighting, we can achieve about 92% accuracies on both drowsiness and non-drowsiness detections. As the detector encounters driver-wearing-glasses scenario, sometimes the reflections in glasses and the glasses frames may cause disturbance of eye openness detection and thus lose a good information from eye-motion detection.

In addition, the performance of *Sunglasses* falls behind very much with other scenarios, although eyes occluded by sunglasses have better visibility

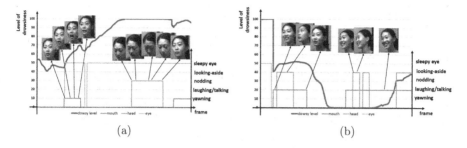

(a)   (b)

Fig. 5. Plots of the processed data aligned with level of drowsiness versus time: (a) Non-drowsy (alert) driver falling asleep example and (b) drowsy driver awake example. Blue curve represents level of drowsiness; below are orange, green and yellow curves represent predictions of mouth- and head- and eye-motions from DBNs, respectively. (Color figure online)

using IR camera compared with RGB camera, the stronger reflections and hindrance deteriorate on eye-contour-point detection. Therefore, to deal with the big challenge on driver-wearing-sunglasses scenario, firstly we design a sunglasses detector to find the region of sunglasses by the image intensity from detected eye points area, and then the sunglasses region is employed gamma correction ($\gamma = 0.4$) to adjust brightness and simple reflection removal on the reflection that would block eye contours. After these improvements, the result of *Sunglasses* is increased 9% which is about 77% accuracy.

The rest two scenarios were experimented in nighttime. The IR videos collected from the chosen camera can still capture clear faces, but facial landmark points are less accurate than in daytime resulting in lower accuracies. The accuracy of the overall scenarios is about 85%, and so do the scores for both drowsiness and non-drowsiness detection.

Figure 5a depicts an example of processing a 60-second video of an alert driver gradually falling asleep by using the proposed algorithm. Orange, green and yellow curves represent the predictions of mouth- and head- and eye-motions from DBNs, respectively, co-existing at every time stamp. As shown, when mouth-DBN first recognizes a yawning cycle the level of drowsiness goes up drastically to about 70%. As the blink rate goes slower and eye closure duration goes longer, the drowsiness percentage keeps growing up to the top 100% until the head-DBN detects a nodding cycle, a high-risk warning should be raised. In contrast, a drowsy driver waking up case illustrated in Fig. 5b shows the sustainability of the proposed system. While no sleepy eye is noticed, the level of drowsiness decrease dramatically. Mouth- and head-DBN detect several laughing/talking motions and head looking aside motions, thus the system determines the drowsiness level of the driver to be below 50%.

5.3 Comparison with Other Solutions

In this experiment, our proposed HTDBN model is compared with baseline model: support vector machine (SVM) on mixed scenarios in the Drowsy Driver Dataset. Before deep learning appears, support vector machine was the most popular technique for data classification, especially the SVM with kernel trick that can efficiently perform a non-linear classification. For a fair comparison, every solution is trained and tested same as the proposed HTDBN solution, yet the only difference is that the determination of the drowsiness state. As for SVM, the probability of drowsiness determines the final drowsiness state, while for HMMs, it is decided by the difference between drowsiness-HMM and nondrowsiness-HMM. Table 2 shows the effectiveness of applying different models (SVM or DBN, HMM) on the proposed framework.

Table 2. The 2nd experiment: comparison between HTDBN and baseline solution on *mixed scenarios* in the drowsy driver detection dataset

	Drowsiness	Non-drowsiness	Accuracy
SVM+SVM	81.16%	74.99%	78.51%
SVM+HMM	81.30%	77.14%	79.43%
DBN+SVM	84.26%	79.20%	82.08%
Ours	**85.39%**	**84.19%**	**84.82%**

The accuracy is apparently lower when SVMs (SVM+SVM solution) are substituted for all detectors in the proposed algorithm. To further discuss the separate effectiveness in different part of the framework, the SVM+HMM solution only replaces DBNs with SVMs in drowsiness-related symptoms, and only HMMs are replaced with a binary-class SVM in drowsy driver detection for DBN+SVM solution. From the results, DBNs play important roles in the overall system which improves 5% accuracy, whilst the usage of HMMs increases 3%. In conclusion, our proposed HTDBN algorithm is more suitable for classification in time series.

5.4 Yawning Detection Performance on YawDD Dataset

Although *Yawning Detection Dataset (YawDD)* [31] is not a sufficient dataset for comprehensive drowsy driver detection because it is only determined by yawning detection. However, to compare with existing approaches, the yawning detector mouth-DBN in our proposed system has been evaluated on the YawDD dataset as well. According to the evaluation scheme from [32], we got 94% yawning detection accuracy with 2% false alarm rate on CASE I *(camera under the mirror)*, and 92% accuracy with 5% false alarm rate on CASE II *(camera on the dash)*. Our results outperform recent methods presented in [31–33], whose accuracies on the better case are 60%, 75%, and 92% but 13% false alarm rate, respectively.

5.5 Computational Complexity

Our driver drowsiness detection system consists of an off-line training phase and on-line detection phase. Though the learning in the network is uninterestingly long, once the model training is finished, with low inference cost, the entire system is able to perform in real-time with MATLAB implementation using a Core i5, 3.1GHz PC with 16GB RAM, at an average speed of 20 fps. More precisely, a single multi-layer feedforward neural networks incurs in linear running time $\mathcal{O}(T)$ and the forward-backward algorithm applied in HMM has time complexity $\mathcal{O}(N^2 T)$, where T is the length of the sequence and N is the number of states.

6 Conclusion

In this paper, we presented a novel HTDBN that utilizes DBNs for learning contextual frame-level representations for drowsiness-related symptoms. By encoding dynamic structure of drowsiness and non-drowsiness information into HMM-based models, the results are robust and promising under different circumstances. The proposed continuous-HMM can model the interactive relations among eyes, mouth and head. Moreover, for performance evaluation, we collected a large drowsy driver detection dataset in which various skin colors, scenarios and lighting conditions are considered. Experimental results on various kinds of scenarios and fusion all together demonstrated the power of the proposed framework in estimating the driver's drowsiness level.

Acknowledgement. The authors would like to thank Qualcomm Technologies Inc. for supporting this research work.

References

1. Bergasa, L., Nuevo, J., Sotelo, M., Barea, R., Lopez, M.: Real-time system for monitoring driver vigilance. IEEE Trans. Intell. Transp. Syst. **7**, 63–77 (2006)
2. World Health Organization: Global status report on road safety 2013: supporting a decade of action: summary. World Health Organization (2013)
3. Wheaton, G., Shults, R.: Drowsy driving and risk behaviors 10 states and Puerto Rico. Online article (2014)
4. National Sleep Foundation: Drowsy driving reduction act of 2015 (2014)
5. Colic, A., Marques, O., Furht, B.: Driver Drowsiness Detection: Systems and Solutions. Springer, Heidelberg (2014)
6. Mercedes-Benz: Attention assist: drowsiness-detection system warns drivers to prevent them falling asleep momentarily. Online article (2008)
7. Teyeb, I., Jemai, O., Zaied, M., Ben Amar, C.: A drowsy driver detection system based on a new method of head posture estimation. In: Corchado, E., Lozano, J.A., Quintián, H., Yin, H. (eds.) IDEAL 2014. LNCS, vol. 8669, pp. 362–369. Springer, Heidelberg (2014). doi:10.1007/978-3-319-10840-7_44
8. Qiang, J., Lan, P., Looney, C.: A probabilistic framework for modeling and real-time monitoring human fatigue. IEEE Trans. Syst. Man Cybern. Part A: Syst. Hum. **36**, 862–875 (2006)

9. Lucey, P., Cohn, J., Kanade, T., Saragih, J., Ambadar, Z., Matthews, I.: The extended Cohn-Kanade dataset (CK+): a complete dataset for action unit and emotion-specified expression. In: IEEE Computer Society Conference on Computer Vision and Pattern Recognition Workshops (2010)

10. Wu, D., Shao, L.: Leveraging hierarchical parametric networks for skeletal joints based action segmentation and recognition. In: IEEE Conference on Computer Vision and Pattern Recognition, pp. 724–731 (2014)

11. Yang, G., Lin, Y., Bhattacharya, P.: A driver fatigue recognition model based on information fusion and dynamic Bayesian network. Inf. Sci. **180**, 1942–1954 (2010)

12. Ji, Q., Zhu, Z., Lan, P.: Real-time nonintrusive monitoring and prediction of driver fatigue. IEEE Trans. Veh. Technol. **53**, 1052–1068 (2004)

13. Mohamed, A., Dahl, G., Hinton, G.: Acoustic modeling using deep belief networks. IEEE Trans. Audio Speech Lang. Process. **20**, 14–22 (2012)

14. Dasgupta, A., George, A., Happy, S., Routray, A.: A vision-based system for monitoring the loss of attention in automotive drivers. IEEE Trans. Intell. Transp. Syst. **14**, 1825–1838 (2013)

15. Alioua, N., Amine, A., Rziza, M.: Drivers fatigue detection based on yawning extraction. Int. J. Veh. Technol. **2014** (2014)

16. Rezaei, M., Klette, R.: Look at the driver, look at the road: no distraction! No accident! In: IEEE Conference on Computer Vision and Pattern Recognition, pp. 129–136 (2014)

17. Smith, P., Shah, M., da Vitoria Lobo, N.: Determining driver visual attention with one camera. IEEE Trans. Intell. Transp. Syst. **4**, 205–218 (2003)

18. Eskandarian, A., Sayed, R.: Analysis of driver impairment, fatigue, and drowsiness and an unobtrusive vehicle-based detection scheme. In: Proceeding of International Conference on Traffic Accidents (2005)

19. Taylor, G., Hinton, G., Roweis, S.: Modeling human motion using binary latent variables. In: Neural Information Processing Systems, pp. 1345–1352 (2006)

20. Hinton, G., Osindero, S.: A fast learning algorithm for deep belief nets. Neural Comput. **18** (2006)

21. Taigman, Y., Yang, M., Ranzato, M., Wolf, L.: Deepface: closing the gap to human-level performance in face verification. In: IEEE Conference on Computer Vision and Pattern Recognition, pp. 1701–1708 (2014)

22. Xiong, X., de la Torre, F.: Supervised descent method and its application to face alignment. In: IEEE Conference on Computer Vision and Pattern Recognition, pp. 532–539 (2013)

23. DeMenthon, F., Davis, L.: Model-based object pose in 25 lines of code. Int. J. Comput. Vis. **15**, 123–141 (1995)

24. Heo, J., Savvides, M.: Gender and ethnicity specific generic elastic models from a single 2D image for novel 2D pose face synthesis and recognition. IEEE Trans. Pattern Anal. Mach. Intell. **34**, 2341–2350 (2012)

25. Yang, X., Tian, Y.: Eigenjoints-based action recognition using naive-bayes-nearest-neighbor. In: IEEE Computer Society Conference on Computer Vision and Pattern Recognition Workshops, pp. 14–19 (2012)

26. Krizhevsky, A., Sutskever, I., Hinton, G.E.: Imagenet classification with deep convolutional neural networks. In: Pereira, F., Burges, C.J.C., Bottou, L., Weinberger, K.Q. (eds.) Advances in Neural Information Processing Systems 25, pp. 1097–1105. Curran Associates Inc, Red Hook (2012)

27. Hinton, G.E.: A practical guide to training restricted Boltzmann machines. In: Montavon, G., Orr, G.B., Müller, K.-R. (eds.) Neural Networks: Tricks of the Trade. LNCS, vol. 7700, 2nd edn, pp. 599–619. Springer, Heidelberg (2012). doi:10.1007/978-3-642-35289-8_32

28. Freund, Y., Haussler, D.: Unsupervised learning of distributions on binary vectors using two layer networks. Technical report, University of California at Santa Cruz, Santa Cruz, CA, USA (1994)

29. Yang, L., Widjaja, B., Prasad, R.: Application of hidden Markov models for signature verification. Pattern Recogn. **28**, 161–170 (1995)

30. Devijver, P.A.: Baum's forward-backward algorithm revisited. Pattern Recogn. Lett. **3**, 369–373 (1985)

31. Abtahi, S., Omidyeganeh, M., Shirmohammadi, S., Hariri, B.: YawDD: a yawning detection dataset. In: Proceedings of the 5th ACM Multimedia Systems Conference, pp. 24–28. ACM (2014)

32. Omidyeganeh, M., Shirmohammadi, S., Abtahi, S., Khurshid, A., Farhan, M., Scharcanski, J., Hariri, B., Laroche, D., Martel, L.: Yawning detection using embedded smart cameras. IEEE Trans. Instrum. Meas. **65**, 570–582 (2016)

33. Zhang, W., Murphey, Y.L., Wang, T., Xu, Q.: Driver yawning detection based on deep convolutional neural learning and robust nose tracking. In: 2015 International Joint Conference on Neural Networks (IJCNN), pp. 1–8. IEEE (2015)

Detection of Driver Drowsiness Using 3D Deep Neural Network and Semi-Supervised Gradient Boosting Machine

Xuan-Phung Huynh, Sang-Min Park, and Yong-Guk Kim[⊠]

Department of Computer Engineering, Sejong University, Seoul, Korea
phunghx@sju.ac.kr, zmqp111@gmail.com, ykim@sejong.ac.kr

Abstract. Detecting drowsiness of the driver with a reliable and confident manner is a challenging task since it requires accurate monitoring of facial behavior such as eye-closure, nodding and yawning. It is even harder to deal with it when she wears sunglasses or scarf, appearing in the data set given for this challenge. One of the popular ways to analyze facial behavior has been using standard face models such as active shape model or active appearance model. These models work well for the frontal faces and yet often stumble for the extreme head pose cases. To handle these issues, we propose an approach based on recent machine learning techniques: first, 3D convolutional neural network to extract features in spatial-temporal domain; secondly, gradient boosting for drowsiness classification; thirdly, semi-supervised learning to enhance overall performance. The highest score from our submissions was 87.46% accuracy, suggesting that this approach has a potential for real application.

1 Introduction

It is known that the major categories of car accidents are divided into distraction and drowsiness of drivers. Drowsy driving refers to when a driver is half-sleeping because of mental and physical fatigue. Currently, more than 30% of deaths caused by car accidents are attributed to drowsy driving. In 2008, National Highway Traffic Safety Administration (NATASHA) estimates that 100,000 police reports on vehicle crashes were direct outcomes of driver drowsiness resulting in 1,550 deaths, 71,000 injuries, and $12.5 billion in monetary losses [1]. For these reasons, many car manufacturers over the world are developing diverse systems that can prevent drowsy driving. To develop a successful driver drowsiness monitoring system, it requires several important components: first a reliable and accurate algorithm by which drowsiness of the driver can be detected; secondly a high-quality driving data set used for training the algorithm; thirdly a high-performing embedded computer.

For the 1st component, since any algorithm for the driver drowsiness detection system is to tell correctly whether she is drowsy and non-drowsy by observing her facial state for a certain period, it normally consists of two parts: the facial state monitoring in the front and the mental state classification in the

C.-S. Chen et al. (Eds.): ACCV 2016 Workshops, Part III, LNCS 10118, pp. 134–145, 2017.
DOI: 10.1007/978-3-319-54526-4_10

back. There are many approaches in modeling facial state in terms of driver drowsiness detection. The conventional approach is to measure the important facial features, such as eye, mouth, and head movement using the metric geometry. The modern approach attempts to utilize the statistical face models such as Active Appearance Model (AAM) and Active Shape Model (ASM). AAM models the face using the shape and color, whereas ASM employs the shape only. It is known that the geometrical approach is often lighter than the statistical one in term of computing cost, although the latter is more reliable than the former. Once the facial states are estimated using one of these models, the system makes an inference whether the driver is drowsy or non-drowsy typically using a classifier. Artificial neural network for the present purpose has been popular, and Markov model has been useful when it needs to make an inference by accounting variation in time-domain [2].

For the 2nd component, manufactures in fact build in-house driving databases and use them in developing their drowsiness detection systems. However, most of them are not available for public use and academic research. Because of such reason, we believe that the present drowsiness driver database will become a milestone in this area because the present dataset appears to be a serious attempt in building a big-data driving database in several aspects. For instance, the number of subjects is large enough having diverse ethnic groups and the illumination conditions cover the day and night cases. Moreover, the subjects wear glasses, sunglasses and scarf. The present dataset contains 36 subjects. It supposes that 18 subjects should be used for training of participants' system, whereas 18 subjects for evaluating and testing. The total dataset consists of more than 1.5 million frames recorded with and without wearing glasses/sunglasses under diverse simulated driving behaviors, such as eye-closing, yawning, nodding, and talking. Overall, the dataset is divided into five categories depending on whether the subject wears either glasses/sunglasses or not: BareFace, Glasses, Night_BareFace, Night_Glasses, and Sunglasses.

When someone would like to employ a successful face models, for instance, ASM, in analyzing facial states of the subjects appearing the present dataset [2], it is expected that so many exception cases will be occurred because of above mentioned problems such as extreme head poses, talking to someone, nodding deeply, and scarf (or sunglasses) wearing. To handle these troubling cases, we thought that first a very powerful face detector is necessary and secondly a non-conventional image processing technique such as deep neural network will be a better option for the present study. Note that recently it is shown that the gaze zone of a driver can be categorized using a deep neural network in real-time [3]. In addition, the present dataset is so huge that it actually fits to the deep learning method. Because of such reasons, we propose an approach in which the facial states estimation consists of two sub-parts: face detection and feature extraction using a deep neural network.

Boosting is a powerful technique for the classification task. For instance, it is well known that AdaBoost algorithm is the most successful face detector so far [5]. Recent advance in boosting research in Machine Learning (ML) area appears to be culminated with Gradient Boosting, especially for the big-data classification tasks as reported in many ML challenges [4,5]. In this study, the Extreme Gradient Booting method is adopted in classifying driver drowsiness [6].

By definition, supervised learning uses labeled data in training, whereas unsupervised learning processes un-labeled data during training. In-between is semi-supervised learning whereby it aims to utilize un-labeled data, in training the supervised learning, to improve its performance. Similar effort has been made in the name of transfer learning. We employ this technique here to improve overall performance of our classifier, i.e. Gradient Boosting Machine.

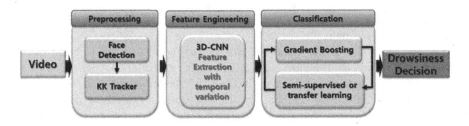

Fig. 1. The system block diagram illustrates three main stages of our system

Thus, our approach consists of three steps as illustrated in Fig. 1:

- Preprocessing: Only the facial area in each frame, rather than the whole image, will be used. Since dataset consists of sequence of images, we will employ a high performing tracker with a Haar-feature face detector in detecting the facial area within the video [7].
- Feature Engineering: The dataset is divided into five categories: BareFace, Glasses, Night_BareFace, Night_Glasses, and Sunglasses. A 3D-CNN has been designed using a deep neural network framework such as Tensorflow framework [8] with Keras library [9]. A sequence of 20 frames is given as input to the network for training as well as testing. The features, accumulated by training, are extracted from the hidden layer of the network and forwarded to the classifier.
- Classification: Gradient boosting machine is adopted for the classification task using XGBoost developed by [6]. To improve overall performance, semi-supervised learning techniques are utilized [10–12].

The rest of the paper is organized as follow. In Sects. 2, 3, and 4, we describe three main parts: preprocessing, feature engineering, and classification respectively. In Sect. 5, we present experimental results. Finally, Sect. 6 summarizes this paper.

2   Face Detection System

The present database basically contains many scenes (or frames) where one subject drives in front of a wheel. Therefore, the upper part of a torso is seen continuously throughout a session. Since it has been decided that we will use only the facial area of the given driver rather than the whole image, a reliable and powerful face detector is required. We use the face detection system that consists of a Haar-feature face detector and a powerful object tracker [7,13]. If one would employ only a face detector, it would fail to detect the face in the 3rd image in Fig. 2, because major portion of her facial features is occluded. However, since the object tracker, in which KCF (Kernelized Correlation Filter) is combined with KF (Kalman Filter), called it KK tracker, is robust against occlusion, it is able to track, i.e. detect the facial area, her deeply nodding head. On the other hand, the face detector also plays an important role when the tracker drifts by some reasons during the tracking, meaning that re-initialization is started by detecting the new face. Within the present detection system, the face detector cooperates with the tracker to improve detection performance.

Fig. 2. The present face detection system, consisting of a face detector and an object (or face) tracker.

3   Feature Engineering

3.1   2D Convolution

In 2D Convolutional Neural Networks (CNNs), convolution layers extract features from feature maps in the previous layer. They perform 2D convolution with 2D filter. For instance, these layers compute the outputs as follows:

$$C(x_{u,v}) = \sum_{s=-k}^{k} \sum_{t=-l}^{l} w(s,t)x_{u-s,v-t} \qquad (1)$$

where w is the filter of size $m x n$; x is input; $k = (m-1)/2$, $l = (n-1)/2$. The convolution filter acts very much like an edge detector. In the sub-sampling layers, 2D max-pooling is performed over the two trailing axes of input x. The output is a set of non-overlapping rectangles, and, for each such sub-region, output is the maximum value. Stacking multiple layers of convolution and max-pooling in an alternating fashion is the main structure of CNN. Back propagation is utilized to learn the parameters of CNN [14].

3.2 3D Convolution

3D convolution layers are used to capture the motion information encoded in the multiple contiguous frames as shown in Fig. 3. Since 2D kernel is only applied on the 2D feature maps to extract the spatial dimensions, 3D kernel is proposed to learn spatiotemporal features. The output of 3D convolutional layer is computed as follows:

$$C(x_{u,v,z}) = \sum_{s=-k}^{k} \sum_{t=-l}^{l} \sum_{r=-q}^{q} w(s,t,r)x_{u-s,v-t,z-r} \qquad (2)$$

where w is the 3D kernel of size $mxnxh$; $q = (h-1)/2$. By this operation, the feature map in the convolution layer is connected to multiple frames in the previous layer to encode motion information. Similar to the case of 2D CNN, max-pooling is utilized to sub-sample the feature maps. The output is a set of non-overlapping cubes. Note that a general design principle of 3D-CNN is similar to 2D-CNN that combines multiple layers of convolution and max-pooling. In addition, the number of feature maps is increased in the late layers by creating different types of features from the same of lower-level feature maps.

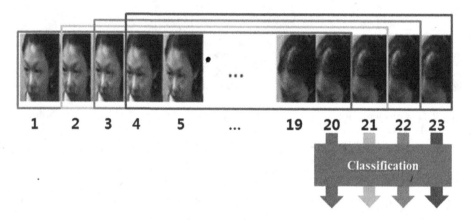

Fig. 3. Sequential frames to 3D CNN. The network accepts 20 frames as its input and extracts features and sends them to the classifier.

3.3 3D CNN Architecture

Figure 4 shows the schematic architecture of our network that uses for extracting features from the sequential frames of the driving dataset. The network consists of 6 convolution layers (Conv.) and 2 fully-connected layers. The input is a sequence of 20 frames as shown in Fig. 3. The max-pooling is implemented in the layer 1, 2, 4 and 6, respectively. It is known that such operation increases invariance property against distortion potentially contained in the input. To improve the generalization error, the dropout operation has been performed in each fully-connected layer with a rate of 0.5. Rectified Linear Units (ReLUs)

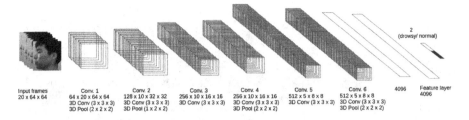

Fig. 4. Schematic architecture of 3D convolutional network. Each convolution layer has its name on the top such as Conv. 1, then the size of output in the 2nd row, 3D convolution with kernel size in the 3rd row, and the max-pooling with its size in the bottom, respectively.

is utilized as activation function in each convolution as well as in hidden layer. Finally, the output layer is built from two hidden layers and it has two units: one of them is drowsy and the other non-drowsy. The errors from the output are propagated back to all fully-connected and convolution layer during the training and it makes changes of weights between layers.

3.4 Details of Learning

We have trained our network for 5 different scenarios separately since the DDD dataset consists of BareFace, Glasses, Night_BareFace, Night_Glasses, and Sunglasses cases. In the network, Nesterov's accelerated gradient is used with the batch size of 32, and the momentum is fixed at 0.9. The weight update formula is given as:

$$V_{t+1} = 0.9 * V_t - \epsilon \nabla f(w_t + 0.9 * V_t)$$
$$w_{t+1} = w_t + V_{t+1} \tag{3}$$

where t is the iteration index, V the momentum variable, ϵ the learning rate, and $\nabla f(.)$ the gradient at w_t, respectively. We use a learning rate for all layers as well as all epochs, which is 0.0015. In addition, we initialize the weights in the form of Glorot weight initialization, also known as Xavier initialization.

3.5 Visualization of Convolutional Layers and Clustering of Features

A 3D-CNN extracts multi-layer deep features from each attribute map. Because the input to the network is given by sequence of 20 frames, the network not only learns the spatial dimensions but also captures the temporal information. Visualization of learned features in each convolution layer is illustrated in Fig. 5. As it is well known, the convolutional filter plays a vital role in extracting important features from the input. Note that the facial features are clearly visible in convolutional layer 1, whereas such features become more abstract in convolution layer 5 and 6.

In order to illustrate the characteristics of features extracted from the output layer, we visualized the feature vectors using t-SNE method, since it is known that it is an efficient tool for visualization of high dimensional data [15]. Figure 6

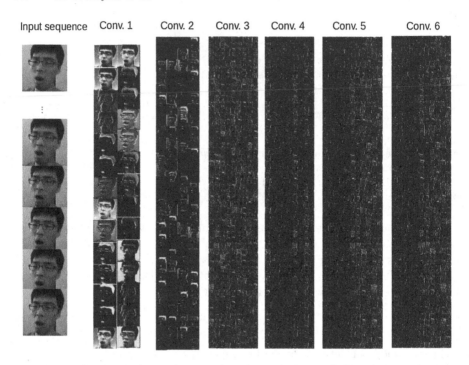

Fig. 5. Visualization of convolution layers for a specific input sequence. Note that feature maps are randomly selected in each layer to illustrate features learned by 3D convolution layers.

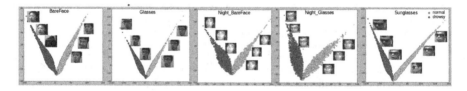

Fig. 6. Visualization of feature layer in our 3D CNN using t-SNE method for different driving scenarios. Red dots represent drowsy faces and green ones normal faces. Note that two clusters are fairly well separated. (Color figure online)

shows five different scenarios and illustrates how the features accumulated from the training data spread out. Indeed, the network is able to separate between drowsy and normal state fairly well for each scenario, although there are some overlaps between two cases as shown in the bottom of each figure.

4   Classification

4.1   Gradient Boosting

The idea of boosting is that a group of weak learners can be modified to be a better model by using them several times in a succession of hypotheses. Each weak leaning refocuses on the examples that the previous ones have found difficult or misclassified. A statistical framework which includes AdaBoost and related algorithms is developed by [4] and called Gradient Boosting Machine (GBM). The objective of GBM is to minimize the loss of the model by adding weak learners using a gradient descent like procedure. This class of algorithms is a stage-wise additive model. Therefore, this framework can be used with arbitrary differential loss functions and expanded the technique beyond binary classification problem to multi-class classification as well as regression.

Gradient Boosting includes three components: (1) a loss function that is to be optimized; (2) a weak leaner to predict; (3) an additive model to add weak learners to optimize the loss function. First, the loss function is differentiable and can be varied depends on the type of problem. We use the logistic binary loss function for binary classification for this challenge. Second, decision trees are utilized as the weak learner. 150 boosted trees are made for each model. Figure 7 show the first tree of the model trained using the glasses scenario.

Trees are constructed in a greedy manner where it selects the best split points based on scores by minimizing the loss. Third, when adding the trees, existing trees in the model are not changed. A gradient descent procedure minimizes the loss function when adding trees. Gradient Boosting is a greedy algorithm and therefore it over-fit a training dataset quickly. There are four ways to regularize the basic gradient boosting: tree constraints, shrinkage, random sampling, and penalized learning. In our experiment, we utilize the Xgboost library [6]. This is a scalable end-to-end tree boosting system, which has been favored among data scientists who have achieved state-of-the-art results in the ML challenges.

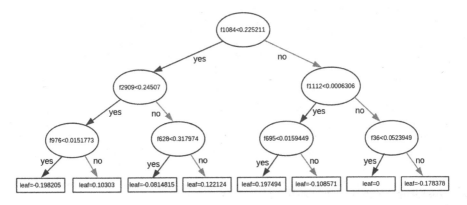

Fig. 7. The first boosted tree of the model trained by the glasses scenario. Each scenario has one specific model with 150 trees. The feature layer has 4096 neural units. The boosted tree considers each unit as a feature and it is named by its index. The leaf node is the logistic loss value.

4.2 Semi-Supervised Learning/Transfer Learning

Given that subjects are independent between training and testing dataset, we employ semi-supervised learning (SSL) to enhance performance of our system. So SSL is the halfway between supervised learning and un-supervised learning. In a similar vein, Transfer Learning (TL) tries to enhance its performance of the learner (or network), either by using labeled data that belongs to different domain with a certain risk, or by labeling a portion of un-labeled data in the same domain using a domain expert.

We have tested two semi-supervised learning, or transfer learning, methods to improve performance of the present system. First, although the testing dataset is un-labeled, or the ground-truth is unknown to the learner, we are able to extract features using our 3D-CNN and then to predict the probability scores using the Gradient Boosting model. Based on these scores, we select top 10% of testing data that have higher probability, and add them to the training dataset. One may call it pseudo-labeling. Then we retrain the Gradient Boosting model with this new training data. Secondly, we have labeled the un-labeled testing data by randomly selecting 2% of it. According to the theory of TL, a domain expert participates at this labeling task. However, note that such labeling process also takes a risk simply because performance can be improved only if the domain expert should have a certain area expertise, otherwise performance can be rather deteriorated.

5   Experiment Results

5.1 System Configuration

We have developed the present system using a GPU cluster platform DIGITS from NVIDIA running Ubuntu 14.04. The hardware specification is as follow:

- CPU: Core i7
- RAM: 64 Gb
- 4 GPUs: Titan X with 12 Gb in memory
- Hard disk: 3 Tb

In addition, we have used diverse open source libraries as follow and programmed with Python:

- Operating System: Ubuntu 14.04
- Python 2.7.6
- Sklearn 0.17.1
- Xgboost 0.4
- Pandas 0.18.1
- Numpy 1.11.1
- Keras 1.0.7
- Tensorflow r0.10
- OpenCV 2.4.9

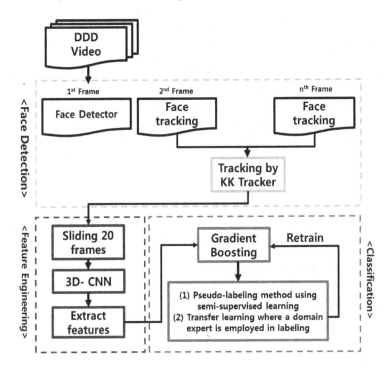

Fig. 8. The workflow of our system consisting of three parts. First, face detection system by combining a Haar-feature face detector and a high-performance object tracker; secondly, feature engineering whereby important features are extracted using a 3D-CNN; thirdly, classification and semi-supervised learning is used.

Figure 8 illustrates the workflow of our system consisting of three parts. The first part is the face detection system where a Haar-feature face detector is combined with a high-performance KK tracker. The second part carries out feature engineering using a 3D-CNN in the spatiotemporal domain. The third part classifies drowsiness of the given driver using gradient boosting and transfer learning/semi-supervised methods. The whole system can work continuously without any interruption as long as the face detector system provides the facial region to the network as shown in [3]. It is believed that the present deep neural network-based system is much reliable than the conventional image processing-base one because the network does not produce any exception as long as we provide input.

5.2   Result

In recent work [2], they have measured independently two facial behaviors such as eye-closing via the ASM face model and head nodding via the head pose estimation method. Then, the final decisions, whether the driver is drowsy or not, are made using a Markov model. Contrast to that study, we do not try to measure facial behaviors separately such as eye-closing, nodding and yawning, even though those ground-truths are actually given. We rather let the network learn such behaviors during the extensive training period using the drowsiness ground-truth value only, showing that our network somehow learns how to make decisions against diverse situations. Table 1 shows the result when randomly chosen 2% of testing dataset are labeled by the domain expert according to the theory of transfer learning. Our second best result comes from the pseudo-labeling process using the semi-supervised learning. It is found that the impact of transfer learning/semi-supervised learning is particularly promising since such method is useful to improve the overall performance of the network.

Table 1. Our best result through the Driver Drowsiness Detection challenge (%)

Scenario	Drowsiness (F1-score)	Non-drowsiness (F1-score)	Accuracy
Bareface	85.93	91.13	89.12
Glasses	92.18	90.36	91.36
Sunglasses	85.36	82.75	84.16
Night_Bareface	88.94	87.64	88.33
Nigh_Glasses	86.45	81.58	84.39
Overall	87.97	86.91	87.46

6   Conclusion

This study presents a new driver drowsiness detection system. For this challenge, we have designed the 3D Convolutional Neural Network to extract key features in term of spatial and temporal domains. The reliable face detection system in this system is important component since all remaining parts are heavily depends on the input, or the facial area of the driver. It is also found that gradient boosting machine is very useful as a binary classifier in making a decision either drowsy or non-drowsy. In addition, we have utilized semi-supervised learning and transfer learning methods to improve performance of the whole system. Given the fact that the DDD dataset contains diverse and tough conditions, we have chosen the deep neural network rather than the standard face model. Results suggest that the network-based system using recent machine learning techniques is a better option because the dataset is the complex and huge big-data, to which the conventional method has difficulty in dealing with.

Acknowledgement. This research was supported by the MSIP (Ministry of Science, ICT and Future Planning), Korea, under the ITRC (Information Technology Research Center) support program (IITP-2016-R2718-16-0017) and the Global IT Talent support program (IITP-2016-R0134-16-1032) supervised by the IITP (Institute for Information & Communications Technology Promotion).

References

1. Diddi, V.K., Jamge, S.: Head pose and eye state monitoring (HEM) for driver drowsiness detection: overview. IJISET-Int. J. Innov. Sci. Eng. Technol. **1**, 504–508 (2014)
2. Choi, I.H., Jeong, C.H., Kim, Y.G.: Tracking a drivers face against extreme head poses and inference of drowsiness using a hidden Markov model. Appl. Sci. **6**, 137 (2016)
3. Choi, I.H., Hong, S.K., Kim, Y.G.: Real-time categorization of driver's gaze zone using the deep learning techniques. In: 2016 International Conference on Big Data and Smart Computing (BigComp), pp. 143–148. IEEE (2016)
4. Friedman, J.H.: Greedy function approximation: a gradient boosting machine. Ann. Stat. **21**, 1189–1232 (2001)
5. Freund, Y., Schapire, R.E.: A desicion-theoretic generalization of on-line learning and an application to boosting. In: Vitányi, P. (ed.) EuroCOLT 1995. LNCS, vol. 904, pp. 23–37. Springer, Heidelberg (1995). doi:10.1007/3-540-59119-2_166
6. Chen, T., Guestrin, C.: Xgboost: a scalable tree boosting system. arXiv preprint arXiv:1603.02754 (2016)
7. Lienhart, R., Maydt, J.: An extended set of Haar-like features for rapid object detection. In: 2002 Proceedings of the International Conference on Image Processing, vol. 1, I-900. IEEE (2002)
8. Abadi, M., Agarwal, A., Barham, P., Brevdo, E., Chen, Z., Citro, C., Corrado, G.S., Davis, A., Dean, J., Devin, M., et al.: Tensorflow: large-scale machine learning on heterogeneous systems, vol. 1 (2015). Software available from https://www.tensorflow.org/
9. Chollet, F.: Keras: Deep learning library for theano and tensorflow (2015)
10. Arnold, A., Nallapati, R., Cohen, W.W.: A comparative study of methods for transductive transfer learning. In: Seventh IEEE International Conference on Data Mining Workshops (ICDMW 2007), pp. 77–82. IEEE (2007)
11. Chapelle, O., Scholkopf, B., Zien, A.: Semi-supervised learning (Chapelle, o. et al., eds.; 2006)[book reviews]. IEEE Trans. Neural Netw. **20**, 542–542 (2009)
12. Shi, X., Fan, W., Ren, J.: Actively transfer domain knowledge. In: Daelemans, W., Goethals, B., Morik, K. (eds.) ECML PKDD 2008. LNCS (LNAI), vol. 5212, pp. 342–357. Springer, Heidelberg (2008). doi:10.1007/978-3-540-87481-2_23
13. Huynh, X.-P., Choi, I.-H., Kim, Y.-G.: Tracking a human fast and reliably against occlusion and human-crossing. In: Bräunl, T., McCane, B., Rivera, M., Yu, X. (eds.) PSIVT 2015. LNCS, vol. 9431, pp. 461–472. Springer, Heidelberg (2016). doi:10.1007/978-3-319-29451-3_37
14. LeCun, Y., Bottou, L., Bengio, Y., Haffner, P.: Gradient-based learning applied to document recognition. Proc. IEEE **86**, 2278–2324 (1998)
15. Maaten, L.V.D., Hinton, G.: Visualizing data using t-SNE. J. Mach. Learn. Res. **9**, 2579–2605 (2008)

MSTN: Multistage Spatial-Temporal Network for Driver Drowsiness Detection

Tun-Huai Shih and Chiou-Ting Hsu[(✉)]

Department of Computer Science, National Tsing Hua University,
Hsinchu, Taiwan
cthsu@cs.nthu.edu.tw

Abstract. Recent survey has shown that drowsy driving is one of the main factors in fatal motor vehicle crashes. In this paper, given only the visual information of the driver, we propose a Multistage Spatial-Temporal Network (MSTN) to efficiently and accurately detect driver drowsiness. The proposed MSTN consists of a spatial CNN, a temporal LSTM, and then followed by a temporal smoothing. Firstly, we use the spatial CNN to effectively extract drowsiness-related features from the face region detected from each video frame. Then, we model the temporal variation of the drowsiness status by feeding a sequence of frame-level features into the Long Short Term Memory (LSTM). Finally, we conduct the temporal smoothing to smooth the predicted drowsiness scores in order to avoid noisy predictions. We evaluate the proposed MSTN using NTHU Drowsy Driver Detection Video Dataset and achieve 82.61% overall accuracy on the testing set.

1 Introduction

Drowsy driving is one of the main factors that cause traffic crashes. Therefore, it is of utmost importance to develop active monitoring systems that help drivers avoid accidents in a timely manner. Because convolutional neural networks (CNN) have shown their powerfulness in many computer vision tasks such as image recognition, object detection, and semantic segmentation, we plan to build a driver drowsiness detection system based on CNN. However, the driver drowsiness status not only depends on spatial appearance, but also depends on the appearance change along time. We must consider both spatial and temporal information in order to make accurate prediction. A recent paper, LRCN [1] proposed an architecture that handles both static input (single image) and dynamic input (video). LRCN consists of a spatial CNN, followed by a temporal network (LSTM). The spatial CNN extracts powerful visual features, while the temporal LSTM models the temporal variation of the extracted features. Inspired by [1], we proposed a Multi-stage Spatial-Temporal Network (MSTN) that also consists of a spatial CNN and a temporal LSTM, but followed by a third stage to perform temporal smoothing. Unlike [1], we do not train the MSTN end-to-end. Instead, we train the MSTN stage-by-stage. Each stage thus only focuses on a single and simpler learning task. We describe the architecture of the MSTN in Sect. 2, and show the experiments in Sect. 3.

© Springer International Publishing AG 2017
C.-S. Chen et al. (Eds.): ACCV 2016 Workshops, Part III, LNCS 10118, pp. 146–153, 2017.
DOI: 10.1007/978-3-319-54526-4_11

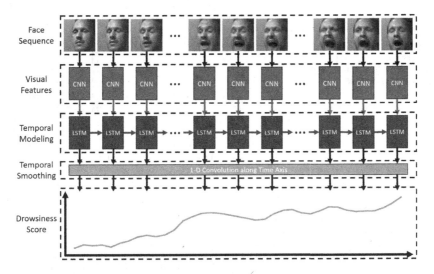

Fig. 1. Overview of the MSTN. It consists of three main stages, where the first stage (red) is the spatial network, and the following stages (blue and green) are temporal networks. (Color figure online)

2 Proposed Method

Figure 1 illustrates the proposed MSTN, which consists of three main stages and each stage corresponds to a single network. In Sect. 2.1, we first describe how we extract frame-level features using a carefully-designed CNN architecture to characterize drivers' drowsiness. Then, we feed a sequence of frame-level features extracted by the previous stage into LSTM to model the temporal variation of drowsiness in Sect. 2.2. Finally, we use a 1-D linear convolution along the time axis to further smooth the predicted drowsiness scores in Sect. 2.3.

2.1 Visual Features

The goal of the first stage is to learn a good feature extractor (denoted as V) to extract drowsiness-related visual features (denoted as F) from a single face image (denoted as I). Because we conduct the feature extractor on every single I, the learned F must be general and robust to various input noises. Here we adopt VGG-16 [6] as our basic model but make several modifications to derive a more general and robust extractor for drowsiness detection.

Figure 2 illustrates the modified VGG-16 architecture V. The original VGG-16 contains 13 conv layers (grouped into conv1–5), 5 pool layers (pool1–5), and 2 fc layers (fc6–7). Because (1) the fc6–7 layers contain a lot of parameters and (2) the input now is a relatively smaller image (64×64 as compared to 224×224), we replace fc6–7 with smaller fc layers. This not only reduces the number of parameters to learn, but also reduces the risk of over-fitting; this modification thus can result in a more general feature extractor.

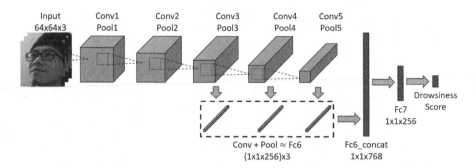

Fig. 2. Feature extractor V, which is a modified VGG-16 with smaller fc layers.

Now, given a $64 \times 64 \times 3$ face image, we first extract not only pool5 but also pool3-4 features to capture discriminative representation. In general, lower-level layers in CNN contain more detailed information, while higher-level layers contain global information. We argue that lower-level layers are also helpful for detecting driver drowsiness. For example, the eyes are relatively small as compared to the entire face, but the last layer in CNN is usually insensitive to such small regions (because of pooling layers). Thus, we believe the lower-level layers are essential to capture such small regions from an entire face image.

For each of the pool3–5 features, we append a 1×1 conv layer, followed by a global max pooling layer to approximate fc6 and result in three vectors of the same length (i.e., 256 in this paper). This approximation has two purposes. One is to reduce the number of parameters. The other is to let the network "automatically" discover drowsiness-related features by pooling operation. We concatenate the pooled vectors and append another small fc layer (fc7) to derive high-level visual features, which is the aforementioned F.

In order to make our feature extractor V more general and also to speed up the training process, we adopt Batch Normalization (BN) [3]. BN normalizes the layer output (or input) by estimating sample mean and variance of the hidden units over mini-batch data. Because BN calculates the sample mean and variance, it will also introduce randomness if a mini-batch consists of different samples. This randomness will reduce the risk of over-fitting (similar to DropOut [2]). In this paper, we place BN before fc6, fc7, and after fc7 (totally 5 BN layers).

Based on the modified VGG architecture, in order to learn drowsiness-related F, we append a prediction layer after fc7 to produce the drowsiness score Y_d. Given Y_d and the ground-truth label $L_d \in \{0, 1\}$, we use SigmoidCrossEntropy-Loss layer in caffe [4] to train V. Using this loss layer, if the drowsiness degree of the input is higher (i.e. more drowsy, or $L_d = 1$), Y_d should be much larger than 0, and vice versa.

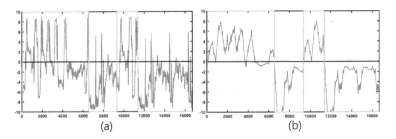

Fig. 3. Temporal smoothing. The orange lines are the drowsiness scores (a) before smoothing and (b) after smoothing. (Color figure online)

2.2 Temporal Modeling

In the previous stage, although we have already trained a model to predict the drowsiness score of each frame, the prediction is merely based on a single frame. Sometimes it is very difficult to determine whether a person is yawning or laughing when only seeing his open mouth. In stage 2, in order to handle this issue, we adopt Long Short Term Memory (LSTM) [7] (i.e., blue cubes in Fig. 1) to model the temporal variation of drowsiness status from a sequence of frame-level features F.

We first extract frame-level features F of all videos by feature extractor V. Note that this step is conducted off-line; i.e., we do not fine-tune V in this stage. Then, for each video, we randomly sample many video clips with the fixed length 50 as our training data. The temporal network we use is a single LSTM layer with 32 hidden units, followed by a new prediction layer (weight-sharing across time) to predict the refined drowsiness score $\hat{Y}_d^t$ of each frame ($t = 1, \ldots, 50$). We still use SigmoidCrossEntropyLoss layer to train the LSTM.

2.3 Temporal Smoothing

In the previous two stages, we use a spatial network (CNN) and a temporal network (LSTM) to predict the drowsiness score of each frame. When testing on the validation set, we find that there still exist some noisy predictions. We have tried several post-processing on the predicted scores, including moving average and mean/median filtering, and can gain up to 5% the improvement of accuracy. This suggests that we can get better performance by "smoothing" the predicted scores. Because the mean filtering is actually a 1-d convolution along the time axis, we should be able to "learn" a network to perform such smoothing instead of using any pre-determined post-processing.

Given the predicted scores $\hat{Y}_d^t$ of all frames, we use a network with a linear 1-d conv layer to predict the smoothed scores $\widetilde{Y}_d^t$. Figure 3(a, b) illustrate the scores before and after temporal smoothing, where the cyan line is the ground-truth labels $L_{d'}$ (mapped to 10), and the orange lines are the drowsiness scores of each frame. After temporal smoothing, some noisy predictions are successfully suppressed.

3   Experiments

3.1   Dataset

We evaluate the proposed MSTN using the NTHU Drowsy Driver Detection (DDD) dataset collected by NTHU Computer Vision Lab. DDD contains 36 subjects of different ethnicities recorded with and without wearing glasses/sunglasses under a variety of simulated driving scenarios, including normal driving, yawning, slow blink rate, falling asleep, burst out laughing, etc., under day and night illumination conditions. The training set contains 18 subjects with 5 different scenarios (BareFace, Glasses, Night_BareFace, Night_Glasses, Sunglasses). Each scenario contains 4 videos, which are combination of drowsiness-related symptoms (yawning, nodding, slow blink rate) and non-drowsiness related actions (talking, laughing, looking at both sides). The evaluation set contains 20 driving videos (from another 4 subjects) with drowsy and non-drowsy status mixed under different scenarios. The testing set contains 70 driving videos (from the remaining 14 subjects).

There are 4 frame-level annotations provided, including drowsiness status (Stillness or Drowsy), eye status (Stillness or Sleepy-eyes), head status (Stillness, Nodding, or Looking aside), and mouth status (Stillness, Yawning, or Talking & Laughing). The annotations of testing set are unavailable.

The videos of Night_BareFace, Night_Glasses scenario were captured at 15 frames per second; BareFace, Glasses, Sunglasses scenario were 30 frames per second. All the videos are grayscale and in 640×480 resolution. No audio information is provided.

3.2   Implementation Details

Data Pre-processing: Given a driving video, we first use Matlab built-in Viola-Jones face detector to detect the driver's face from all frames. We then crop the detected face bounding boxes (along with some boundary pixels) and resized the cropped face regions to a fixed size (66×66 for training, and 64×64 for testing). Because the frame rate of most videos is 30 fps and there is only subtle difference between adjacent frames, we sub-sample the video frames by a factor of 3 and input the face sequence in the frame rate of 10 fps to the proposed MSTN. We will up-sample the predicted scores back to the original length later. After subsampling, because the videos in DDD are grey-scale, we replicate the single-channel images to become 3-channel images so as to generalize the proposed method to handle either color or grey-scale input videos.

Label Pre-processing: Among the 4 annotations, drowsiness and eye have two statuses (0 or 1), while head and mouth have three statuses (0, 1, or 2). We do not use SoftmaxLoss for head and mouth. Instead, we "map" the statuses [0, 1, 2] into the labels [0.5, 1, 0]. We observe that, among the three statuses of head, i.e., Stillness (0), Nodding (1), and Looking aside (2), status 1 is the one that most relates to the drowsiness, while the status 2 is the least related one. Therefore, we set status 1 to have label 1 and status 2 to have label 0. We set the Stillness status to have label 0.5, because its drowsy degree lies between status 1 and status 2. After the mapping, we use SigmoidCrossEntropyLoss for all the annotations during training.

Environment: We conduct the implementation in a single Win10 PC with Intel Core i7 6700 CPU, NVIDIA GTX 960 GPU, 16GB DDR4 memory, and 3TB HDD.

Stage 1 Training: We use caffe [4] to fine-tune V from the pre-trained VGG-16 [6] by fixing conv1-2 weights for simplification. Weights of the newly-added layers are randomly initialized by Gaussian. We perform random cropping and horizontal flipping of the input images as run-time data augmentation. Because we adopt BN [3], we also shuffle the training data list to ensure randomness. For the solver, we choose Adam [5] with mini-batch size 64, fixed learning rate 0.0005, weight decay 0.0005, and momentum 0.9 & 0.99. We train V up to 50000 iterations, taking less than 4 hours.

Stage 1 Testing: We discard the prediction layer of V and extract drowsiness-related features F of each frame. F will be used in the next stage.

Stage 2 Training: We randomly sample video clips with fixed length 50 (equivalent to 5 s) from training set as the input of LSTM. Note that the input features are already extracted in stage 1 (and thus fixed). For the solver, except for a larger learning rate (0.005) and batch size 8 (8 clips), all other settings are the same as stage 1. We train the LSTM up to 10000 iterations, taking less than 5 min.

Stage 2 Testing: Although the training clip length (T) of the LSTM is 50, we have tried different T during testing. In addition, to get more smoothed prediction, we perform "overlapped" testing. For example, given a video with 100 frames and T = 50, we feed 3 video clips (frame index 1–50, 26–75, and 76–100) into the LSTM. The stride size is set to T/2, and the scores of overlapped frames are simply averaged.

Stage 3 Training: Given the stage 2 scores (drowsiness, eye, head, and mouth) of all videos from the "evaluation" set, we uniformly sample many clips with length K + 200 and stride 200, where K = 201 is the kernel size of the 1-D convolution. We feed all score clips into a temporal smoothing network with a single 1-D linear conv layer to produce the smoothed scores. Because we do not perform zero-padding, the output length will be $401 - (201 - 1) = 201$. The learning rate is fixed to 0.001, and the weight decay is set to 0. We train this network up to 1000 iterations, taking less than 2 min.

Stage 3 Testing: Given the stage 2 scores of a video with length N, we first pad 200 scores to both head and tail of the video (resulting in length N + 200). Then, we crop N clips with length 201 (equals to K) centered on the original N frames, and feed them into the temporal smoothing network to get the smoothed scores of each frame. Although it is possible to feed a longer clip to reduce the testing time, we choose to feed a clip with the same length as K to handle different video lengths. Finally, because the input video is sub-sampled previously, we up-sample the scores back to the original video length using 3× linear interpolation.

Table 1. Average accuracies (%) on the evaluation set.

	Drowsiness	Eye	Head	Mouth
Stage 1	75.14	74.68	91.35	88.74
Stage 2, T = 40	76.66	75.46	93.08	90.81
Stage 2, T = 50	76.76	75.68	93.03	90.86
Stage 2, T = 60	77.07	76.14	93.06	90.82
Stage 2, T = 120	77.60	76.67	93.08	90.84
Stage 2, T = 180	77.81	76.97	93.06	90.88
Stage 3, K = 201	**85.52**	**85.11**	**94.74**	**93.06**

Table 2. Average F1 scores and accuracies (%) on the testing set.

Scenario/author	Drowsiness F1-score	Nondrowsiness F1-score	Accuracy/rank
Bareface	90.08	92.49	91.45
Glasses	83.59	81.56	82.64
Sunglasses	86.92	82.46	85.02
Night-Bareface	81.81	80.98	81.41
Night-Glasses	73.46	75.40	74.47
Yong-Guk Kim	**87.97**	**86.91**	**87.46 (1st)**
Ours, overall	82.82	82.38	82.61 (2nd)
Jie Lyu	75.62	75.17	75.40 (3rd)

3.3　Results on DDD Evaluation Set

Table 1 lists the average accuracies among 20 videos of the evaluation set. The spatial network V in the first stage already achieves 75.14% accuracy of drowsiness even though its prediction is merely based on a single frame. The temporal network LSTM in the second stage models the temporal variation of the drowsiness status, and thus improves the accuracy of drowsiness to 77.81%. It is worth noting that a longer clip length T during testing achieves higher accuracies. Finally, after adopting temporal smoothing with kernel size 201 in the third stage, we achieve 85.52% accuracy of drowsiness.

3.4　Results on DDD Testing Set

Table 2 lists the average F1 scores and accuracies on the testing set. We also list the results of top three teams participating the competition this year. In terms of accuracy, the proposed MSTN works pretty well under the Bareface scenario (91.45%), but is 16.98% worse when under the Night-Glasses scenario (74.47%). The underlying reason might be the poor lighting environment or the occlusion caused by the glasses. In terms of F1 score, the balanced F1 scores among all scenarios show that the proposed MSTN does not make biased prediction. Finally, with the 82.61% overall accuracy, we won the second place in the ACCV DDD 2016 challenge.

Table 3. Time complexity (sec) of the proposed MSTN, where stage 0 loads and extracts driver faces from the entire video. 100 is the video length (in second).

Stage 0	Stage 1	Stage 2	Stage 3
41.29/100	7.32/100	0.3/100	0.65/100

3.5 Time Complexity

Table 3 lists the time complexity of the proposed MSTN. Given a 100-s video (3000 frames), the data pre-processing step (stage 0) takes about 41 s to extract 3000 raw face images from the video. After $3\times$ sub-sampling, the first stage takes 7.32 s to extract visual features of the 1000 faces. The remaining two stages (including $3\times$ linear up-sampling) take about 1 s to predict the drowsiness scores with length 3000. Without considering the data pre-processing step, the proposed MSTN only takes about 8% time of the entire video to produce the results.

4   Conclusion

In this paper, we proposed a Multistage Spatial-Temporal Network (MSTN) for driver drowsiness detection. The spatial network V effectively extracts robust drowsiness-related frame-level features, and the temporal network LSTM then models the temporal variation of the drowsiness status. Finally, a temporal smoothing network with a single 1-D linear convolution smooths the predicted scores from previous stage. Experiments show that the proposed MSTN is both accurate and efficient, and is able to be applied on real-world applications.

References

1. Donahue, J., Hendricks, L.A., Guadarrama, S., Rohrbach, M., Venugopalan, S., Saenko, K., Darrell, T.: Long-term recurrent convolutional networks for visual recognition and description. In: CVPR (2015)
2. Hinton, G.E., Srivastava, N., Krizhevsky, A., Sutskever, I., Salakhutdinov, R.R.: Improving neural networks by preventing coadaptation of feature detectors. arXiv:1207.0580 (2012)
3. Ioffe, S., Szegedy, C.: Batch normalization: accelerating deep network training by reducing internal covariate shift. arXiv:1502.03167 (2015)
4. Jia, Y., Shelhamer, E., Donahue, J., Karayev, S., Long, J., Girshick, R., Guadarrama, S., Darrell, T.: Caffe: convolutional architecture for fast feature embedding. arXiv:1408.5093 (2014)
5. Kingma, D., Ba, J.: ADAM: a method for stochastic optimization. arXiv:1412.6980 (2014)
6. Simonyan, K., Zisserman, A.: Very deep convolutional networks for large-scale image recognition. In: ICLR (2015)
7. Zaremba, W., Sutskever, I.: Learning to execute. arXiv:1410.4615 (2014)

Driver Drowsiness Detection System
Based on Feature Representation Learning
Using Various Deep Networks

Sanghyuk Park$^{(\boxtimes)}$, Fei Pan, Sunghun Kang, and Chang D. Yoo

School of Electrical Engineering, KAIST, Guseong-dong,
Yuseong-gu, Dajeon, Republic of Korea
{shine0624,feipan,sunghun.kang,cd_yoo}@kaist.ac.kr

Abstract. Statistics have shown that 20% of all road accidents are fatigue-related, and drowsy detection is a car safety algorithm that can alert a snoozing driver in hopes of preventing an accident. This paper proposes a deep architecture referred to as deep drowsiness detection (DDD) network for learning effective features and detecting drowsiness given a RGB input video of a driver. The DDD network consists of three deep networks for attaining global robustness to background and environmental variations and learning local facial movements and head gestures important for reliable detection. The outputs of the three networks are integrated and fed to a softmax classifier for drowsiness detection. Experimental results show that DDD achieves 73.06% detection accuracy on NTHU-drowsy driver detection benchmark dataset.

1 Introduction

Over the years, various safety-related driving assistant systems have been proposed to reduce the risk of car accidents, and statistics have shown fatigue to be a leading cause of car accidents. In fact, the American Automobile Association released a figure in 2010 that 17% of all fatal crashes in the USA could be attributed to tired drivers. This seems to be a global trend. In Germany, several studies conducted by Volkswagen AG in 2005 indicate that 5–25% of all collisions are caused by driver falling asleep. Failing concentration can impair steering behavior and reduce reaction time, and studies show that drowsiness increases the risk of collision by several factors. These figures indicate the need for a reliable driver drowsiness detection algorithm.

Drowsiness detection is studied by monitoring vehicle-based measurements, behavioral measurements, and physiological measurements. Vehicle-based measurements are from steering wheel movements, driving speed, brake patterns, and standard deviation of lane positions [8–12]. Behavioral measurements are obtained from driver eye/face movement using a camera. Physiological measurements such as heart rate, electrocardiogram (ECG) [7], electromyogram (EMG) [24], electroencephalogram (EEG) [6] and electrooculogram (EOG) [25] can be used to monitor drowsiness. Subjective measurements based on questionnaires and electrophysiological measures of sleep can also be made but it is

© Springer International Publishing AG 2017
C.-S. Chen et al. (Eds.): ACCV 2016 Workshops, Part III, LNCS 10118, pp. 154–164, 2017.
DOI: 10.1007/978-3-319-54526-4_12

generally difficult to obtain drowsiness feedback from a driver in a real driving situation. There are reasons for and against using each measurement, and currently it is not clear which measurement is most reliable and cost effective. This paper is focused on detecting drowsiness by monitoring facial features and head movements of the driver.

It is assumed that drowsiness will manifest as rapid and constant blinking, nodding or head swinging, and frequent yawning [13,15–18]. PERCLOS (percentage of eyelid closure over the pupil over time) is considered a reliable measure for predicting drowsiness and has been incorporated in commercial products such as Seeing Machines and Lexus. Other facial movements such as inner brow rise, outer brow rise, lip stretch, jaw drop and eye blink have also been known to be markers for drowsiness. Until recently, most vision-based drowsiness detection has relied on hand-crafted features for monitoring facial and head movements. In general, hand-crafted features have shown limited effectiveness in real-world scenarios which might include drivers wearing sunglasses and large variation in illumination. On the other hand, features learned based on deep learning have been more effective in real-world scenarios. Choi et al. [14] developed a gaze zone detection algorithm based on features learnt using a convolutional neural network (CNN). Based on the learnt features, support vector machine (SVM) is used to predict drivers gaze zone. Dwivedi et al. [12] used a 3-layered CNN to learn facial features of detected face region from the input image. The outputs of the last layer are considered as the extracted features. On the basis of these features, the softmax classifier was trained and used for drowsiness prediction.

In this paper, we propose a deep architecture referred to as deep drowsiness detection (DDD) network in detecting driver drowsiness from input video. The proposed DDD network learns appropriate features for the task and predicts drowsiness of the driver. We consider both RGB video as well as optical flow as inputs. The proposed DDD network consists of three deep networks: AlexNet [1], VGG-FaceNet [2] and FlowImageNet [3]. Given an image sequences, the AlexNet is fine tunned to learn features related to drowsiness. The VGG-FaceNet is trained to learn facial feature related to drowsiness which is robust to genders, ethnicity, hair style and various accessories adornment. FlowImageNet takes dense optical flow image that is extracted from consecutive image sequences and is trained to learn behavior features related to drowsiness such as facial and head movements. Each three networks are independently fine-tunned for multi-class drowsiness classification given the following four classes: non-drowsiness, drowsiness with eye blinking, nodding and yawning. The softmax outputs of each networks are averaged for final classification. Before obtaining the final softmax output, fully connected (fc) 7 layer features of a block frames are concatenated for classification. We refer to this architecture as independently-averaged architecture (IAA). For comparison, the three networks are also integrated such that their fc7 layer features are concatenated, and based on this concatenated feature, input videos are classified into one of four classes. We refer to this architecture as feature-fused architecture (FFA). The proposed algorithm is evaluated on NTHU-driver drowsiness detection benchmark video dataset. The prediction results are presented in terms of detection accuracy. Experimental results show that DDD achieves 73.06% detection accuracy on NTHU-drowsy driver detection benchmark dataset.

The rest of this paper is organized as follows. Section 2 describes the proposed DDD network in detail. Section 3 presents experimental results comparing and analyzing various algorithms. Finally, Sect. 4 concludes the paper.

2   Proposed Deep Drowsiness Detection (DDD) Network

In this section, we present the proposed DDD network to detect driver drowsiness from complicated driving scenarios under varying circumstance. The proposed DDD network consists of two processes: learning feature representations and ensemble detection. During feature representation learning, we use three different networks: AlexNet, VGG-FaceNet and FlowImageNet. During drowsiness detection, we use two different ensemble strategies: independently-averaged architecture (IAA) and feature-fused architecture (FFA). The framework of DDD network for drowsiness detection using FFA is shown in Fig. 1.

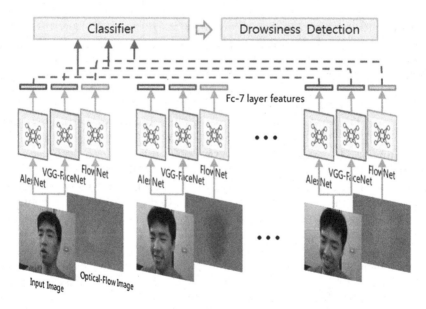

Fig. 1. The framework of deep drowsiness detection (DDD) network for drowsiness detection using feature-fused architecture (FFA).

2.1   Image Feature Representation Learning Based on AlexNet

To extract image feature which is robust to various backgrounds and environment changes (*i.e.,* indoor/outdoor, day/night) from the input image sequences, we adopt a pre-trained AlexNet model. A 8-layered AlexNet showed that deep

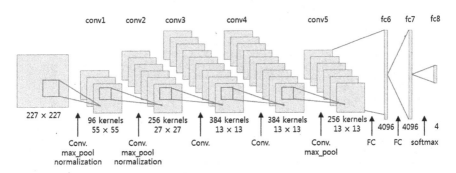

Fig. 2. The example architecture of AlexNet model for DDD network.

convolutional neural networks can significantly outperform other methods for the task of large scale image classification. This Alexnet consists of 5 convolution layers and 3 fc layers which has 60 million parameters and 650,000 neurons, and this model is trained with 1.2 million images for 1000 categories classification. For the proposed DDD networks, we fine-tunned AlexNet to classify multi-class drowsiness. The architecture and parameters of AlexNet model for DDD network are summarized in Fig. 2.

Each input image is down-sampled to size of 227 × 227. Training dataset is constructed by subtracting the mean value from pre-trained model, and AlexNet is fine-tunned using the drowsiness detection training dataset. The first convolutional layer filters the 227 × 227 × 3 input image with 96 kernels of size 11 × 11 × 3 and stride of 4 pixels. The second convolutional layer takes as input the (after normalized and pooled layers) output of the first convolutional layer and filters it with 256 kernels of size 5 × 5 × 48. The third, fourth, and fifth convolutional layers are connected to one another without any intervening pooling or normalization layers. The third convolutional layer has 384 kernels of size 3 × 3 × 256 connected (after normalized and pooled layers) to the outputs of the second convolutional layer. The fourth convolutional layer has 384 kernels of size 3 × 3 × 192, and the fifth convolutional layer has 256 kernels of size 3 × 3 × 192. The fully-connected layers have 4096 neurons. The output of the last fc layer is fed into a 4-way softmax layer which produces a distribution over the 4 class labels such as non-drowsy state, drowsy state with eye blinking, drowsy state with head nodding, and drowsy state with mouth yawning.

2.2   Facial Feature Representation Learning Based on VGG-FaceNet

To extract facial feature representation which is robust to facial characteristics (*i.e.*, genders, ethnicities) from the input image sequences, we adopt a pre-trained VGG-FaceNet model. A 16-layered VGG-FaceNet model was trained on various celebrity faces and evaluated on faces recognition task from the Labeled Faces in the Wild and YouTube faces datasets. The VGG-FaceNet consists of 13 convolution layers and 3 fc layers based on VGG-Very-Deep-16 CNN

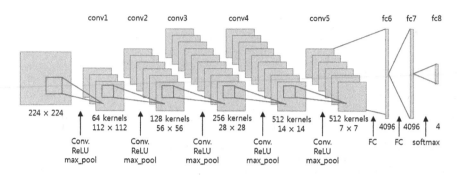

Fig. 3. The example architecture of VGG-FaceNet model for DDD network.

architecture, and this model is trained with 2.6 million images for 2,622 people recognition. The architecture and parameters of VGG-FaceNet model for DDD network are summarized in Fig. 3. Due to the deeper structure of VGG-FaceNet, we describe shorter version of VGG-FaceNet in Fig. 3. Similar to AlexNet training, we trained VGG-FaceNet with multi-class drowsiness classification about drowsy statements in fine-tuning manner. During the training, we down-sampled the images with a fixed resolution of 224×224, and the training dataset is constructed in the same way in AlexNet training. The training image dataset is passed through a stack of many convolutional layers, max pooling layers, Rectified Linear Unit (ReLU) activation function. Then these stacked layers are followed by three fc layers. The first two fc layers have 4096 neurons each, and the output of the last fc layer is fed into a 4-way softmax which produced a distribution over 4 class labels in the same way as training AlexNet.

2.3 Behavior Feature Representation Learning Based on FlowImageNet

To extract behavior feature representation which is related to movement patterns about drowsy states such as face and head gestures from the input image sequences, we adopt a pre-trained FlowImageNet model. A 8-layered FlowImageNet model consists of 5 convolution layers and 3 fc layers, and this model is used for video activity recognition task using UCF101 dataset [5]. The architecture and parameters of FlowImageNet model for DDD network are summarized in Fig. 4. As training AlexNet, FlowImageNet is trained using multi-class drowsiness classification about drowsy statements in fine-tuning manner. During training, we down-sampled the images with a fixed resolution of 227×227, and the training dataset is constructed in the same form as training AlexNet. Dense optical flow was calculated using [4] from consecutive image sequences and transformed into two channels of flow images by scaling and shifting x and y flow values to a range of $[-128, +128]$. A third channel for the flow image was created by calculating the magnitude of flows. The training image dataset is also passed through a stack of many convolutional layers, ReLU activation function,

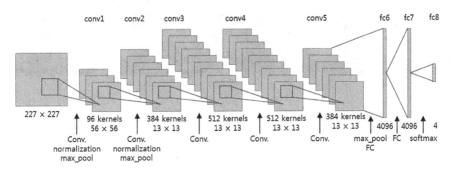

conv1   conv2   conv3   conv4   conv5   fc6 fc7 fc8

227 × 227   96 kernels   384 kernels   512 kernels   512 kernels   384 kernels   4096 4096   4
            56 × 56   13 × 13   13 × 13   13 × 13   13 × 13
            Conv.   Conv.   Conv.   Conv.   Conv.   max_pool FC softmax
            normalization normalization   FC
            max_pool   max_pool

Fig. 4. The example architecture of FlowImageNet model for DDD network.

max pooling layers and normalization. A stack of various layers is followed by three fc layers. The first two fc layers have 4096 neurons each, and the output of the last fc layer is also fed into a 4-way softmax which produces a distribution over the 4 class labels in the same way as training AlexNet and VGG-FaceNet model.

2.4 Ensemble Detection Using DDD Network

The outputs of the three networks are combined to make single prediction. Ensemble model is well known to increase accuracy on various machine learning tasks [19–21], such as voting, averaging, stacking classifiers with regressors, and so on. For the proposed DDD network, we adopt two different fusion strategies: independently-averaged architecture (IAA) and feature-fused architecture (FFA). During IAA, the probability distributions of each network output for multi-class classification are integrated, and average probabilities are used to determine the driver drowsiness. During FFA, the three networks are also integrated such that their fc7 layer features are concatenated, and based on this concatenated feature, input video are classified into one of four classes using SVM [22].

3 Experiments

In this section, we provide competitive experimental results using proposed DDD network on drowsy driver detection video dataset. Due to the lack of previous benchmark performance on this dataset, we compare with the performance of several well-known classification algorithms such as variants of CNNs and LRCN [3].

3.1 Drowsy Driver Detection Video Dataset

To evaluate proposed DDD network, we use NTH Drowsy Driver Detection (NTHU-DDD) video dataset[1]. This video dataset contains 36 subjects including

---

[1] http://cv.cs.nthu.edu.tw/php/callforpaper/2016_ACCVworkshop/.

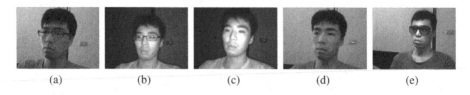

Fig. 5. Example frames of NTHU-DDD video dataset with different situations: (a) wearing glasses, (b) wearing glasses at night, (c) bare face at night, (d) bare face and (e) wearing sunglasses

Fig. 6. Example frames of same situation (night bareface) and different behaviors (mixing drowsy and non-drowsy state) from 1 video clip.

Fig. 7. Example frames of different situations (night wearing glasses, night bareface, wearing glasses, waering sunglasses, bareface) and same behavior (drowsy state) from 5 video clips.

different people, both genders, different ethnicities, which is in 5 situations as shown in Fig. 5. Each situation contains at least two behaviors about drowsy states, such as slow blinking, nodding, and yawning as shown in Fig. 6. The total dataset consists of train dataset, evaluation dataset, and test dataset. The train dataset consists of 360 video clips (722,223 frames) of 18 subjects. The evaluation dataset consists of 20 video clips (173,259 frames) of 4 subjects and test dataset consists of 70 clips (736,132 frames) of 14 subjects. During training and evaluation, each frame is binary labeled: drowsy or non-drowsy. The ground-truth label for test dataset is not publicly available yet. The dataset includes different physical attributes including variety in skin tone, fatigue, facial structure, clothes and hair styles. Thus, the algorithm for drowsiness detection should consider robustness and efficiency in all circumstances as shown in Fig. 7. The videos are in 640 × 480 pixels, 30 frames per second AVI format without audio.

3.2 Drowsiness Detection Experiments

The proposed DDD network aims to classify each frame in videos based on feature representation learning via different type of deep networks. Due to the lack of ground truth label of test dataset, we substituted evaluation dataset for test dataset. We train each convolutional neural network of DDD using 90% of train dataset, and validate proposed DDD network using the last 10% of train dataset. Firstly, train images are extracted 1 frame from every 10 frames in the videos. To improve robustness and generalization of our model, we adopted data augmentation based on pixel intensity. Each image was augmented by using intensity adjustment with 15%, 30%, −15%, −30%, and normalized into a range of [0, 255]. These training images resized to desired network input sizes are fed into each CNN. Further, training images are randomly fed into a multilayer CNN. The output of the last layer composed of 4 classes. All networks are trained using general fine tuning manner with modification of last fc8 and softmax layers. All weights are learnt via back-propagation and they are token as the learnt feature representations which are convolved with input images to produce the final feature representation to classify drowsy state. In our experiment, the output of the fc7 layer are considered as the trained feature representations. The feature representations are fed into the last softmax classifier and trained. Once the classifier has been trained, the whole images of evaluation dataset are tested on the trained classifier. We implement proposed DDD network using the MatConvNet package [23]. The batch size was set to 10, momentum to 0.9, and training was regularized by weight decay 10^{-4}. The learning rate was initially set to 10^{-3}, and then decreased by a factor of 10 when the validation set accuracy stopped improving.

Due to the lack of state-of-the-art performance on this dataset, we compared composed algorithm with the performance of several well-known multiclass classification algorithm such as variants of CNNs and LRCN. For this, we fine-tunned each AlexNet, VGG-FaceNet, FlowImageNet and LRCN using same training dataset and ground truth label, independently. For the human-level test results, all the extracted image sequences from the evaluation vdieo dataset were labeled manually as drowsiness state or non-drowsiness state by 5 human experts (doctoral students in computer vision area).

We evaluated the performance of individual CNN based models, LRCN, and proposed DDD network. The results are shown in Tables 1 and 2. In our experiments, DDD-IAA showed better results compare with DDD-FFA. For the DDD-IAA, we combine the outputs of three models of DDD by averaging their soft-max class probabilities. This improves the performance due to complementarity of the models. The average accuracies for different subjects were 70.81% (DDD-FFA) and 73.06% (DDD-IAA) which are higher than other CNNs including AlexNet, VGG-faceNet, FlowImageNet, LRCN as shown in Table 1. Because we ensembles the several features and sent into classification to yield outputs, which is robust to detect various drowsiness situation. Although these accuracies are lower than human experts's results but better than variants of previous CNN based models. Also, proposed DDD network showed better accuracies for different situation as shown in Table 2.

Table 1. Drowsiness detection accuracies for different subjects (%) on evaluation dataset

ID	Human	AlexNet [1]	VGG-FaceNet [2]	FlowImageNet [3]	LRCN [3]	DDD-FFA	DDD-IAA
004	73.37	61.12	66.20	65.92	52.65	78.26	66.87
022	83.42	80.30	77.80	53.36	77.31	78.64	86.27
026	85.07	56.68	66.12	59.33	58.86	68.16	69.00
030	81.47	65.62	61.28	67.40	63.15	58.16	70.11
Average	80.83	65.93	67.85	61.50	62.99	70.81	73.06

Table 2. Drowsiness detection accuracies for different situations (%) on evaluation dataset

Situations	Human	AlexNet [1]	VGG-FaceNet [2]	FlowImageNet [3]	LRCN [3]	DDD-FFA	DDD-IAA
Bareface	82.04	70.42	63.87	56.33	68.75	79.41	69.83
Glasses	78.83	61.63	70.53	61.61	61.73	74.10	75.93
Sunglasses	80.89	70.20	57.00	67.57	71.47	61.89	69.86
Night-bareface	82.54	64.69	73.75	66.82	57.39	70.27	74.93
Night-glasses	79.87	62.70	74.10	55.17	55.63	68.37	74.77
Average	80.83	65.93	67.85	61.50	62.99	70.81	73.06

4   Conclusion

This paper proposes a deep architecture referred to as deep drowsiness detection (DDD) network for learning effective features and detecting drowsiness given an input image of a driver. Previous approaches could only make decisions based on carefully hand-crafted features such as eye blinks and head gestures for detecting driver drowsiness. Deep network based feature representation learning approaches have been providing an automated and efficient set of learned features which help us to classify the driver as drowsy or non-drowsy very accurately. Especially, model ensemble are fusion strategies improves the performance due to complementarity of the models. Experimental results show that DDD achieves 73.06% detection accuracy on NTHU-drowsy driver detection benchmark dataset.

Acknowledgement. This work was partly supported by Institute for Information & communications Technology Promotion (IITP) grant funded by the Korea government (MSIP) (No. B0101-16-0307, Basic Software Research in Human-level Lifelong Machine Learning (Machine Learning Center)) and supported by the National Research Foundation of Korea (NRF) grant funded by the Korea government (MSIP) (No. 2010-0028680).

References

1. Krizhevsky, A., Sutskever, I., Hinton, G.E.: Imagenet classification with deep convolutional neural networks. In: NIPS, pp. 1097–1105 (2012)

2. Parkhi, O.M., Vedaldi, A., Zisserman, A.: Deep face recognition. In: BMVC, vol. 1, p. 6 (2015)
3. Donahue, J., Anne Hendricks, L., Guadarrama, S., Rohrbach, M., Venugopalan, S., Saenko, K., Darrell, T.: Long-term recurrent convolutional networks for visual recognition and description. In: CVPR, pp. 2625–2634 (2015)
4. Brox, T., Bruhn, A., Papenberg, N., Weickert, J.: High accuracy optical flow estimation based on a theory for warping. In: Pajdla, T., Matas, J. (eds.) ECCV 2004. LNCS, vol. 3024, pp. 25–36. Springer, Heidelberg (2004). doi:10.1007/978-3-540-24673-2_3
5. Soomro, K., Zamir, A.R., Shah, M.: UCF101: a dataset of 101 human actions classes from videos in the wild. arXiv preprint arXiv:1212.0402 (2012)
6. Li, W., He, Q.C., Fan, X.M., Fei, Z.M.: Evaluation of driver fatigue on two channels of EEG data. Neurosci. Lett. **506**, 235–239 (2012)
7. Patel, M., Lal, S.K.L., Kavanagh, D., Rossiter, P.: Applying neural network analysis on heart rate variability data to assess driver fatigue. Expert Syst. Appl. **38**, 7235–7242 (2011)
8. Mattsson, K.: In vehicle prediction of truck driver sleepiness. Master's thesis, Luleå University of Technology, vol. 107 (2007)
9. Boyle, L.N., Tippin, J., Paul, A., Rizzo, M.: Driver performance in the moments surrounding a microsleep. Transp. Res. Part F: Traffic Psychol. Behav. **11**, 126–136 (2008)
10. Friedrichs, F., Yang, B.: Drowsiness monitoring by steering and lane data based features under real driving conditions. In: 2010 18th European Signal Processing Conference, pp. 209–213 (2010)
11. Forsman, P.M., Vila, B.J., Short, R.A., Mott, C.G., Dongen, H.P.: Efficient driver drowsiness detection at moderate levels of drowsiness. Accid. Anal. Prev. **50**, 341–350 (2013)
12. Dwivedi, K., Biswaranjan, K., Sethi, A.: Drowsy driver detection using representation learning. In: 2014 IEEE International Advance Computing Conference (IACC), pp. 995–999 (2014)
13. Lee, S.J., Jo, J., Jung, H.G., Park, K.R., Kim, J.: Real-time gaze estimator based on driver's head orientation for forward collision warning system. IEEE Trans. Intell. Transp. Syst. **12**, 254–267 (2011)
14. Choi, I.H., Hong, S.K., Kim, Y.G.: Real-time categorization of driver's gaze zone using the deep learning techniques. In: International Conference on Big Data and Smart Computing (BigComp), pp. 143–148 (2016)
15. Singh, M., Kaur, G.: Drowsy detection on eye blink duration using algorithm. Int. J. Emerg. Tech. Adv. Eng. **2**, 363–365 (2012)
16. Saito, H., Ishiwaka, T., Okabayashi, S.: Applications of driver's line of sight to automobiles-what can driver's eye tell. In: 1994 Proceedings of Vehicle Navigation and Information Systems Conference, pp. 21–26 (1994)
17. Horng, W.B., Chen, C.Y., Chang, Y., Fan, C.H.: Driver fatigue detection based on eye tracking and dynamk, template matching. In: 2004 IEEE International Conference on Networking, Sensing and Control, vol. 1, pp. 7–12 (2004)
18. Smith, P., Shah, M., da Vitoria Lobo, N.: Monitoring head/eye motion for driver alertness with one camera. In: ICPR, p. 4636 (2000)
19. Polikar, R.: Ensemble based systems in decision making. IEEE Circuits Syst. Mag. **6**, 21–45 (2006)
20. Rokach, L.: Ensemble-based classifiers. Artif. Intell. Rev. **33**, 1–39 (2010)
21. Opitz, D., Maclin, R.: Popular ensemble methods: an empirical study. J. Artif. Intell. Res. **11**, 169–198 (1999)

22. Fan, R.E., Chang, K.W., Hsieh, C.J., Wang, X.R., Lin, C.J.: LIBLINEAR: a library for large linear classification. J. Mach. Learn. Res. **9**, 1871–1874 (2008)
23. Vedaldi, A., Lenc, K.: Matconvnet: convolutional neural networks for matlab. In: Proceedings of the 23rd ACM International Conference on Multimedia, pp. 689–692 (2015)
24. Sahayadhas, A., Sundaraj, K., Murugappan, M.: Electromyogram signal based hypovigilance detection. Biomed. Res. **25**, 281–288 (2014)
25. Chieh, T.C., Mustafa, M.M., Hussain, A., Hendi, S.F., Majlis, B.Y.: Development of vehicle driver drowsiness detection system using electrooculogram (EOG). In: 1st International Conference on Computers, Communications, Signal Processing with Special Track on Biomedical Engineering, pp. 165–168 (2005)

Representation Learning, Scene Understanding, and Feature Fusion for Drowsiness Detection

Jongmin Yu[1], Sangwoo Park[1], Sangwook Lee[2], and Moongu Jeon[1(✉)]

[1] Department of Electronical Engineering and Computer Science,
GIST, Gwangju, South Korea
jm.andrew.yu@gmail.com, {swpark,mgjeon}@gist.ac.kr
[2] Department of Information Communication Engineering,
Mokwon University, Daejeon, South Korea
slee@mokwon.ac.kr

Abstract. We propose a novel drowsiness detection method based on 3D-Deep Convolutional Neural Network (3D-DCNN). We design a learning architecture for the drowsiness detection, which consists of three building blocks for representation learning, scene understanding, and feature fusion. In this framework, the model generates a spatio-temporal representation from multiple consecutive frames and analyze the scene conditions which are defined as head, eye, and mouth movements. The result of analysis from the scene condition understanding model is used to auxiliary information for the drowsiness detection. Then the method subsequently generates fusion features using the spatio-temporal representation and the results of the classification of scene conditions. By using the fusion features, we show that the proposed method can boost the performance of drowsiness detection. The proposed method demonstrates with the NTHU Drowsy Driver Detection (NTHU-DDD) video dataset.

1 Introduction

Drowsiness detection is a critical function of the smart vehicle system for road safety. A lot of people are killed, or significantly injured because drivers falling asleep at the road in every year. According to the report of the national highway traffic safety administration in USA, 100,000 police-reported crashes are closely related to driver fatigue each year. This causes in an estimated 1,550 death, 71,000 injuries, and 12.5 billion in monetary losses [1]. These figures may be the tip of the iceberg, since currently it is difficult to attribute crashes to drowsiness. Therefore, it is essential issues to develop a detection method for driver drowsiness in order to prevent car accidents.

In previous, diverse techniques have been proposed and developed for detecting driver drowsiness. The techniques normally can be divided into three approaches. The first approaches are to detect physiological cues [2–5]. The methods evaluate drivers biomedical signals such as heart rate, brain activity, and pulse rate. Furthermore, they can ensure great detection accuracy, which helps

© Springer International Publishing AG 2017
C.-S. Chen et al. (Eds.): ACCV 2016 Workshops, Part III, LNCS 10118, pp. 165–177, 2017.
DOI: 10.1007/978-3-319-54526-4_13

to reduce drowsiness driving accidents. For example, a fuzzy mutual-information-based wavelet packet transform model was developed to extract drowsiness-related information from a set of electroencephalogram, electrooculogram, and electrocardiogram signals [2]. However, the main disadvantages of these methods are that physiological equipment must be attached to a body of the driver and more expensive than other methods. Therefore, they are hard to use and other devices that detect the driver are essential in order to be used in reality such as sensor and camera.

The second approaches are to detect driving behaviour [6–10]. The methods based on drivers driving act detect their drowsiness by monitoring the variation of lateral position, vehicle speed, steering, breaking, acceleration, and gear change with sensors. Ersal et al. proposed a framework to classify driving behaviour [6]. The method is shown that the driving-behavior-based scheme with support vector machines (SVM) assists systematically classify whether the driver is asleep or not. However, a variety of data must be necessary to use the techniques to get good detection accuracy because every person has a different personality and vehicle and drives a different road such as driving habit, car model, and road condition, therefore, the rule which is a reference can't be exactly established.

The third approaches are to detect visual features. Nowadays, many computer vision approaches have been developed to be used in the real world for detecting drowsiness driver [11–17]. The methods analysis facial feature extracted from the face image of drivers. If drivers are especially tired, they have different facial information as usual such as eye blinking, yawning, eye and head moving, facial expression and so on. Using the special features, these approaches can classify driving condition as drowsiness or non-drowsiness. To detect these facial feature informations, a standard camera is normally installed on the dashboard. Except that the camera must be set up to see the face of the driver, it does not cause annoyance to the driver because there is no attached device to them and depend on the individuals driving characteristics. Thus, they can better classify condition of the driver. Garcia et al. developed system is composed of three stages [11]. The first stage conduct pre-processing and the second stage processes information of pupil. The third stage computes PERCLOS from eyes closure (EL) information. Mbouna et al. proposed visual analysis eye state and head pose for monitoring of a vehicle driver with single camera without the use of source of light [12]. Wang et al. presented the visual analysis based on combination eye state and the state of a vehicle drivers mouth to solve the issue of wearing glasses [13]. Dwivedi et al. proposed convolutional neural network in order to make facial features such as blink rate, eye closure, yawning, eye brow shape [18].

In recent, there are a lot of state-of-the-art methods based on the deep learning approach proposed for solving various computer vision challenge such as object detection [19], recognition [20], action recognition [21], scene classification, and image segmentation [22]. The key of the success is that using deep learning approach, a rich and discriminative representation can be extracted via multi-layer nonlinear system [23]. Therefore, it is reasonable to expect that detecting drowsiness in videos can also benefit from deep learning method. In our proposed framework, we employ the paradigms of the convolutional neural network (CNN) and multi-layer fully connected neural network (a.k.a., deep neural network) to extract useful spatio-temporal representation for the scene understanding and the drowsiness detection.

We proposed a novel learning framework with 3D deep convolutional neural network (3D-DCNN), that is conceptually appropriate to combine the scene understanding knowledge to the drowsiness detection method. In this framework, 3D-DCNN is used to extract spatio-temporal representation from the multiple consecutive frames, and the scene understanding model encodes the extracted representation to the scene condition vector. The fusion model generates a feature map which is specified to the result of the scene understanding model using the spatio-temporal representation, then by using the fusional feature, the framework detects the drowsiness of driver. Our main contribution is the learning framework that can train the four building blocks which include the 3D-DCNN, the scene understanding model, fusion model, and the detection model simultaneously, even though spatio-temporal representation using single input.

The rest of the paper is organized as follows. In Sect. 2, The detail configuration of the proposed architecture is described. Training and inference procedures are represented in Sect. 3. Section 4 shows an experimental results and Sect. 5 concludes this paper.

2   Architecture

The proposed learning framework for the driver drowsiness detection is based on four main components which are composed on 3D-DCNN f_d, the scene understanding model f_{su}, the fusion model f_{fu}, and the detection model f_{det}. Figure 1

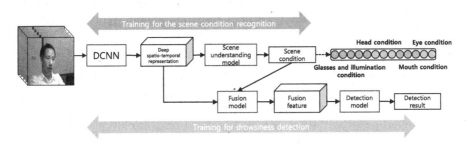

Fig. 1. The architecture of the proposed framework. The red boxes denote the models and the black boxes define features that are extracted from the models. Given the spatio-temporal representation from 3D-DCNN, the classification model for the scene condition predicts both the status of glasses and illumination conditions and the condition of facial elements such as eye, mouth and motion of head. The outputs of the scene understanding model and spatio-temporal representation are subsequently fed into the fusion model, which generate fusion feature that can represent a condition-specified representation. By using the fusion feature, we determine a status of drowsiness. In the training phase, we first train the 3D-DCNN and the classification model, and leverage the predicted condition vector and the spatio-temporal representation from 3D-DCNN and the classification model to train the fusion model and the detection model. (Color figure online)

illustrates an overall architecture of the proposed framework. First, the deep convolutional neural network extracts the spatio-temporal representations from input data for capturing appearance and motion information in a video clip simultaneously. Second, the extracted spatio-temporal representation directly apply to the scene understanding model. By using the understanding model, we estimate the scene conditions that related with the state of glasses and motion of facial elements (i.e., eye, mouth), that can help us to understand the motion of driver. Then, In order to detect the sleepiness of drivers, the spatio-temporal representations and the results of the scene understanding agglomerate together to form scene condition specified features. In last, The detection model based on the deep neural network determines a state of driver drowsiness using the fusion feature. Once the proposed learning framework is trained, the scene understanding model and the drowsiness detection model shares the input structurally. In the following, we describe the proposed architecture in details.

2.1 3D Deep Convolutional Neural Network

First, we describe a method for learning the spatio-temporal representation using 3D-DCNN. The objective of the learning of spatio-temporal representation is discovering a rich and discriminative feature that can represent information of both appearance and motion. Sleepiness is accompanied by various visual and physiological symptoms such as change in the shapes of eyelids and mouth. In this work, we adopted 3D-DCNN to interpret these symptoms spatially and temporally.

Let x denotes a training video clip with $R^{W \times H \times T}$ dimensions where W, H, and T are the width, the height, and the temporal length respectively. Given input video clip x, the 3D-DCNN extracts a spatio-temporal representation as

$$\alpha = f_d(x; \theta_d), \quad \alpha \in R^{W_\alpha \times H_\alpha \times D_\alpha} \tag{1}$$

where θ_d is the parameter for the DCNN, and α is a learnt spatio-temporal representation that is activation values of the hidden units in the last convolutional layer of 3D-DCNN. W_α, H_α, and D_α denote the width, height, and depth of the spatio-temporal representation. 3D-DCNN based on convolutional neural network consists of six convolutional layers and two pooling layers. Figure 2 shows the architectural detail for 3D-DCNN. In order to capture a temporal feature, we employ 3D local receptive field suggested by [24]. An activation value of hidden units using 3D local receptive field can be defined as

$$\alpha = \rho\left[\sum_{i}^{W_r} \sum_{j}^{H_r} \sum_{k}^{D_r} (v_{i,j,k} w_{i,j,k} + b_r)\right] \tag{2}$$

where α is activation value of hidden unit, and v, w, and b_r are the input value, the weight, and bias respectively. W_r, H_r, and D_r denote the width, the height, and the depth of 3D local receptive field, ρ is an activation function for the

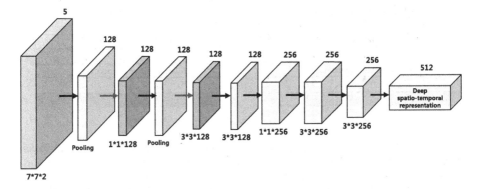

Fig. 2. The illustration of the architecture detail of the 3D-DCNN. The green box denotes input data, the grey and blue boxes represent convolutional layer and pooling layer respectively. Numbers that are located in the upside of the boxes denote the depth of each layer, and numbers of under the boxes denote the dimension and structure of the local receptive field. The black arrows represent that a convolution operation between adjacent layers and the red arrows denote that a pooling operation. (Color figure online)

convolution layer. We employ the Rectified Linear Units (ReLUs) [25] to the proposed 3D-DCNN.

3D local receptive field can capture the spatial feature and temporal feature simultaneously, whereas 2D local receptive field only can extract the spatial feature. The extracted representations are used to understand the scene conditions and generate fusion feature for drowsiness detection.

2.2 Scene Understanding

Given input data x, we assume that each video clip is associated with a scene condition labels and a detection labels, which are given by either ground-truth (in training dataset) or predictions (in inferencing). The scene condition contains the three categories of the physiological element and one category for the status of glasses and illumination, and each category is defined by using a one-hot vector to represent states of facial element and condition for glasses and illumination. In this work, we used four category: (1) condition of glasses and illumination, (2) head, (3) mouth, (4) eye. The condition labels $\mathcal{L}_s$ are generated by concatenating these one-hot vectors. The learning procedure of the scene understanding model is similar to the back propagation algorithms [26]. The model estimates a label for the scene condition from given the spatio-tempora representation $\boldsymbol{\alpha}$, then computes the difference between the predicted labels and annotation labels in order to train the parameters of the network of the scene understanding model. The prediction of the condition is written by

$$\mathcal{L}_c = f_{su}(\boldsymbol{\alpha}; \theta_{su}), \quad \mathcal{L}_c \in R^{L \times 1} \tag{3}$$

where $\mathcal{L}_c$ is a predicted label associated to input data x, and L is a size of the label. θ_{su} is the parameters of the fully connected network (Fig. 3(a)) in the

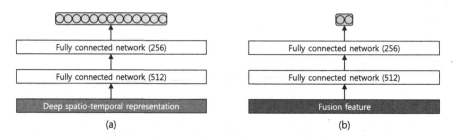

Fig. 3. The illustration of the structures of the multi-layer fully connected networks (a.k.a., deep neural network) which are used to the scene understanding model and the detection model. (a) is a network for the scene understanding model, and (b) is a network for the detection model. The numbers in parentheses represent the number of hidden units in each layer. The green and orange boxes define the outputs of each networks. (Color figure online)

scene understanding model. In this work, The glasses and illumination conditions which includes glasses, noglasses, night-glasses, night-noglasses and sunglasses are defined by 5-dimensional vector. The properties of head and mouth motions represented as 3-dimensional vectors, the change of eye is defined by 2-dimensional vector. Therefore, scene condition labels are generated as 13-dimensional vector. Given the spatio-temporal representation as input, the scene understanding model is trained to minimize the objective funsion represented as follow

$$\min_{\theta_d, \theta_{sc}} \sum_i E_{su}(\mathcal{L}_c, \hat{\mathcal{L}}_c). \tag{4}$$

where $\hat{\mathcal{L}}_c \in R^{L \times 1}$ denotes an annotation of input data, and E_{su} denotes mean-square error function between the annotation and predicted results. The spatio-temporal representation and the result of the scene understanding model are then used to generate a fusion feature to detect the drowsiness of driver, which is discussed in the following subsections.

2.3 Fusion Model

Given a representation extracted from 3D-DCNN $\alpha \in R^{W_\alpha \times H_\alpha \times D_\alpha}$ and its associated condition labels $\mathcal{L}_c^{'} \in R^{L \times 1}$, the objective of our fusion model is to discover a set of condition specified feature map β. Where each element of β describes the information that helps to understand for a video clip. Our fusion model is defined as follow

$$\beta = f_{fu}(\alpha, \mathcal{L}_c; \theta_{fu}) \tag{5}$$

$$v_i = \frac{exp(\beta_i)}{\sum_j exp(\beta_j)} \tag{6}$$

where β_i represents i-th element of the unnormalized fusion feature, v_i is i-th element of the normalized fusion feature. To specified the representation pay attention to only a scene condition, we normalize β to v using a softmax function attempted in [22,27].

The objective our fusion model is to learn a set of condition-specified representation from given spatio-temporal representation α extracted from 3D-DCNN and its associated condition labels $\mathcal{L}_c$. We employ multiplicative interactions [29] between the representation and the condition labels to generate conditions-specified feature vector v using the proposed fusion model f_{fu}. We adopted the training method that is suggested by [28], for learning the parameters represented by a 3-way tensor to capture the correlation between the spatio-temporal representation and the condition labels. The method is formulated as follow

$$\beta = W_{fu}(W_{fea}\alpha \otimes W_{sc}\mathcal{L}_c) + b_{fu}. \tag{7}$$

where $b_{fu} \in R^{d \times 1}$ is the bias of the fusion model, and $\otimes$ denotes element-wise matrix multiplication the weights are given by $W_{fu} \in R^{M \times d}$, $W_{fea} \in R^{d \times W_\alpha H_\alpha D_\alpha}$ and $W_{sc} \in R^{d \times L}$, where M and d denote the number of hidden unit in the fusion model. Hong et al. observes that using element-wise multiplication interaction between the feature map gives results by capturing high-order dependency between relevant features [29]. Intuitively, v represents a condition-specific feature defined over all spatio-temporal representation. The generated feature are then used to detect the drowsiness of driver, which is discussed in next subsection.

2.4   Drowsiness Detection

The fusion model described in the previous subsection generates a set of condition-specific feature v, which provide useful information for the drowsiness detection. The next step of the proposed framework is to determine whether drivers are sleepy or not by using the given fusion features. As same as the network for the scene understanding model, we put an additional fully connected neural network (Fig. 3(b)) on top of the fusion model as follow:

$$o_{det} = f_{det}(v; \theta_{det}). \tag{8}$$

where o_{det} denote the output of detection model, θ_{det} is the model parameters. An optimization scheme for both f_{fu} and f_{det} operate under the detection objective. Using detection annotation associated with fusion feature, our detection network is trained to minimized the detection loss as follows:

$$\min_{\theta_f, \theta_{det}} \sum_i E_{det}(o_{det}, \hat{o}_{det}) \tag{9}$$

where $\hat{o}_{det}$ is a target value associated to input data x, and E_{det} denote the cost function for the fully connected network in detection model. We employ a cross-entropy loss function for E_{det}. Since training requires the predictions for the scene condition and spatio-temporal representation, the objective function is optimized with all models.

3   Training and Inference

Combining Eqs. (3) and (8), the overall objective function is defined by

$$\min_{\theta_d,\theta_{sc},\theta_f,\theta_D} \sum_i (E_{su}(\mathcal{L}_c, \hat{\mathcal{L}}_c) + \lambda E_{det}(o_D, \hat{o}_D)) \tag{10}$$

where λ assigned for balancing between the scene understanding model and the detection model. During training the framework, we used the condition labels for the first term in Eq. (9), and predicted conditions are used to second term. Note that it allows to us joint optimization for the four main building blocks of the proposed framework. We can regularize it to avoid finding a poor local optimized solution for the drowsiness detection results by jointly training with the scene understanding objective. After training, we remove loss function for the condition classification in Eq. (3), because it is required only in training to learn spatio-temporal representation and condition classification model.

For detection from input video clip using the trained framework, we first apply 3D-DCNN to extract spatio-temporal representation. Then we classify the scene condition and construct fusion feature from the classification model and the fusion model using the spatio-temporal representation. The final detection result is obtained by comparing the maximum activation value in the output of the detection model.

4   Experiments

4.1   Training Dataset

We perform drowsiness detection to verify the strength of our framework using NTHU Drowsy Driver Detection Dataset (NTHU-DDD Dataset). The training dataset composed of videos which are recorded in various driving condition (i.e., a gender of drivers, glasses or no glasses, sunglasses, illumination conditions). Each frame of videos vectorization to diverse categories that are associated with the drowsiness and the status of facial elements of drivers such as eye, mouth, and movement of head. Figure 4 shows an example snapshot of NTHU-DDD dataset.

To train the proposed framework, we divide videos into sets of a small video clip that contains the five consecutive frames, and assign an annotation for scene condition and the state value of drowsiness. An annotation for the scene condition denotes two characteristics: (1) the information for the facial elements such as eye, mouth and movement of head, (2) recording condition that had been pre-classified by directories such as 'glasses', 'non-glasses', 'night glasses, 'night non-glasses', and 'sunglasses'.

We employ a concept of IOU (intersection over union) [30], since the video clip can contain multiple categories and it may cause confusion. Therefore, we assign the value that occupying more than 50% of the value of frame-level annotation as the annotation value of clip. Figure 5 shows the concept of IOU used in our

Fig. 4. The example snapshots of NTHU Drowsy Driver Detection Dataset (NTHU-DDD Dataset). Diverse conditions of the dataset including various subjects with different skin tone, a presence of glasses, gender, illumination condition that contains a colour sensor and an infrared sensor.

Fig. 5. The illustration for the concept of temporal IOU. Left and right sides of the arrow denote annotations for each frame and video clip respectively.

experiment. In our experiment, each clip consists of five consecutive frames, and we determine the annotation value which is observed more than 3 in each clip. We resize image frame with the width of 224 pixels and height of 224 pixels. Thus, the size of the video clip for training is $224 \times 224 \times 5$. During training the framework, we adopted two phase of the randomization for the dataset for preventing an over-fitting problem. Initially, we divided the video into a set of the clip and rearrange the order of the clip randomly. In second, we group a set of the clip using the glasses and illumination condition category, then pick a clip at random from each grouped dataset as training data.

4.2   Experimental Results

The proposed framework is tested on the evaluation dataset of NTHU-DDD dataset. The evaluation dataset consists of 20 videos that contain various driving situations. The videos in this dataset are provided with multiple annotations for the scene condition and the drowsiness detection. We evaluated the scene understanding model and the drowsiness detection model respectively. The scene understanding model is evaluated by using validation accuracy. The quantitative evaluation results of the detection model are averaged over all video for which have equal glass and illumination categories (glasses, non-glasses, night-glasses, night-non-glasses, sunglasses).

The result are evaluated using quantitative evaluation method. Quantitative evaluation is carried out using the F-measure, which is an assessment of test accuracy, F-measure is harmonic mean of precision and detection rate, where precision and recall are defined as follow:

$$precision = \frac{TP}{TP + FP} \qquad (11)$$

$$Detectionrate(DR) = \frac{TP}{TP + FN} \qquad (12)$$

$$F\text{-}measure = \frac{2 \times Precision \times DR}{Precision + DR} \qquad (13)$$

where TP (true positive) is the number of correctly identified as drowsiness state, FN (false negative) is the number of drowsiness detection result incorrectly identified as non-drowsiness state, FP (false positive) is the number of non-drowsiness detection result incorrectly identified as drowsiness state, and TN (true negative) is the number of correctly identified as non-drowsiness state. Table 1 shows the validation accuracy for the scene understanding model using the evaluation dataset. The scene understanding model seems the validation accuracy of more than 88% across to all condition labels. Table 2 shows the

Table 1. Validation accuracies of the scene understanding model using the evaluation dataset in NTHU-DDD dataset.

Scenario	Glasses and illumination	Head	Mouth	Eye
Bare face	0.99	0.97	0.98	0.84
Glasses	0.94	0.91	0.94	0.73
Sunglasses	0.97	0.96	0.77	0.66
Night bare face	0.99	0.92	0.94	0.72
Night glasses	0.96	0.93	0.80	0.89
Average	0.968	0.938	0.886	0.762
Total average				0.888

Table 2. F-measures and accuracies for the evaluation dataset in NTHU-DDD dataset. It is averaged for the five categories about the glasses and illumination conditions.

Scenario	Drowsiness	Non-drowsiness	Accuracy
Bare face	0.758	0.751	0.751
Glasses	0.776	0.691	0.723
Sunglasses	0.723	0.694	0.709
Night bare face	0.681	0.649	0.684
Night glasses	0.692	0.633	0.683
Average	0.726	0.689	0.712

F-measure for the proposed framework for the drowsiness detection. The results of the detection show that our proposed framework achieves the average accuracy of 0.712. However, the detection results in night illumination categories that contain the night-glasses and night-non-glasses categories, are lower than other categories. It could be interpreted that the learnt model have been biased to specific categories because the ratio of amounts of data in each category is under unbalanced. The proposed framework is implemented with the Tensorflow library. A prediction operates at frame rates close to real-time (24–32 fps on GPU).

5   Conclusion

In this paper, we have proposed a learning framework for drowsiness detection without any eye, head, and face detection methods. The proposed framework exploits extra scene condition prediction to improve drowsiness detection on the dataset. The proposed framework that consists of 3D-DCNN, classification model, fusion model, and detection model, is appropriate to detect sleepiness of driver at various wearing condition and illumination conditions. Previous approaches could only make decisions based on extra methods such as facial element detection and extra sensor information (i.e., heart beat). The proposed framework provides an automated and efficient feature learning which helps us to classify the scene conditions and the drowsiness of driver. The scheme was tested on a NTHU-DDD dataset. Both quantitative and qualitative result show that the proposed framework can provide the drowsiness detection scheme for practical systems.

However, the proposed framework embeds two main drawbacks that can make significant degradation in performance. Firstly, a direct application of fusion feature to the detection model may cause a critical problem, because the activation of hidden units tends to be sparse due to the softmax operation (a.k.a., Eq. (3)) and may lose information during an extraction of fusion feature. And it could be caused a model that over-fitted to an incorrect solution. Secondly, the performance of the framework is closely related with the training dataset, and the proposed framework trained by NTHU-DDD dataset may not be suitable to detect the drowsiness in situations that are much different to training dataset. In future work, we devise a new model which can overcome the problem of feature sparsity in the feature fusion model, and design a new data argumentation method for the lack of training data.

Acknowledgment. This work was supported by Institute for Information & communications Technology Promotion (IITP) grant funded by the Korea government (MSIP) (No. B0101-15-0525, Development of global multi-target tracking and event prediction techniques based on real-time large-scale video analysis) and Center for Integrated Smart Sensors as Global Frontier (CISS-2013M3A6A6073718).

References

1. Schroeder, P., Meyers, M., Kostyniuk, L.: National survey on distracted driving attitudes and behaviors–2012. Technical report (2013)
2. Khushaba, R.N., Kodagoda, S., Lal, S., Dissanayake, G.: Driver drowsiness classification using fuzzy wavelet-packet-based feature-extraction algorithm. IEEE Trans. Biomed. Eng. **58**, 121–131 (2011)
3. Patel, M., Lal, S., Kavanagh, D., Rossiter, P.: Applying neural network analysis on heart rate variability data to assess driver fatigue. Expert Syst. Appl. **38**, 7235–7242 (2011)
4. Tran, Y., Craig, A., Wijesuriya, N., Nguyen, H.: Improving classification rates for use in fatigue countermeasure devices using brain activity. In: 2010 Annual International Conference of the IEEE Engineering in Medicine and Biology, pp. 4460–4463. IEEE (2010)
5. Papadelis, C., Chen, Z., Kourtidou-Papadeli, C., Bamidis, P.D., Chouvarda, I., Bekiaris, E., Maglaveras, N.: Monitoring sleepiness with on-board electrophysiological recordings for preventing sleep-deprived traffic accidents. Clin. Neurophysiol. **118**, 1906–1922 (2007)
6. Ersal, T., Fuller, H.J., Tsimhoni, O., Stein, J.L., Fathy, H.K.: Model-based analysis and classification of driver distraction under secondary tasks. IEEE Trans. Intell. Transp. Syst. **11**, 692–701 (2010)
7. Yang, J.H., Mao, Z.H., Tijerina, L., Pilutti, T., Coughlin, J.F., Feron, E.: Detection of driver fatigue caused by sleep deprivation. IEEE Trans. Syst. Man Cybern.-Part A: Syst. Hum. **39**, 694–705 (2009)
8. Liu, C.C., Hosking, S.G., Lenné, M.G.: Predicting driver drowsiness using vehicle measures: recent insights and future challenges. J. Saf. Res. **40**, 239–245 (2009)
9. Takei, Y., Furukawa, Y.: Estimate of driver's fatigue through steering motion. In: 2005 IEEE International Conference on Systems, Man and Cybernetics, vol. 2, pp. 1765–1770. IEEE (2005)
10. Wakita, T., Ozawa, K., Miyajima, C., Igarashi, K., Katunobu, I., Takeda, K., Itakura, F.: Driver identification using driving behavior signals. IEICE Trans. Inf. Syst. **89**, 1188–1194 (2006)
11. Garcia, I., Bronte, S., Bergasa, L.M., Almazán, J., Yebes, J.: Vision-based drowsiness detector for real driving conditions. In: 2012 IEEE Intelligent Vehicles Symposium (IV), pp. 618–623. IEEE (2012)
12. Mbouna, R.O., Kong, S.G., Chun, M.G.: Visual analysis of eye state and head pose for driver alertness monitoring. IEEE Trans. Intell. Transp. Syst. **14**, 1462–1469 (2013)
13. Wang, P., Shen, L.: A method of detecting driver drowsiness state based on multi-features of face. In: 2012 5th International Congress on Image and Signal Processing (CISP), pp. 1171–1175. IEEE (2012)
14. Minkov, K., Zafeiriou, S., Pantic, M.: A comparison of different features for automatic eye blinking detection with an application to analysis of deceptive behavior. In: 2012 5th International Symposium on Communications Control and Signal Processing (ISCCSP), pp. 1–4. IEEE (2012)
15. Panning, A., Al-Hamadi, A., Michaelis, B.: A color based approach for eye blink detection in image sequences. In: 2011 IEEE International Conference on Signal and Image Processing Applications (ICSIPA), pp. 40–45. IEEE (2011)
16. Kurylyak, Y., Lamonaca, F., Mirabelli, G.: Detection of the eye blinks for human's fatigue monitoring. In: 2012 IEEE International Symposium on Medical Measurements and Applications Proceedings (MeMeA), pp. 1–4. IEEE (2012)

17. Suzuki, M., Yamamoto, N., Yamamoto, O., Nakano, T., Yamamoto, S.: Measurement of driver's consciousness by image processing-a method for presuming driver's drowsiness by eye-blinks coping with individual differences. In: 2006 IEEE International Conference on Systems, Man and Cybernetics, vol. 4, pp. 2891–2896. IEEE (2006)
18. Dwivedi, K., Biswaranjan, K., Sethi, A.: Drowsy driver detection using representation learning. In: 2014 IEEE International Advance Computing Conference (IACC), pp. 995–999. IEEE (2014)
19. Ren, S., He, K., Girshick, R., Sun, J.: Faster R-CNN: towards real-time object detection with region proposal networks. In: Advances in Neural Information Processing Systems, pp. 91–99 (2015)
20. Girshick, R., Donahue, J., Darrell, T., Malik, J.: Region-based convolutional networks for accurate object detection and segmentation. IEEE Trans. Pattern Anal. Mach. Intell. **38**, 142–158 (2016)
21. Du, Y., Wang, W., Wang, L.: Hierarchical recurrent neural network for skeleton based action recognition. In: Proceedings of the IEEE Conference on Computer Vision and Pattern Recognition, pp. 1110–1118 (2015)
22. Qi, X., Li, C.G., Zhao, G., Hong, X., Pietikainen, M.: Dynamic texture and scene classification by transferring deep image features. Neurocomputing **171**, 1230–1241 (2016)
23. Xu, D., Ricci, E., Yan, Y., Song, J., Sebe, N.: Learning deep representations of appearance and motion for anomalous event detection. arXiv preprint arXiv:1510.01553 (2015)
24. Tran, D., Bourdev, L., Fergus, R., Torresani, L., Paluri, M.: Learning spatiotemporal features with 3D convolutional networks. In: 2015 IEEE International Conference on Computer Vision (ICCV), pp. 4489–4497. IEEE (2015)
25. Krizhevsky, A., Sutskever, I., Hinton, G.E.: Imagenet classification with deep convolutional neural networks. In: Advances in Neural Information Processing Systems, pp. 1097–1105 (2012)
26. Le Cun, B.B., Denker, J.S., Henderson, D., Howard, R.E., Hubbard, W., Jackel, L.D.: Handwritten digit recognition with a back-propagation network. In: Advances in Neural Information Processing Systems. Citeseer (1990)
27. Xu, K., Ba, J., Kiros, R., Cho, K., Courville, A., Salakhutdinov, R., Zemel, R.S., Bengio, Y.: Show, attend and tell: neural image caption generation with visual attention. arXiv preprint arXiv:1502.03044, **2**, 5 (2015)
28. Memisevic, R.: Learning to relate images. IEEE Trans. Pattern Anal. Mach. Intell. **35**, 1829–1846 (2013)
29. Hong, S., Oh, J., Han, B., Lee, H.: Learning transferrable knowledge for semantic segmentation with deep convolutional neural network. arXiv preprint arXiv:1512.07928 (2015)
30. Farfade, S.S., Saberian, M.J., Li, L.J.: Multi-view face detection using deep convolutional neural networks. In: Proceedings of the 5th ACM on International Conference on Multimedia Retrieval, pp. 643–650. ACM (2015)

Joint Shape and Local Appearance Features for Real-Time Driver Drowsiness Detection

Jie Lyu[1(✉)], Hui Zhang[2(✉)], and Zejian Yuan[1(✉)]

[1] Institute of Artificial Intelligence and Robotics, Xi'an Jiaotong University, Xi'an,
China
jiejielyu@outlook.com, yzejian@gmail.com
[2] Shenzhen Forward Innovation Digital Technology Co. Ltd., Shenzhen, China
jet.zhang@forward-innovation.com

Abstract. In this paper, we propose a framework to detect driver drowsiness from video sequences for an advanced driver assistance system. Our method extracts the effective facial descriptors to describe the drowsiness based on face alignment, and classifies the driver facial states via random forest (RF), finally short-term voting and long-term correlation are applied to output smooth results with long-term memory. In particular, the proposed descriptors can encode both shape and local appearance by the located facial landmarks, and utilize the information from multiple frames to enhance the reliability. The classification and alignment based on RF structure are very efficient for drowsiness detection. Our system can obtain 94% accuracy on our F-DDD dataset and 88.18% accuracy on the evaluating set of NTHU-DDD dataset, meanwhile, the implementation achieves 22 FPS for 640×480 videos.

1 Introduction

Driver drowsiness has been one of the main causes of road accidents. According to a report from World Health Organization (2013) [1], 1.24 million people die on roads every year, and about 6% of those accidents happen because of driving in drowsy states.

Drowsiness is defined as the lack of sleep [2], normally involves lowering head, yawning and closing eyes. There are three typical types of methods to detect driver drowsiness including: physiology [3], behavior [4] and vehicle based [5]. And our approach is based on facial behaviors captured by images. Our system is shown in Fig. 1. A camera, located in the front of the vehicle, gets the frontal face images of the driver. Thus, this embedded system detects driver drowsiness without contact. And it alarms tired driver before abnormal vehicle trajectory appears. Especially, this embedded device is able to take images in the evening with safely active near-infrared light.

It still meets some challenges to effectively detect the driver drowsiness from images. The illumination in the driving room is always inconstant and the shadows will easily cause large difference in the same state. The typical normal states are defined as opening eyes, blinking and turning head, which are easy to be

C.-S. Chen et al. (Eds.): ACCV 2016 Workshops, Part III, LNCS 10118, pp. 178–194, 2017.
DOI: 10.1007/978-3-319-54526-4_14

Fig. 1. Driver Drowsiness Detection System (D^{3}S) and driver's frontal face image. The system is installed in the front of the vehicle to capture driver frontal face images, shown in the right, and to analyse drowsy states. Our embedded device is presented in the red box. (Color figure online)

confused with drowsy states. Particularly, the driver wearing glasses may cause some difficulties, and the videos in scenarios of night are hard to extract discriminative representation due to the blurry quality of images. Our method can accurately locate landmarks and make decisions in an unified random forest structure to achieve high speed. Meanwhile, it is robust for illumination and partial occlusion. Especially, we propose a novel feature encoding appearance and spatio-temporal information. The appearance representation is extracted like HOG, which ensures illumination invariance. And the complex facial states are effectively described by landmarks. In addition, the temporal information is significant to recognise the facial behaviors in a certain period of time.

The main contributions include: (1) we propose a framework on the multi-frame, and unify face alignment and classification into a random forest structure, (2) we design an efficient feature to distinguish facial states, and (3) we build a dataset named Forward Driver Drowsiness Detection (F-DDD), which includes 291 clips.

2   Related Work

Drowsiness detection is a special kind of expression recognition. In the last few years, lots of people studied facial expression. [6–8] proposed some methods to recognise facial expression from a single image. [9–11] proposed to use spatial-temporal information from image sequences. Jung et al. [12] used deep learning to recognise expressions in videos. But never did they detect driver drowsiness.

Several papers detect the driver drowsiness by estimating head pose [13,14] or calculating the state of eyes [15]. But these states don't cover typical drowsy states. [5] proposed a framework to detect driver drowsiness in which it combined vehicle based method and head pose estimation together, while the system is too complex to maintain. [3] presented a method to detect multiple drivers' emotion from physiological signals. However, detecting physiological signals is not convenient.

Face alignment is the first component of our system. [16,17] proposed to detect facial landmarks under occlusion. And [18] detected the landmarks in the wild. But those methods are not optimised for Driver Drowsiness Detection independently. Ren et al. [19] proposed a fast and high-precision approach to face alignment, which is adopted by us.

[20] studied facial expression recognition from near-infrared video sequences, using LBP-TOP features. But the method only considers appearance information. Our system can also capture near-infrared videos, moreover our method can be easily extended to the near-infrared videos in evening. Particularly, we take appearance and spatio-temporal information into consideration.

In recent years, some researchers have proposed methods to detect driver drowsiness. Shirakata et al. [21] utilized the change of pupil diameter to detect the imperceptible drowsiness, while it is not convenient for driver to carry such a system to extract the information of the pupil and it is almost impossible to observe the pupil of driver with active infrared light at night. Nakamura et al. [22] coded the representation by considering the wrinkles change and distance of each part and then estimated the drowsiness-degree by k-NN. Nevertheless, the method cannot satisfy real-time and causal requirement. Akrout and Mahdi [23] presented a spatio-temporal features for detecting driver drowsiness state and lack of concentration. After locating the iris and two eyelids via the circular hough transform, they decide the states of driver by measuring the eye-blinking and estimating the 3D head pose. However, the hough transform cannot give a precise location of iris and eyelids under complex practical driving room. And based on facial points location which is not robust to the shadow and large facial deformation, their system cannot work in the reality.

3   Our Approach

The eyes and mouth are the most important parts to judge whether a driver is exhausted or not, thus their locations should be obtained at first. Ren et al. [19] proposed a model based on the residual of facial shape and implemented with random forests regression.

Given a facial image I and its initial shape S^{k-1} at stage k, the approach predicts the shape residual ΔS^k between current shape and the true shape from Eq. (1), supposing that the feature mapping function Φ^k and regression weight W^k exist. The approach can obtain facial shape close to true shape by iterating over Eq. (1), from an input image I and an initial shape S^0.

$$\Delta S^k = W^k \Phi^k(I, S^{k-1}) \tag{1}$$

The method learns a set of binary feature mappings via random forests and solves the weighted matrix by the support vector regression.

Especially, We select specific 51 facial landmark points S, which indicate key points' locations of key parts, to attain the robustness for large difference among facial shapes. Moreover, we also retrain this model from the combining dataset, which includes the existed datasets and our facial images of drivers, to locate driver facial landmark points with high precision.

Based on those landmark points, we extract shape and local appearance features and judge the driver's drowsy state by random forest classifier followed by a short-term voting and long-term correlation, which is shown as Fig. 2.

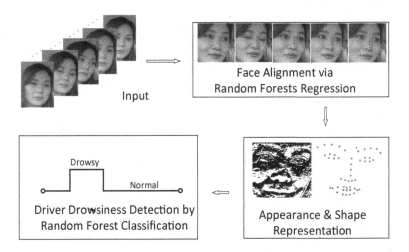

Fig. 2. The framework of the Driver Drowsiness Detection

3.1 Shape and Local Appearance Features (SLAFs)

A shape feature is designed to describe face poses and an appearance feature is built to represent local region states. Furthermore, we combine Shape Feature and Appearance Feature together on the multi-frame to form a long vector $\mathbf{x}$ which includes appearance and spatio-temporal information, shown in Fig. 3.

Shape Feature. Given the facial shape S^t on the t-th frame I^t, the shape feature mapping $\varPhi_s$ is proposed to extract descriptor from the global face and each local part shape S_i^t. We get the representation of the shape $\mathbf{x}_{s,i}$ in every part, R→Right eye, L→Left eye, M→Mouth, G→Global, by the specific mapping $\varPhi_s$ as Eq. (2).

$$\mathbf{x}_{s,i}^t = \varPhi_s(S_i^t), i \in \{R, L, M, G\} \tag{2}$$

This mapping $\varPhi_s$ calculates the relative shape of each part, which is defined as the differences between absolute shape S_i^t and the center of the corresponding shape $\overline{S_i^t}$. And it normalizes this relative shape by the width of shape in x coordinate. Specifically, it is defined as Eq. (3).

$$\varPhi_s(S_i^t) = \frac{S_i^t - \overline{S_i^t}}{\max\limits_{p \in S_i^t}(x_p) - \min\limits_{p \in S_i^t}(x_p)}, \tag{3}$$

where x_p denotes the x coordinate of a point p in the shape S_i^t. In our method, the global shape is represented with a 102-dimensional vector $\mathbf{x}_{s,G}^t$, shapes of eyes with 12-dimensional vectors $\mathbf{x}^t_{s,L}$ and $\mathbf{x}^t_{s,R}$ respectively and shape of mouth with a 40-dimensional vector $\mathbf{x}^t_{s,M}$.

Local Appearance Feature. Given the location of every part from its corresponding key points, appearance feature mapping Φ_a is proposed to extract descriptor from each local region of input frame I^t. This mapping Φ_a crops the local regions and calculates the corresponding gradient images at first, then divides each gradient image into 2×2 blocks and divides the orientation into 8 directions, and finally builds a normalized 32-dimensional vector by calculating the statistics histogram of gradient orientations.

This specific mapping function Φ_a is applied on three parts, Left eye (L), Right eye (R) and Mouth (M). The mapping extracts the appearance representation $\mathbf{x}_{a,j}$ of this part, which is shown as Eq. (4).

$$\mathbf{x}^t_{a,j} = \Phi_a(I^t, S^t_j), j \in \{L, R, M\} \tag{4}$$

In our method, these three local parts are represented with 32-dimensional vectors $\mathbf{x}^t_{a,L}$, $\mathbf{x}^t_{a,R}$ and $\mathbf{x}^t_{a,M}$ respectively.

SLAFs on the Multi-Frame. After extracting local appearance feature and shape feature, a normalized descriptor $\mathbf{x}^t$ is built by mapping the two representations together, which is shown as Eq. (5). Namely, this mapping outputs a 262-dimensional descriptor including a 166-dimensional shape feature and a 96-dimensional local appearance feature from each frame.

$$\mathbf{x}^t = [\mathbf{x}^t_{s,G}; \mathbf{x}^t_{s,L}; \mathbf{x}^t_{s,R}; \mathbf{x}^t_{s,M}; \mathbf{x}^t_{a,L}; \mathbf{x}^t_{a,R}; \mathbf{x}^t_{a,M}] \tag{5}$$

Given the length of the sliding temporal windows w, which is an odd number, we get a long vector $\mathbf{x}^t_w$ by putting the SLAFs of those frame together. Specifically, it is defined as Eq. (6). When w is equal to 5, $\mathbf{x}^t_w$ is a 1310-dimensional vector. Normally, we normalize the vector with L2 norm and then output a representation $\mathbf{x}$ of these frames.

$$\mathbf{x}^t_w = [\mathbf{x}^{t-\frac{w}{2}}; \ldots; \mathbf{x}^t; \ldots; \mathbf{x}^{t+\frac{w}{2}}] \tag{6}$$

The procedure of extracting the vector $\mathbf{x}^t_w$ is shown as Fig. 3 in detail, from which we know that the descriptor encodes appearance and spatio-temporal information. It is significant to maintain the effectiveness for a certain period of time and robustness for shadow and worse locating precision.

3.2   Random Forest Classification

We apply a random forest on classifying the feature descriptor $\mathbf{x}$, calculated by Eq. (6) and normalized. The random forest randomly combines the Shape Feature and Local Appearance Feature on the multi-frame. By this way, the input vector $\mathbf{x}$ would be mapping into a very highly dimensional feature to

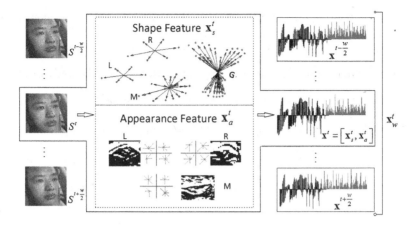

Fig. 3. The flow diagram of the method for extracting Shape and Local Appearance Features.

improve performance while calculated with lower consumption in every weak learner.

A sample set $\{\mathbf{x}_i, y_i,\}, y_i \in \{0,1\}, i = 1, 2, \ldots, n$, created from a dataset, is essential for training the classifier, where 0 indicates normal state and 1 indicates drowsy state. This paper sets a simple linear classifier as every branch node of the random forest whose input vector is randomly chosen from a long vector $\mathbf{x}$. Each tree classifier divides training sample set into left child and right child in the branch nodes, and calculates the probability of each class in the leaf nodes. Thus, the leaf nodes of r-th tree store the probability of c class $p_r(c|\mathbf{x})$, $c \in \{0,1\}$.

When a sample $\mathbf{x}$ reaches the leaf node, the tree classifier would output the probability of every class $p_r(c|\mathbf{x})$ in this leaf node. In binary classification, tree classifier outputs the probability of the positive class $p_r(c = 1|\mathbf{x})$ under an input vector $\mathbf{x}$. Thus the output of the random forest is the probability of drowsy state $p(c = 1|\mathbf{x})$, which is shown as Eq. (7), where T is the forest size [24].

$$p(c = 1|\mathbf{x}) = \frac{1}{T} \sum_{r=1}^{T} p_r(c = 1|\mathbf{x}) \tag{7}$$

We take drowsy states as positive samples, represented with 1, and normal states as negative samples, represented with 0. If given the probabilistic threshold of the positive τ^*, we get the output label y of an input $\mathbf{x}$ from Eq. (8).

$$y = \begin{cases} 0, p(c = 1|\mathbf{x}) < \tau^* \\ 1, p(c = 1|\mathbf{x}) \geq \tau^* \end{cases} \tag{8}$$

Because of tree structure, the random forest can achieve high speed when applied. Due to the combination of lots of weak learners, the random forest fits complicated practical situations with high accuracy. In our system, face align-

ment and classification are based on the random forest structure. This guarantees
the running speed and detection accuracy.

3.3   Short-Term Voting and Long-Term Correlation Based on Video

The results Y that classifier outputs include noise and short-term memory. To
avoid wrong detection and gain long term memory, we present a method as the
post processing step after classification. The procedure of short-term voting and
long-term correlation is illustrated as Fig. 4.

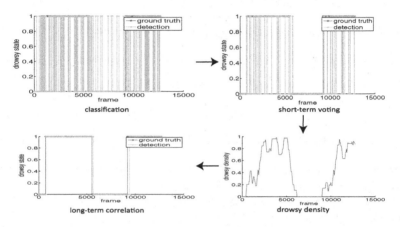

Fig. 4. The illustration shows the procedure of short-term voting and long-term correlation.

Short-Term Voting. A smooth filter is applied to obtain short-term memory,
which decides the state in a frame by voting according to the states of the current
frame and the former several frames. This operation can remove some isolated
false drowsy states and false normal states. Thus relative smooth drowsiness
states are given by this filter with the factor of short-term voting T. Theoretically,
short-term voting result $Y_s(t)$ in t-th frame can be presented by Eq. (9), in which
$Y(t-k)$ means the result of classification in $(t-k)$-th frame, V_t is the number
voting for drowsiness, and $\delta(\cdot)$ is the binary indicating function.

$$
V_t = \sum_{k=0}^{k=T-1} \delta(Y(t-k) = 1),
$$
$$
Y_s(t) = \begin{cases} 0, V_t < T/2 \\ 1, V_t \geq T/2 \end{cases}
$$

(9)

Long-Term Correlation. We define drowsiness density ρ, which is the convolution between a filter F with a specific length L and smooth drowsiness states
Y_s, as the degree of driver drowsiness. Furthermore, it is given by Eq. (10), in

which, $*$ represents convolutional operation, the elements of filter F are all $1/L$, which can also be different value to achieve better performance.

$$\rho = F * Y_s \tag{10}$$

Given a threshold of density τ_ρ, we can calculate drowsy states $Y_l(t)$ in the t-th frame by binary function as Eq. (11).

$$Y_l(t) = \begin{cases} 0, \rho(t) < \tau_\rho \\ 1, \rho(t) \geq \tau_\rho \end{cases} \tag{11}$$

For the long-term memory, a relative larger number should be selected as the specific length L, thus we can obtain some clips with drowsy state. To avoid missing some gaps between clips, we get the correlation between clips and the corresponding gap by computing the ratio of the gap length to the clips length, and fill in some gaps existing strong correlation with adjacent drowsy clips.

4   Experiments

We build a dataset named Forword Driver Drowsiness Detection (F-DDD) to analyse the effects of those parameters of our method (JSLA) which attains 94% accuracy on testing set of F-DDD dataset under the best parameters configuration. A dataset named NTHU Drowsy Driver Detection (NTHU-DDD) is combined with F-DDD to train several available classification models. Finally, the performance of the proposed method is evaluated on the evaluating and testing set of the NTHU-DDD dataset. The system achieves 88.18% accuracy on the evaluating set, and 75.45% on the testing set. Meanwhile, our method can satisfy real-time performance on 640×480 videos.

4.1   Our Dataset

For the lack of existed publicly available dataset with low latency, a dataset and corresponding feature set are built to analyse the effects of those parameters. Figure 5 shows these three kinds of drowsy states defined as lowering head, yawning and closing eyes and three kinds of normal states defined as opening eyes, blinking and swing head.

Data Collection and Labeling. We take 19 videos with 5 min long and 1920×1080 resolution from 15 men and 3 women. Normally, a conclusion whether driver is tired or not can only be drawn after a certain period of time. We label the starting and ending frames of every drowsy state, as well as normal states. To ensure the consistency between training and application, it is necessary to reserve some normal frames at the start and the end of every positive video piece as transitional states.

Data Statistics. We cut every video into pieces with 640×480 resolution according to the labeled frames, and build a database which includes 291 video

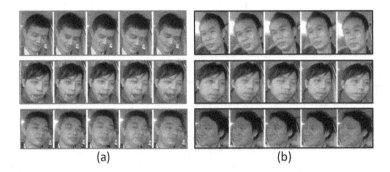

Fig. 5. Three kinds of typical positive samples and three kinds of typical negative samples on images sequence. (a) Lowering head (1st), yawning (2nd) and closing eyes (3rd). (b) Opening eyes (1st), blinking (2nd) and swing head (3rd).

pieces with different numbers of frames. Table 1 shows the ratios of three kinds of drowsy states in the corresponding positive sample set. Closing eyes is the most frequent tired state and yawning is the least frequent one.

Table 1. The ratios of three kinds of drowsy states in the corresponding positive sample set. Note that every video may include several states.

Positive set	Lowering head	Yawning	Closing eyes	Videos
Testing	59.3%	7.4%	63.0%	27
Training	29.9%	7.3%	81.8%	137

Evaluation Criteria. We present a criterion to judge whether the classifier works well or not. In binary classification, we usually define the ACC (accuracy) as the ratio of the number of true predicted labels n and the number of total samples N. Namely, $ACC = n/N$.

4.2 Effects Analysis

Extracting features from our F-DDD database by presented approach, we have obtained 258 positive samples and 653 negative samples on the testing set, similarly 787 positive samples and 2366 negative samples on the training set.

Effects of Random Forest Parameters. The effects of changing parameters, such as forest size T, tree depth d and every weak learner dimensions D, are shown as Fig. 6, and we draw some conclusions about the binary classification as follows:

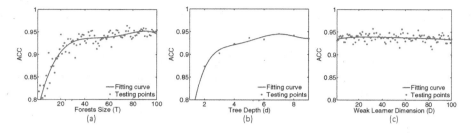

Fig. 6. The effects of changing parameters on random forest classifier. (a) $T = 1-100$, $d = 7$ and $D = 48$. (b) $T = 58$, $d = 1-9$ and $D = 48$. (c) $T = 58$, $d = 7$ and $D = 1-100$.

Forest Size. The accuracies almost keep invariant if the forest size is more than thirty, thus we tend to choose the number of trees near 40 to improve the accuracy as well as speed.

Depth of Trees. The accuracy is bad when the tree depth is less than four, and it is also not good when the tree depth is more than seven because of overfitting, so that we would better to choose the depth equal to seven.

Dimension of Weak Learners. The accuracy does not have noticeable differences when changing the weak learner dimensions, and we make the number of dimensions close to 30 to make sure the randomness is enough.

Probabilistic Threshold. The best probabilistic threshold τ^* should be chosen to judge if the output of the classifier is positive or negative, according to Eq. (12).

$$\tau^* = \arg \max_{\tau \in (0,1)} ACC(\tau) \tag{12}$$

By changing the threshold from 0 to 1 by 0.01 step length, we get the best threshold 0.26.

Effects of Sliding Windows. Changing w the length of sliding windows, the latency and accuracy will change accordingly. If w is too small, such as equal to 3, the features will have no discernment between blinking and closing eyes. Naturally, the wrong detection increases. While w is too large, such as more than 9, the latency can be more than 1 s, which cannot satisfy our requesting for 0.5 s delay. Considering accuracy and latency simultaneously, we set w to 5.

Comparison of SF and LAF. By changing the parameters, we have knowledge of that Local Appearance Feature (LAF) has more importance than Shape Feature (SF) from Fig. 7(a)–(c). According to Fig. 7(d), we know that the accuracy of LAF is higher than that of SF with respectively the best parameters. And the accuracy SLAF including shape and appearance information is just a little higher than that of the two.

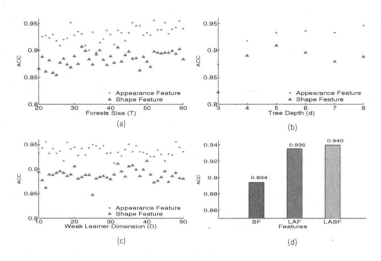

Fig. 7. The comparison of accuracy between the Shape Feature and the Local Appearance Feature (LAF). (a) $T = 20-60$, $d = 7$ and $D = 30$. (b) $T = 50$, $d = 3-8$ and $D = 30$. (c) $T = 50$, $d = 7$ and $D = 10-50$. (d) Under the best respective parameters of three features.

4.3   Evaluating on NTHU-DDD Dataset

After training models on the dataset consisting of F-DDD and NTHU-DDD, we evaluate our method (JSLA) on the evaluating and testing set of NTHU-DDD dataset. Because of some problems in NTHU-DDD dataset, such as too long latency and too much wrong labels of eyes' states, we need to relabel the drowsiness label to satisfy our low latency requirement.

NTHU-DDD Dataset. The training set of NTHU-DDD involves 18 volunteers including 10 men and 8 women, and each volunteer acts as driver in five scenarios listed as glass, no glasses, glasses at night, no glasses at night and sunglasses. In every scene, the volunteer records four videos with different content, which are non-sleepy, sleepy states combining three typical states, blinking with nodding and yawning. Each video has four annotation files recording the states of drowsiness, eyes, head and mouth. The labels of each part are given as Table 2.

Table 2. The labelled states of each part on NTHU-DDD dataset

Label	Sub-category			
	Drowsiness	Eyes	Head	Mouth
0	Normal	Normal	Normal	Normal
1	Drowsy	Sleepy	Nodding	Yawning
2	-	-	Looking aside	Talking & Laughing

Figure 8 illustrates some typical states in a frame of NTHU-DDD dataset respectively. Drowsiness states are related to not only those typical states, but also long-short-term memory. For example, blinking includes closing eyes states in some frames, but it cannot be labelled as drowsiness state according to the context. A particularly obvious character of NTHU-DDD is the long latency, which means that some frames with typically normal states between two drowsiness states are labelled as drowsiness even if the normal time is a few seconds continuously.

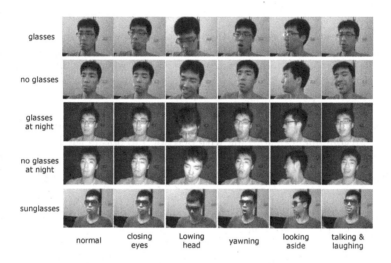

Fig. 8. The illustration of some typical states in a frame of NTHU-DDD dataset respectively.

Selecting Parameters of Long-Term Correlation. The specific length L and the threshold of drowsy density τ_ρ are the most important parameters affecting the accuracy of long-term correlation. Figure 9(a) shows that the accuracy is sensitive to the threshold of drowsy density, while the best thresholds are different in those videos. Therefore, a self-adapting threshold should be applied to the long-term correlation. In a single video, the mean value of drowsy density $\bar{\rho}$ can indicates the entire drowsy level, which can guide us to select a better threshold which is related to the character of this video. Figure 9(b) turns out that selecting the threshold as the half of $\bar{\rho}$ can achieve better stable accuracy under changing the length of long-term filter.

4.4   System Level Performance

With the well-tuned parameters, we evaluate the performance of the system on the evaluating set and testing set of NTHU-DDD dataset, in which the resolution of those videos are 640×480. The evaluating set include 20 videos and another

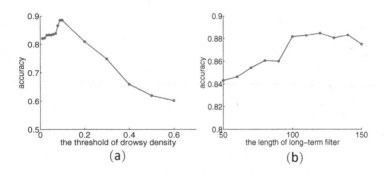

Fig. 9. The accuracy under changing the threshold of drowsy density and the length of long-term filter. (a) changing the threshold of drowsy density τ_ρ under $L = 120$, (b) changing the length of long-term filter L under $\tau_\rho = \overline{\rho}/2$, where $\overline{\rho}$ is the mean value of drowsy density ρ of a video.

70 videos are included in the testing set. To optimise the rate and speed of face detection, we combine face tracking with detection to obtain the location of face by applying the tracking method in [25] to our system. For large variation of head pose, other approaches [26–28] can also be selected as the tracking module.

Accuracy. On the evaluating set, the system can achieve 88.18% accuracy, and some details are shown as Table 3.

Table 3. The accuracies of all scenarios on the evaluating set of NTHU-DDD dataset.

Scenario	No glasses	Glasses	Sunglasses	Night-no glasses	Night-glasses	Overall
Accuracy (%)	95.65	96.35	83.35	69.74	95.74	88.18

On testing set, the system can obtain 75.40% accuracy, and the accuracy of all scenarios are presented as Table 4, which is the result of the ACCV Workshop on Driver Drowsiness Detection from Video 2016.

The accuracies in glasses and no glasses scenarios are almost the best on both the evaluating and testing sets, which indicates that the system can work stably in day time and implies that our F-DDD dataset including only day-time samples can improve the performance of the system in day time. In night scenarios, the system performs unsteadily, and two main reasons lead to this result: one is that in active infrared scenarios the discernibility of the representation is not enough, and the other one is that the quantity of samples in night scenarios is much too small to train a better classifier. In sunglasses scenario, the accuracies are almost the worst, of which the reason is that the states of eyes with sunglasses cannot be distinguished by the classifier and it can only output judgment according to

Table 4. The accuracies of all scenarios on the testing set of NTHU-DDD dataset are the results of the ACCV Workshop on Driver Drowsiness Detection from Video 2016.

Scenario	Drowsiness	Non-drowsiness	Accuracy
No glasses	84.47%	84.14%	84.31%
Glasses	87.41%	86.20%	86.83%
Sunglasses	53.49%	66.38%	60.97%
Night-no glasses	86.65%	82.27%	84.77%
Night-glasses	59.40%	60.61%	60.01%
Overall	75.62%	75.17%	75.40%

the states of mouth and head, while mouth and head cannot represent all drowsy states. Overall the system works well in day-time scenarios but just passably in night scenarios, and In sunglasses scenario, it can only give judgment of driver drowsiness according to yawning and lowering head.

Speed. Table 5 shows the time consumption of every component, and the entire system achieves real-time performance with a core of i7 CPU, at a speed of about 22 FPS.

Table 5. The average time consumption of every component on video: Face Detection (FD), Face Tracking (FT), Face Alignment (FA), Extracting Feature (EF) and Classification (Cla). Others (O) includes reading and writing frames, cropping regions and so on. For the entire system, the speed achieves about 22 FPS.

Component	FD & FT	FA	EF	Cla	O	Total
Time (ms)	14.6	26.8	1.7	0.2	2.1	45.4

From Table 5, we know that face detection/tracking and face alignment cost the most time in a processing cycle. In practical, the time of face detection and tracking can be optimized to less than 10 ms, and the speed of face alignment, which is related to the number of landmark points, can be accelerated by decreasing the number of points without losing the accuracy.

Results on Videos. Processing some videos by our driver drowsiness detection system, we present the results which includes both the situation where the system works well and some bad samples as Fig. 10, from which we know that our system can detect general driver drowsiness robustly, but it is not stable for driver's face with serious deformation on account of that the landmark regressor cannot ensure the locating precision due to the lack of such training samples. At night driving with active infrared light, the driver drowsiness can also be detected by the system generally, while wrong detections may happen if the driver's state is amphibolous such as talking and laughing with closing eyes.

Figure 10(c) illustrates that the system can also detect driver drowsiness well with active infrared at night, even if the driver face cannot be observed because of large variation of head pose. While some failed detections shown as Fig. 10(d) still happen as the result of the inconsistent states of eyes and mouth.

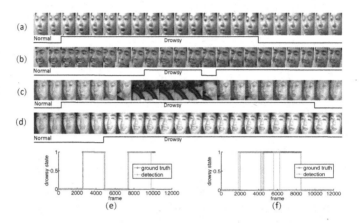

Fig. 10. The results of the Driver Drowsiness Detection System. (a) and (c) show the correct detection results given by classifier and short-term voting, while (b) and (d) present the false drowsy states, but the scenarios of (c) and (d) are at night. (e) illustrates a better consequence by long-term correlation. In the contrary, (f) shows a worse case.

5   Conclusion

We introduce the serious problems caused by driving with drowsiness and propose a framework to detect driver drowsiness based on facial behaviors. Shape and Local Appearance Features are proposed to represent the facial states on the multi-frame. The approach to extracting features has taken shape and appearance information into consideration. And our experiments show that Local Appearance Feature is more important than Shape Feature because that Local Appearance Feature includes partial significant information of Shape Feature. Random Forest Classifier is used to unify face alignment into random forest structure to ensure the speed. We build a dataset named Forward Driver Drowsiness Detection (F-DDD) including 291 video clips to combine with the training set of NTHU-DDD and train a better classification model. The consequences of the experiments on the NTHU-DDD dataset show that the proposed method can fulfil the real-time requirement and achieve about 75.40% accuracy on the testing set and 88.18% accuracy on the evaluating set. What's more, the method can obtain 94% accuracy on our F-DDD dataset.

Acknowledgement. This work was supported by the National Basic Research Program of China (No. 2015CB351703), the National Key Research and Development Program of China (No. 2016YFB1001001), and the National Natural Science Foundation of China (No. 61573280).

References

1. World Health Organization: Global Status Report on Road Safety 2013: Supporting a Decade of Action: Summary. World Health Organization, Geneva (2013)
2. Colic, A., Marques, O., Furht, B.: Driver Drowsiness Detection - Systems and Solutions. Springer Briefs in Computer Science. Springer, Heidelberg (2014)
3. Wang, J., Gong, Y.: Recognition of multiple drivers' emotional state. In: ICPR, 8–11 December 2008, pp. 1–4 (2008)
4. Smith, P., Shah, M., da Vitoria Lobo, N.: Determining driver visual attention with one camera. IEEE Trans. Intell. Transp. Syst. **4**, 205–218 (2003)
5. Rezaei, M., Klette, R.: Look at the driver, look at the road: no distraction! no accident! In: CVPR, 23–28 June 2014, pp. 129–136 (2014)
6. Liu, W., Song, C., Wang, Y.: Facial expression recognition based on discriminative dictionary learning. In: ICPR, 11–15 November 2012, pp. 1839–1842 (2012)
7. Chew, S.W., Lucey, S., Lucey, P., Sridharan, S., Conn, J.F.: Improved facial expression recognition via uni-hyperplane classification. In: CVPR, 16–21 June 2012, pp. 2554–2561 (2012)
8. Liu, P., Han, S., Meng, Z., Tong, Y.: Facial expression recognition via a boosted deep belief network. In: CVPR, 23–28 June 2014, pp. 1805–1812 (2014)
9. Sun, Y., Yin, L.: Facial expression recognition based on 3D dynamic range model sequences. In: Forsyth, D., Torr, P., Zisserman, A. (eds.) ECCV 2008. LNCS, vol. 5303, pp. 58–71. Springer, Heidelberg (2008). doi:10.1007/978-3-540-88688-4_5
10. Drira, H., Amor, B.B., Daoudi, M., Srivastava, A., Berretti, S.: 3D dynamic expression recognition based on a novel deformation vector field and random forest. In: ICPR, 11–15 November 2012, pp. 1104–1107 (2012)
11. Wang, Z., Wang, S., Ji, Q.: Capturing complex spatio-temporal relations among facial muscles for facial expression recognition. In: CVPR, 23–28 June 2013, pp. 3422–3429 (2013)
12. Jung, H., Lee, S., Yim, J., Park, S., Kim, J.: Joint fine-tuning in deep neural networks for facial expression recognition. In: ICCV, 7–13 December 2015, pp. 2983–2991 (2015)
13. Geng, X., Xia, Y.: Head pose estimation based on multivariate label distribution. In: CVPR, 23–28 June 2014, pp. 1837–1842 (2014)
14. Demirkus, M., Precup, D., Clark, J.J., Arbel, T.: Probabilistic temporal head pose estimation using a hierarchical graphical model. In: Fleet, D., Pajdla, T., Schiele, B., Tuytelaars, T. (eds.) ECCV 2014. LNCS, vol. 8689, pp. 328–344. Springer, Heidelberg (2014). doi:10.1007/978-3-319-10590-1_22
15. Shi, T., Liang, M., Hu, X.: A reverse hierarchy model for predicting eye fixations. In: CVPR, 23–28 June 2014, pp. 2822–2829 (2014)
16. Burgos-Artizzu, X.P., Perona, P., Dollár, P.: Robust face landmark estimation under occlusion. In: ICCV, 1–8 December 2013, pp. 1513–1520 (2013)
17. Wu, Y., Ji, Q.: Robust facial landmark detection under significant head poses and occlusion. In: ICCV 2015, 7–13 December 2015, pp. 3658–3666 (2015)
18. Zhu, X., Ramanan, D.: Face detection, pose estimation, and landmark localization in the wild. In: CVPR, 16–21 June 2012, pp. 2879–2886 (2012)

19. Ren, S., Cao, X., Wei, Y., Sun, J.: Face alignment at 3000 FPS via regressing local binary features. In: CVPR, 23–28 June 2014, pp. 1685–1692 (2014)
20. Taini, M., Zhao, G., Li, S.Z., Pietikäinen, M.: Facial expression recognition from near-infrared video sequences. In: ICPR, 8–11 December 2008, pp. 1–4 (2008)
21. Shirakata, T., Tanida, K., Nishiyama, J., Hirata, Y.: Detect the imperceptible drowsiness. SAE Int. J. Passeng. Cars - Electron. Electr. Syst. 3, 98–108 (2010)
22. Nakamura, T., Maejima, A., Morishima, S.: Detection of driver's drowsy facial expression. In: 2nd IAPR Asian Conference on Pattern Recognition, ACPR 2013, Naha, Japan, 5–8 November 2013, pp. 749–753 (2013)
23. Akrout, B., Mahdi, W.: Spatio-temporal features for the automatic control of driver drowsiness state and lack of concentration. Mach. Vis. Appl. 26, 1–13 (2015)
24. Criminisi, A., Shotton, J., Konukoglu, E.: Decision forests: a unified framework for classification, regression, density estimation, manifold learning and semi-supervised learning. Found. Trends Comput. Graph. Vis. 7, 81–227 (2012)
25. Danelljan, M., Häger, G., Khan, F.S., Felsberg, M.: Accurate scale estimation for robust visual tracking. In: British Machine Vision Conference, BMVC 2014, Nottingham, UK, 1–5 September 2014 (2014)
26. Chen, D., Yuan, Z., Wu, Y., Zhang, G., Zheng, N.: Constructing adaptive complex cells for robust visual tracking. In: Proceedings of the IEEE ICCV, pp. 1113–1120 (2013)
27. Chen, D., Yuan, Z., Hua, G., Wu, Y., Zheng, N.: Description-discrimination collaborative tracking. In: Fleet, D., Pajdla, T., Schiele, B., Tuytelaars, T. (eds.) ECCV 2014. LNCS, vol. 8689, pp. 345–360. Springer, Heidelberg (2014). doi:10.1007/978-3-319-10590-1_23
28. Chen, D., Yuan, Z., Hua, G., Wang, J., Zheng, N.: Multi-timescale collaborative tracking. IEEE TPAMI (2016). doi:10.1109/TPAMI.2016.2539956

Workshop on Meeting HCI with CV

3D Pose Estimation of a Front-Pointing Hand Using a Random Regression Forest

Dai Fujita and Takashi Komuro$^{(\boxtimes)}$

Graduate School of Science and Engineering, Saitama University, Saitama, Japan
komuro@mail.saitama-u.ac.jp

Abstract. In this paper, we propose a method for estimating the 3D poses of a front-pointing hand from camera images to realize freehand pointing interaction from a distance. Our method uses a Random Regression Forest (RRF) to realize robust estimation against environmental and individual variations. In order to improve the estimation accuracy, our method supports the use of two cameras and integrates the distributions of the hand poses for these cameras, which are modeled by the Gaussian mixture model. Moreover, tracking of the hand poses further improves the estimation accuracy and stability. The results of performance evaluation showed that the root mean square error of the angle estimation was 4.10°, which is accurate enough to expect that our proposed method can be applied to user interface systems.

1 Introduction

With the progress of hardware performance and computer vision technology, user interface (UI) systems that recognize users' hand gestures from camera images are now being commercialized. Using hand gestures, users can interact with systems from a distance without having to use special devices. However, many of the existing hand gesture UIs have problems; for example, they can only recognize some gesture commands, and they require large hand motions, which sometimes makes users feel tired.

Pointing interaction from a distance by using hand pointing gestures enables various input operations with small hand motions. Leap Motion [1] is a device that can recognize various kinds of hand poses accurately, including the 3D poses of a pointing hand. However, it can recognize a hand only near the sensor, but it cannot recognize a hand from a distance.

In order to realize interaction with a display from a distance in an ordinary indoor environment, it is desirable to attach a camera or cameras to the display and recognize a front-facing hand that is pointing toward the camera/cameras. However, it is difficult to recognize the poses of a front-facing hand since its appearance exhibits few distinctive features.

Electronic supplementary material The online version of this chapter (doi:10. 1007/978-3-319-54526-4_15) contains supplementary material, which is available to authorized users.

C.-S. Chen et al. (Eds.): ACCV 2016 Workshops, Part III, LNCS 10118, pp. 197–211, 2017.
DOI: 10.1007/978-3-319-54526-4_15

There has been some research on robust hand recognition methods based on machine learning [2,3]. However, they only recognize a limited number of hand shapes and are not sophisticated enough to recognize the poses of a pointing hand. On the other hand, methods using a Random Regression Forest (RRF) for estimating the poses of a specific object (such as a face, a body, or a car) have shown their effectiveness lately [4–7].

In this paper, we propose a method for recognizing a front-pointing hand with sufficiently high accuracy to be used for freehand pointing interaction from a distance. The pointing directions of a hand are estimated using an RRF to realize robust estimation against environmental and individual variations. In order to improve the estimation accuracy, our method supports the use of two cameras and utilizes the difference in appearance between the two camera images. Moreover, tracking of the hand pose based on a probability model is performed to further improve the estimation accuracy and stability.

2   Related Work

2.1   Hand Pose Recognition

The shape of a hand is variable and it is difficult to define the orientation unless the pose is limited. Therefore, there have been few studies on methods for estimating only the orientation of a hand. On the other hand, estimation of arbitrary hand poses having a high degree of freedom (DOF) in joint motions has been studied.

For instance, Oikonomidis et al. have proposed a method for estimating the 26 DOF poses of a hand by formulating the estimation as an optimization problem using the hand's appearance and constraints on the hand poses [8]. Using the hand images with various poses generated by CG, Keskin et al. have constructed a learning model based on a Random Decision Forest that classifies each pixel into the parts of the hand [9]. The results were used to estimate the 3D poses of a hand, in combination with estimation of joint positions.

Tompson et al. have proposed a method for estimating the poses of a hand by extracting feature points on a hand using a convolutional neural network and solving the inverse kinematics by optimization [10]. Sharp et al. proposed a method that roughly classifies hand poses using a modified model of a Random Decision Forest and then estimates detailed hand poses using particle swarm optimization [11].

These methods have high estimation accuracy, but are not sufficiently specialized for recognizing a pointing hand. In addition, many of these methods use depth images for estimation, but depth cameras are expensive and are not yet widely used in consumer electronics. Therefore, we believe that hand recognition methods using only color images are still useful.

There have been studies on systems that can recognize a pointing hand using multiple cameras installed in the environment. Schick et al. have developed a system that recognizes pointing interaction at a distance using two cameras to extend touch input on a large display [12]. Hu et al. have developed a system that estimates hand pointing directions accurately by tracking the hand using two cameras [13]. These systems use cameras installed in the environment, such as on a wall or a ceiling. This makes it difficult to move the system, and also, camera calibration is required after installation. Therefore, it is difficult to use such systems for casual use like a webcam. Moreover, the cameras view the hand from different directions, which makes recognition easier, but limits the operating space to the area that all the cameras can see.

To overcome these limitations, it is desirable to attach a camera or cameras to the display and to recognize a front-facing hand that is pointing toward the camera/cameras.

2.2 Random Regression Forest (RRF)

A Random Regression Forest (RRF), a kind of random forest proposed by Breiman [14,15], is one of the ensemble learning models for regression. An RRF consists of many regression trees. A regression tree predicts a continuous value for the input data (feature) by repeated branching from the root node until reaching a leaf node. Though a regression tree has the problem that overfitting often occurs, the RRF avoids the overfitting problem by constructing multiple regression trees with low correlation and that are trained using a set of randomly sampled training data.

RRFs are becoming widely used in computer vision and are also being applied to pose estimation. Fanelli et al. have proposed a method for estimating the position and orientation of a face using multivariate regression with an RRF [4]. This method approximates the output of each regression tree by a normal distribution and realizes high accuracy by excluding the output with low confidence based on the variance of the distribution. Girshick et al. have proposed a method for estimating the joint positions of a human body by using an RRF in which the output of each regression tree is a mixture of normal distributions and by estimating the body pose at 200 fps [5]. Hara and Chellappa have proposed a new RRF training method that performs optimal node splitting in the output space of the data, and verified the performance by pose estimation of faces and cars [6]. Zhen et al. improved the performance of multivariate regression by applying the features that were trained using manifold learning to an RRF, and applied the method to pose estimation of faces.

3 Pose Estimation of a Pointing Hand Using an RRF

Our method estimates the pose of a hand that is pointing toward the camera and that is detected in a camera image captured from a distance of a few meters. Since the pose of the hand is limited to that of a pointing hand, the pose is determined only from the pointing direction. We define the pointing direction by a yaw angle θ (horizontal rotation) and a pitch angle ϕ (vertical rotation) that are represented in Euler angles, as shown in Fig. 1a. In this study, we do not consider roll rotation.

Our method supports the use of not only a single camera but also two cameras placed at the same height and with parallel optical axes. By using two cameras that are separately placed, the appearance difference of a pointing hand can be utilized. The appearances greatly differ depending on the camera position, as shown in Fig. 1b and c. Even when a hand is overlapping the user's face when viewed from one camera, which makes it difficult to recognize the hand, or when an index finger, which is an important cue for recognizing the pointing direction, is not clearly seen from one camera, it is expected that the other camera can compensate for this. Therefore, the use of two cameras can improve the accuracy more than just obtaining sensor information twice.

| (a) | (b) left image | (c) right image |

Fig. 1. (a) Definition of hand pose and image examples. The pose is determined from a 2-DOF pointing direction. (b)(c) Example of appearance difference depending on the camera position. In the right image, a hand is overlapping the user's face, and most of the index finger is not seen. In contrast, in the left image, the hand is clearly seen.

Before estimating the pose of a pointing hand using our method, it is necessary to detect the pointing hand in an image. We conducted a preliminary experiment to detect a pointing hand using a simple SVM-based method and obtained a detection rate of 87.3%. In this study, we assume that the pointing hand is already detected in the image.

3.1   Pose Estimation from a Monocular Image

First, we describe the pose estimation algorithm in the case of using a single camera. A feature vector v extracted from the image of a pointing hand is the input to the algorithm. In reference to Fanelli et al.'s method [4], we use an RRF that outputs a multivariate probability distribution. By regression using the RRF, the distribution $p(x|v)$ of the pointing direction $x = (\theta, \phi)^{\mathrm{T}}$ is obtained. The RRF used for estimation is constructed by training in advance.

Training of RRF. A set of training data C that consists of N data samples is used for training. The i-th data sample $(v^{(i)}, x^{(i)})$ $(i = 1, \ldots, N)$ is a combination of the feature vector $v^{(i)}$ and the pointing direction vector $x^{(i)} = (\theta^{(i)}, \phi^{(i)})^{\mathrm{T}}$.

The distribution of pointing directions in a training dataset C, $p(x|C)$ is assumed to follow the bivariate normal distribution

$$p(x|C) = \mathcal{N}(x; \mu, \Sigma) = \frac{1}{\sqrt{\det(2\pi\Sigma)}} \exp\left\{-\frac{1}{2}(x - \mu)^{\mathrm{T}} \Sigma^{-1} (x - \mu)\right\}, \quad (1)$$

where $\mathcal{N}(\boldsymbol{x}; \boldsymbol{\mu}, \Sigma)$ is the probability density of the normal distribution with the mean $\boldsymbol{\mu}$ and the covariance matrix Σ at the point $\boldsymbol{x}$.

The training dataset at a node j, C_j, is divided into the following two sets by thresholding:

$$
\begin{aligned}
C_{j,\mathrm{L}} &= \{(\boldsymbol{v}, \boldsymbol{x}) \mid (\boldsymbol{v}, \boldsymbol{x}) \in C_j, v_i > \lambda\} \\
C_{j,\mathrm{R}} &= \{(\boldsymbol{v}, \boldsymbol{x}) \mid (\boldsymbol{v}, \boldsymbol{x}) \in C_j, v_i \le \lambda\}
\end{aligned}
\tag{2}
$$

Here, i and λ are the split function parameters that represent an index of an element in the feature vector and a threshold, respectively, and v_i is the i-th element in the feature vector $\boldsymbol{v}$. The optimal split parameters at a node j, i_j^* and λ_j^*, are determined from randomly chosen parameters i and λ so as to maximize the objective function $I(C_j, i, \lambda)$.

Information gain

$$
I(C_j, i, \lambda) = H(C_j) - \sum_{k \in \{\mathrm{L}, \mathrm{R}\}} \frac{|C_{j,k}|}{|C_j|} H(C_{j,k})
\tag{3}
$$

is used as the objective function.

Here, $H(C)$ is the differential entropy of the normal distribution that a training dataset C follows and is written as

$$
H(C) = - \int p(\boldsymbol{x}|C) \ln p(\boldsymbol{x}|C) \, \mathrm{d}\boldsymbol{x} = \frac{1}{2} \ln(\det(2\pi e \Sigma))
\tag{4}
$$

and $|C|$ shows the number of data samples included in the dataset C.

The stopping criteria for node splitting is when the tree reaches to a certain depth $D_{\max}$ or when the number of data samples in the training dataset becomes less than a certain value. At a terminal node j, the output probability distribution is determined as follows using Eq. (1)

$$
p(\boldsymbol{x}|\boldsymbol{v}_j) = p(\boldsymbol{x}|C_j) = \mathcal{N}(\boldsymbol{x}; \boldsymbol{\mu}_j, \Sigma_j)
\tag{5}
$$

where $\boldsymbol{v}_j$ is a feature vector that will reach the node j.

To avoid degeneracy of the distribution and overestimation of the distribution, we set the lower limit of a variance $v_{\min}$. When $\det(\Sigma_j) < v_{\min}$, the covariance matrix is replaced by $\Sigma_j = \mathrm{diag}(\sqrt{v_{\min}}, \sqrt{v_{\min}})$. In the experiment described later, we set $v_{\min} = 25$.

Regression using an RRF. When a feature vector $\boldsymbol{v}$ is input to the RRF, each regression tree i outputs a distribution $p_i(\boldsymbol{x}|\boldsymbol{v}) = \mathcal{N}(\boldsymbol{x}; \boldsymbol{\mu}_i, \Sigma_i)$ as in Eq. (5), and the posterior distribution of the pointing direction,

$$
p(\boldsymbol{x}|\boldsymbol{v}) = \frac{1}{T} \sum_{i=1}^{T} \mathcal{N}(\boldsymbol{x}; \boldsymbol{\mu}_i, \Sigma_i),
\tag{6}
$$

is calculated by averaging the output distributions from all trees. T is the number of regression trees. This output distribution follows a mixture of normal distributions.

3.2   Pose Estimation from Stereo Images

In this section, we extend the pose estimation algorithm described above to the two-camera (stereo) case. The basic algorithm is common to both the monocular and stereo cases, and in particular, the RRF model is the same.

We assume that the pointing hand has already been detected in both the left and right input images, and that the image window in the left image, $W_L = (x_L, y_L, s_L)$ and that in the right image, $W_R = (x_R, y_R, s_R)$, have been obtained. Here, $(x_{L/R}, y_{L/R})$ [pixel] is the center position of the window in the left/right image, and $s_{L/R}$ [pixel] is the width of the window. The origin of the image coordinate system is located at the center of the image. v_L and v_R are the feature vectors that are extracted from the left and right image windows, respectively. The flow of pose estimation in the stereo case is shown in Fig. 2.

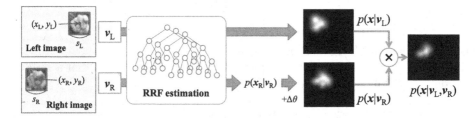

Fig. 2. Flow of pose estimation in the stereo case. The distributions of hand pointing directions in the left and right images are obtained using the same RRF as in the monocular case. The pointing direction seen from the right camera is transformed to that seen from the left camera, and then, the two distributions are integrated.

First, the distributions of hand pointing directions in the left and right images are obtained using the same RRF as in the monocular case:

$$p(\boldsymbol{x}_L | \boldsymbol{v}_L) = \frac{1}{T} \sum_{i=1}^{T} \mathcal{N}(\boldsymbol{x}_L; \boldsymbol{\mu}_{i,L}, \Sigma_{i,L}), \quad p(\boldsymbol{x}_R | \boldsymbol{v}_R) = \frac{1}{T} \sum_{i=1}^{T} \mathcal{N}(\boldsymbol{x}_R; \boldsymbol{\mu}_{i,R}, \Sigma_{i,R}), \quad (7)$$

where $\boldsymbol{x}_L$ is the pointing direction seen from the left camera, and $\boldsymbol{x}_R$ is that seen from the right camera.

To unify the coordinate systems, the pointing direction seen from the right camera is transformed to that seen from the left camera. The yaw angle is changed by

$$\Delta\theta = \text{atan2}(x_L x_R + f^2, (x_L - x_R)f), \quad (8)$$

where x_L and x_R are the x coordinates of the center positions of the left and right image windows, and f [pixel] is the focal length of the cameras. From here, we abbreviate $\boldsymbol{x} = \boldsymbol{x}_L$.

Using $\Delta\theta$, we obtain the distribution of the pointing directions seen from the right image that is transformed to that seen from the left image:

$$p(\boldsymbol{x}|\boldsymbol{v}_\mathrm{R}) = p\left(\boldsymbol{x}_\mathrm{R} + \begin{pmatrix} \Delta\theta \\ 0 \end{pmatrix} \middle| \boldsymbol{v}_\mathrm{R}\right) = \frac{1}{T}\sum_{i=1}^{T}\mathcal{N}(\boldsymbol{x}; \boldsymbol{\mu}'_{i,\mathrm{R}}, \Sigma_{i,\mathrm{R}}) \tag{9}$$

Here, we set $\boldsymbol{\mu}'_{i,\mathrm{R}} = \boldsymbol{\mu}_{i,\mathrm{R}} + (\Delta\theta, 0)^\mathrm{T}$.

Next, the two distributions are integrated and the distribution of the pointing directions, $p(\boldsymbol{x}|\boldsymbol{v}_\mathrm{L}, \boldsymbol{v}_\mathrm{R})$, given both feature vectors $\boldsymbol{v}_\mathrm{L}$ and $\boldsymbol{v}_\mathrm{R}$, is obtained. We assume that $\boldsymbol{v}_\mathrm{L}$ and $\boldsymbol{v}_\mathrm{R}$ are conditionally independent given $\boldsymbol{x}$, i.e. $p(\boldsymbol{v}_\mathrm{L}, \boldsymbol{v}_\mathrm{R}|\boldsymbol{x}) = p(\boldsymbol{v}_\mathrm{L}|\boldsymbol{x})p(\boldsymbol{v}_\mathrm{R}|\boldsymbol{x})$, and that the prior distribution $p(\boldsymbol{x})$ follows a uniform distribution. Then, using the Bayes' theorem, integration is realized as follows.

$$p(\boldsymbol{x}|\boldsymbol{v}_\mathrm{L}, \boldsymbol{v}_\mathrm{R}) \propto p(\boldsymbol{x}|\boldsymbol{v}_\mathrm{L})p(\boldsymbol{x}|\boldsymbol{v}_\mathrm{R}) \tag{10}$$

$$\propto \left(\sum_{i=1}^{T}\mathcal{N}(\boldsymbol{x}; \boldsymbol{\mu}_{i,\mathrm{L}}, \Sigma_{i,\mathrm{L}})\right)\left(\sum_{i=1}^{T}\mathcal{N}(\boldsymbol{x}; \boldsymbol{\mu}'_{i,\mathrm{R}}, \Sigma_{i,\mathrm{R}})\right). \tag{11}$$

Due to integration, the uncertainty of the estimation decreases, as shown in the scatter plots in Fig. 2.

3.3   Pose Tracking Using Multi-frame Images

The proposed method described above uses a single-frame image/images for estimation. In this section, we improve the estimation accuracy and stability by performing pose tracking based on Bayesian estimation as shown in Fig. 3. First we show the algorithm in the monocular case and then we explain that it can easily be extended to the stereo case.

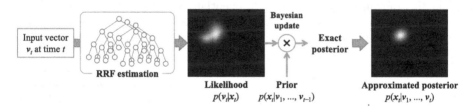

Fig. 3. Flow of tracking the pose of a pointing hand ($t > 1$). The posterior distribution is calculated from: the prior distribution predicted from the pose estimate in frame $t - 1$; and the RRF output in frame t. The posterior distribution is approximated by a normal distribution for successive tracking.

Probability Model of the Pointing Direction. Let $\boldsymbol{x}_t$ be the pointing direction in frame t, and $\boldsymbol{v}_t$ be the feature vector extracted from the detected image window in frame t. The initial frame $t = 1$ is the frame in which the pointing hand is first detected. We assume that the estimate of the current pose, $p(\boldsymbol{x}_t|\boldsymbol{v}_1,\ldots,\boldsymbol{v}_t)$, follows the following normal distribution:

$$p(\boldsymbol{x}_t|\boldsymbol{v}_1,\ldots,\boldsymbol{v}_t) = \mathcal{N}(\boldsymbol{x}_t; \boldsymbol{\mu}_t, \Sigma_t) \tag{12}$$

This assumption is the same as in the Kalman filter. The proposed tracking method is an extension of the Kalman filter with the likelihood $p(\boldsymbol{v}_t|\boldsymbol{x}_t)$ being a mixture of normal distributions.

Prediction of Prior Distribution. We assume that the pointing direction, $\boldsymbol{x}_t$, follows a Markov process and varies with time on the random walk model:

$$\boldsymbol{x}_t = \boldsymbol{x}_{t-1} + \varepsilon_t, \varepsilon_t \sim \mathcal{N}(\boldsymbol{0}, \Sigma_{\mathrm{d}}) \tag{13}$$

The prior distribution of the current pose is predicted using the known estimate in frame $t-1$ as follows:

$$p(\boldsymbol{x}_t|\boldsymbol{v}_1,\ldots,\boldsymbol{v}_{t-1}) = \mathcal{N}(\boldsymbol{x}_t; \boldsymbol{\mu}_{t-1}, \Sigma_{t-1} + \Sigma_{\mathrm{d}}). \tag{14}$$

Calculating the Posterior Distribution. The predicted prior distribution is updated using the observation $\boldsymbol{v}_t$ to obtain the posterior distribution $p(\boldsymbol{x}_t|\boldsymbol{v}_1,\ldots,\boldsymbol{v}_t)$. From the Bayes' theorem and assuming uniform distributions for $\boldsymbol{x}_t$ and $\boldsymbol{v}_t$, the likelihood can be written as $p(\boldsymbol{v}_t|\boldsymbol{x}_t) \propto p(\boldsymbol{x}_t|\boldsymbol{v}_t)$. Therefore, the likelihood becomes proportional to the output distribution from the RRF shown in Eq. (6), and the posterior distribution can be written as

$$p(\boldsymbol{x}_t|\boldsymbol{v}_1,\ldots,\boldsymbol{v}_t) \propto \left(\sum_{i=1}^{T} \mathcal{N}(\boldsymbol{x}_t; \boldsymbol{\mu}_{t,i}, \Sigma_{t,i}) \right) \mathcal{N}(\boldsymbol{x}_t; \boldsymbol{\mu}_{t-1}, \Sigma_{t-1} + \Sigma_{\mathrm{d}}). \tag{15}$$

The obtained posterior distribution becomes a mixture of normal distributions. If this exact posterior distribution is used as the prior in the next frame, the number of components increases exponentially with time [16]. Therefore, the posterior distribution is approximated by a normal distribution by putting the mean and covariance matrix of the posterior distribution into Eq. (12), and it is regarded as the estimate $p(\boldsymbol{x}_t|\boldsymbol{v}_1,\ldots,\boldsymbol{v}_t)$ in frame t.

Extension to the Stereo Case. The pose tracking method in the monocular case described above can easily be extended to the stereo case. Let $\boldsymbol{v}'_t = \{\boldsymbol{v}_{t,\mathrm{L}}, \boldsymbol{v}_{t,\mathrm{R}}\}$ be the combination of left and right feature vectors observed in frame t. The extension to the stereo case is realized by replacing $\boldsymbol{v}_t$ with $\boldsymbol{v}'_t$.

4   Accuracy Evaluation

We conducted an experiment to measure the performance of the proposed method. First, we visualized the estimation results and distributions for subjective evaluation, and then we evaluated the accuracy of pose estimation for objective evaluation. We assumed that the pointing hand was already detected, and that the image window in each camera image was obtained.

4.1   Conditions and Dataset

We set the range of recognizable hand angles to

$$-30° \leq \theta < 30°, -15° \leq \phi < 35°, \tag{16}$$

which is thought to be a sufficiently wide range for pointing interaction from a distance. The range of yaw angles was wider than that of the pitch angles to compensate for the shift of the yaw angle, which reduces the range where the information from the two cameras can be integrated.

We set the size of image windows to 40×40, and used Histograms of Oriented Gradients (HOG) [17] as the feature. The cell size was 8×8 pixels, the block size was 2×2 cells, and the number of dimensions was 576. The number of regression trees, T, in the RRF was 40, and the maximum depth of the trees was 18.

In the tracking, the covariance of the process noise, Σ_d, was set to a constant multiple of an identity matrix $\Sigma_d = \mathrm{diag}(\sigma^2_{\mathrm{pnoise}}, \sigma^2_{\mathrm{pnoise}})$. We set $\sigma_{\mathrm{pnoise}} = 5$ in the monocular case, and $\sigma_{\mathrm{pnoise}} = 3.75$ in the stereo case, which are the values that minimized the root mean square error (RMSE) of the estimated angles in the accuracy evaluation described later.

We used a desktop PC (Intel Core-i7 4770 CPU) for calculation. The proposed method was implemented with parallel processing in a multi-threaded environment.

The pointing hand dataset used for training the RRF and evaluating the estimation performance was constructed by collecting data from 14 participants (2 female, 12 male) who were university students in their twenties.

The images were captured under the fluorescent lighting conditions as in a usual indoor environment, and all the images were captured in the same environment. The participants were asked to point at various positions that were indicated by a pointing stick, and the images of the pointing hand were captured by two cameras (Firefly MV, Point Grey Research Inc.) with 8 mm focal length lenses, which were placed at a distance of 2 m with a baseline of 40 cm. The participants moved their hand in a space inside a steel rack, and the ground truth data was collected by measuring the pointing directions from above using a Kinect sensor.

To control individual differences of the pointing hand shapes, we imposed three restrictions: (1) the index finger of a right hand should be used for pointing, (2) when changing the pointing direction, the whole hand should move without moving the index finger, and (3) the back of the hand should always face upward.

As a result of collecting data, we obtained 20,471 source images whose pointing directions were within the angular range in Eq. (16). By shifting the source images pixel by pixel in eight directions, we obtained a training dataset that consisted of 184,239 pointing hand images. Examples of the pointing hand images are shown in Fig. 4.

As an evaluation dataset, we used 4,111 pairs of stereo images, whose pointing directions in both the left and right images were within the above angular range. When evaluating images of a participant, the images of that participant were excluded from the training dataset.

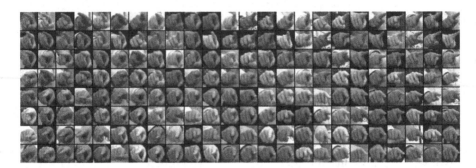

Fig. 4. Examples of hand pointing images for training. Hands with various shapes and poses are included.

It took about ten minutes to train the RRF using all the training data, and the constructed RRF consumed 38.9 MB of memory.

4.2 Subjective Evaluation

By changing the conditions such as the number of cameras and the use of tracking, we visualized the estimation results and the distributions for evaluation data of a participant. For evaluation in the monocular case, the left images of the stereo images in the evaluation dataset were used. In the case without tracking, the mode of the distribution was used to obtain a point estimate of the pose. In the case with tracking, the mean of the posterior distribution was used.

First, the estimation results in the monocular case without tracking are shown in Fig. 5. The white box frame indicates the region of an image window that is set in advance. The arrow indicates the estimated pointing direction. In frames $F = 60$, 178, and 316, the actual and estimated pointing directions were almost the same, and it seems that the estimation was successful. However, in frame $F = 385$, the actual pointing direction was downward but the estimated direction was toward the front, which was apparently wrong.

Figure 6 shows the probability distribution for each image in Fig. 5. In frames $F = 60$ and 316, the probability density of the output distribution was concentrated in a certain region, and the estimate was near the ground truth. In frame

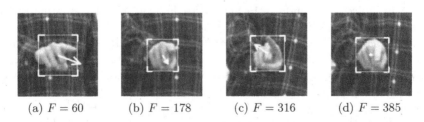

(a) $F = 60$ (b) $F = 178$ (c) $F = 316$ (d) $F = 385$

Fig. 5. Estimation result in the monocular case without tracking. F is the frame number.

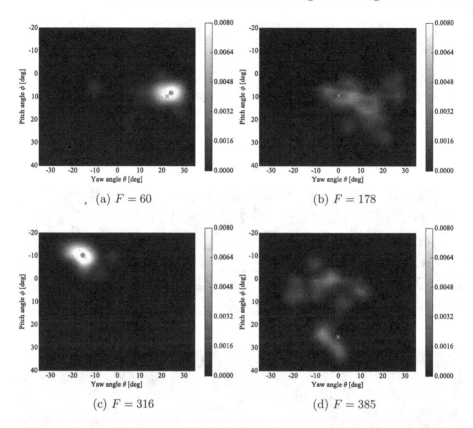

(a) $F = 60$

(b) $F = 178$

(c) $F = 316$

(d) $F = 385$

Fig. 6. Pose distributions for the images in Fig. 5. The red double circle indicates the points with the maximum probability density, the yellow cross mark indicates the ground truth, the blue points are grid points at $10°$ intervals, and the green point indicates the origin. (Color figure online)

$F = 178$, the distribution was ambiguous, and there was a difference of about $7°$ between the estimate and the ground truth. In frame $F = 385$, the distribution was more ambiguous and there was a difference of more than $20°$. However, in both frames $F = 178$ and 385, there was a certain level of density around the ground truth, and it is thus expected that correct estimation is possible by using more information.

Next, the estimation results in frames $F = 178$ and 385 in the stereo case with tracking are shown in Fig. 7, and the probability distributions are shown in Fig. 8. The estimation accuracy was improved in both frames because integration of the left and right distributions and use of the prior distribution improved the estimation accuracy, as shown in Fig. 8 (c) and (f).

The proposed method including detection ran roughly at 10 fps in the stereo case with tracking.

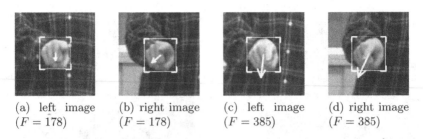

(a) left image
($F = 178$)

(b) right image
($F = 178$)

(c) left image
($F = 385$)

(d) right image
($F = 385$)

Fig. 7. Estimation result in the stereo case with tracking.

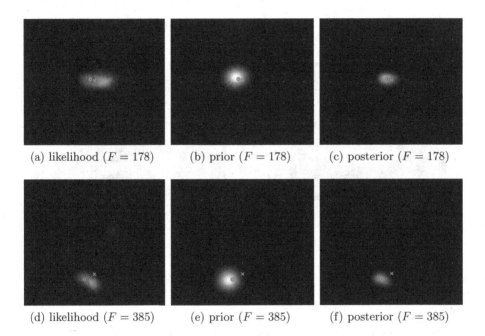

(a) likelihood ($F = 178$)

(b) prior ($F = 178$)

(c) posterior ($F = 178$)

(d) likelihood ($F = 385$)

(e) prior ($F = 385$)

(f) posterior ($F = 385$)

Fig. 8. Pose distributions for the images in Fig. 7. The red double circle indicates the mean of the posterior distribution. (Color figure online)

4.3   Evaluation of Estimation Accuracy

For objective evaluation of the proposed method, the estimation accuracy was measured under various conditions, including the number of cameras, the use of tracking, and different parameter values. The indices of measurement were (1) the mean absolute error (MAE), (2) the median absolute error (Median AE), and (3) the root mean square error (RMSE) of angle estimation, and (4) the success rate, which is the percentage of estimations whose error was below ϵ [deg]. The error was measured using the angle between two vectors, namely, the estimate and the ground truth, and yaw and pitch angle errors were measured when needed.

Table 1. Mean absolute error (MAE), median absolute error (Median AE), and root mean square error (RMSE) of angle estimation, success rate, which is the percentage of estimations whose error was below $\epsilon = 10$ [deg], and mean computation time for one item of input data [μs] with the proposed method (not including feature extraction).

Condition	σ_{pnoise}	MAE	Median AE	RMSE	Success rate	Computation time
Stereo + tracking	3.75	3.37	2.90	4.10	98.3%	427.0
Stereo	-	3.66	3.01	4.70	97.1%	381.5
Monocular + tracking	5	4.24	3.62	5.29	95.6%	79.2
Monocular	-	5.06	3.76	7.42	91.6%	164.5

Table 2. Yaw and pitch angle errors of the proposed method, expressed as (yaw, pitch) in degrees.

Condition	σ_{pnoise}	MAE	Median AE	RMSE
Stereo + tracking	3.75	(2.63, 1.75)	(2.09, 1.40)	(3.47, 2.32)
Stereo	-	(2.80, 1.94)	(2.12, 1.51)	(3.83, 2.85)
Monocular + tracking	5	(3.24, 2.21)	(2.56, 1.62)	(4.36, 3.16)
Monocular	-	(3.77, 2.77)	(2.68, 1.77)	(5.76, 4.84)

The results of error measurement with a varying number of cameras and the use of tracking are shown in Table 1, and the results of measuring the yaw and pitch angle errors are shown in Table 2. These results showed that the use of two cameras greatly improved the accuracy, and that the use of tracking also contributed to accuracy improvement. The MAE of 3.37° and the RMSE of 4.10° may not be sufficient for detailed interaction, but we expect that our method can be used for rough interaction at a distance using hand pointing. The computation time for one item of input data was less than 1 in all the cases, which is fast enough for realtime operation.

4.4   Comparison with Existing Methods

The yaw and pitch MAEs in the monocular case without tracking shown in Table 2 were similar to those reported in other studies [4,6,7]. However, the experimental conditions, such as the type of camera (RGB or depth), the target type (face or hand), the DOFs and their range, and the dataset, were largely different, and we cannot compare the values as they are. The biggest difference between our method and other existing methods is that our method obtains a mixture of normal distributions as an output of the RRF and uses it for estimation.

Therefore, in order to compare our method with existing methods under the same conditions, we measured the estimation accuracy of the method that

Table 3. Accuracy evaluation of existing methods and our method in the monocular case without tracking [deg]. The kernel radius r having the smallest RMSE was selected.

Method	MAE (yaw, pitch)	Median AE	RMSE (yaw, pitch)
Simple averaging	6.21 (4.26, 3.75)	4.79	7.91 (5.78, 5.55)
Mean-shift ($r = 18$)	5.03 (3.70, 2.80)	3.90	7.05 (5.45, 4.63)
Our method	5.06 (3.77, 2.77)	3.76	7.42 (5.76, 4.84)

calculates the arithmetic mean of the means of the output distributions from the regression trees, which is the standard technique widely used in RRFs, and the method that removes outliers in the outputs using the mean-shift algorithm (Epanechnikov kernel), which is the technique used in [4]. In both methods, the outputs having a determinant of the covariance matrix of over 300 were excluded as outliers in order to improve the accuracy.

The evaluation results of existing methods and our method in the monocular case without tracking are shown in Table 3. In the case of simple averaging, all the statistics of the errors were worse than those of our proposed method, which shows that the use of a probability distribution improved the estimation accuracy. In comparison with the method using the mean-shift algorithm, our proposed method showed almost the same performance. However, the method using the mean-shift algorithm cannot be extended to integration of information from two cameras and pose tracking using the probability distribution. The biggest advantage of our proposed method is its versatility in integrating distributions.

5 Conclusion

In this paper, we have proposed a method for recognizing a pointing hand to realize freehand pointing interaction from a distance. Our method estimates the pointing direction of the hand using a Random Regression Forest (RRF), which outputs bivariate probability distribution. In order to improve the estimation accuracy, our method integrates the distributions of the hand poses seen from two cameras, and performs tracking of the hand poses using multi-frame images. The results of a performance evaluation showed that the root mean square error of the angle estimation was 4.10°, which is accurate enough to expect that our proposed method can be applied to user interface systems.

Future work will include further improving the estimation accuracy, expanding the operating space by using cameras with a wider viewing angle, and developing a user interface applications using the pointing hand recognition.

References

1. Leap Motion: https://www.leapmotion.com/
2. Kölsch, M., Turk, M.: Robust hand detection. In: Sixth IEEE International Conference on Automatic Face and Gesture Recognition, pp. 614–619 (2004)

3. Song, J., Sörös, G., Pece, F., Fanello, S.R., Izadi, S., Keskin, C., Hilliges, O.: In-air gestures around unmodified mobile devices. In: 27th Annual ACM Symposium on User Interface Software and Technology, pp. 319–329 (2014)
4. Fanelli, G., Gall, J., Gool, L.V.: Real time head pose estimation with random regression forests. In: IEEE Conference on Computer Vision and Pattern Recognition, pp. 617–624 (2011)
5. Girshick, R., Shotton, J., Kohli, P., Criminisi, A., Fitzgibbon, A.: Efficient regression of general-activity human poses from depth images. In: IEEE International Conference on Computer Vision, pp. 415–422 (2011)
6. Hara, K., Chellappa, R.: Growing regression forests by classification: applications to object pose estimation. In: Fleet, D., Pajdla, T., Schiele, B., Tuytelaars, T. (eds.) ECCV 2014. LNCS, vol. 8690, pp. 552–567. Springer, Heidelberg (2014). doi:10.1007/978-3-319-10605-2_36
7. Zhen, X., Wang, Z., Yu, M., Li, S.: Supervised descriptor learning for multi-output regression. In: IEEE Conference on Computer Vision and Pattern Recognition, pp. 1211–1218 (2015)
8. Oikonomidis, I., Kyriazis, N., Argyros, A.: Efficient model-based 3D tracking of hand articulations using kinect. In: Proceedings of the British Machine Vision Conference, pp. 101.1–101.11 (2011)
9. Keskin, C., Kıraç, F., Kara, Y.E., Akarun, L.: Hand pose estimation and hand shape classification using multi-layered randomized decision forests. In: Fitzgibbon, A., Lazebnik, S., Perona, P., Sato, Y., Schmid, C. (eds.) ECCV 2012. LNCS, vol. 7577, pp. 852–863. Springer, Heidelberg (2012). doi:10.1007/978-3-642-33783-3_61
10. Tompson, J., Stein, M., Lecun, Y., Perlin, K.: Real-time continuous pose recovery of human hands using convolutional networks. ACM Trans. Graph. **33**, 169:1–169:10 (2014)
11. Sharp, T., Keskin, C., Robertson, D., Taylor, J., Shotton, J., Kim, D., Rhemann, C., Leichter, I., Vinnikov, A., Wei, Y., Freedman, D., Kohli, P., Krupka, E., Fitzgibbon, A., Shahram, I.: Accurate, robust, and flexible real-time hand tracking. In: 33rd Annual ACM Conference on Human Factors in Computing Systems, pp. 3633–3642 (2015)
12. Schick, A., van de Camp, F., Ijsselmuiden, J., Stiefelhagen, R.: Extending touch: Towards interaction with large-scale surfaces. In: ACM International Conference on Interactive Tabletops and Surfaces, pp. 117–124 (2009)
13. Hu, K., Canavan, S., Yin, L.: Hand pointing estimation for human computer interaction based on two orthogonal-views. In: 20th International Conference on Pattern Recognition, pp. 3760–3763 (2010)
14. Breiman, L.: Random forests. Mach. Learn. **45**, 5–32 (2001)
15. Criminisi, A., Shotton, J., Konukoglu, E.: Decision forests: a unified framework for classification, regression, density estimation, manifold learning and semi-supervised learning. Found. Trends Comput. Graph. Vis. **7**, 81–227 (2012)
16. Ali-Löytty, S., Niilo, S.: Gaussian mixture filter in hybrid navigation. In: European Navigation Conference, pp. 831–837 (2007)
17. Dalal, N., Triggs, B.: Histograms of oriented gradients for human detection. In: IEEE Computer Society Conference on Computer Vision and Pattern Recognition, vol. 1, 886–893 (2005)

Fingertips Tracking Algorithm for Guitarist Based on Temporal Grouping and Pattern Analysis

Zhao Wang[⊠] and Jun Ohya

Department of Modern Mechanical Engineering, Waseda University,
Shinjuku, Japan

Abstract. This paper proposes a temporal Grouping and pattern analysis-based algorithm that could track the fingertips of guitarists during their guitar playing towards the actualization of the automatic guitar fingering recognition system. First a machine learning-based Bayesian Pixel Classifier is used to segment the hand area on the test data. Then, the probability map of fingertip is generated on the segmentation results by counting the voting numbers of the Template Matching and Reversed Hough Transform. Furthermore, a temporal Grouping algorithm, which is a geometry analysis for consecutive frames, is applied to removal noise and group the same fingertips (index finger, middle finger, ring finger, little finger). Then, a data association algorithm is utilized to associate 4 tracked fingers (index finger, middle finger, ring finger, little finger) with their correspondent tracked results frame by frame. Finally, particles are distributed only between the associated fingertip candidates to track the fingertips of guitarist effectively. The experimental result demonstrates that this fingertip tracking algorithm is robust enough for tracking fingertips (1) without any constrains such us color marker; (2) under the complex contexts, such us complicated background, different illumination conditions, (3) with the high tracking accuracy (mean error 3.36 pixels for four fingertips).

1 Introduction

String instruments with fixed fret positions such as guitar offer multiple alternative positions, i.e. string-fret combinations where a given note can be played. Each position can furthermore be played by any of the four left hand fingers and considering that multiple notes are played at the same time, it becomes apparent that there exists a large number of possible fingering alternatives for a single set of notes [1]. Therefore, fingering is a cognitive process that maps each note on a music score to a fingered position on some instrument. Fingering involves several competences: (i) musical analysis (including both structural and aesthetical issues), for the interpretation of the notes in input, (ii) physical constraints, posed by the instrument where the notes have to be played, (iii) bio-mechanical constraints, which characterize the possible figures of the hand [2].

© Springer International Publishing AG 2017
C.-S. Chen et al. (Eds.): ACCV 2016 Workshops, Part III, LNCS 10118, pp. 212–226, 2017.
DOI: 10.1007/978-3-319-54526-4_16

The idea of applying computer vision-based expert systems toward the guitar fingering problem is found in [1,3] which reports a number of rules used by expert guitarists in decision making and proposes the optimum path paradigm as a suitable method for applying these criteria and calculating the optimal fingering solution. The reason why researchers did this fingering problem by using computer vision method instead of audio method is that as we mentioned before, a large number of possible fingering alternatives exist just for a single set of note, which means even though audio signal may sound right, but the real fingering is wrong, and because of this, we must analyze whether the user of our system is right or wrong from computer vision's method. As we know, learning how to play guitar is very interesting, but difficult, because it involves many techniques: reading scores, pressing strings, sweeping, slipping and so on so forth. Going to a guitar teacher is one of the most useful and effective methods to learn, but it is very difficult to go very frequently such as every day. However, if a computer at home can autonomously teach how to play the guitar, it is very convenient, and the progress could be very fast. Such an automatic guitar teaching system should be able to teach many techniques such as the ones mentioned above. In order to achieve the system, this paper focuses on one fundamental but significant module: specifically, fingertip tracking as we mentioned before, analyzing the fingering of guitarist is the problem of the optimum path of the fingertips, it is not only about whether guitarist play the right note, press the right place, but also whether he moves the fingers in a right manner.

Motokawa and Saito [4] built a system called Online Guitar Tracking that supports a guitarist by using augmented reality. This is done by showing visual aid information (i.e. the virtual fingers model) on a real stringed guitar and this becomes an aid to learning to play the guitar. Online Guitar Tracking uses augmented reality to detect the guitar so that the player can learn how to hold the strings of the guitar by overlapping the player's hand onto a manual model. Joseph Scarr and Richard Green [5] proposed an algorithm that uses a markerless approach to successfully locate a guitar fretboard in a webcam image, normalise it and detect the individual locations of the guitarist's fretting fingers. Preliminary testing of this system shows that it is more accurate at note recognition. Burns and Wanderley [6] detected the positions of fingertips for the retrieval of the guitarist fingering without markers. They used the circular Hough transform to detect fingertips. By fixing a camera on the guitar neck, the guitar neck and the camera are relatively static, but the fixed camera brings inconveniences to guitar players. Kerdvibulvech and Saito [7–9] proposed a series of novel approaches to detect the position of the player's fingers such as stereo cameras are used to compute the 3D positions of fingers using the color markers attached on fingertips, and template matching is used to localize the fingertip positions and etc. All of the research fixed some problem in guitar fingering teaching such as chord recognition, guitar neck tracking and finger detection. But it is obvious that some research works researched computer vision-based methods for supporting guitarists to improve their skills. However, there are also limitations in their research: (1) some of them use inconvenience tool, such us color markers, neck-fixed camera and ARTag; [4,6,7] (2) Instead of tracking the guitar neck and fingers to obtain the comprehensive fingering of guitarist, some of them only detect guitar neck [5,6] or finger [5,6,8] in certain frames so their work cannot analyze the real fingering.

Compared with other multiple target tracking problems such as pedestrian tracking or vehicle tracking, there are several difficulties in guitarist's fingertips tracking: (1) each fingertip has similar features such as shape and color compared with other fingers; while, every pedestrian or vehicle may differ at its appearance in the same scene; (2) during the playing of guitar, the fingertips move without any regular pattern, they could be at any places on the fretboard of guitar based on the score; while, pedestrian or vehicle's movement always follows some rules such us moving at a nearly constant speed in a straight line or at a nearly constant acceleration, which makes the tracking easy to be estimate; (3) self-occlusions of fingertips happens during the guitar play. Our work aims at solving all these difficulties in guitar fingering recognition by detecting the fingertip candidates, which share the nearly same appearance and shape with multiple weighted features, applying a reasonable data association method to find the correspondent fingertips in consecutive frames, and tracking the fingertips.

Section 1 introduces our research background and some related work, Sect. 2 presents the overview of our fingertip tracking system, Sect. 3 covers the detailed algorithm of our fingertip tracking module, Sect. 4 describes the experimental results, analyzes the accuracy of our fingertip tracking module, Sect. 5 gives the conclusions of the research and discusses the future work.

2   Overview of the Proposed Method

As depicted in Fig. 1 (module is shown in rectangle; data is shown in circle), first, the input of this fingertip tracking module is the output of another guitar neck tracking system done by our previous work [11]. Generally speaking, that guitar neck tracking system includes three steps: (1) guitar neck detection; (2) guitar neck tracking; (3) normalization of the guitar neck, and this system tracks guitar neck frame by frame by tracking the feature points detected on the fretboard of guitar neck. After inputting the tracked guitar neck result, first a machine learning-based Bayesian Pixel Classifier [8,12] is used to discriminate the hand area to acquire the hand segmentation area. Then, the probability map of fingertip is generated on segmentation results by counting the voting numbers of the template matching or Reversed Hough Transform. Here, both of template matching and reversed Hough Transform are used as weighted features to extract the fingertip candidates. On the generated probability map, the higher intensity pixels indicate the pixels with higher probability to be the fingertips. Furthermore, a temporal Grouping algorithm, which is a geometry analysis for buffer images (10 adjacent frames), is applied to removal noise and group the same fingertips (index finger, middle finger, ring finger, little finger) on the successive frames, and the ROIs are drawn only at the position of the clustered fingertip candidates. Then, a data association algorithm is utilized to associate

4 tracked fingers (index finger, middle finger, ring finger, little finger) with their correspondent trajectories frame by frame. We apply this data association algorithm by defining four pattern for tracked fingertips movement during the whole process: the active pattern, new track pattern, vanished track and consecutive noise pattern [9], and all the tracked trajectories of fingertip are fitted into these fours patterns in order to solve the problem such as self-occlusion mentioned in Sect. 1. Finally, we use particle filter to track the fingertips of guitarist in case of complex and volatile hand gesture change by continuously distributing particles within the associated ROIs of fingertips on every frame.

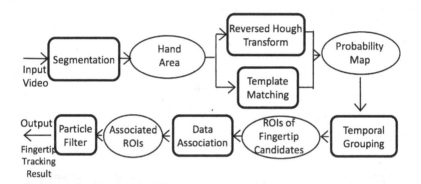

Fig. 1. Diagram of our proposed method

3 Fingertip Tracking Module

3.1 Input of Fingertip Tracking Module

As mentioned in Sect. 2, the input of our fingertip tracking module is the output of an another guitar neck tracking system published by Wang and Ohya [10]. Generally, this guitar neck tracking system applied an optical-flow based tracking method to track the guitar neck, and they also developed a solution which can recover occluded feature points by analyzing geometric relationship between feature points. The output of this guitar neck tracking system is shown in Fig. 3, the red rectangle in Fig. 2 shows the tracking result of the guitar neck, and Fig. 3 shows the projection result of guitar neck tracking: no matter how the guitarist shakes the guitar during the play, the guitar neck area would always be projected to the center of a new image sequence with the resolution of 1300 pixels by 300 pixels. The reason why we do the projection process is to analyze the fingering (the green contours in Fig. 3) in the following fingering analysis process.

3.2 Hand Segmentation by Bayesian Pixel Classifier [8, 12]

One of the most difficult problems in hand segmentation hand area cannot be discriminated from the whole image by simply specifying a color threshold, because

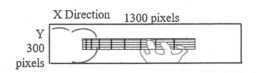

Fig. 2. The conceptual image of the input of our fingertip tracking module in the guitar teacher system. (Color figure online)

Fig. 3. The projection of the neck tracking result. (Color figure online)

the skin color varies a lot under different illuminations. Instead of manually setting up a threshold, a Bayesian Classifier [8,12] is adopted to help to separate skin-color pixels from non-skin-color pixels.

During the off-line learning process, first hand areas are manually depicted in the training data set that only the hand areas appear in their original colors, while non-hand areas are black. During the on-line testing process, every pixel in a testing image is scanned to calculate the possibility to be the hand-color pixel. The segmentation process is shown from Figs. 4, 5, 6 and 7. The detailed algorithm is shown in [8,12].

Fig. 4. An example of test data

Fig. 5. Testing result of Bayesian pixel

Fig. 6. An example of test data

Fig. 7. Testing result of Bayesian pixel

3.3   Probability Map Obtained by Template Matching and Reversed Hough Transform

After the hand segmentation, the next step is to localize the position of each fingertip on every frame because as we mentioned in Sect. 1, the evaluation of guitarist's fingering is actually the problem of analyzing the optimal path of each fingertip. Therefore, in order to analyze the fingering, first we have to localize the position of each fingertip candidate from our segmentation result.

The related works detect fingertips mainly using two methods, (a) semi-circle detection [4,6], because fingertips share the semi-circle shape compared with other parts of hand (b) template matching [7,8], with the predefined fingertip models, we can match the segmentation result with those models to localize the position of fingertips. Our algorithm detect candidates of the fingertips from the hand segmentation result by applying both reversed Hough Transform and template matching, and both of them are weighted, because based on our test results, this combined weighted method works much better than any single one. The formula is shown as follows:

$$Fin_{x,y} = \alpha T_{sum(x,y)} + (1 - \alpha)R_{sum(x,y)} \tag{1}$$

where, $Fin_{x,y}$ indicates the fingertips detection result at pixel (x, y) in the segmentation result. $T_{sum(x,y)}$ and $R_{sum(x,y)}$ indicate the result of template matching and reversed Hough Transform result at pixel (x, y) respectively, a indicate the weight. From Eq. (1), the candidates of fingertips are acquired based on two weighted parts, one is the result of template matching, and the other is the result of reversed Hough transform.

For the part of reversed Hough Transform, after segmenting the hand area (Fig. 9), we test the probability that every pixel in the hand segmentation result is the center of a set of circles (Fig. 10). The counts of intersected pixels between the set of circles and the hand contour are saved in an accumulator that has the same size with the Hand Segmentation result. Figure 12 shows the way to save the count of intersections. A new accumulator that has same size as the hand segmentation result is prepared to record the count. In Fig. 12, each cell indicates every single pixel; the red points indicate the position of the fingertips; every pixel records the max count of intersection when a set of circles (radius varies from 15 to 25) is drawn as this pixel to be the center of the set of circles. From Fig. 11, it is clear that the pixels around fingertips own higher count than other pixels, because fingertips have semi-circle shape, and fingertips contours have more intersection pixels with circles. The $R_{sum(x,y)}$ in Eq. (1) is calculate as follow

Fig. 8. Input image   **Fig. 9.** Hand segmentation

Fig. 10. Reversed Hough Transform   **Fig. 11.** Pixels intersect at fingertip

$$\sum_{r=15}^{25} \sum_{x',y'} \left[(x', y') \epsilon l \bigcap (x - x')^2 + (y - y')^2 = r \right] \tag{2}$$

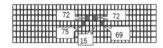

Fig. 12. A same size image recording the count of the intersection. (Color figure online)

where, (x', y') is a searched circle centered at coordinates (x, y) with the radius of r, l is the set of hand contour pixels (Fig. 8).

For Template Matching, we build our work based on Kerdvibulvech's work [9]. We use six templates to localize the fingertip candidate's position considering the variance of fingertip during the guitar playing. The reason why we use six template models to fit our segmentation result is that the fingertip shape can be approximated with a semicircular shape while the other hand parts are roughly straight. The template models are shown in Fig. 14. The $T_{sum(x,y)}$ in Eq. (1) is calculate as follow:

$$T_{sum(x,y)} = \sum_{i=0}^{N_0} \frac{\sum_{x',y'} [T(x', y') - H(x + x', y + y')]^2}{\sqrt{\sum_{x',y'} (T(x', y')^2) \sum_{x',y'} H(x + x', y + y')^2}} \tag{3}$$

where $T(x, y)$: a searched template at coordinates (x, y) and $H(x, y)$ is the current frame where the search is running, N_0 is the number of the fingertip models for template matching. Finally, the normalized result of $Fin_{(x,y)}$ in Eq. (1) (Fig. 13):

Fig. 13. Hand segmentation result

Fig. 14. Six templates with different sizes and orientations of fingertipsl

$$Fin_{Normal} = \frac{Fin_{x,y}}{Fin_{x_{max},y_{max}}} * 255 \tag{4}$$

where (x_{max}, y_{max}) is the position of the largest template matching result on the frame The result of fingertip localization called as probability map [9] is shown as a new image with the same resolution as the hand segmentation result depicted in Fig. 15. In the probability map, the intensity of each pixel indicates the probability of the same coordinate pixel in hand segmentation to be the fingertip (higher intensity equals higher probability to be the fingertip).

3.4   Obtaining Fingertip Candidates by Temporal Grouping

The probability map cannot always generate good results that accurately locate positions of fingertips; sometimes noises also own higher probability, as shown in

Fig. 15. Probability map based on template matching and reversed Hough Transform

Fig. 15. In order to remove noises and group the same fingertips in consecutive probability maps, we implement a Grouping algorithm [6] by focusing on the consecutive frames on every single frame. Figure 16 shows a conceptual image of the temporal grouping algorithm. The gray and the red areas in the left side indicate the pixels that have a higher probability to be fingertips (the red circles indicate noises). On consecutive frames (Fig. 16), the same color lined fingertip candidates are linked into a group, because in time sequence, (for instance, at frame t, we focus on 3 previous frames from Frame t to Frame t-2), these grouped candidates have close Euclid distance (smaller than 10 pixels) in the consecutive frames. On the other hand, when candidates come from noise, as red circle in the left side, no candidates on consecutive frames can be grouped with it even though it has a high probability. Finally, we generate the ROIs (rectangles with different colors in Fig. 16) only at the positions of the clustered fingertips. We use the temporal grouping algorithm on every frame in order to (1) generate the ROIs for clustered fingertip candidates; (2) remove the high probability noise which has the high probability while cannot be clustered in consecutive frames.

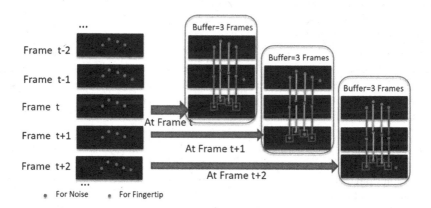

Fig. 16. Conceptual image of clustering algorithm. (Color figure online)

3.5   ROI Association and Fingertip Tracking

In this multiple objects tracking problem, we have to find the correspond fingers in consecutive frame in order to decide where to distribute the particles. In this fingertips tracking algorithm, we generate ROIs in temporal Grouping algorithm to distribute particles within the associated ROIs. But, in real guitar playing

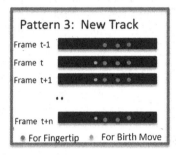
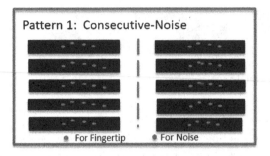
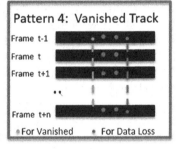
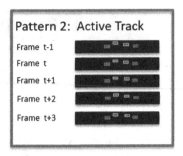

Fig. 17. Conceptual images of four patterns. (Color figure online)

situation, the problem such as finger-self-occlusion, finger-frame-out, noise etc. happen frequently, which means sometimes, we cannot find the correspondent ROIs in consecutive frames. Therefore, it is hard to decide how to distribute the particles to track the fingertip later.

In related work, Chutisant [9] and Ng [10] solved this data association by monitoring two vector frame by frame: (1) Track-to-region vector (2) Region-to-track vector. They tried to find the data association in consecutive frames for these multiple targets tracking problem by calculating these two vectors to find the relationship between the ROIs detected on the current frame and the trajectories tracked before. But their work cannot solve the problem mentioned before, for example their work cannot recover the track after finger-self-occlusion happens.

The four pattern are shown in Fig. 17. In our work, we fix the problem by analyzing and classifying all the generate ROIs (in Temporal Grouping) into 4 patterns: (1) consecutive-noise pattern, (2) new track pattern, (3) vanished track (4) active pattern in order to find the correspondent ROIs in consecutive frames and distribute particles only within the associated ROIs.

(1) Consecutive-Noise Pattern. In order to discriminate from noise (randomly appeared in Temporal Grouping algorithm), we name this type of noise as consecutive-noise, because this type of noise appears nearly at the same positions on consecutive frames and it cannot be removed in temporal Grouping algorithm. As an example of consecutive-noise pattern shown in the up-left image sequences of Fig. 17, on the adjacent frames from Frame t-1 to Frame

t+n, the noise is detected from Frame t, lasts for n frames, and disappear on Frame t+n. This noise pattern obviously cannot be removed based on temporal Grouping, because this noise pattern lasts long enough (n frames) so the Group classifies the noise as the true fingertip. Therefore, when we detected this noise pattern, if it lasts equal or shorter than n frames, we classify them into the consecutive noise pattern.

On the other hand, sometimes at the position of only one fingertips we detect two fingertips because of noise as shown in the up-left image sequence of Fig. 17, we call this pattern as duplicate pattern. We solve this pattern by setting a threshold N, and if within the area of N, two fingertips are detected, we remove the one that has lower probability (low intensity in probability map) to be the fingertip.

(2) New Track Pattern. This pattern is used to solve the problem such us frame-in or newly started fingertips tack as shown in Fig. 10, and is similar to the noise pattern. As shown in Fig. 17, if the new target appears, and lasts for more than n frames, we classify them as the new track pattern.

(3) Vanished Tack Pattern. This pattern is used to solve the problem such us self-occlusion and frame-out. As shown in Fig. 17, from Frame t, two fingertips candidates in probability map disappear (the orange track and the blue track), but from Frame t+n, the blue track recovers and the fingertip candidate starts to show up around the candidate detected at Frame t−1, obviously we cannot classify them into new track because of data loss at the fingertips detection based on reversed Hough Transform and template matching. On the other hand, the orange track cannot find any candidates from Frame t to Frame t+n, we classify them into the vanished track because of frame-out problem during the guitar playing.

(4) Active Pattern. This pattern, which indicates on every frame, all the candidates of fingertips are grouped based on Grouping, and all the tracks associate the candidates of fingertips detected on the current frame.

Figure 18 shows the diagram of four patterns-based processing. On any current Frame N, we denote that there are total M_N ROIs were generated in temporal grouping algorithm. ROIs need to be associated with their predecessors (ROIs on previous frame). First, based on the comparison between M_N and M_{N-1} (the number of ROIs generate in previous frame), (1) if $M_N = M_{N-1}$, we determine each ROI belong to active pattern by calculating the Euclid distance between each two associated ROIs on adjacent frames. If all the distances $d(M_N) \doteq d(M_{N-1})$ between each two ROIs (with the shortest distance in adjacent frame) are smaller than the threshold, we consider these ROIs belong to active Pattern. On the other hand, if the distance of any two ROIs (with the shortest distance in adjacent frame) is larger than threshold, we consider these two ROIs belong to Noise Pattern (the ROI on latter frame) and Vanished Pattern (the ROI on former frame). (2) If $M_N > M_{N-1}$, which indicates that we generate more ROIs than previous frame, we determine this pattern belongs to noise pattern or new track pattern based on their life span l, if the surplus ROIs (because of MN > MN-1) on the frame last more than l frames, we

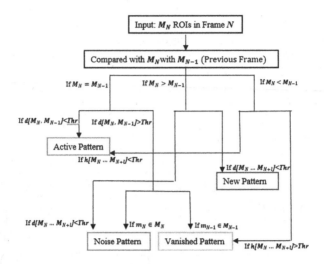

Fig. 18. Four patterns-based processing

consider these ROIs as new track, or as the noise pattern. (3) If $M_N < M_{N-1}$, which indicates that we generate less ROIs than previous frame, we determine this pattern belongs to active pattern or vanished pattern also based on their life span l. Still, if the missing ROIs (because of $M_N < M_{N-1}$) on the frame keep missing more than l frames, we consider these ROIs as vanished pattern, or as the active pattern, because of temporary fingertip missing, such us self-occlusion. So we associate the ROIs on every frame based this method shown in Fig. 18 in order to find the correspond ROIs on consecutive frames, and distribute the particles only within the associated ROIs to track the fingertips. As we mentioned before, because of the diversity and variability (the self-occlusion, frame-in, high-speed-movement and etc.) of the hand of the guitarist during the guitar playing, it is very hard to track the fingertips of guitarist. Meanwhile, each fingertip has similar feature such as shape and color compared with other fingers which makes the tracking problem become more difficult compared with other multiple targets tracking problem. By classifying ROIs to these four pattern of ROIs on every frame, we can associate the correspondent ROIs on consecutive frames, and distribute the particles only between the associated ROIs to track the fingertips of guitarist effectively. The particle filter is totally implemented on the work of Ng [10].

4   Evaluation

In this experiment, we used the guitar playing data collected from four experimental participants, and four videos were taken for each participant, and the duration of each video lasts nearly 20–30 s. Totally, 16 videos (5500 frames) were used to test the validity of the fingertips tracking system.

Fig. 19. The examples of experimental input and condition (textured clothing, complicated background, dim illumination, different participants)

All the test data are taken with iPhone front camera with the compressed resolution of 480 pixels by 300 pixels. All the videos were taken under different illumination conditions (daylight, incandescent lamp and etc.) with complicated background and different subject's clothing, as shown in Fig. 19. The system used for testing was a Macbook Pro machine with a 2.7 GHz i7 processor and an Nvidia GeForce 650 m GT graphics card. The algorithm was implemented in C++, using Xcode 6.4 and version 2.8 of the OpenCV library.

In these 16 test video, we let the participants play two musical pieces: (1) Symmetrical Exercise and (2) Scales (24 majors or Minor because it has been proved these two daily guitar excises help the players to develop their stamina and rhythm abilities, meanwhile developing valuable muscle memory in their picking hand, and no matter novice or professional player yielded significant positive results by practicing these two pieces of notes.

Table 1 shows us the comparison of fingertips tracking accuracy between the related work and our work. The ground truth is manually observed by human eye, while the tracking result is obtained by the fingertips tracking system. With the mean error 7.25, 2.2, 1.75, 2.25 pixels for fore finger, middle finger ring finger and little finger respectively, it is obvious that our work is accurate enough for tracking the fingertip of guitarist during guitar playing.

Table 1. The comparison of tracking accuracy (mean error) between related work and our work (accuracy for tracking four fingertips: 5500 frames)

Tracking error	Forefinger	Middle finger	Ring finger	Little finger	Mean
Our work	7.25	2.2	1.75	2.25	3.36
Our previous [11]	10.2	7.4	5.56	6.15	7.4
Kerdvibulvech [8]	17.15	13.25	10.83	6.71	11.98
Italic					

Figures. 20, 21, 22 show the vision comparison between our work and our previous method [11] for a same image sequence with 518 consecutive frames. From Fig. 15 we can figure out compared with the related work, our tracking result (dotted colorful line) tracks the real fingertips (full line) more accurately.

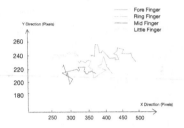

Fig. 20. Ground truth trajectory of fingertips

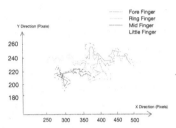

Fig. 21. The comparison between our work and ground truth (dotted line indicates tracking result)

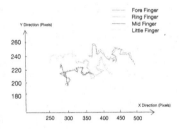

Fig. 22. The comparison between our previous method [11] and ground truth

5   Conclusion

This paper proposed a fingertips tracking algorithm within the automatic guitar fingering teaching system. First a machine learning-based Bayesian Pixel Classifier is used to segment the hand area on the test data Then, the probability map of fingertip is generated on segmentation results by counting the voting numbers of the Template Matching and Reversed Hough Transform, and on the probability map, the higher intensity pixels indicate the pixels with higher probability to be the fingertips. Furthermore, a Grouping algorithm, which is a geometry analysis for buffer images (10 adjacent frames), is applied to removal noise and group the same fingertips (index finger, middle finger, ring finger, little finger) on the successive frames, and based on the Grouping results, we draw the ROIs only at the position of the clustered fingertip candidates. Then, a data association algorithm is utilized to associate 4 tracked fingers (index finger, middle finger, ring finger, little finger) with their correspondent tracked result frame by frame. We apply this data association by defining four pattern for fingertips movement during the whole process: the active pattern, new track pattern, vanished track pattern and consecutive-noise pattern, and every tracked trajectory of fingertip is fitted into these fours patterns in order to solve the problem such us self-occlusion and frame-out mentioned in Sect. 1. Finally, we used particle filter to track the fingertip of guitarist in case of complex and volatile hand gesture change by continuously distributing the particles between associated ROIs of fingertips generate before.

The experiment result demonstrates that this fingertip tracking algorithm is accurate enough for tracking fingertips (1) under the complex contexts, such as complicated background, different illumination conditions; (2) with the high tracking accuracy (the mean error 7.25, 2.2, 1.75, 2.25 pixels for fore finger, middle finger, ring finger and little finger respectively).

In the future, First, machine-learning-based fingering evaluation system is very necessary. The features that can comprehensively and thoroughly express the fingering of guitar and hand need to be selected. Also, the classifier such as HMM model will be used to discriminate good fingering or not in the future.

If the research mentioned before could be finished in the future, the fingering of guitarist could be analyzed comprehensively and thoroughly.

References

1. Radisavljevic, A., Driessen, P.: Path difference learning for guitar fingering problem. In: Proceedings of the International Computer Music Conference, vol. 28 (2004)
2. Radicioni, D., Lombardo, V.: Guitar fingering for music performance strings. **40**(45) (2005)
3. Sayegh, S.I.: Fingering for string instruments with the optimum path paradigm. Comput. Music J. **13**(3), 76–83 (1989)
4. Motokawa, Y., Saito, H.: Support system for guitar playing using augmented reality display. In: Proceedings of the Fifth IEEE and ACM International Symposium on Mixed and Augmented Reality (ISMAR 2006), pp. 243–244. IEEE Computer Society (2006)
5. Scarr, J., Green, R.: Retrieval of guitarist fingering information using computer vision. In: 25th International Conference on Image and Vision Computing New Zealand (IVCNZ), Queenstown, New Zealand, 8–9 November 2010, pp. 1–7 (2010). ISSN2151-2191
6. Burns, A.: Visual methods for the retrieval of guitarist fingering. In: Proceedings of the Conference on New Interfaces for Musical Expression, Paris, France, pp. 196–199 (2006). ISBN: 2-84426-314-3
7. Kerdvibulvech, C., Saito, H.: Real-time guitar chord estimation by stereo cameras for supporting guitarists. In: Proceedings of 10th International Workshop on Advanced Image Technology 2007 (IWAIT 2007), Bangkok, Thailand, pp. 147–152, January 2007
8. Kerdvibulvech, C., Saito, H.: Guitarist fingertip tracking by integrating a Bayesian classifier into particle filters. Int. J. Adv. Hum.-Comput. Interact. (AHCI), 10 p. (2008). Hindawi Publishing Corporation. ISSN 1687-5893

9. Kerdvibulvech, C., Saito, H.: Markerless guitarist fingertip detection using a Bayesian classifier and a template matching for supporting guitarists. In: Proceedings of the 10th ACM/IEEE Virtual Reality International Conference, VRIC 2008, Laval, France (2008)

10. Ng, W., Li, J., Godsill, S., Vermaak, J.: A hybrid approach for online joint detection and tracking for multiple targets. In: Proceedings of IEEE Aerospace Conferences, pp. 2126–2141 (2005)

11. Zhao, W., Ohya, J.: Research on detecting and tracking the guitar neck and guitarist's fingers from a video sequence. EI (Electron. Imaging), D-11-12, San Francisco (2016)

12. Kovac, J., Peer, P., Solina, F.: Human skin color group for face detection, vol. 2. IEEE (2003)

13. http://www.guitarworld.com/three-steps-shred-fundamental-daily-practice-techniques-about-15-minutes

Intuitive Pointing Position Estimation for Large Scale Display Interaction in Top-View Depth Images

Hye-mi Kim, Daehwan Kim, Yong Sun Kim, and Ki-Hong Kim[✉]

Electronics and Telecommunications Research Institute, Daejeon, Korea
kimgh@etri.re.kr

Abstract. In this paper, we propose an intuitive pointing position estimation method for large scale display interaction in top-view depth images. The depth sensor is mounted above the users' head in order to avoid the sensor occluding the display. In order to estimate the pointing position, we detect the user's head and estimate the position of the user's eye. To calculate the center of the head, we propose a head segmentation method. We use an iterative binary partitioning method and a one-to-one correspondence method to detect and track the hands, respectively. The 3D positions of the head and hands were converted to the real world coordinates and the pointing position was estimated on the eye-hand ray intersecting with the large screen. Experimental results show that we improve the head detection rate applying our head segmentation method. Also, we calculate the pointing direction accuracy and the proposed method has a good performance compared with conventional methods even in dark environments.

1 Introduction

As displays become larger in size and resolution, a user interface for large scale displays is needed. It is difficult to use a mouse on large scale displays because it is painstaking to find the mouse cursor. Furthermore, large scale displays provide immersive environments; thus, the user interface should also provide them. The best interface for users that provides an immersive environment is free hand or full body interactions that do not require auxiliary devices to be worn.

In using large display interfaces, direct manipulation through pointing and selecting remains the dominant interaction paradigm in conventional user interfaces [14]. The estimated pointing position corresponds to the position of the mouse pointer and selecting the indicated object corresponds to clicking the mouse button. However, using body gestures, the selecting gesture can be easily implemented via pausing with one's arm outstretched, whereas implementation of an intuitive pointing interface is comparatively complicated. The conventional pointing gesture recognition method is counterintuitive; that is, the user's outstretched hand is regarded as the mouse pointer in the predefined area.

In this research, we propose the intuitive pointing position estimation method without any instruction to the users to point at the target. The rest of this

© Springer International Publishing AG 2017
C.-S. Chen et al. (Eds.): ACCV 2016 Workshops, Part III, LNCS 10118, pp. 227–238, 2017.
DOI: 10.1007/978-3-319-54526-4_17

Section, we discuss the related works and the approach of our method. We represent the head and hand detection method in Sects. 2 and 3, respectively. In Sect. 4, the pointing position estimation method is represented and the experimental results are shown in Sect. 5. Conclusion and future works are drawn in Sect. 6.

1.1 Related Works

Humans usually use their fingers or hands when they point at something far apart from themselves. Specifically in [13], Taylor reports users tend to point to the target by placing their fingertip just lateral to the eye target line. However, in most of the conventional methods which treat pointing estimation, only finger or arm direction is used for pointing. In [2], they used an adapted Adaboost cascade detector for hand gesture detection and an Active Appearance Model to estimate the pointing direction from stereo views. In [14], Vogel and Balakrishnan estimated the pointing direction using only the direction of the index finger and Lee and Lee [5] also recognized that the user is pointing or not only with the index finger. Schick et al. [10] found the voxels that belong to the arms of the users based on 3D reconstruction using standard RGB cameras.

Several researches tracked not only hands but also the head to determine the pointing direction [1,6,8,15]. Ball et al. [1] performed interaction with a large scale display using a wireless Gyration GyroMouse for direction. To track physical navigation in 3D space, they used a VICON vision-based system to track the users' head. However, as mentioned above, additional devices are inconvenient to wear and it is not intuitive to use them although they can enhance the accuracy of the pointing estimation. Nikel and Stiefelhagen tracked the head and hands using a fixed-baseline stereo camera based on the disparity map and the skin color map [6]. Pateraki et al. estimated a pointed target by fusing information regarding hand pointing gestures and the pose of the user's head [8]. However, they only considered whether several targets in the environment were correctly identified rather than measuring the accuracy of the pointing estimation precisely.

1.2 Proposed Approach

For the large scale display, a flat panel display, such as TV, can be used. However, there is a limit to the TV screen size. While several televisions can be used to compose a large scale grid panel display, there are bezels between flat panel displays and the images cannot be visualized seamlessly.

In this paper, a large screen with projectors is used to implement the immersive display. The projection on the screen becomes clearer as the location becomes darker. While the color information changes significantly according to the illumination conditions as depicted in Fig. 1(a) and (c), the depth of the images retain the 3D features as depicted in Fig. 1(b) and (d).

In order to estimate the precise pointing position even though it is dark, the user's head and hands are detected and tracked using the depth information without the color information. As a RGBD sensor to capture the depth

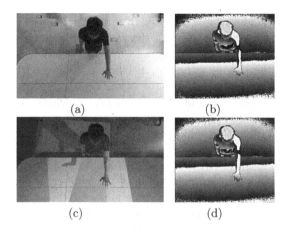

Fig. 1. A color image (a) and a depth image (b) when the light is on. A color image (c) and a depth image (d) when the light is off. (Color figure online)

images, Kinect was used. Kinect was mounted above the user's head pointing toward the ground in order to avoid the sensor occluding the large scale display and disrupting the user's concentration on the screen. Microsoft Kinect SDK provides skeleton tracking results when the sensor is positioned in front of the user. In this paper, we introduce the head and hand detection method in depth images captured from the Kinect mounted above the user's head and estimate the pointing position by the eye-hand ray instead of the index finger direction or the arm direction.

2   Head Detection and Segmentation

First, the holes in the depth images captured from the commercial RGBD sensor are filled. However, if the obtained depth values were changed significantly as a result of the hole filling process, accurate pointing position estimation is impossible. In order to minimize changes in the original depth values, a median filter was used. As a result, the original accurate depth values obtained from the sensor were retained and the holes were filled.

2.1   Feature Extraction

In [12], Tang et al. proposed a feature, the Histogram of Oriented Normal Vectors (HONV), designed specifically to extract local 3D geometric characteristics for the purpose of object recognition with a depth sensor. Since the surface of an object can be described using the orientation of its tangent plane at every surface point, the local 3D geometry characteristics can be represented as a local distribution of the normal vector orientation on the tangent plane. The normal vector orientation can be represented as an ordered pair of azimuthal angle and zenith angle which can be easily computed from the gradients of the depth image. HONV is formed as a concatenation of the local histograms of azimuthal angle and zenith angle and is used as a feature for object detection tasks.

In this paper, the shape of the search block was changed from a rectangle to a circle in order to optimize the shape of the search block for enhancing the head detection rate: the shape of the human head is closer to a circle than a rectangle. When the shape of the search blocks are rectangles, the pixels that represent the background or shoulders can be included in the block and this noise affects the histogram, which forms the features for detection. In order to normalize the histogram of the cells with different shapes, the L1 norm is used through dividing every histogram value into the number of pixels in each cell. The detection window is divided into m by n non-overlapped cells. For each cell, the orientation of the normal vector at each pixel is quantized and voted into a 2D histogram $I \times J$ feature vectors from each cell applying bilinear interpolation for avoiding boundary effects. We concatenate the HONV feature from each cell and get the final feature representation of the detection window.

2.2   Head Detection and Segmentation

These feature vectors are used in conjunction with the class labels to train the linear support vector machine (SVM) classifier. The depth pyramid is computed in order to detect the various sizes of the unspecified users. First, the body part is detected in which the head is detected in order to operate the overall system in real time. The window size of the head is fixed as 33×33.

For the post bounding box selecting, we find the block in which the brightest depth pixel is included. The pixel of the head always has higher pixel values than the shoulders when users are standing and pointing at the target in front of the display.

In order to calculate an accurate center position of the head in the depth image, the head area should be segmented with the shoulder and background removed. The distance between the closest pixel to the sensor and every pixel in the block is calculated. If this distance is lower than the threshold, the pixel is assigned to the head area. The average of each pixel assigned to the head area is calculated, and this is the three-dimensional head position. The seed point is set to the closest pixel from the sensor instead of the center of the head candidate block in order to calculate an accurate center of the head, not only when the head is well-detected but also when it is not detected properly. This is discussed in more detail in Sect. 5.1. The threshold of the distance from the closest pixel is set to 20 cm.

3   Hand Detection and Tracking

Most depth-based body part detection methods use the random forest algorithm based on the feature of the relative difference of the distance between two points [11]. Because such detection methods use the whole body points to obtain the

Fig. 2. The hand detection procedure based on human silhouette analysis. (a) Human silhouette, (b) two points selection among the human silhouette points, (c) inflection points found by using iterative binary partitioning method, (d) removal of concave points along z-direction of cross products, (e) hand detection results by the geodesic calculation.

relative difference feature, they are not appropriate for detecting partial body parts from a specific perspective.

The proposed hand detection and tracking algorithm locates the end points that create a convex shape in the human silhouette point set, and then associates the end points in the previous frame with the end points in the current frame. Figure 2 depicts the hand detection procedure based on human silhouette analyses.

3.1   Hand Detection

The hand detection is performed through finding the end points among the human silhouette points. The end-point location process has three stages. First, it locates all inflection points in the human silhouette points. Second, it removes the concave points from the inflection points and only retains the convex points. Third, it selects the end points from the remaining convex points.

In order to easily locate all inflection points in the input silhouette points, the iterative binary partitioning method (IBPM) that automatically selects the inflection points from an ordered point set is used. The overall process of locating the inflection points is as follows.

– Draw a line connecting the initial and final points in the ordered point set.
– Calculate the length values of all normal vectors from each point on the ordered point set to the line.
– Select the point with the maximal length value and add the selected point to the inflection point set.
– Divide the ordered point set into two point sets at the selected point.
– Iterate above stages until the point sets are no longer divisible.

Figure 3 presents an example of locating the inflection points using the proposed iterative binary partitioning method. In order to remove the concave points from the located inflection points, the Z direction of the cross product between the front and rear points of each inflection point is used. The clockwise ordered point set and the right-hand rule is used. If the Z direction is negative, the point has a convex shape; otherwise, it has a concave shape. Therefore, it obtains the

Left point set	Right point set

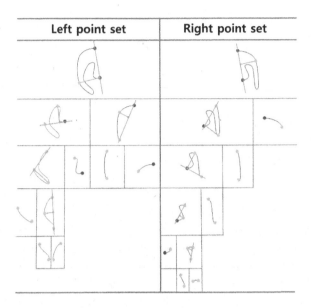

Fig. 3. Iterative binary partitioning process based on inflection points.

convex points through removing the inflection points (concave points) with a positive sign from the located inflection points.

In order to only select the end points from the convex points, the length of the geodesic path from the head position point to each convex point is used. The convex points with a length value more than the constant are regarded as the hand positions.

Finally, the hand center position is corrected using the mean-shift with a 2D disc filter. The filter weight is 2D Gaussian. In order to limit large movements of the hands, we empirically fix the maximal size of the movement.

3.2 Hand Tracking

Hand tracking uses a one-to-one correspondence method based on the distance between $t-1$ and t hand positions. It tracks hands through preferentially selecting a hand pair with the minimal distance between the hand positions in consecutive frames.

In addition, when it calculates the distance between two hands, the t hand position is calibrated because hands move quickly and randomly. It is assumed that the hand motions are approximated by the constant velocity model and the correct t hand position through adding the difference value between the $t-2$ hand position and the $t-1$ hand position.

4   Pointing Position Estimation

In Sects. 2 and 3, the 3-dimensional head position and the hand position of the user in the depth sensor coordinates positioned above the user's head. From these positions, the cross point on the eye-hand ray intersecting with the large screen in the real world coordinates are calculated and the user's pointing position on the large scale display is estimated.

4.1   Coordinate Transformation

The display position can be obtained by measuring the distance from the origin in the real world coordinates. However, the 3D position of the head and hand uses Kinect coordinates. Therefore, these points should be converted to real world coordinates points. In order to resolve this, the pattern board on which there are four distinguishable and predetermined points is placed on the floor as depicted in Fig. 4. Using SVD, we calculate the rotation and translation matrix R and t converting the head and the hand points from the Kinect coordinates to the real world coordinates. A point in the real world coordinates P' is given by

$$P' = \mathrm{R}P + \mathrm{t} \tag{1}$$

where P is a point in the Kinect coordinates.

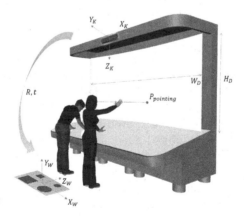

Fig. 4. 3D world coordinates, pattern board, and a point on eye-hand ray intersecting with the screen.

4.2   Eye-Hand Ray Generation

In Sect. 2, the three-dimensional head center position was obtained; to be clear, the center position is the top of the user's head. However, as Taylor reported in [13], the pointing direction is not the head-hand line, but the eye-hand line. Therefore, the head position should be corrected to the eye position. When standing and pointing at the target in front of the screen, the eyes are always lower than the top of the head. Thus, the Y value is reduced by the average distance between the top of the head and the eye. The average distance is set to 10 cm.

As depicted in Fig. 4, the eye and hand positions in the Kinect coordinates are converted into the two points in world coordinates and the position is obtained on the eye-hand ray intersecting with the screen whose size, resolution, and relative position from the origin in world coordinates are measured. We calculate the transformation matrix H_p from measured information. The pointing position p' on the screen is given by

$$p' = H_p p \tag{2}$$

where p is a (x, y) position of $P_{pointing}$ on the plane in the world coordinates and p' is a (u, v) position of $P_{pointing}$ in the screen image.

5   Experimental Results

For the head detection, we take a sliding window approach over the depth image pyramid. Each detection window is divided into 11×11 pixel cells. The window consists of the non-overlapped 3×3 cells. For 2D histogram of the zenith angle and azimuth angle, 9×6 2D bins are used. We build 3 layer depth pyramid and the scale factors are 0.4, 0.5 and 0.6. The size of the display is $4\,m \times 2\,m$ and the user is 1.7 m apart from the screen. The display has 2540×1080 pixels. Kinect is positioned above the user's head.

5.1   The Performance of Head Detection

For real-time processing, the stride of the sliding window is set to 11, which is the same as the cell size. Due to the wide stride, the accurate head detection can fail. Therefore, the accurate head region was calculated using the local information near the head center in Sect. 2.2, even though the head was not detected. Furthermore, this process exhibited good performance for the head detection failure as well as the failure due to the wide stride, and it extracted the center of the head, which was not detected by the SVM classifier.

In Fig. 5, the pink rectangles indicate the head candidates and the yellow dots are the seeds for the head area segmentation. The head detection result of the user on the left was not perfect due to the stride of the sliding window method, and the shoulder of the user on the right was detected instead of the head. Nevertheless, the yellow points were extracted as seeds for the head segmentation, and the precise head center points were obtained from the seeds in both cases as indicated by the green dots.

For head detection, 602 head images and 2267 non-head images were obtained. For the SVM classifier learning, the OpenCV library was used. In Fig. 6, the final head segmentation results are indicated by the purple areas and the centers of the heads are indicated by the green dots. The head detection rate from the SVM classifier was 90.3%, and after implementing the proposed head

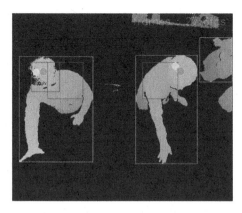

Fig. 5. The results of the head candidates (the pink rectangles), the highest pixels from the ground (the yellow dots), the final head centers (the green dots) (Color figure online)

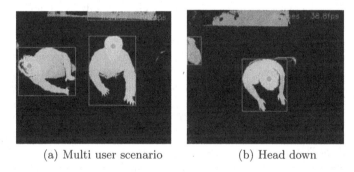

(a) Multi user scenario   (b) Head down

Fig. 6. Head area detection results and the center of the head (Color figure online)

segmentation method, the rate increased to 98.1%. According to the experiments, it can be confirmed that the proposed head segmentation method improves the detection performance by 7.8%.

5.2 The Accuracy of the Pointing Position Estimation

When using a mouse, the user moves the mouse after confirming the current position of the mouse pointer. In contrast, for intuitive pointing interfaces, the user can point at the target directly and instantly. In the experiments, in order to confirm the intuitiveness of the proposed method, users did not receive visual feedback, including the current pointing position on the screen, which corresponds to the mouse pointer when pointing at the target. Eleven participants stood 175 cm from the screen and pointed at the yellow dots on the screen. The average pointing error was 13.18 cm and the average error degree between the estimated pointing position on the eye-hand ray and the target position on the screen was $4.3°$.

Table 1. Pointing direction accuracy

Methods	Estimation error
Jojic et al. [3]	15 cm
Richarz et al. [9]	10°
Park and Lee [7]	7.2°
Our method	13.18 cm/4.3°

The proposed method was compared with conventional intuitive pointing estimation methods [3,7,9]. In [3], the spatial resolution was approximately 15 cm at a distance of 3 m from the screen. The accuracy was slightly lower than the proposed method, but the distance from the user to the screen was farther. However, in [3], visual feedback was also provided for precise pointing and it increased the achievable target resolution to approximately 5 cm. The visual feedback improved the pointing estimation precision, but users were likely to adjust their hand positions subtly due to the visual feedback. Therefore, it disrupts the user pointing at the target immediately and intuitively. In [7], the performance of the proposed method was measured for two pointing gestures. One gesture was a large pointing gesture performed with the user's full arm extended and the other was a small pointing gesture reduced to the user's forearm and hand movement only. In the current study, only large pointing gestures were considered and the accuracy of the pointing position estimation of the proposed method was better than that of Park and Lee's method. As described in Table 1, the performance of the proposed method was good compared with the conventional methods.

The screen was divided into 36 sections, and the pointing estimation accuracy corresponding to the distance between the user and the target on the screen was analyzed. As seen in Fig. 7, the pointing estimation error tended to be lower at the center of the screen, i.e. close to the user, and higher as it moved further from the center. This resulted from the distance between the user and the target increasing as the target on the screen became further from the center. As a result, the position error on the eye-hand ray intersecting with the screen could be relatively large at the side of the screen, even if the estimated hand or head position had similar errors in a different position.

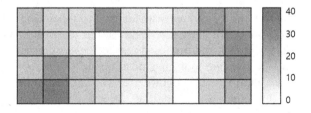

Fig. 7. The pointing position estimation error according to the section on the screen.

6   Conclusions and Future Work

In this paper, a system that allows intuitive pointing interactions in dark environments using only depth information without color information was introduced. The sensor is positioned above the user's head in order to avoid the users seeing the sensor when they are interacting because that situation can interrupt the users concentration on the screen. In order to estimate the pointing position, the user's head was detected and the position of the user's eye was estimated. As feature vectors for the head detection, the HONV features were enhanced in order to remove the noise and to obtain refined feature vectors. For classification, the linear SVM classifier was used. In order to calculate the center of the head, a head segmentation method was proposed; this method improved the head detection performance by 7.8%. The iterative binary partitioning method and one-to-one correspondence method were used to detect and track the hands, respectively. The 3D positions of the head and hands were converted to the real world coordinates and the pointing position was estimated on the eye-hand ray intersecting with the large screen. In the experiments, the average pointing error was 13.18 cm and the average error degree between the estimated pointing position on the eye-hand ray and the target position on the screen was 4.3°. Compared with the conventional methods, the proposed method has good performance, but its significant feature is that it works in dark environments.

As discussed in Sect. 4, in order to implement the intuitive pointing interface in large scale displays, the size of the display and the relative position to the origin in space are required. This causes inconveniences in that the size, resolution, and relative position should be measured repeatedly whenever the display is replaced or the position of the display or sensor is changed. In the future, using an auxiliary device as referred to in [4], the problem of the display replacement or position change will be resolved.

Acknowledgement. This work was supported by the ICT R&D program of MSIP/IITP. (15501-15-1016, Instant 3D object-based Join & Joy content technology supporting simultaneous participation of users in remote places and enabling realistic experience).

References

1. Ball, R., North, C., Bowman, D.A.: Move to improve: promoting physical navigation to increase user performance with large displays. In: Proceedings of the SIGCHI Conference on Human Factors in Computing Systems, pp. 191–200 (2007)
2. Hu, K., Canavan, S., Yin, L.: Hand pointing estimation for human computer interaction based on two orthogonal-views. In: International Conference on Pattern Recognition, pp. 3760–3763. IEEE (2010)
3. Jojic, N., Brumitt, B., Meyers, B., Harris, S., Huang, T.: Detection and estimation of pointing gestures in dense disparity maps. In: IEEE International Conference on Automatic Face and Gesture Recognition, pp. 468–475 (2000)

4. Kim, J.S., Park, J.M.: Sensor-display registration for 3D physical user interaction using a flat-panel display. In: International Conference on Pattern Recognition, pp. 1675–1680 (2014)
5. Lee, D., Lee, S.: Vision-based finger action recognition by angle detection and contour analysis. ETRI J. **33**, 415–422 (2011)
6. Nickel, K., Stiefelhagen, R.: Pointing gesture recognition based on 3D-tracking of face, hands and head orientation. In: Proceedings of the 5th International Conference on Multimodal Interfaces, pp. 140–146 (2003)
7. Park, C.B., Lee, S.W.: Real-time 3D pointing gesture recognition for mobile robots with cascade HMM and particle filter. Image Vis. Comput. **29**, 51–63 (2011)
8. Pateraki, M., Baltzakis, H., Trahanias, P.: Visual estimation of pointed targets for robot guidance via fusion of face pose and hand orientation. Comput. Vis. Image Underst. **120**, 1–13 (2014)
9. Richarz, J., Scheidig, A., Martin, C., Muller, S., Gross, H.M.: A monocular pointing pose estimator for gestural instruction of a mobile robot. Int. J. Adv. Robot. Syst. **4**, 139–150 (2007)
10. Schick, A., van de Camp, F., Ijsselmuiden, J., Stiefelhagen, R.: Extending touch: towards interaction with large-scale surfaces. In: Proceedings of the ACM International Conference on Interactive Tabletops And Surfaces, pp. 117–124 (2009)
11. Shotton, J., Fitzgibbon, A., Cook, M., Sharp, T., Finocchio, M., Moore, R., Kip-man, A., Blake, A.: Real-time human pose recognition in parts from single depth images. In: 2011 IEEE Conference on Computer Vision and Pattern Recognition, pp. 1297–1304. IEEE (2011)
12. Tang, S., Wang, X., Lv, X., Han, T.X., Keller, J., He, Z., Skubic, M., Lao, S.: Histogram of oriented normal vectors for object recognition with a depth sensor. In: Lee, K.M., Matsushita, Y., Rehg, J.M., Hu, Z. (eds.) ACCV 2012. LNCS, vol. 7725, pp. 525–538. Springer, Heidelberg (2013). doi:10.1007/978-3-642-37444-9_41
13. Taylor, J.L., McCloskey, D.: Pointing. Behav. Brain Res. **29**, 1–5 (1988)
14. Vogel, D., Balakrishnan, R.: Distant freehand pointing and clicking on very large, high resolution displays. In: Proceedings of the 18th Annual ACM Symposium on User Interface Software and Technology, pp. 33–42 (2005)
15. Yoo, B., Han, J.J., Choi, C., Yi, K., Suh, S., Park, D., Kim, C.: 3D user interface combining gaze and hand gestures for large-scale display. In: CHI 2010 Extended Abstracts on Human Factors in Computing Systems, pp. 3709–3714 (2010)

Investigating Size Personalization for More Accurate Eye Tracking Glasses

Yi-Yu Hsieh, Chia-Chen Liu, Wei-Lin Wang, and Jen-Hui Chuang[(✉)]

National Chiao Tung University, Hsinchu, Taiwan
jchuang@cs.nctu.edu.tw

Abstract. Personalized eyewear frame could improve the accuracy of eye tracking. To obtain the personalized frame size (temple length), we propose a new measuring instrument that consists of (i) the hardware, a 3D printed trial frame which has marks but no scales, and (ii) the software, a vision-based measurement which is view invariant. The vision-based measurement has accuracy and precision that are both 0.02 cm, while the trial frame can achieve a precision of 0.17 cm for secure wearing. Moreover, dispersion up to 2.56 cm is obtained among the personalized frame sizes for just a fairly small group of users, indicating the importance of having such a personalized measurement system.

1 Introduction

Eye tracking glasses provide a new vision for discoveries in marketing [1], usability [2], medical [3], psychology [4], and human performance [5]. To improve the accuracy of eye tracking, camera calibration is required after the eye tracking glasses is positioned on the face [6,7]. Ideally, the position between the eye tracking glasses and the face should be fixed before and after calibration. An accuracy of 0.5° [8] is achieved by one of the available eye tracking glasses with no help from personalized frame size to fix the position [9–13]. Better accuracy can be achieved if the eye tracking glasses have personalized frame size (temple length) to enable more secure wearing. To obtain such a frame size, we propose a new measuring instrument which consists of (i) the hardware, a 3D printed trial frame which has marks but not scales, and (ii) the software, a vision-based measurement which is view-invariant. It is shown that the vision-based measurement can achieve high accuracy and precision, with both within 0.02 cm, while the trial frame will enable a secure wearing, with a precision of 0.17 cm, if more curved frame tails (ear hooks) are adopted. Moreover, dispersion up to 2.56 cm is obtained among the personalized frame sizes for a fairly small group of users, indicating the importance of having such a personalized measurement.

2 Trial Frame

In this section, we describe the design procedure of proposed trial frame, and for each step, its design concepts.

© Springer International Publishing AG 2017
C.-S. Chen et al. (Eds.): ACCV 2016 Workshops, Part III, LNCS 10118, pp. 239–248, 2017.
DOI: 10.1007/978-3-319-54526-4_18

First, we take inspiration from a trial frame developed by Hasegawa Bicoh [14]. The trial frame has adjustable temple size, which is inherited by later steps of designing our trial frame. The adjustable temple size is measured by the scale printed on the trial frame. Screws are used to fix the temple size, so that the user can read the scale after they take off the trial frame.

In the next step, we adopt an emulated approach for the instrumentation using a pair of sports glasses which has adjustable temple size. The structure of the sports glasses is simpler than that of the trial frame developed in [14]. Specifically, the sports glasses has no scale and screw. Without the scale, we try to measure the adjustable temple size from its image captured by a camera. However, we cannot measure the temple size directly from the image unless the image distortion due to perspective projection is well calibrated. Therefore, we measure the view-invariant cross-ratio among four dark blue stripes marked on the temple, along a line going through centers of the stripes.

We conduct a usability test for the above measurement, and find several problems. First, the four dark blue stripes on the black frames of sports glasses are hard to detect in low-light scenes, due to the low contrast between the two colors. Second, the adjustable part of the temple is prone to be occluded by user's hair, for the adjustable part is located near the end of the temple. Third, the line going through centers of the four stripes might not exist due to imaging and computation errors.

Finally, we introduce the proposed trial frame. The color of the proposed trial frame is white, and two light green rectangles with adjustable distance are marked on each temple. Obviously, the rectangles (and the four vertical edges) can be detected more robustly under perspective projection compared with the four strips used earlier. Moreover, the light green color gives a greater intensity in pictures of low-light scenes than the dark blue color, and has a good contrast versus the white temple. The boundary of the adjustable part (marked with one of the rectangle) of temple is now located nearer to the root of the temple, making it less likely to be occluded by the user's hair. Besides, we design two swappable temples with different curvature at the two ends of the proposed trial frame, so that we can compare them in terms of consistency in the wearing process in later experiments. Furthermore, the proposed trial frame is 3D printed, so that it is available to consumers.

3   Vision-Based Measurement

In this section, we propose a vision-based measurement. We show a flowchart of the vision-based measurement in Fig. 1.

First, the user adjust the temple size and take a picture of the two rectangular marks on the trial frame. From the picture we detect the two rectangle marks. Then we find the four vertical edges of the two rectangle marks. Next, we find the center of each of the four vertical edges, and draw a regression line of the four centers. Then we project the four centers onto the regression line, and derive four collinear points. Therefore, the distance between the two collinear points in the middle is the distance between the two marks.

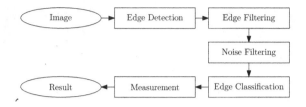

Fig. 1. Flowchart of proposed vision-based measurement

3.1 Edge Detection

To find the four vertical lines from the four vertical edges of the two rectangular marks, we utilize edge detection through image gradient. To derive the vertical edges, we calculate the image gradient in the horizontal direction with

$$Grad(x,y) = I(x, y-1) - I(x, y+1), \tag{1}$$

where $Grad(x,y)$ is the image gradient at the location (x,y), and I is the image intensity at the location (x,y). To calculate the image gradient of the RGB image, we calculate (1) for of R, G, and B-channels, respectively.

As mentioned in the previous section, we design the trial frame to be white, while the rectangular marks to be green. Therefore, the four vertical lines locate at the white-green or green-white edges, as shown in Fig. 2a. Next, we visualize the white-green edges as red edges, and green-white edges as blue edges, as shown in Fig. 2b.

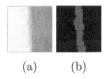

(a)   (b)

Fig. 2. (a) A detected white-green edge is visualized as (b) a red edge. (Color figure online)

3.2 Edge Filtering

We might detect edges that are not white-green or green-white; thus in this subsection, we filter out these edges. The filtering is done by checking whether there are all white pixels on either the left or right side of the edge, and whether there are all green pixels on the other side, as shown in Fig. 3.

After edge filtering, there may still be some false edges. Since the false edges are distributed randomly, we remove them based on the analysis of connected components. We derive the connected components by connecting white-green and green-white edges, as Fig. 4 shows. If the size of the connected component is too small, the corresponding edges are regarded as noises and removed.

With the edge points derived, we categorize them into four categories based on their x-coordinates. The results are shown in Fig. 5.

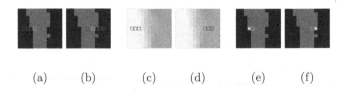

(a)   (b)   (c)   (d)   (e)   (f)

Fig. 3. Edge filtering remove edges that are not white-green or green-white: (a)(b) For an edge, (c)(d) check its neighboring pixels whether they are white or green. (e)(f) If yes, then we highlight the leftmost and rightmost pixels of the edge; otherwise, we remove the edge. (Color figure online)

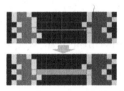

Fig. 4. A green connected component is derived by connecting the white-green (shown as red) and green-white (shown as blue) edges. (Color figure online)

Fig. 5. The edge points are categorized into four categories.

3.3   Measurement

For each of the four categories derived from the last subsection, we find its center point. Then, we project the four centers to four points a, b, c and d on a regression line. Next, we can use their cross-ratio and some known distances to measure the temple size. The cross-ratio r of a, b, c and d along the regression line is defined in the following Eq. 2.

$$r = \frac{|\overline{ab}||\overline{cd}|}{|\overline{bc}||\overline{ad}|},\qquad(2)$$

where $|\overline{ab}|, |\overline{cd}|, |\overline{bc}|, |\overline{ad}|$ are the lengths of line segments in the image space. The cross-ratio $R = r$ of the four centers A, B, C and D of the vertical edges of rectangular marks, due to projective invariant, is defined as

$$R = \frac{|\overline{AB}||\overline{CD}|}{|\overline{BC}||\overline{AD}|},\qquad(3)$$

where $|\overline{AB}|, |\overline{CD}|, |\overline{BC}|, |\overline{AD}|$ are the lengths of line segments in the scene space. Only the length $|\overline{BC}|$ is unknown since it is the adjustable distance between the two rectangular marks. The lengths $|\overline{AB}|$ and $|\overline{CD}|$ are known since they are the width of the rectangular marks, and the length $|\overline{AD}|$ is the sum of $|\overline{AB}|$, $|\overline{BC}|$ and $|\overline{CD}|$. Therefore, we can rearrange (3) as

$$R = \frac{|\overline{AB}||\overline{CD}|}{|\overline{BC}|(|\overline{AB}| + |\overline{BC}| + |\overline{CD}|)}. \tag{4}$$

Use quadratic formula, we derive the solution of the distance $|\overline{BC}|$ as

$$|\overline{BC}| = \frac{-R(|\overline{AB}| + |\overline{CD}|) + \sqrt{(|\overline{AB}| + |\overline{CD}|)^2 R^2 + 4R|\overline{AB}||\overline{CD}|}}{2R}, \tag{5}$$

which in term gives the temple size.

4   Experimental Results

In this section, we conduct experiments on the proposed measuring instrumentation to examine (i) the accuracy and precision of the software, vision-based measurement, (ii) the precision of the hardware, trial frame, and (iii) the dispersion of the measured object, personalized frame size.

4.1   Accuracy and Precision of Vision-Based Measurement

We review the precision of another frame measurement. The frame measurement is composed of a trial frame with adjustable temple [14]. The adjustable temple has a linear scale that shows the temple size. The linear scale has a unit of 0.1 cm, which guarantees a precision of 0.05 cm. On the other hand, the proposed vision-based measurement can achieve a precision of 0.02 cm, as presented in the following.

We describe the procedure to calculate accuracy and precision. We are given a measured value and an actual value. First, we derive the absolute error, the difference between the measured value and the actual value. Then, we derive the mean and standard deviation of the absolute error. Lastly, we define the accuracy and precision as the above mean and standard deviation, respectively.

Following is the procedure to derive the measured value and actual value. We place the trial frame on a flat surface. Then we set the actual value of frame size using a ruler which has an unit of 0.1 cm. Next, we capture an image using a smartphone camera, then derive the measured value through the visual-based measurement.

We conduct the experiment in different setup, e.g. different actual values including 0.28 cm and every 0.5 cm within the range from 0.5 to 4.5 cm, and with camera rotated about the X or Y-axis or translated along the Z-axis, as shown in Fig. 6.

Table 1(a) shows all the actual values and measured values in this experiment. Table 1(b) shows the absolute errors derived from the actual values and measured values. We also show the absolute errors in Fig. 7. One can see that even the largest absolute error, 0.06 cm, is less than the unit of the ruler, 0.1 cm. From the absolute errors, we derive the accuracy and precision both are 0.02 cm.

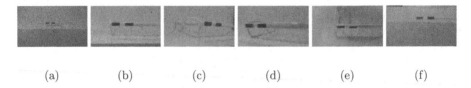

(a)   (b)   (c)   (d)   (e)   (f)

Fig. 6. Camera poses setup in the experiment (a) translated along the Z-axis in the positive direction, (b) in the negative direction, (c) rotated about the Y-axis in the clockwise direction, (d) in the counter-clockwise direction, (e) rotated about the X-axis in the clockwise direction, (f) in the counter-clockwise direction.

4.2   Precision of Wearing a Trial Frame

The precision of wearing a trial frame is affected by the temple curvature of trial frame. We define the temple curvature as the reflex angle at the end of the temple. In this subsection, we conduct experiments on two temples with different temple curvatures.

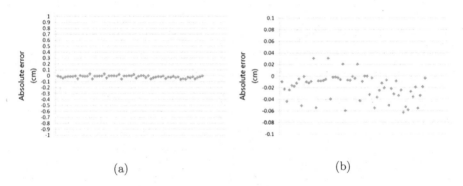

(a)   (b)

Fig. 7. The absolute error of vision-based measurement shown in (a) larger and (b) smaller scale.

We create two swappable temples for the proposed trial frame, each has an unique temple curvature, as shown in Fig. 8. For each of the examinees, we conduct experiments on the two temples of different curvatures, and we record the derived temple size. In the next several days, we conduct subsequent experiments of the same kind on the same examinee until we obtain ten measured values for each of the temple of unique curvature. We ensure a time period of at least one day between two experiments, so that the measured results are independent from each other.

For the ten measured values for each trial frame of unique temple curvature, we calculate their mean and standard deviation. Table 2 lists the measured results for the same examinee, where we represent the mean of the ten measured results for each of the right/left temple of unique curvature as AVG, and the

Table 1. (a) Actual values and, for each actual value, the measured values. The measured values are derived from 60 input images of the trial frame that has translation along the positive(+)/negative(−) direction of the Z-axis, or rotation in the clockwise (CW)/counter-clockwise (CCW) direction about the X-axis/Y-axis. (b) Actual values and absolute errors.

(a)

Actual value (cm)	Measured value (cm)					
	Transformation		Rotation			
	Z-axis		Y-axis		X-axis	
	+	−	CW	CCW	CW	CCW
0.28	0.27	0.26	0.26	0.24	0.26	0.26
0.50	0.49	0.49	0.50	0.45	0.50	0.49
1.00	0.99	0.99	1.03	0.95	0.99	0.99
1.50	1.49	1.50	1.53	1.46	1.50	1.50
2.00	1.99	2.00	2.02	1.94	1.99	1.99
2.50	2.59	2.50	2.52	2.46	2.49	2.50
3.00	2.97	2.99	3.00	2.94	2.96	2.97
3.50	3.48	3.49	3.49	3.45	3.48	3.47
4.00	3.97	3.97	3.98	3.94	3.95	3.94
4.50	4.46	4.50	4.48	4.44	4.47	4.48

(b)

Actual value (cm)	Absolute error (cm)					
	Transformation		Rotation			
	Z-axis		Y-axis		X-axis	
	+	−	CW	CCW	CW	CCW
0.28	−0.01	−0.02	−0.02	−0.04	−0.02	−0.02
0.50	−0.01	−0.01	0.00	−0.05	0.00	−0.01
1.00	−0.01	−0.01	0.03	−0.05	−0.01	−0.01
1.50	−0.01	0.00	0.03	−0.04	0.00	0.00
2.00	−0.01	0.00	0.02	−0.06	−0.01	−0.01
2.50	0.00	0.00	0.02	−0.04	−0.01	0.00
3.00	−0.03	−0.01	0.00	−0.06	−0.04	−0.03
3.50	−0.02	−0.01	−0.01	−0.05	−0.02	−0.03
4.00	−0.03	−0.03	−0.02	−0.06	−0.05	−0.06
4.50	−0.04	0.00	−0.02	−0.06	−0.03	−0.02

standard deviation as SD. One can see that standard deviation is greater for the temples with small curvature in Fig. 9, we visualize the deviations of the ten measured values from the corresponding AVGs.

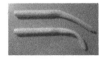

Fig. 8. Tails of trial temple with larger curvature (top) and smaller curvature (bottom).

Table 2. Measured values, mean, and standard deviation derived from ten experiments for left/right temples with unique curvature for the same examinee

Curvature	Side	Measured value (cm)										AVG	SD
Small	Left	1.55	1.35	2.28	2.23	1.38	2.52	1.09	2.18	2.14	1.26	1.80	**0.52**
	Right	1.57	1.59	2.47	2.46	1.31	2.90	1.00	2.42	2.76	1.39	1.99	**0.68**
Large	Left	1.79	2.00	1.55	2.14	1.90	2.27	2.06	2.15	2.08	1.89	1.98	**0.21**
	Right	1.94	1.85	1.75	2.03	1.53	2.08	1.62	1.77	2.06	1.68	1.83	**0.19**

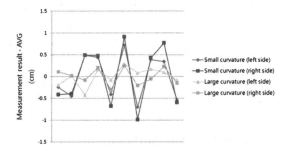

Fig. 9. The deviation from the mean values of left/right temple with large (red) / small (blue) curvatures (Color figure online)

Table 3. The mean of standard deviation for the temples of small and large curvature examined on examinees A–I

Examinee		A	B	C	D	E	F	G	H	I
Standard deviation (cm)	Small	0.60	0.13	0.32	0.12	0.32	0.22	0.19	0.22	0.57
	Large	0.17	0.09	0.15	0.14	0.32	0.10	0.13	0.22	0.18

We conduct the aforementioned experiment on nine examinees A, B,..., I. In Table 3, we show the mean of the standard deviation (MSD) of the measured values for the left and right temples. One can see that the MSD values for temples with small curvature can be 3.5 times larger than those with larger curvatures.

4.3   Dispersion of Personalized Frame Size

According to the results derived in Subsect. 4.2, the trial frame with larger temple curvature is more accurate. Therefore, in this section, we use the trial frame with larger temple curvature to measure the frame size of thirty examinees. The measured frame sizes are shown in Table 4. We also plot the measured value in Fig. 10a, and show its distribution in Fig. 10b.

Table 4. Measured values for thirty examinees

Measured value (cm) per examinee														
1.86	0.95	2.04	1.27	0.99	2.06	1.13	1.29	0.91	1.61	1.62	1.11	1.35	1.73	1.79
1.15	0.70	1.76	1.02	0.43	1.72	2.99	1.34	0.89	0.96	1.50	2.29	1.40	1.08	1.26

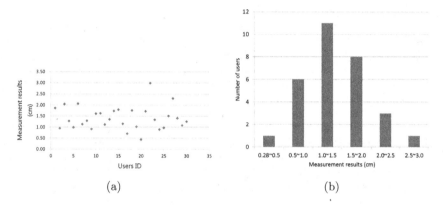

(a)   (b)

Fig. 10. (a) The measured values of frame sizes for different people and (b) their distribution

From Table 4, we see the largest and smallest frame sizes are 2.99 cm and 0.43 cm, respectively. Therefore, the range of the dispersion of the frame size is 2.56 cm.

Despite a truly small number of people considered in this experiment, such a value is significant in terms of affecting the stability in wearing an eye tracking device.

5   Summary

In this paper, we propose a new measuring instrument to obtain personalized size information (temple length) for eyewear frames. The measuring system consists of a 3D printed trial frame which has marks but not scales, and a vision-based measuring scheme which is view-invariant. The vision-based measurement has accuracy and precision that are both 0.02 cm, while the trial frame can achieve a precision of 0.17 cm for secure wearing if more curved frame tails (ear hooks) are adopted. Moreover, dispersion up to 2.56 cm is obtained among the personalized frame sizes for a small group of 30 users, indicating the importance of having such a personalized measurement system to improve the accuracy of eye tracking as well as other eyewear applications.

References

1. Lindström, A., Berg, H., Nordfält, J., Roggeveen, A.L., Grewal, D.: Does the presence of a mannequin head change shopping behavior? J. Bus. Res. **69**, 517–524 (2016)
2. Cave, A.R., Blackler, A.L., Popovic, V., Kraal, B.J.: Examining intuitive navigation in airports. In: Proceedings of Design Research Society Conference, vol. 1, pp. 293–311 (2014)
3. O'Meara, P., Munro, G., Williams, B., Cooper, S., Bogossian, F., Ross, L., Sparkes, L., Browning, M., McClounan, M.: Developing situation awareness amongst nursing and paramedicine students utilizing eye tracking technology and video debriefing techniques: a proof of concept paper. In: Proceedings of International Emergency Nursing, vol. 23, pp. 94–99 (2015)
4. Gomes, A., Oh, H., Chisik, Y., Che, M.: Ilha Musical: a CAVE for nurturing cultural appreciation. In: Proceedings of International Conference on Interaction Design and Children, pp. 232–235 (2012)
5. Marshall, S.P.: What the eyes reveal: measuring the cognitive workload of teams. In: Duffy, V.G. (ed.) ICDHM 2009. LNCS, vol. 5620, pp. 265–274. Springer, Heidelberg (2009). doi:10.1007/978-3-642-02809-0_29
6. Tobii: Camera calibration for Tobii Pro glasses 2. https://youtu.be/Gl_UVvFTuxU?t=7m55s
7. SMI: Camera calibration for SMI eye tracking glasses. https://youtu.be/scODR342dJA?t=22s
8. SMI: Specifications of SMI eye tracking glasses 2 wireless. http://www.eyetracking-glasses.com/fileadmin/user_upload/documents/smi_etg2w_flyer_naturalgaze.pdf
9. Tobii: Tobii Pro glasses 2. http://www.tobiipro.com/product-listing/tobii-pro-glasses-2/
10. SMI: SMI eye tracking glasses 2 wireless. http://www.eyetracking-glasses.com/products/eye-tracking-glasses-2-wireless/technology/
11. ASL: ASL Mobile Eye-XG glasses. https://imotions.com/hardware/eye-tracking-glasses/
12. Kassner, M., Patera, W., Bulling, A.: Pupil: an open source platform for pervasive eye tracking and mobile gaze-based interaction. CoRR abs/1405.0006 (2014)
13. EyeControl: Eyecontrol. http://www.eyecontrol.co.il/
14. Hasegawa Bicoh: Trial frame. http://www.bicoh.co.jp/products1002.html

HeadPager: Page Turning with Computer Vision Based Head Interaction

Zhenyu Tang[✉], Chenyu Yan, Sijie Ren, and Huagen Wan

College of Computer Science and Technology, Yuquan Campus,
Zhejiang University, 38 Zheda Rd., Hangzhou 310027, China
roy.james0717@gmail.com

Abstract. As computers become a general tool for reading texts, pictures and other kinds of contents, human-computer interactions in the reading process are worth studying for improving user's reading experience. One issue with reading on computers is that user's hands are often occupied in turning pages, which prevents them from performing certain tasks well. It is possible to free user's hands from physical input devices by designing new input methods. We present a prototype system that uses a camera to monitor user's head, and when the user makes head gestures, the system helps the user turn pages on the currently active document. We evaluate the usability of our system through experiment and questionnaire. Feedback from users indicates that our system is helpful in hand-occupying situations.

1 Introduction

Page turning on electronic devices is a common and simple operation, it serves readers during their scanning, browsing and reading. Thus the design principle of its interface is to be as clear and simple as possible that shall not disturb the reading procedure.

Nowadays, the most used methods for page turning around the world are by mouse clicking and scroll, a touch pad, or a touch screen. The common feature of the three methods is that at least one human hand is required to directly operate on certain devices. Though these methods are easily learned by users, they are not necessarily the best option under all circumstances. Specifically, users may not want to or cannot use their hands to turn pages in certain scenarios, such as when their hands are occupied in other activities including playing instruments or cooking while they still need to look at the screen, when their hands are injured, when they do not wish to expose hands in a cold winter, and when they are too tired to raise up hands. These users might consider alternative methods of turning pages, and thus new methods for hand-free interaction are in demand.

Current methods for hand-free interaction include eye control [1], voice control [2] and head gesture control [3,4]. Eye trackers are widely applied in user study, but the technology is based on statistical assumptions, from which the result is not so accurate in unusual individuals. When applied to high-end intelligent mobile phone, most users report this new function to be unusable. Also,

© Springer International Publishing AG 2017
C.-S. Chen et al. (Eds.): ACCV 2016 Workshops, Part III, LNCS 10118, pp. 249–257, 2017.
DOI: 10.1007/978-3-319-54526-4_19

reading itself is eye-occupying, using eyeballs to control can be ambiguous to reading devices. Voice control is also a hot topic in artificial intelligence. However, it is bothering to use voice while reading since this activity usually requires a quiet environment. Head gesture control is a relatively better alternative for hand-free control in reading, since it relates to no major pitfall mentioned above. So we decide to apply a computer vision driven approach, on controlling page turning by head movement.

2   Recognized Tasks

Basically, we aim to free users' hands from turning pages. With head movement being our top choice, proper gestures are to be designed that are intuitive to understand and easy to accomplish.

Secondly, we need to provide fluent page turning experience. Due to different reading speed and pausing habit, meaning how many lines does one read before the next turning operation, the interface should allow users to page up and down freely with clear gestures and get easy access to speed settings.

Thirdly, we should give clear guidance. The interface should provide guidance for the user that could be easily understood and remembered.

3   Experimental Design

3.1   Interaction Gestures

Conventional methods track the user's head by facial recognition, and define several operations by facial expressions such as mouth opening and stretching [4]. These methods can perform robustly if properly implemented, however, some operations defined are often not so intuitive as to let the user make direct connections between interaction inputs and outputs. It might also cause the user to feel uncomfortable when making facial expressions frequently. Besides, in the task of page turning, the user might not want to alter the position of mouse pointer so we do not tend to affect mouse movements.

Naturally, we try to connect the user's head location to page turning signals. We assume that most readers are likely to keep their heads near the straight up central position, since tilting head while long-time reading is generally a bad habit. Then we decide that, when users lean their head to their left, pages will be turned upward; when users lean to their right, pages will be turned downward, as shown in Fig. 1. Note that we do not use information of when people lower or raise up their heads, because those actions are normal movements during reading and nothing should be triggered accordingly. In addition, we relate the speed of page turning with the extent to which the user's head deviates from the original central position. If one wants to turn pages faster than usual, he only needs to lean his head to the left most area within the range of detection.

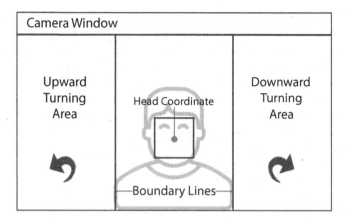

Fig. 1. Partitioning of the camera frame. Paging signal is triggered when the head coordinate passes boundary lines. Boundary lines are adjustable.

3.2 Designed Tasks

We are curious to know how will our system affect the effectiveness of task completion under certain conditions. We suppose that when users need to spare one or both of their hands while reading, our system might be helpful. Thus we design the following two experiment tasks.

1. **Writing task:** We prepare several question sets consisting of elementary level single choice math questions. Adjacent questions are properly separated by blank lines in a single document. In this task, the user needs to answer each question by writing choices on the answer sheet. The user will be forced to turn pages and write down answers interchangeably.
2. **Typing task:** We show users a document containing one paragraph of an essay with frequent line breaks, and offer them with a modern smart phone to type on. The user's task is to type the complete paragraph without line breaks on the smart phone with both hands holding the phone horizontally. Also, the user must switch between turning pages and typing.

In order to make comparisons between traditional mouse paging method and our head paging method, experiments must be run in pairs. The experiment comprises four tasks, two in each paging method.

4 Developed Prototype

4.1 Algorithm and Implementation

The minimum requirement for our system is a modern computer with Windows system installed, and an internal or external camera. The reader should maintain moderate reading distance from the screen and the camera as well. We developed our prototype for head movement controlled page turning and name it as *HeadPager*.

The core algorithm for human face detection is mainly based on HAAR cascade classification with OpenCV 2.4.1 [5] and facial landmark tracking [6]. Since there is no need for high-precision head localization, we do not call face detection in every camera frame. Instead, we first run face recognition to locate the user's head, and then detect facial landmarks within the head bounding box and use the center landmark to update the head position. For most frames, we only track facial landmarks in a neighborhood area around the detected center in the previous frame. After this modification, the computational cost is greatly reduced and the algorithm can easily run in high frame rate. Also, to enhance the stability of tracking, we update the new position by a time dependent decay factor to utilize information from previous detections. We partition the 2-D screen coordinate to three sections with two vertical boundary lines. For example, two lines could be placed at 30% frame width to the left and right fringe of the frame. When the user's head passes through a boundary, the corresponding paging signal is triggered. For paging speed control, additional boundary lines could be conveniently set. The actual paging action is realized with system level command (mouse wheel and keyboard simulation) that works on any applications.

In order to ensure the generality of our software, we add a calibration utility to it. The calibration serves to adopt the page turning criteria to a specific user, by sampling the current static position of the user's head during several frames and defining the sampled average position as the new central position. The calibration process is recommended when the user does not constantly sit in the middle of camera frames.

However, through our tests, it is also possible that the tracking fails during some consecutive frames. In such cases, we immediately re-initialize with whole face detection to continue the interaction. The workflow described above is depicted in Fig. 2.

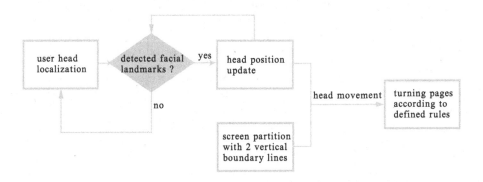

Fig. 2. Workflow of our algorithm.

4.2 User Interface

For the simplicity of our software, the user interface will not contain any implicit function that is not directly relevant to user's operations. Thus, we only allow users to start/stop the page turning, change the turning style, adjust range of movement, call a calibration to recenter their heads and view the help information. And the software should run in background until the user terminates it.

5 Usability Evaluation

5.1 Participants and Apparatus

We invited 10 participants (4 females), ranging from 20 to 26, with a mean age of 21.7 (SD = 2.0). All experiment was conducted under the same environment to rule out the influence of unrelated variables. The study and questionnaire took each subject approximately 20 min without interruption to complete.

5.2 Experiment Materials

For writing tasks, we programmatically generated simple math questions about addition and subtraction of integer numbers between 0 and 100. Each question set has 30 such questions. In each screen shot of one document, the user can only see two questions and their choices. We guarantee that each question has and only has one correct answer.

For typing tasks, we manually edited essays from analytical writing samples of Graduate Record Examinations (GRE). Each document contains approximately 90 English words. In each screen shot of one document, the user can see two lines of sentences that have approximately 10 words. The smart phone used is an iPhone 6s plus and we require participants to not use auto-completion while typing.

A questionnaire is designed for our experiment. We follow the framework of [7], participants first answer a question concerning their expectations about our system, and after their actual experience with it, they answer the same question concerning their past experience. The before and after questions have the same meaning but are rephrased properly. All rating scale used in our questionnaire follows a 7-point Likert scale.

5.3 Pilot Experiment

In order to assert that our experiment is fair and feasible, a pilot experiment is run by three participants before recruiting more people, and we do not use data from these three participants in formal analysis.

In the experiment, we first let participants complete writing tasks on 3 question sets, and then asks them to do typing tasks on 3 essay paragraphs of similar length. We record the time spent on each task and observe that it takes almost the same amount of time to complete tasks with materials of the same type.

Additionally, we discover that our system does not work well under back-light illumination since that makes human faces unrecognizable. So we always maintain a suitable lighting environment and require users to keep a moderate distance from the camera to be detected.

5.4   Procedure

Before the experiment, we first briefly introduce our study to the participants. Then we ask them to fulfill a short questionnaire about their general information, handedness, computer usage, and expectation of our system. Each user is allowed several minutes to get familiar with the system.

During the experiment, the experimenter gives clear instructions before each task begins. The order of four tasks are randomly chosen for each participant. Time for completing each task is recorded by the experimenter. Right after completing each task, participants need to rate the fluency (whether the task often interrupted) and comfortableness (whether they feel dizzy) completing this task.

After the experiment, participants are asked to fulfill another questionnaire concerning their feelings about the easiness of learning, comfortableness of using, and willingness of future usage for our system. There is also one last open question asking for their suggestions to our system that they can answer in text form.

5.5   Results

All tasks are run on the same computer, in the same well-lit room, without any accidental situations. All participants are right-handed and report no former experience with non-contact paging techniques. 60% of the participants report a daily computer reading usage of 3–5 h, while others read on computers for 1–3 h. Half of them report that sometimes they feel uncomfortable or inconvenient to spare one hand on computer while reading.

Task Completion Time: Figure 3 shows the time distribution for completing tasks among all participants. We run paired t-tests with $\alpha = 0.05$ to find out if there is statistically significant difference between task completion time with different paging methods. For writing tasks, there is no significant difference in time with mouse scroll method and head paging method. But for typing tasks, there is significant difference in time on the level $p < 0.05$, which means using head paging method is more efficient than using mouse for typing tasks.

We were expecting that in both tasks head paging would outperform mouse method, which is not true for writing tasks. The reason might be that in writing tasks, we observe that most participants are able to hold the mouse and pen at the same time so that they do not need to drop the pen and can switch back to writing immediately. This indicates that their hands are not fully occupied

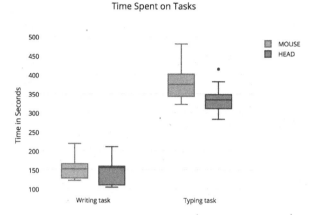

Fig. 3. Time spent on four tasks measured in seconds.

in writing tasks. But in typing tasks, if they are going to scroll the page using mouse, they must put their right hand away from the phone. The pose restoration takes some time but head paging enables them to keep both hands on the phone throughout the task.

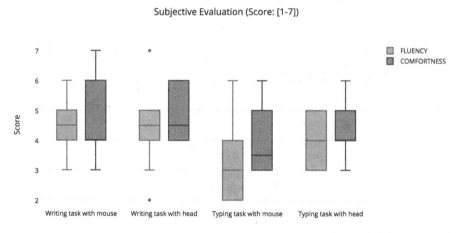

Fig. 4. Subjective evaluation of fluency and comfortableness. The higher the value, the better performance a method has, where 7 = very fluent and very comfortable.

Subjective Results: We would also like to track if there is any difference between participants' perceived effectiveness and our quantitative measure. Figure 4 shows the result of participants rating. Although we see some increase of rating using head paging instead of mouse paging in some cases, statistical tests indicate no significant difference in any pair.

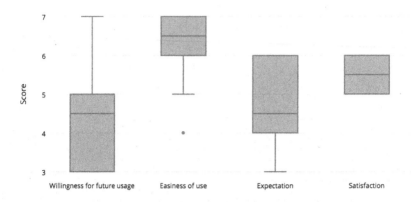

Fig. 5. Subjective evaluation of overall performance of the head paging system.

Figure 5 shows a general evaluation of the head paging system from participants. Most ratings are on the positive side (>3) of the scales. Note that category
Expectation and *Satisfaction* are comparable since they mean the same thing
although are asked before and after the experiment. We run a Mann-Whitney U
test and discover that there is significant difference on the level $p < 0.05$ for the
two ratings, meaning that participants have better remarks for our system after
their trial.

6   Conclusions and Future Work

The results from our experiment indicate our system have positive effects in
assisting users completing tasks that involve page turning and occupy both
hands. This makes sense as head interaction frees their hands from unnecessary interruptions and thus enhances task efficiency.

Though our head paging system helps in some tasks, it is not significantly
more preferred by users. This is interesting because besides efficiency, there are
other aspects that users care about such as easiness and precision of control.
Some answers given by users to the last open question has given us insights
on further development of the software. For instance, one participant says the
gestures are not flexible enough, because users might want to lean their head only
to check their cellphone next to the computer but not to trigger page turning.
This issue could be resolved by refining interaction gestures. One solution is to
encode vertical head position information with horizontal information we already
used to create more simple-to-learn page turning rules. Another promising choice
would be to utilize the help of gaze input to enable more reliable and efficient
interactions.

The system we presented in this work uses a standard setup that is easily found on any modern computers. So a more complete and robust further system could be built based on our prototype to help people turn pages.

References

1. Gips, J., Olivieri, P.: EagleEyes: an eye control system for persons with disabilities. In: The Eleventh International Conference on Technology and Persons with Disabilities, pp. 1–15 (1996)
2. Rourke, M., Clough, R., Brackett, P.: Method and means of voice control of a computer, including its mouse and keyboard. US Patent 6,668,244 (2003)
3. Fu, Y., Huang, T.S.: hMouse: head tracking driven virtual computer mouse. In: 2007 IEEE Workshop on Applications of Computer Vision, WACV 2007, p. 30. IEEE (2007)
4. Tu, J., Tao, H., Huang, T.: Face as mouse through visual face tracking. Comput. Vis. Image Underst. **108**, 35–40 (2007)
5. Itseez: Open source computer vision library (2015). https://github.com/itseez/opencv
6. Uřičář, M., Franc, V., Hlaváč, V.: Detector of facial landmarks learned by the structured output SVM. In: Csurka, G., Braz, J. (eds.) VISAPP 2012: Proceedings of the 7th International Conference on Computer Vision Theory and Applications, vol. 1, pp. 547–556. SCITEPress—Science and Technology Publications, Setúbal (2012)
7. Jokinen, K., Hurtig, T.: User expectations and real experience on a multimodal interactive system. In: INTERSPEECH (2006)

Exploring Manipulation Behavior on Video See-Through Head-Mounted Display with View Interpolation

Chun-Jui Lai[1]([⊠]), Ping-Hsuan Han[2], Han-Lei Wang[1], and Yi-Ping Hung[1,2]

[1] Department of Computer Science and Information Engineering,
National Taiwan University, Taipei, Taiwan
r03922084@csie.ntu.edu.tw
[2] Graduate Institute of Networking and Multimedia,
National Taiwan University, Taipei, Taiwan

Abstract. Video see-through HMD mixes the real and virtual world, and users can have a good experience on virtual part, but the real part captured by cameras still have some problem, especially the distance perception. In this paper, we try to remove the error due to the distance between cameras and users. We use depth image-based rendering algorithm to re-compute the true distance of the scene, and render the correct image to the user. And we use multiple cameras with different viewpoints to reduce the occlusion areas.

In order to analyze the effect brought to the users, we implement the device with recent HMD and depth camera, and design an experiment to compare with human eyes and recent video see-through device.

1 Introduction

By using a head-mounted display (HMD), we can have an immersive virtual reality (VR) experience. But users cannot see any information from the real world, it is very inconvenient and dangerous if we split VR and real world completely. To solve this problem, video see-through HMD can acquire images from real environment and present them into the HMD, then, we could build a mixed reality (MR) or augmented reality (AR) system. However, how to append and calibrate cameras on HMD for recovering real environment is still a research issue.

Currently, there are some products had appended single or stereo cameras on their devices. HTC VIVE has a single camera in front of its device. OVRVISION Pro [1] proposed to append dual cameras on current HMD to capture left and right images. Even though we have single or stereo images from real world, due to the difference of viewpoints, images captured by cameras are different to what eyes see (as shown in Fig. 1). So we need to use the correct distance to re-compute the true corresponding between real world and user eyes.

View interpolation algorithm can use images from multiple views to compute 3D information in the scene, and use the 3D information to interpolate a virtual view. By using the RGB-D image, we could also calculate the 3D information

© Springer International Publishing AG 2017
C.-S. Chen et al. (Eds.): ACCV 2016 Workshops, Part III, LNCS 10118, pp. 258–270, 2017.
DOI: 10.1007/978-3-319-54526-4_20

of real environment, and project it into a virtual view. Then we could use these methods to get the view from position of user eyes. Nevertheless, there is still another problem, occlusion, need to be solved.

There are some disocclusion inpainting algorithms attempt to solve occlusion problem, and can get results close to the truth. However, due to the lack of scene information, no matter how does the algorithm work, they cannot recover true information of real world. Appending more cameras with different positions to cover all the scene may solve the root cause of this problem. But the number of cameras we can append on HMD is limited. So how to configure the cameras to minimize the occlusion areas is also an important issue.

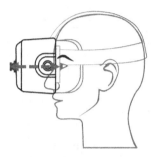

Fig. 1. Correspondence between human eyes and video see-through cameras

2   Related Work

In order to see the outside world while using the HMD, see-through HMD can satisfy this requirement. There are two methods to achieve see-through HMD, optical see-through and video see-through HMD.

There have been many optical see-through HMD devices, such as Google glass, Epson BT-200, Microsoft HoloLens, etc. But there are some difficulties to need to overcome, especially in the calibration of the user view. In order to match the virtual object to real world, calibration of user eyes and the line of sight is needed. And the calibrated parameter should be personalized, so the calibration method need to be simple and easily accessible.

Compare to the optical see-through HMD products, video see-through HMD products are fewer. But every smartphone can be a video see-through HMD by using a Google Cardboard. However, to achieve stereo vision, stereo or depth cameras are necessary. AR-Rift [2] and Steptoe et al. [3] proposed to append two webcam on current HMD to capture left and right images, and it is similar to the OVRVISION Pro. However, the distance perception is wrong, and users are easily giddy when using the HMD.

Shenchang et al. [4] proposed the view interpolation algorithm to solve the low framerate in computer graphic. In virtual world, the camera parameters and positions are known and ideal, even the depth image is available. Zitnick et al. [5] proposed to use eight calibrated camera to interpolate the real view between these cameras. However, the stereo reconstruction is needed, so the computation speed is very slow if we want to get better depth images.

To skip the stereo reconstruction step, we can use depth camera to acquire the depth image directly. Fehn et al. [6] proposed the depth image-based rendering (DIBR) algorithm to use a pair of color and depth image to synthesize the left and right views for stereo vision. However, the occlusion would occur, so disocclusion inpainting is necessary. Because the disocclusion inpainting algorithm use the neighboring color information of occlusion areas, if the occlusion is too large, the result would be messy. Smolic et al. [7] proposed to use multiple RGB-D images to generate stereo view for 3D display. Because there are multiple reference views, the virtual can be generate with smaller occlusion, and the result would be better. Ndjiki-Nya et al. [8] proposed a better method to improve the hole filling algorithm by analyzing the texture around the occlusion areas. But this hole filling algorithm need several to tens of minutes for a frame according to the patch size and occlusion area size, and it is difficult to achieve real-time rendering.

3   Method

We use DIBR [6,7,9] algorithm to interpolate the virtual views from the color and depth images. Different to common DIBR method, we need to achieve real time rendering for HMD usage, so we use DirectCompute to accelerate DIBR algorithm. And we simplified the algorithm to easily implement it on GPU. There are four steps in our implement, as shown in Fig. 2.

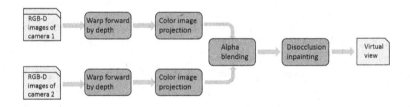

Fig. 2. Flowchart of DIBR algorithm implemented in this paper

3.1   Warp Forward Reference by Depth Image

We consider the W * H image as $(W-1) * (H-1)$ tightly arranged squares, and we split each square into two triangles (as shown in Fig. 3). And we can use camera parameter to warp the triangles from cameras to virtual views for left and right eyes according to follow formula.

$$\begin{bmatrix} u' \\ v' \\ 1 \end{bmatrix}_{virtual} = A_{virtual} * [R \mid T]^{ref}_{virtual} * Depth * A^{-1}_{ref} * \begin{bmatrix} u \\ v \\ 1 \end{bmatrix}_{ref}$$

$$A : Camera\ Matrix$$

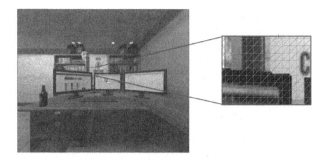

Fig. 3. Squares and triangles in pixel-wise.

3.2 Color Image Projection and Interpolation

After warping forward, we can get the correspondence between reference viewpoints and virtual viewpoint. If the depth of any vertices on the triangle is 1.05 times greater or 0.95 times smaller than any other two vertices, we consider the triangle as an occlusion area because the difference of depth is too large.

As shown in Fig. 4, because the warped vertices of triangles (red points in virtual texture) may not be integer points, we consider the bounding box (green rectangle in virtual texture) which contains integer points inside triangle. And we use Barycentric coordinates to check that if the integer points inside the bounding box are also inside the warped triangle or not. For the points inside the warped triangle, we apply Barycentric interpolation to compute the color.

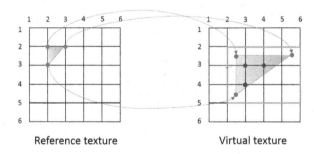

Reference texture   Virtual texture

Fig. 4. Warped triangle from reference texture to virtual viewpoint. (Color figure online)

3.3 Alpha Blending

Different to the case in [6], we may have more than two reference cameras in some camera configurations, each point on virtual texture may have more than two corresponding points. Therefore, we use warped depth map to compare the corresponding points. In our case, the virtual viewpoints are always behind the reference cameras, so the foreground would not be occluded by the background.

If the difference of depth between corresponding points are very large, we could simply use the nearest point. If there are two or more near corresponding points, we average the color of these points with the same weight. (Fig. 5)

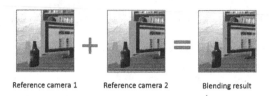

Reference camera 1   Reference camera 2   Blending result

Fig. 5. Blinding result, green areas are occlusion regions (Color figure online)

3.4   Disocclusion Inpainting

To reduce computation loading, we use a simple way to implement disocclusion inpainting algorithm instead of the texture analysis method proposed in [8]. In our case, because the cameras are in front of the user, we can ensure that the occlusion area occurred when background is occluded by foreground. Therefore, for each pixel in occlusion area, we consider the horizontal line and vertical line cross the pixel and boundary of occlusion area. There are four intersection points of the two lines and the boundary, and we take then as ends of the pixel. If the depth of left end and right end are the same, we consider the line is on a plane, so we inpaint the pixel by color of left end and right end. It is similar if up end and right end have the same depth. Otherwise, we use the end with large depth to inpaint the pixel. (Fig. 6)

Fig. 6. Results before and after disocclusion inpainting. (a) Occlusion. (b) Disocclusion inpainting. (1) Part of original result. (2) Corresponding depth of (1). (Color figure online)

4   Hardware Implementation

In [10], there is a simulation showing that view interpolation is achievable for HMD, so we attempt to use current RGB-D camera and HMD device to implement our method. And we calibrate the cameras and get the intrinsic and extrinsic matrices. We process the depth image to solve the problem when mapping the depth image to color camera, and apply the camera parameters to present the virtual view results on HMD. Finally, we also do a user study to evaluate the effect of our method.

4.1   Hardware Configuration

We use Oculus Rift Development Kit 2 (DK2) as the based HMD, RealSense F200 as the RGB-D camera. Because infrared devices would interfere to each other and the RealSense F200 use infrared to compute depth image, we can only use single depth camera for implementation. We use OVRVISION Pro as the reference of user eyes, and compare to our method. Because the framerate of RealSense is about 30 fps, the upper limit of our method is also 30 fps. However, ignoring the input limitation, our method can achieve over 60 fps for a pair of 800 * 800 left and right images on a Nvidia GTX980 GPU. Figure 7 show the appearance of our device from the front view.

Fig. 7. The device we used to implement our method.

4.2   Camera Calibration and Preprocessing

We use Zhangs calibration method [11] to calibrate the color camera of RealSense and OVRVISION cameras, and get the intrinsic parameters. In order to warp the depth image from the depth camera to the color camera of RealSense, we use QueryInvUVMap function and get the inverse UV map provided by RealSense SDK, so we can directly get the mapping correspondence from depth image to color image. Following equation shows the transformation from depth to color image of RealSense.

$$u = InvUVMap[i, j].x * depth_{width}$$
$$v = InvUVMap[i, j].y * depth_{height}$$

[i, j] is the pixel position of color image, and [u, v] is corresponding pixel position of depth image.

After calibrating the intrinsic of the cameras, we need to calibrate the extrinsic parameter of the cameras. We let the upper left inner corner as the origin of chessboard coordinate system, the horizontal axis as X-axis and the vertical axis as Y-axis. Because we had known the square size of the chessboard about 25 mm, we can set the coordinate system of the chessboard in millimeter, as show in Fig. 8(c).

We let OVRVISION Pro and RealSense F200 capture chessboard simultaneously (Fig. 8). Then we can find the inner corners of the chessboard, and get pixel positions of the inner corners. And we had set the coordinate of all inner corner, according to follow equation:

$$\begin{bmatrix} u \\ v \\ 1 \end{bmatrix} = Intrinsic * [R \mid T] * \begin{bmatrix} X \\ Y \\ Z \\ 1 \end{bmatrix},$$

$$R : rotation\ matrix, T : translation\ matrix$$

For each inner corner, we can get two equations. Because rotation matrix has 3 degrees of freedom and translation vector also has 3 degrees of freedom. There are totally 6 degrees of freedom of the extrinsic. Hence, if we have three or more inner corners, we can get at least six equations and solve the rotation matrix and translation vector. So we can get the extrinsic matrix from chessboard coordinate to camera coordinate.

Figure 9 shows the correspondence between chessboard and OVRVISION and RealSense camera. We set the coordinate of RealSense color camera as global coordinate on HMD. By follow equation,

$$[R \mid T]_{OVRVISION}^{Realsense} = [R \mid T]_{OVRVISION}^{Chessboard} * ([R \mid T]_{RealSense}^{Chessboard})^{-1}$$
$$= [R \mid T]_{OVRVISION}^{Chessboard} * [R \mid T]_{Chessboard}^{RealSense}$$
$$[R \mid T]_{B}^{A} : extrinsic\ matrix\ from\ A\ to\ B$$

we can compute the extrinsic matrix from RealSense color camera to OVRVISION left and right camera.

To setting the virtual view position for users, we consider the left and right cameras of OVRVISION are just in front of user eyes, so we only need to decide the distance between OVRVISION and user eyes. According to the extrinsic calibrated by mentioned method, the distance between OVRVISION and RealSense color camera in Z-axis is about 25 mm. We measure the distance between RealSense color camera and the lens as about 70 mm. And the diameter of the lens we measured is about 40 mm, and we have known the FOV of the lens is about 106, so we can compute the focus of the lens is about 15 mm. Therefore, the distance between camera center of OVRVISION Pro and user eyes is about 110 mm. Figure 10 shows the distance correspondence.

Due to the hardware design of depth camera of RealSense, there will be some occlusion in original depth image (Fig. 11(a)). After warping the depth image to the viewpoint of color camera, the warped depth image cannot complete match the color image, especially in near object (Fig. 11(d)).

We find that the occlusion only influences right part of near objects, so we pad the right part of objects inside 50 cm with about 10 pixels. And the depth image can match the color image more completely (Fig. 11(f)).

After applying the method implemented in Sect. 3, we can get stereo images for left and right eyes. Due to the noise of depth camera and the large occlusion

Fig. 8. Images for calibrating OVRVISION Pro and color camera of RealSense F200. (a) Left camera and (b) right camera of OVRVISION Pro. (c) RealSense F200 and chessboard coordinate (Color figure online)

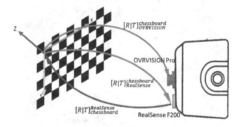

Fig. 9. The correspondence between chessboard, OVRVISION and RealSense.

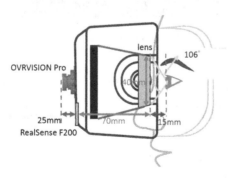

Fig. 10. Distance between OVRVISION Pro to user eyes.

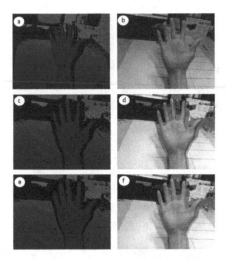

Fig. 11. RealSense images. (a) Original depth image. (b) Original color image. (c) Warped depth image. (d) Blending result of (b) and (c). (e) Padding depth image. (f) Blending result of (b) and (e). (Color figure online)

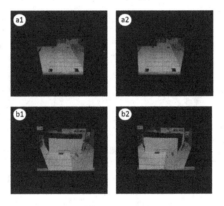

Fig. 12. Results presented in HMD. (1) Left view. (2) Right view. (a) Table plane. (b) Near objects.

of near object, we do not apply disocclusion inpainting in the implementation for better visual effect. Figure 12 shows the result presented in the HMD.

4.3 Experiment

There are three modes in our experiment. Mode 1 is without wearing the HMD, and Mode 2 is wearing the HMD with cropped OVRVISION view, and Mode 3 is wearing the HMD with view interpolation of RealSense images (our method). The difference between these modes is only the view presented to users. Figure 13 shows the view of left eyes that users can see in the different mode.

Fig. 13. Users left view in our experiment. (a) Mode 1: without wearing the HMD. (b) Mode 2: wearing the HMD with cropped OVRVISION view. (c) Mode 3: our method.

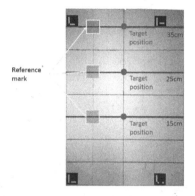

Fig. 14. The distance sheet used in our experiment. (Color figure online)

In each mode, there are three tasks with distance 15 cm, 25 cm, 35 cm, and we require users to put the 3 cm * 3 cm * 3 cm cube on the request target position. We use a 3 cm * 3 cm blue square as the reference mark, and draw red lines in the distance of 5 cm, 10 cm, 35 cm, as shown in Fig. 14. We randomly assign the task and ensure each task is assigned ten times in each mode.

In each task, the participant is requested to see the target position of requested distance on the distance sheet. When the participant has determined the target distance, then he/she need to close his/her eyes and put the cube on the position he/she thinks to be the correct position. Then we capture the position of the cube and remove it, so the participant does not know where he/she put the cube.

To measure the position of the cube and the distance errors, we use ArUco marker detection [12,13] to recognize the markers. Figure 15 shows the setting of markers. We put the markers with id 1–4 around the sheet, and set the lower left corner of marker with id 1 as the origin. Then we measure the distance between the markers. The marker size is 26 mm, so we can set all the coordinate of the four corners of the four markers. Similar to the extrinsic calibration, if we have calibrated the intrinsic of the camera, we can compute the transformation between the camera and the marker coordinate system, and the position of camera on the coordinate system.

We paste a marker on the cube, so we can also detect the cube position on the image by the marker detection. According to the following equation:

$$\begin{bmatrix} X \\ Y \\ Z \end{bmatrix} = [R \mid T]_{plane}^{camera} * (Z_{cam} - 30\,mm) * Intrinsic^{-1} * \begin{bmatrix} u \\ v \\ 1 \end{bmatrix},$$

Z_{cam} : the z coordinate of the camera in marker coordinate system.

We can compute the 3D position of the marker and the Y-axis of the cube center. Then we can compute the distance in Y-axis between target position and cube center.

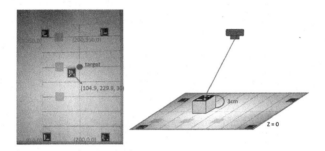

Fig. 15. The coordinate of distance sheet, and the detected marker of the cube.

4.4   Result and Discussion

There are 12 participants in out experiment. Figure 16 shows the box-plots of our results, each box-plot shows the distance error in different task. Total result means that we consider the distance error of all the task.

According to the experiment results, the average of distance error in the mode 1 that users do not wear the HMD is about the same in each task. It means the distance perception of users is approximately consistency in these tasks without wearing the HMD. However, the distance error increases when the target distance increases if users wear the see-through HMD no matter in mode 2 or 3. There may be some factors that affect user to determine the distance when the distance of target increases.

According to Fig. 16, we can find that the distribution of mode 2 that users wear the HMD with the camera view is obviously different to mode 1 and 3. By the Tukeys test, we can find that mode 2 is indeed significantly different to mode 1 in all task. Although mode 3 that users wear the HMD with interpolated view is only significantly different to mode 2 in task 1' and total, comparing to mode 1, the difference between mode 2 and mode3 is larger. So we should increase our sample to clarify this part.

In the feedbacks of 12 participants, users may feel the view is oppressive to eyes in mode 2 because the view is too close and large. And the view of mode 3 is far but close to the view without using the HMD. Hence, compare to the mode 2, mode 3 is closer to mode 1 according to the experiment result and participant feedbacks.

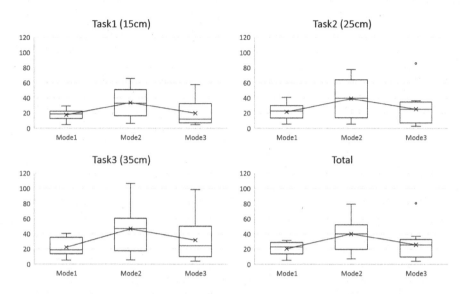

Fig. 16. Box-plot of results.

Table 1 shows the averages and standard deviation of each task in each mode, repeated-measures ANOVA, and the pairwise Tukeys test.

Table 1. Statistics of distance error (mm)

Task	Mode1	Mode2	Mode3	RM-ANOVA	Tukeys test
Task1	17.7(6.67)	34.0(19.3)	19.9(17.8)	0.015*	1–2, 2–3
Task2	21.7(10.5)	39.4(25.3)	25.4(22.6)	0.035*	1–2
Task3	22.4(12.7)	46.9(30.9)	32.0(27.7)	0.007**	1–2
Total	20.6(8.70)	40.2(23.4)	25.8(20.5)	0.005**	1–2, 2–3

$\cdot p < 0.1, *p < 0.05, * * p < 0.01$

5   Conclusion and Future Work

Users may lose the sense of distance when using video see-through HMD without any processing on the view, and even get dizzy because the view changes more than original perspective. Hence, we proposed to use view interpolation to solve this problem.

According to our experiment results, view interpolation is indeed helpful to reduce the difference of sense of distance when using the video see-through HMD. By the simulation results [10], if we have multiple RGB-D cameras with wide

FOV, we can interpolate stereo views close to the views of user eyes. It is possible to implement on current devices, and provide better visual effect to the users.

However, due to the limit of current hardware, we could only use single RGB-D camera with small FOV. There is some problem necessary to overcome for reaching the results close to the ideal case. One is that the FOV of current RGB-D cameras is not large enough for HMD usage, and the other is that only single depth image is available for each frame due to the interference between IR devices.

To eliminate these problems, we will try to use multiple color cameras with large FOV and apply the stereo reconstruction algorithm, and then we can get images with large FOV and more depth image. In the other hand, we will also try to design a cameras system which can make IR depth cameras work alternatively, then we could get multiple depth and color images. By these attempt, we could reach the results of the ideal case in the future.

References

1. OVRVISION. http://ovrvision.com/entop/
2. AR-RIFT. http://willsteptoe.com/post/66968953089/ar-rift-part-1
3. Steptoe, W., Julier, S., Steed, A.: Presence and discernability in conventional and non-photorealistic immersive augmented reality. In: IEEE International Symposium on Mixed and Augmented Reality (ISMAR). IEEE (2014)
4. Chen, S.E., Williams, L.: View interpolation for image synthesis. In: Proceedings of the 20th Annual Conference on Computer Graphics and Interactive Techniques. ACM (1993)
5. Zitnick, C.L., Kang, S.B. Uyttendaele, M., Winder, S., Szeliski, R.: High-quality video view interpolation using a layered representation. ACM Transactions on Graphics (TOG), vol. 23, no. 3. ACM (2004)
6. Fehn, C.: Depth-image-based rendering (DIBR), compression, and transmission for a new approach on 3D-TV. Electronic Imaging 2004. International Society for Optics and Photonics (2004)
7. Smolic, A., Muller, K., Dix, K., Merkle, P., Kauff, P., Wiegand, T.: Intermediate view interpolation based on multiview video plus depth for advanced 3D video systems. In: 15th IEEE International Conference on Image Processing. IEEE (2008)
8. Ndjiki-Nya, P., Koppel, M., Doshkov, D., Lakshman, H., Merkle, P., Muller, K., Wiegand, T.: Depth image-based rendering with advanced texture synthesis for 3-D video. IEEE Trans. Multimedia 13(3), 453–465 (2011)
9. Zinger, S., Luat, D., de With, P.H.N.: Free-viewpoint depth image based rendering. J. Vis. Commun. Image Represent. 21(5), 533–541 (2010)
10. Lai, C.-J., Han, P.-H., Hung, Y.-P.: View interpolation for video see-through head-mounted display. In: ACM SIGGRApPH Posters. ACM (2016)
11. Zhang, Z.: A flexible new technique for camera calibration. IEEE Trans. Pattern Anal. Mach. Intell. 22(11), 1330–1334 (2000)
12. Garrido-Jurado, S., Muoz-Salinas, R., Madrid-Cuevas, F.J., Marn-Jimnez, M.J.: Automatic generation and detection of highly reliable fiducial markers under occlusion. Pattern Recogn. 47(6), 2280–2292 (2014)
13. Garrido-Jurado, S., Muoz-Salinas, R., Madrid-Cuevas, F.J., Medina-Carnicer, R.: Generation of fiducial marker dictionaries using mixed integer linear programming. Pattern Recogn. 51, 481–491 (2016)

Workshop on Human Identification for Surveillance (HIS) Methods and Applications

Multi-cue Information Fusion for Two-Layer Activity Recognition

Yanli Ji[1]([⊠]), Jiaming Li[2], Hong Cheng[1], Xing Xu[1], and Jingkuan Song[3]

[1] University of Electronic Science and Technology of China, Chengdu, China
yanliji@uestc.edu.cn
[2] State University of New York at Stony Brook, New York, USA
[3] Columbia University, New York, USA

Abstract. Human activities involve complex multi-cue information. We propose a multi-cue information fusion based two-layer recognition approach for visual activity recognition. On the bottom layer, we learn features of body motion, interactive objects and scenes related with activities using deep networks. On the top layer of recognition, we fuse multi-cue information for activity recognition. In our experiments, we evaluate the performance of each single-cue information and various combinations of multi-cue information in activity recognition. We evaluate the effectiveness of two fusion methods, a linear support vector machine (SVM) classifier and a fully connected network. Experimental results illustrate that scene and body motion provide larger contributions for activity recognition, and recognition by fusing multi-cue information achieves 3%–12% higher MAcc than using single-cue information in the CCV database. Compared with state-of-the-art works, our approach achieves high level results both in CCV and UCF-101 databases.

1 Introduction

Visual activity recognition has wide applications in surveillance, video search, etc. Because of the effect from light, cloth color, person number, various activity categories etc., activity recognition is still a big challenging problem. In traditional frameworks, activity recognition mainly relies on human centroid motion features [1,2]. However, human activities also involve complex context information except motion, e.g. interactive objects, scenes. Recognizing human activities by fusing various context information is beneficial to applications of event detection in surveillance and human daily life understanding.

The effectiveness of each single-cue information has been certified in previous works. Obviously, human centroid motion is a critical information in human activities. For motion based activity recognition, recurrent neural networks (RNNs) with Long Short-Term Memory (LSTM) efficiently learn feature representations and model long-term temporal dependencies automatically in activity sequences [3–6]. LSTM is a good choice to learn motion features in activity sequences. Objects that involved in human activities are most frequently used as context information in activity recognition. Li et al. [7] presented a mutual

© Springer International Publishing AG 2017
C.-S. Chen et al. (Eds.): ACCV 2016 Workshops, Part III, LNCS 10118, pp. 273–285, 2017.
DOI: 10.1007/978-3-319-54526-4_21

context model to jointly model objects and human poses in human-object inter-action activities. Using the context model, object detection provides a strong prior for better human pose estimation, while human pose estimation improves the accuracy of detecting the objects that interact with human. Moreover, Baldassano et al. [8] studied on how the brain builds complex relationships between human pose and object identity into a new percept, and indicated that human brain encoded human-object interactions as more than the sum of their parts. Therefore, combining related objects, human centroid motion and interaction relationships of them are very important for accurate activity recognition. Furthermore, scenes where activities occur is another crucial information in activity recognition. For example, people usually play basketball in a basketball court, play football in a grassland and put birthday cakes on tables. Russakovsky et al. [9] introduced the Imagenet database and gave a review on current approaches for image classification. It can be used for scene feature learning. With information of multiple context subjects, an efficient approach is required to fuse these features for activity recognition.

There are many approaches fusing multiple information for image caption generation [10,11]. For activity recognition, Venugopalan et al. [12] trained LSTM model by fusing visual features and optical flow on video-sentence pairs of human action and learned to associate a sequence of video frames to a sequence of words in order to generate a description of the event video. Ordonez et al. [13] presented a generic deep framework for activity recognition based on convolutional and LSTM recurrent units. Rohrbach et al. [5] learned robust visual classifiers to distinguish verbs, objects, and places in a video. Then using discriminant scores obtained by visual classifiers (SVM) as new input features, captions of videos were generated using LSTM. The advantage of using deep networks for fusion is that deep networks learn fusion weights automatically. In addition, Wu et al. [14] use three CNN-LSTM nets to model spatial, short-term motion and audio clues respectively, then used logistic regression to train the best class-specific fusion weights for streams to generate the final predictions using weighted streams. Weighted combination is another way for information fusion. To find a proper fusion method, we evaluate the performance of deep networks and linear combination models on fusing multi-cue information for activity recognition.

In this paper, we propose a multi-cue information fusion based two-layer activity recognition approach. Figure 1 shows the diagram of the framework. We realize activity recognition in two layers. On the bottom layer, deep networks, e.g. Convolutional Neural Networks (CNNs) and LSTMs, are used to learn features of context subjects, i.e. human centroid motion, interactive objects and scenes, and generate a primary determination score for each cue. On the top layer, we use multi-cue fusion approach to fuse features of multiple context subjects for activity recognition. We evaluate the effectiveness of two frequently used fusion methods, a linear SVM classifier and a fully connected network. The major contributions of this paper are (1) Proposing a two-layer activity recognition approach, which includes bottom-layer feature learning and top-layer multi-cue fusion; (2) Using fusion approaches to fuse multi-cue information for activity recognition.

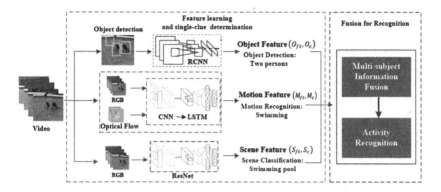

Fig. 1. Two-layer activity recognition based on multi-cue information fusion.

The paper is organized as follows. Section 3 explains our proposed framework in detail. In Sect. 4, we illustrate experiments and analyses of the proposed approach. Finally, a conclusion is given in Sect. 5.

2 Related Work

RNN-LSTM network is frequently applied for human action analysis and video caption because of its effectiveness on sequence learning. Grushin et al. [3] demonstrated that LSTM's performance remains robust even as experimental conditions deteriorate, i.e. lower quantities of training data, poorer video quality and shorter available input data sequences. Ng et al. [15] modeled the video as an ordered sequence of frame, and employed a recurrent neural network that uses Long Short-Term Memory (LSTMs) cells to learn motion features in the frame sequence. Donahue et al. [4] presented Long-term Recurrent Convolutional Networks (LRCNs), a class of models that unifies the state-of-the-art in visual and sequence learning. Using LRCNs, the work showed good performance on activity recognition, image description and video description. For skeleton based action recognition, Mahasseni et al. [6] and Zhu et al. [16] used Regularizing Long Short Term Memory (RLSTM) to model skeleton sequences and achieved high recognition accuracy.

There have been lots of works using deep networks for information fusion [5,12–14]. Moreover, Ibrahim et al. [17] presented a two-layer deep temporal model for group activity recognition. In their work, LSTM model was designed to represent action dynamics of individual people in a sequence and another LSTM model was designed to aggregate person-level information for whole activity understanding. Therefore, LSTM network is a good choice for multiple information fusion in activity recognition. Except that, feature fusion is also performed at a middle-level way. Du et al. [18] presented an end-to-end two-layer RNN for skeleton based action recognition. They divided the human skeleton into five parts according to human physical structure, and then separately feed them to five subnets. As the number of layers increases, the representations extracted by the subnets are hierarchically fused to be the inputs of higher layers. In addition, Ma et al. [19] presented the Reduced Analytic Dependency Model (RADM) to fuse action features in score level. Liu et al. [20] used a sample-specific late fusion method to propagate the fusion weights, where positive samples have higher fusion scores than negative samples. Linear fusion is a traditional but efficient method. In this paper, we evaluate deep networks and linear fusion methods to select a proper fusion method for multi-cue information fusion based activity recognition.

3   Two-Layer Activity Recognition

This section explains the proposed two-layer activity recognition. In the bottom layer processing, deep networks are used to learn features of multiple cues, and generate a determination score for each cue. On the top layer, we use multi-cue information fusion approach for activity recognition.

3.1   Motion Feature Learning

Donahue et al. [4] presented a long-term recurrent convolutional networks (LRCN) approach to learn spatial-temporal information of activities. Using LRCN networks, we learn motion features and primary determination scores of activity categories in human activities. The LRCNs combines CNNs and RNNs to learn spatial-temporal information of activities. AlexNet [21] is used in the LRCNs for feature learning. Finally, a LSTM network is adopted for activity sequence learning. In the approach, both RGB and optical flow information are considered and are processed independently in two channels. Finally, the approach fuses the two channels by weighted sum the determination scores of RGB and optical flow images.

In our approach, giving the network a strong initialization to accelerate training speed, we finetune the revealed pre-trained model with video clips of 16 frames. We also use two individual channels, one for RGB frames and the other for optical flow. Optical flow calculation follows paper [22], and then we convert optical flow data to optical flow images by centering optical flow values around 128 and multiplying by a scalar to make flow values fall between 0 and 255. LRCNs is used to learn motion features in activities and to generate primary determination scores which indicate activity categories. Determination scores of activity categories are recorded as M_c, and features of activities on higher layer of LSTM are recorded as M_{ft}.

3.2   Interactive Object Detection

Li et al. [7,8] released works on human-object interaction, and certified that interactive objects was beneficial for human activity recognition. To determine objects that the human interacting with in activities, we use R-CNN [23] to detect object categories and object numbers. For the end, we firstly generate category-independent region proposals using Binarized Normed Gradients (BING) [24] for objectness estimation, which produces only a small number of proposals and has high computational efficiency. Secondly, a Convolutional Neural Network (CNN) is utilized to learn object features from these proposals, and a SVM classifier is trained to score extracted features of each proposal for object detection following the framework [23]. To avoid detecting an object repeatedly, a greedy non-maximum suppression for each class is absolutely necessary.

In our approach, object detection is a middle step for activity recognitions. To estimate contributions from objects in activity recognition, we learn features of objects, O_{ft}, by CNN, and also obtain determination scores of objects, O_c, which are scored by SVM. Finally, we integrate object features and objects scores respectively with other information for activity recognition.

3.3 Scene Classification

Scene classification is performed based on the deep residual network, ResNet, proposed by He et al. [25], which is the winner in the ILSVRC 2015 classification task. To avoid overfitting and acceleration, we finetune the pre-trained 50-layer ResNet network. The pre-trained ResNet is trained on the ILSVRC-2012 database containing 1.2M images [9]. Although ResNet is much deeper than any other networks, it is proven that this net is faster and far less complex than any previous networks such as VGG nets [26] and GoogLeNet [27].

For finetuning, we separate scenes of videos to several categories in an activity database, and construct a target vector containing the same element number with the category number to label each scene category. To represent scenes in activities, we also extract learned features S_{ft} on high layer of ResNet and keep final determination results S_c for information fusion in activity recognition.

3.4 Multi-cue Information Fusion

On the top layer of the proposed approach, we use fusion methods to integrate multi-cue information, i.e. motion, interactive objects and scenes, for activity recognition. We evaluate two major fusion methods for activity recognition, a linear SVM classifier and a fully connected network. Diagrams of two methods are shown in Fig. 2.

To realize the evaluation in an impartial situation, we do not add weights to multi-cue features. Using a linear SVM classifier (Fig. 2(a)), we combine features or determination scores of motion, objects and scenes into a long vector and use it as an input feature for classifier training. In a fully connected network (Fig. 2(b)), we input combined feature vectors to the network for training and recognition.

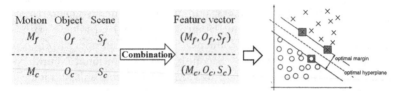

(a) Linear fusion using linear SVM.

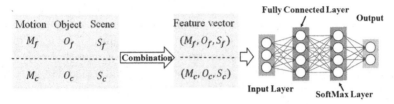

(b) Fusion using fully connected network.

Fig. 2. Multi-cue information fusion for recognition.

4   Experiments and Results

4.1   Databases

We evaluate the multi-cue fusion based recognition approach on the Columbia Consumer Video (CCV) database [28] and the UCF-101 database [29].

CCV database contains 9317 videos captured from YouTube which are annotated into 20 categories. Since this paper discusses human activity recognition, we pick out those videos with obvious topic of human activities except video categories of Dog, Cat and Birds, and totally 17 categories, i.e. Playing Basketball, Playing Baseball, Playing Soccer, Ice Skating, Skiing, Swimming, Biking, Graduation, Birthday, Wedding Ceremony, Wedding Dance, NonMusic Performance, Parade, Beach, Playground, etc. Then we separate the selected video set into training and testing sets, respectively. In the selected video set, 60 video samples are used for training and the left samples for testing. Here, Mean accuracy (MAcc) is calculated to evaluate the overall performance on selected video set.

UCF-101 database contains large-scale videos of human activities. It totally consists of 13,320 video clips categorized into 101 action categories, i.e. Baby Crawling, Playing Cello and Typing, etc. Compared with the CCV database, qualities of video clips in UCF-101 database are much better, with less video jitter and rotating. In our experiments, we adopt the whole UCF-101 database to evaluate our approach. Similarly, we split it into two parts for training and testing tasks and compute the MAcc to measure the performance. We refer to the work [29] to split training and testing set.

4.2   Activity Recognition Using Motion Information

We adopt LRCN to learn motion features from RGB images and optical flow, and obtain determination scores for activity classification. We fine-tune the network using selected activity videos in the CCV and UCF-101 databases. In our experiments, the learning rate starts from $R = 10^{-3}$ for both of RGB images and optical flow. But the the learning rate decreases to $10^{-1}R$ after every 10,000 iterations for RGB images, and decreases to $10^{-1}R$ after every 20,000 iterations for optical flow images. Training with LRCN for RGB images stops at 30,000 iterations and stops at 70,000 iterations for optical flow images. Iteration number in our experiment is larger than [4], which produces a smaller loss and improves recognition accuracy.

We list the motion based recognition results in the CCV and the UCF-101 databases in Table 1. We do experiments using single-cue information for activity recognition in order to evaluate motion features, thus we simply use a linear SVM classifier for recognition. In Table 1, we show results of recognition using RGB images, using optical flow and fusing RGB images and optical flow, respectively. The experiment shows that RGB information performs much better than optical flow in activity recognition. Obviously, fusing RGB images and optical flow performs better than using single feature for activity recognition.

Table 1. Activity recognition results using motion information.

Database	RGB (MAcc %)	Optical flow (MAcc %)	RGB and optical flow (MAcc %)
CCV	69.3	37.5	71.3
UCF-101	86.7	14.0	87.9

4.3   Object Detection in Activities

Combining the region proposal approach, BING [24], and R-CNN [23], we detect objects and count their number in activity videos. Since categories and numbers of objects varies greatly in different activity videos, it is impossible to use O_{ft} to represent interactive objects in activity videos. Therefore, we simply use the determination score O_c for object representation, which indicates the category and the number of each object in an activity video. Statistic results show that an average accuracy of 67.3% is obtained for object detection in the CCV database. Using O_c for activity recognition, we only get a result of 17.9% (MAcc) in the CCV database. In the UCF-101 database, the activity recognition result is 6.7% (MAcc) using O_c independently. Results of activity recognition using object information independently are very poor in both the two databases because the relative dependency between objects and activities is not very high. The other reason is that the framework in [23] does not cover all the objects in our experiments, which leads to detection error. Enlarging the object data space is our future work.

4.4   Scene Classification

Based on the visual characters of places where activity occurs, we manually define scene categories in the CCV and the UCF-101 databases. In the CCV database, we separate scenes into 11 categories, and define them to be Beach, Church, Swimming Pool, Beach, Desk, Fence, Grassland, Basketball Court, Pathway, Playground, Snowfield and Stage. Similarly, we define 14 scene categories in the UCF-101 database. The categories are Basketball court, Floor, Snow field, Arena, Billiard table, Desk, Fence, Grassland, Gymnasium, Pathway, Sky, Wall, Water and No Scene. Furthermore, we define an empty category, No Scene, to label the situation that foreground objects take up nearly all the places on video frames, so that it is difficult to understand scenes.

We fine-tune the deep residual network, ResNet [25] for scene classification. In the experiment of fine-tuning, we start stochastic gradient descent solver at a learning rate of $R = 10^{-2}$ with a mini-batch size of 4. Momentum is set to 0.9. Similarly, the learning rate decreases to $10^{-1}R$ after every 10,000 iterations. The training stops after 60,000 iterations.

Using fine-tuned ResNet, we obtain a MAcc of 90.4% for 11-category scene classification in the CCV database and a MAcc of 79.1% for 14-category scene classification in the UCF-101 database. We test the contribution of scene information on activity recognition. Results are shown in Table 2. Recognition is realized by a linear SVM classifier. Experiments illustrate that scene information makes contribution in activity recognition, and obviously feature S_{ft} performs much better than S_c in both the CCV and the UCF-101 databases. Therefore, feature S_{ft} is selected to represent scene information in following experiments of multi-cue fusion for activity recognition.

Table 2. Results of activity recognition using scene information.

Database	Feature S_{ft} (MAcc %)	Score S_c (MAcc %)
CCV	80	66.9
UCF-101	62.8	33.1

4.5   Multi-cue Information Fusion for Activity Recognition

We test various combination modes for activity recognition. Table 3 shows us all experiment results obtained in the CCV database. We show experiment results of the UCF-101 database in Table 4.

Compared with using single-cue information for activity recognition, recognition by fusing multi-cue information always reaches higher accuracies. Figure 3 illustrates the result clearly. The same conclusion can also be made in the UCF-101 database. Recognition by fusing multi-cue information achieves 3%–12% higher MAcc than using single-cue information in the CCV database,

Table 3. Results of activity recognition by fusing multi-cue information in the CCV database.

Multi-subject information	Fused features	Fusion method	MAcc (%)
Motion+Object	M_c, O_c	Linear SVM	71.9
Motion+Scene	M_c, S_c	Linear SVM	75.6
Motion+Scene	M_c, S_{ft}	Linear SVM	80.2
Motion+Scene	M_{ft}, S_{ft}	Linear SVM	80.5
Motion+Scene	M_{ft}, S_c	Linear SVM	71.5
Motion+Scene+Object	M_c, S_c, O_c	Linear SVM	77.3
Motion+Scene+Object	M_c, S_c, O_c	Fully connected net	79.3
Motion+Scene+Object	M_c, S_{ft}, O_c	Linear SVM	80.5
Motion+Scene+Object	M_c, S_{ft}, O_c	Fully connected net	**83.1**
Motion+Scene+Object	M_{ft}, S_{ft}, O_c	Linear SVM	**82.7**
Motion+Scene+Object	M_{ft}, S_{ft}, O_c	Fully connected net	**82.9**

Table 4. Results of activity recognition by fusing multi-cue information in the UCF-101 database.

Multi-subject information	Fused features	Fusion method	MAcc (%)
Motion+Object	M_c, O_c	Linear SVM	89.9
Motion+Scene	M_c, S_c	Linear SVM	90.4
Motion+Scene	M_c, S_{ft}	Linear SVM	73.7
Motion+Scene	M_{ft}, S_{ft}	Linear SVM	80.9
Motion+Scene	M_{ft}, S_c	Linear SVM	88.1
Motion+Scene+Object	M_c, S_c, O_c	Linear SVM	**90.7**
Motion+Scene+Object	M_c, S_c, O_c	Fully connected net	**91.1**
Motion+Scene+Object	M_c, S_{ft}, O_c	Linear SVM	75.2
Motion+Scene+Object	M_c, S_{ft}, O_c	Fully connected net	77.1
Motion+Scene+Object	M_{ft}, S_{ft}, O_c	Linear SVM	80.7
Motion+Scene+Object	M_{ft}, S_{ft}, O_c	Fully connected net	81.2

and achieves 4%–30% higher MAcc in the UCF-101 database. Moreover, for multi-cue fusion, we find that the fully connected network performs better than other methods, and Linear SVM also provides an encouraging fusion effectiveness. Using the fully connected network, we obtain the highest MAcc of 83.1% in the CCV database, and 91.1% in the UCF-101 database.

In the CCV database, the feature fusion mode of M_{ft}, S_{ft}, O_c outperforms other fusion modes except the fusion mode of M_c, S_{ft}, O_c. In the UCF-101 database, M_c, S_c, O_c outperforms all other feature fusion modes. It illustrates that

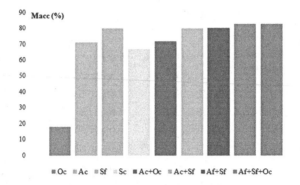

Fig. 3. Comparison of activity recognition using single-cue and multi-cue information.

scene feature S_{ft} contributes largest for activity recognition in the CCV database, and scene information has tight relationship with Activities. But in the UCF-101 database, motion information contributes larger than scene information. Comparing results in Tables 1 and 2, we can also obtain the same conclusion.

4.6 Comparing with State-of-the-Art Works

We compare our proposed approach with state-of-the-art works in activity recognition. Comparisons in the CCV and the UCF-101 databases are shown in Tables 5 and 6, respectively. The comparison shows that our proposed approach fusing multi-cue information outperforms most of previous visual recognition approaches in two databases except Wu et al. [14]. Using visual information, recognition results in our approach are 0.9% (MAcc) lower than [14] in the CCV database, and 1.1% lower than [14] in the UCF-101 database. The comparison indicates that our approach reaches state-of-the-art level.

Table 5. Comparing with state-of-the-art works in the CCV database.

Approaches	Information	MAcc (%)
Jiang et al. [28]	Visual	59.5
Xu et al. [30]	Visual	60.3
Ma et al. [19]	Visual	63.4
Jhuo et al. [31]	Visual + Audio	64.0
Liu et al. [20]	Visual	68.2
Wu et al. [14]	Visual	84
Ours	Visual	**83.1**

Table 6. Comparing with state-of-the-art results in the UCF-101 database.

Approaches	Information	MAcc (%)
Donahue et al. [4]	Visual	82.9
Srivastava et al. [32]	Visual	84.3
Simonyan et al. [33]	Visual	88.0
Ng et al. [15]	Visual	88.6
Lan et al. [34]	Visual	89.1
Wu et al. [14]	Visual	92.2
Ours	Visual	**91.1**

5 Conclusion

Human activities involve complex multi-cue information and relative relationships. It is not appropriate to use motion information for activity recognition. In this paper, we proposed a multi-cue information fusion based two-layer recognition approach for activity recognition. On the bottom layer, we learn features of human centroid motion, and objects, scenes which relate to activities using deep networks. On the top layer of recognition, we test fusion methods, a SVM classifier and a fully connected network, to fuse multi-cue information for activity recognition. In experiments, we evaluate contributions of each single-cue information and various combinations of multi-cue information for activity recognition, and also evaluate two fusion methods. Experiments illustrate that scene and motion contribute larger on activity recognition, and recognition by fusing multi-cue information achieves 3%–12% higher MAcc than single-cue information in the CCV database, and achieves 4%–30% higher MAcc than single-cue information in the UCF-101 database. Compared with previous works, our approach outperforms most visual recognition approaches, and achieves state-of-the-art level in both of the CCV and the UCF-101 databases. Nevertheless, it is possible to improve the performance of our approach. One possible direction in future is to improve the effectiveness of each single-cue information for activity recognition. Moreover, developing a multi-cue fusion model to fuse relative relationships of subjects is the other research direction.

Acknowledgement. This research is supported by the Natural Science Foundation of China (NSFC) under grant No. 61305043 and grant No. 61673088.

References

1. Poppe, R.: A survey on vision-based human action recognition. Image Vis. Comput. **28**, 976–990 (2010)
2. Chaquet, J.M., Carmona, E.J., Fernndez-Caballero, A.: A survey of video datasets for human action and activity recognition. Comput. Vis. Image Underst. **117**(6), 633–659 (2013)

3. Grushin, A., Monner, D., Reggia, J., Mishra, A.: Robust human action recognition via long short-term memory. In: Proceedings of International Joint Conference on Neural Networks (IJCNN) (2013)
4. Donahue, J., Hendricks, A.L., Guadarrama, S., Rohrbach, M., Venugopalan, S., Saenko, K., Darrell, T.: Long-term recurrent convolutional networks for visual recognition and description. In: Proceedings of CVPR (2015)
5. Rohrbach, A., Rohrbach, M., Schiele, B.: The long-short story of movie description. Pattern Recognit. **9358**, 209–221 (2015)
6. Mahasseni, B., Todorovic, S.: Regularizing long short term memory with 3D human-skeleton sequences for action recognition. In: Proceedings of CVPR (2016)
7. Yao, B., Fei-Fei, L.: Recognizing human-object interactions in still images by modeling the mutual context of objects and human poses. IEEE Trans. Pattern Anal. Mach. Intell. Arch. **34**, 1691–1703 (2012)
8. Baldassano, C., Beck, D.M., Fei-Fei, L.: Human-object interactions are more than the sum of their parts. Cerebral Cortex 1–13 (2016)
9. Russakovsky, O., Deng, J., Su, H., Krause, J., Satheesh, S., Ma, S., Huang, Z., Karpathy, A., Khosla, A., Bernstein, M., Berg, A.C., Fei-Fei, L.: Imagenet large scale visual recognition challenge. In: arXiv:1409.0575v3 (2015)
10. Young, P., Lai, A., Hodosh, M., Hockenmaier, J.: From image descriptions to visual denotations: new similarity metrics for semantic inference over event descriptions. Trans. Assoc. Comput. Linguist. **2**, 67–78 (2014)
11. Srivastava, N., Salakhutdinov, R.: Multimodal learning with deep boltzmann machines. J. Mach. Learn. Res. **15**, 2949–2980 (2014)
12. Venugopalan, S., Rohrbach, M., Donahue, J., Mooney, R., Darrell, T., Saenko, K.: Sequence to sequence-video to text. In: Proceedings of ICCV (2015)
13. Ordnez, F.J., Roggen, D.: Deep convolutional and lstm recurrent neural networks for multimodal wearable activity recognition. Sensors **16**, 1–25 (2016)
14. Wu, Z., Jiang, Y., Wang, X., Ye, H., Xue, X., Wang, J.: Fusing multi-stream deep networks for video classification. In: arXiv:1509.06086 (2015)
15. Ng, J.Y., Hausknecht, M., Vijayanarasimhan, S.: Beyond short snippets: deep networks for video classification. In: Proceedings of CVPR (2015)
16. Zhu, W., Lan, C., Xing, J., Zeng, W., Li, Y., Shen, L., Xie, X.: Co-occurrence feature learning for skeleton based action recognition using regularized deep LSTM networks. In: Proceedings of AAAI (2016)
17. Ibrahim, M.S., Muralidharan, S., Deng, Z., Vahdat, A., Mori, G.: A hierarchical deep temporal model for group activity recognition. In: Proceedings of CVPR (2016)
18. Du, Y., Wang, W., Wang, L.: Hierarchical recurrent neural network for skeleton based action recognition. In: Proceedings of CVPR (2015)
19. Ma, A.J., Yuen, P.C.: Reduced analytic dependency modeling: robust fusion for visual recognition. Int. J. Comput. Vis. **109**, 233–251 (2014)
20. Liu, D., Lai, K., Ye, G., Chen, M., Chang, S.: Sample-specific late fusion for visual category recognition. In: Proceedings of CVPR (2013)
21. Krizhevsky, A., Sutskever, I., Hinton, G.E.: Imagenet classification with deep convolutional neural networks. In: Proceedings of NIPS (2012)
22. Liu, C.: Beyond Pixels: Exploring New Representations and Applications for Motion Analysis. Massachusetts Institute of Technology, Massachusetts (2009)
23. Girshick, R., Donahue, J., Darrell, T., Malik, J.: Rich feature hierarchies for accurate object detection and semantic segmentation. In: Proceedings of CVPR (2014)

24. Cheng, M., Zhang, Z., Torr, P.: BING: binarized normed gradients for objectness estimation at 300fps. In: Proceedings of CVPR (2014)
25. He, K., Zhang, X., Ren, S., Sun, J.: Deep residual learning for image recognition. In: arXiv:1512.03385 (2015)
26. Simonyan, K., Zisserman, A.: Very deep convolutional networks for large-scale image recognition. In: Proceedings of ICLR (2015)
27. Szegedy, C., Liu, W., Jia, Y., Sermanet, P., Reed, S., Anguelov, D., Er-han, D., Vanhoucke, V., Rabinovich, A.: Going deeper with convolutions. In: Proceedings of CVPR (2015)
28. Jiang, Y., Ye, G., Chang, S., Ellis, D., Loui, A.: Consumer video understanding: a benchmark database and an evaluation of human and machine performance. In: Proceedings of ACM ICMR (2011)
29. Soomro, K., Zamir, A., Shah, M.: Ucf101: a dataset of 101 human actions classes from videos in the wild. In: CRCV-TR-12-01 (2012)
30. Xu, Z., Yang, Y., Tsang, I., Sebe, N., Hauptmann, A.: Feature weighting via optimal thresholding for video analysis. In: Proceedings of ICCV (2013)
31. Jhuo, I., Ye, G., Gao, S., Liu, D., Jiang, Y., Lee, D., Chang, S.: Discovering joint audio-visual codewords for video event detection. Mach. Vis. Appl. **25**, 33–47 (2014)
32. Srivastava, N., Mansimov, E., Salakhutdinov, R.: Unsupervised learning of video representations using LSTMs. In: CoRR (2015)
33. Simonyan, K., Zisserman, A.: Two-stream convolutional networks for action recognition in videos. In: Proceedings of NIPS (2014)
34. Lan, Z., Lin, M., Li, X., Hauptmann, A.G., Raj, B.: Beyond gaussian pyramid: multi-skip feature stacking for action recognition. In: CoRR (2014)

Piecewise Video Condensation for Complex Scenes

Yingying Chen[1,2(✉)], La Zhang[1,2], Jinqiao Wang[1,2], and Hanqing Lu[1,2]

[1] National Laboratory of Pattern Recognition,
Institute of Automation, Beijing, China
{yingying.chen,la.zhang,jqwang,luhq}@nlpr.ia.ac.cn
[2] University of Chinese Academy of Sciences, Beijing, China

Abstract. Video synopsis or condensation provides an efficient way to video storage and browsing. Lots of improvements have been made for boosting the speed or improving the condensed quality, which have shown promising results. However, most of the existing approaches cannot effectively deal with complex scenes, such as sudden changes or background object movement. In this paper, we propose a robust video condensation approach for complex scenes. A video segmentation method is designed to analyze the background complexity and divide the input video into several segments. The advantage is two-fold: one is to judge the complexity of backgrounds; the other is to generate a piecewise background image for each segment. Then, we adopt a divide-and-conquer strategy for video condensation. We keep the original video segments for complex backgrounds while maximally condense the other segments. Next, we introduce a feedback scheme and a selective diffusion strategy to keep the integrity of foreground objects, followed by a sticky trajectory method to remove noisy fragments and reduce blinking effect. Furthermore, an adaptive truncation strategy is introduced to raise the condensation ratio and improve the visual quality. Experimental results demonstrate the effectiveness of our approach.

1 Introduction

Nowadays millions of surveillance cameras are installed for abnormal incidents, criminal evidences detection and traffic management etc. The world witnesses a large amount of video data recorded for security purposes every day. Browsing and indexing activities in these abundant videos are a time consuming and boring work for viewers. To alleviate the burden for video browsing and searching effectively, many approaches of video condensation were proposed, such as fast forwarding [1–3], video summarization [4], video montage [5], video synopsis [6–8] and ribbon carving [9].

The goal of video condensation is to shorten the original videos with minimum information loss. Video synopsis, first proposed by Peleg and his colleagues [6–8], showed better performance on controlling the loss of information than other previous video abstraction approaches. It mainly involves three steps:

© Springer International Publishing AG 2017
C.-S. Chen et al. (Eds.): ACCV 2016 Workshops, Part III, LNCS 10118, pp. 286–300, 2017.
DOI: 10.1007/978-3-319-54526-4_22

(1) Extract moving objects from the original video to constitute the basic processing unit "tube" (a tube is a spatio-temporal sequence of a moving object); (2) Generate background images by shifting temporal median window over original frames; (3) Rearrange extracted tubes and densely stitch them into background images. Online video condensation (OVC) proposed an online framework [10–12], by transforming the tube rearrangement into a stepwise optimization problem. However, these approaches ignored the complexity of surveillance scenes. As illustrated in Fig. 1, in these complex scenes such as background sudden changes or continuous background object movement, moving objects are difficult to be extracted completely. Moreover, it is also hard to generate proper background images for tube stitching. For example, when the elevator door is open or the horizontal sliding door is open in Fig. 1(a), they are all judged as foreground. The condensation results are shown in Fig. 1(b), where we can see that the visual effects are not acceptable when the moving doors are directly stitched into the backgrounds. Therefore, this kind of background complexity analysis is critical to improve the quality of video condensation. To deal with this problem, we adopt a video segmentation method to estimate the background complexity and divide the input video into segments. For these complex segments such as door open in Fig. 1(a), we keep the original video segments into the synopsis. For the other segments, we maximally condense the content and concatenate condensed results with the complex segments.

(a)   (b)

Fig. 1. Examples of complex backgrounds. (a) Original videos: elevator, sliding door. (b) The condensation results with online video synopsis.

For complex scenes, the integrity of moving objects and the continuity of object trajectories are also important factors for the condensation quality. Therefore, to adapt various changes in complex scenes, we introduce a self-adaptive background modeling approach based on sample consensus with a flexible feedback scheme to automatically adjust the model parameters. Besides, a selective diffusion method is employed to overcome the problems like incomplete foregrounds or false detections brought by intermittent moving objects. For the continuity of object trajectories and reducing blinking effect in a condensed video for better visual effects, sticky tracking was proposed in [11] to merge nearest object cubes before tracking. This method reduces the blinking effect caused by occlusions between objects, however, it also sticks patches caused by background noise and fragments of other objects. Therefore, we argue that it is more reasonable that we generate trajectories by concatenating moving objects in consecutive frames then stick trajectories with overlapping objects. In this way, not only the blinking effect is reduced but also fragments with noise and other objects are removed.

Above all, in this paper we propose a piecewise condensation approach based on the analysis of scene complexity. Based on the background complexity analysis, we divide the input video into several segments. Then we utilize a self-adaptive background modeling approach with a feedback scheme and a selective diffusion strategy to keep the integrity of foreground objects, followed by a sticky trajectory strategy to remove noisy fragments and reduce blinking effect. Finally, we employ an adaptive truncation approach to make the condensed video more compact. The contributions of this paper are summarized as follows:

- We propose a piecewise video condensation approach for complex scenes by dividing the input video into different segments based on the analysis of background complexity and adopting a divide-and-conquer condensation strategy.
- To keep the integrity of moving objects and the continuity of object trajectories, we introduce a self-adaptive background modeling approach with a flexible feedback and selective diffusion scheme.
- We put forward a sticky trajectory strategy to remove noisy fragments and reduce blinking effect.
- We employ an adaptive truncation approach in the process of piecewise optimization to make the condensed video more compact.

2   Related Work

There has been an increasing interest in video presentation and summarization for along time, which is critical for video storage, browsing and indexing.

For video summarization, key frames were usually selected to form a new representative image. Based on maximum frame discrepancy strategies, key frames were usually selected in these approaches. Fast forwarding [1] or video skimming [2] was one efficient video browsing solution, its idea was selecting some representative frames as key frames to replace the whole video, the remaining frames were skipped. For this technique, Choosing the frames with high interest or high activity adaptively adaptively was not an easy task. Some adaptive methods for choosing key frames were proposed [3], but the biggest problem was the big loss of information, especially the fast activities during the dropped frames.

Some researchers generated new images using regions of interest (ROI) beyond the whole frames. For example, video mosaic [13] was a synthetic representation by stitching successive video frames, covering more comprehensive information than a single key-frame. Another typical work was video collage [14]. A video sequence was compacted to get a single image by seamlessly arranging ROIs on a given canvas. Storyboards [15] and narratives [16] represented the course of events by a static image with an explicit temporal cue.

Video montage [5], analyzed both the spatial and temporal information, extracted the informative portions in input videos and condensated them together, its condensate rate was high but caused visual unpleasant and the total loss of context. The ribbon carving method condensed video through removing the ribbons without activities in every frame [9]. It achieved low condensation ratio and also lost a lot of context information. In dynamic video narrative [17], all duplication specific objects were seamlessly stitched into the background video according to its time axis. In terms of a high condense rate, video synopsis [7,8] had made a big success and attracted the attention of many researchers. Feng et al. [10] proposed an online method, in which tubes were filled in a spatio-temporal volume one by one like playing a Tetris game. However, motion structure was not considered, as well as the time consistency of tubes in their method. Huang et al. [18] regarded the synopsis video generation problem as a maximum a posteriori (MAP) estimation problem, where the appearing frames of object instances chronologically rearranged in real time according to an online updated synopsis table. In [11], the optimization problem of tube rearrangement was transformed into a stepwise optimization problem and used Graphic Processing Unit (GPU) and multicore technique to further improve the speed.

3   Overview

As illustrated in Fig. 2, the proposed approach includes video segmentation, background image generation, piecewise condensation, adaptive truncation and object stitching. Firstly, based on the background complexity analysis, we divide the input video into several segments. Then we introduce sample consensus model like ViBe [19] to extract moving objects with a feedback scheme and a selective diffusion strategy to keep the integrity of foreground objects, followed by a sticky trajectory method instead of sticky tracking [10] to remove

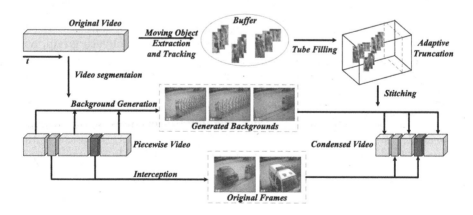

Fig. 2. Overview of piecewise video condensation.

noisy fragments and reduce blinking effect. Furthermore, an adaptive truncation strategy is introduced to raise the condensation ratio and improve the visual quality. And object tubes are stitched into the generated backgrounds with a modified Poisson editing [7]. In the followings, we will introduce these stages in details.

4 Video Segmentation Based on Background Complexity

The complexity of the background in videos can be reflected by temporal changes of background median. Videos are segmented depending on the difference among temporal medians of video clips. The segment process is shown in Fig. 3.

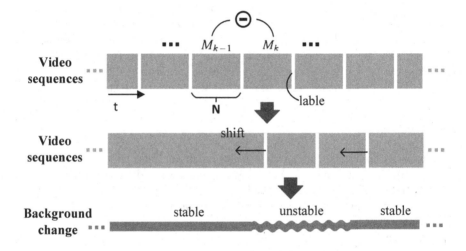

Fig. 3. Video segmentation based on background change.

Firstly, video sequences are divided into several groups in time order, each group contains N frames. Temporal medians are computed in each group, one temporal median frame M_k corresponds to one group k. The last frame id of each group is selected as a label to each group. Let M_{k-1} and M_k be two neighboring temporal median frames, L_{k-1} and L_k are their labels respectively. We compute the frame difference between (in Fig. 3, $\ominus$ stands for this process) each pair of temporal median frames in time order. We use D_k to represent the distance between M_{k-1} and M_k. If D_k exceeds the threshold D_{thre}, we consider the background changes a lot, L_k will be recorded, otherwise, M_k will be updated for next computation $\ominus$:

$$\begin{cases} M_k = (1 - \alpha)M_{k-1} + \alpha M_k \\ \alpha = \lambda D_k / H_{thre} \end{cases} \tag{1}$$

where α is the updating rate ($\alpha \in [0, 1]$), the updating process can help detect some slow changes of background (i.e. illumination variation). According to the distribution of recorded labels in the video, the whole video is segmented into stable periods and unstable periods. Since the temporal median method has hysteretic effects, we set the labels to shift $1.5N$ before. The temporal window size N decides the sensitivity of background change detection. We have tried different values of N on lots of videos and found that $N = 300$ is a proper value choice. Figure 4 shows background changes reflected by temporal median difference method on different videos and foreground areas measure the distance D_k.

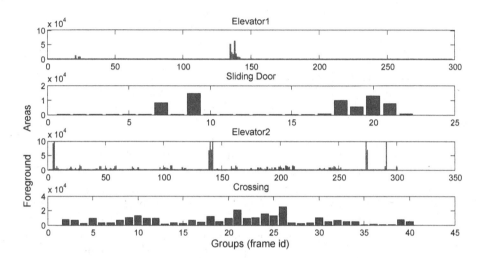

Fig. 4. Background changes reflected by temporal median difference in different complex scenes.

5   Objects Extraction and Sticky Trajectory

5.1   Objects Extraction

To the frames with stable background, we adopt an improved background modeling approach with self-adaptive ability illustrated in Fig. 5. Based on the frames of input video sequences, firstly, we build a sample based background model. Like Vibe [20], each pixel model is represented by a sequence of historical samples based on sample consensus. The background model, named B, is formed by a series of pixel models, each of which contains a set of N recent background samples:

$$B(x) = \{B_1(x), B_2(x), \ldots, B_N(x)\} \tag{2}$$

In our method, we fixed $N = 35$ to strike a balance between accuracy and speed.

$$S_t(x) = \begin{cases} 1 \; if \#\{dist(I_t(x), B_n(x)) < R, \forall n\} < \#min \\ 0 \qquad\qquad otherwise \end{cases} \tag{3}$$

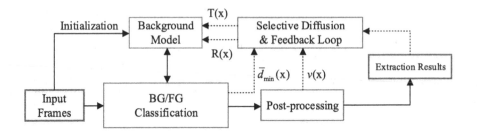

Fig. 5. Moving object detection.

where S_t is the segmentation result, $dist(I_t(x), B_n(x))$ measures the distance between a given background sample and corresponding current observation. R is the distance threshold and $\#min$ is the minimum number of matches required for a background classification. We set $\#min$ to 2 as it was demonstrated in [20].

To adapt various changes in complex scenes, a flexible feedback scheme is presented to automatically adjust the model parameters. Decision threshold R and update rate T are two most important parameters in this modeling process. we consider R and T as two pixel-level state variables. Two frame-size maps are defined to store the current value of R and T. In feedback loops, they are decided by recursive moving average map $\bar{d}_{min}$ and blinking pixels accumulators v. The recursive moving average map $\bar{d}_{min}$ is a measure of background dynamics, it is calculated by the distance between samples and current observations.

Then, a selective diffusion method is employed to overcome the problems like incomplete foregrounds or false detections brought by intermittent moving objects. For those intermittent motionless foreground objects, the pixels in their area are different from surrounding background pixels, so we prevent the diffusion from surrounding background to the foreground. For those sudden moving of background objects, the background area that they leave behind is similar with surrounding background, so we accelerate the diffusion from surrounding background to the foreground. So far, we extract moving foreground objects from video sequences frame by frame.

Because the context information in generated background images is corresponding to the original frames which is used to extract tubes, bounding boxes of extracted objects can be used for computing energy cost and object stitching directly, without any additional segmentation method like graph cut [11], which can accelerate the condensation process.

5.2   Sticky Trajectory

After moving objects extraction, we concatenate moving objects in consecutive frames together to obtain object trajectories. Although many tracking methods have been proposed, those methods may be difficult to suit for video condensation. The reason is that the break of object trajectory will cause blinking effect

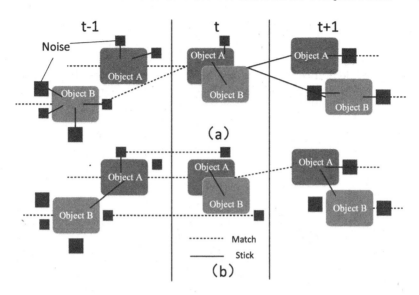

Fig. 6. Comparison between sticky tracking [11] and sticky trajectory. (a) Sticky tracking, (b) Sticky trajectory.

in condensed video. For example, if there are two objects in a video: object A and B. When a part of object A is occluded by object B at frame t, object A will loses this part in condensed video at some point. If worse, object A is occluded by object B at frame t totally, then object A will disappear abruptly and then appear again in the view. To reduce blinking effect in a condensed video for better visual effects, sticky tracking was proposed in [11] shown in Fig. 6(a). In [11], if occlusions happen to two or more object tubes, they will be merged into a single tube, as if they are sticking together. The key point is to launch merging before matching. This method reduces the blinking effect caused by occlusions between objects, but it also sticks noise caused by dynamic background and fragments of other objects. These noise also increases the number of foreground objects, which will take more time for optimization and reduce the compression ratio.

Therefore, instead of sticky tracking [11], we adopt a sticky trajectory approach to not only reduce the blinking effect but also remove fragments with noise and other objects. As shown in Fig. 6(b), different with [11], we generate trajectories by concatenating moving objects in consecutive frames at first. Then, we remove the noise trajectories that are very short because noise often abruptly appears and disappears. Finally, we stick trajectories that have overlapping objects at the some point. In this way, trajectories are generated before sticky process, which benefits removing noisy fragments caused by dynamic background and other objects easily. In addition, the compression ratio is increased with better visual effects.

6　Piecewise Condensation

For the segmented videos with stable backgrounds and unstable backgrounds, we adopt a divide-and-conquer strategy for video condensation. In the following steps, we detail the background image generation and energy minimization:

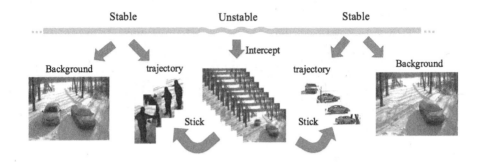

Fig. 7. Background image generation and piecewise condensation.

Background Generation. As shown in Fig. 7, based on the video segmentation, object extraction and sticky trajectory, background images can be generated respectively using temporal median method in each segment with stable background. Tubes from different segments will be stitched into corresponding background images, so as to keep the context information completely.

Energy Minimization. A divide-and-conquer strategy is adopted for video condensation. We keep the original video segments with unstable backgrounds while maximally condensing the other segments. In order to achieve visually pleasing condensed results, the problem of rearrangement of tubes is formulated as an energy function, visual overlap and lost of information are defined as energy cost [7]. Therefore, the task of video synopsis is to solve a problem of energy cost minimization. We use online strategy to transform the global cost minimization to a stepwise optimization, to make condensation faster and ensure low memory cost. Let B denote the set of tubes, the stepwise optimization is solved by a greedy algorithm [10,21]:

$$l_i^* = \arg\min_{l_i} E(l_i)$$
$$s.t. E(l_i) = E_a(l_i) + \sum_{j \in B} E_c(l_i | l_j) \tag{4}$$

where $l = \{l_i\}_{i=1}^{|B|}$ is the set of play start time of tubes, every tube T_i have one corresponding l_i. The $E_a(l_i)$ named activity cost measures the cost of extracting tube T_i from original video and stitching it into a generated background at time l_i. We mainly consider the collision cost between each of the two tubes $E_c(l_i, l_j)$. In online framework, tubes are filled in a condensed space one by one, so simply regard $E_c(l_i, l_j) = E_c(l_i|l_j)$. The greedy algorithm may lead to decreasing of condensation ratio because it condenses the video on finding locally optimal solution in each synopsis clip. The adaptive truncation could decrease this kind of impact.

6.1 Adaptive Truncation

Tubes extracted from original video will be rearranged to give new time labels. Figure 8 is the top view of tube filling in condensation space. Figure 8(a) shows the traditional way of tube filling [7]. Tube division is adopted to improve the condensation ratio but causes blinking effect in condensed video (object appears and disappears suddenly in the middle of frames). For a high visual quality, instead of segmenting tube, we truncate condensation space to improve the condensation ratio. As shown in Fig. 8(b), because tubes are always with different

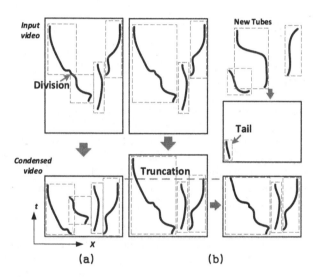

Fig. 8. Adaptive truncation.

temporal length, the compactness in the latter part is very low. In the process of adaptive truncation, the former space with higher compactness will be condensed first. Tubes already filled in the final volume are truncated into two parts, i.e., the body parts and tails. When new tubes come, the former tails are filled into the condensation space firstly, then they will be optimized together with new coming tubes. The temporal truncated location in condensation space can be estimated by mean length approximately:

$$T_{tra} = \frac{1}{n} \sum_{i=1}^{n} (l_i + L_i) \qquad (5)$$

where l_i is the start time label of tube T_i, and L_i is its length. The adaptive truncation breaks one-off process of tube filling into several steps and makes the condensed video more compact in each step. That would help to alleviate the problem brought by greedy optimization through considering a part of future tubes. Moreover, the discontinuity of tubes between condensation space is avoided [21]. Figure 9 shows the comparison results between condensation with adaptive truncation (AT) and without adaptive truncation. Condensation result with AT shows more compact.

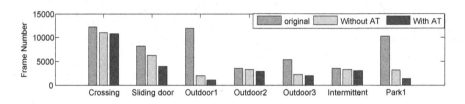

Fig. 9. Comparison results with adaptive truncation and without truncation.

7   Experiments

To evaluate the performance of the proposed approach, we carried experiments on two video datasets. One is a public dataset from [11], composed by 9 videos captured from indoor and outdoor scenes. Another one was collected by ourselves, including 9 complex scenes ("intermittent" is captured by moving camera). Table 1 presents the results of our approach. The condensation ratio (CR) denotes the frame number ratio between condensed videos and original videos, and AoMU, PoMU are abbreviations of average of memory usage and peak of memory usage, respectively.

Condensation Ratio. As shown in Table 1, the lowest condensation ratio is 1.16 while the highest one is 32.4. The lower condensation ratio generally results from complex background changes, which are truncated as unstable video segments.

Speed. As shown in Table 1, the speed decreases with the increase of the the pixel resolution. For the video sequence with solution (320×240), the processing speed is about 100 fps. For high resolution (740×576) video sequences, the process speed still has about 41 fps.

Memory Usage. For those high resolution (740×576) video sequences, the memory usage peak of our system is lower than 2.0 GB.

Subjective Evaluation. The robustness of video condensation can be reflected by the subjective evaluation of condensation quality for different scenes. Two criteria are proposed for evaluating the visual quality of condensed video, including visual pleasing and comprehensible:

Table 1. Our condensation results on 18 videos.

Video	Resolution	#Frame (Num)	Speed (fps)	CR	AoMU (MB)	PoMU (MB)
Overpass [11]	320 × 120	23950	103.1	20.91	150	167
Exit [11]	320 × 240	81538	100.6	18.36	293	313
Garden [11]	320 × 240	33826	100.9	13.68	280	284
Outdoor [11]	320 × 240	138583	98.4	2.99	316	348
Park1 [11]	352 × 288	10221	109.9	7.70	372	374
Passage [11]	352 × 288	51041	98.9	10.01	377	395
Street [11]	704 × 576	100114	42.5	8.48	1465	1530
Staircase [11]	704 × 576	46109	41.5	3.76	1466	1545
T-junction [11]	704 × 576	637470	43.6	32.40	1493	1899
Elevator1	352 × 288	89992	93.2	15.00	371	374
Elevator2	352 × 288	90001	94.0	3.47	401	439
Crossing	740 × 576	12266	42.2	1.13	1508	1959
Slidingdoor	352 × 288	8244	107.1	2.10	378	390
Outdoor1	480 × 360	11999	82.8	11.46	660	716
Outdoor2	724 × 416	3495	74.2	1.21	1082	1085
Outdoor3	352 × 288	5323	102.3	2.70	385	410
Irondoor	352 × 288	90001	97.3	2.93	385	410
Intermittent	568 × 376	3500	40.22	1.16	762	771

1. Visual pleasing: Do you think this synopsis is comfortable for viewers? You can score based on the following aspects: Overlap, blinking effect and object completeness.
2. Comprehensible: Can you infer the original object behavior information from this synopsis?
3. Overall Satisfied: Do you think the synopsis overall is comfortable for viewers?

We set the score scale as 1–5 for these two criteria, where the score below 3 means the visual quality of this condensed result is bad and the score of condensed results with accepted quality is above 4. We invited 36 participants to score the synopsis results. All the participants have strong background knowledge in video surveillance, they were requested to watch the original videos first, then compare the condensed results with original videos and give their scores on two aspects: visual pleasing and comprehensible. We compared average scores by two criteria between our method and online video condensation (OVC) method [11]. The statistical results of subjective feedbacks on two datasets are illustrated in Fig. 10. The score of our approach is close to OVC in dataset1 [11] because most videos have relatively simple backgrounds. But in the dataset of complex scenes, our method achieves better performance on visual quality.[1] The scenes

---

[1] http://pan.baidu.com/s/1i329jzn.

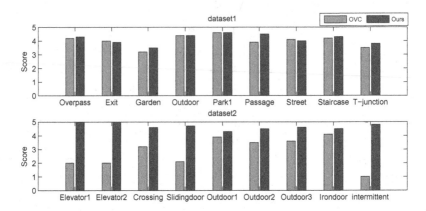

Fig. 10. Comparison results of Subjective evaluation between our method and OVC [11].

Table 2. Comparison results with the state-of-art methods.VS: Video Synopsis, RC: Ribbon Carving, CR: Condensation Ratio, OVC: Online Video Condensation.

Method	Speed	Memory	CR	Blinking effect	Robustness
VS [7]	10 fps	High	User	High	Low
RC [22]	Slow	Huge	Low	-	-
OVC [10]	100 fps+(GPU)	Low	High	Low	Low
Ours	40 fps+(CPU)	Low	High	Low	High

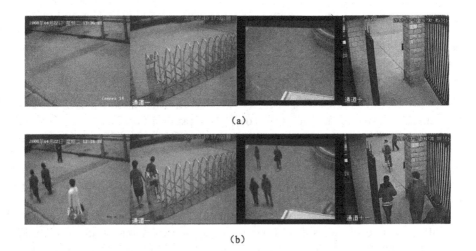

(a)

(b)

Fig. 11. Example frames of condensation video in four different scenes, which are outdoor3, Slindingdoor, outdoor1, Irondoor, from left to right respectively. (a) One frame in input video, (b) One frame in our condensation video.

"intermittent" captured by intermittent moving camera also can be condensed with a high visual quality by our approach.

Table 2 gives a summary of comparison between our approach and other state-of-art video condensation methods. It shows that compared with video synopsis [7] and Ribbon carving [10], our method has faster speed with lower memory, better visual quality. We cannot directly compare our processing speed with [11], since they run in GPU. But compared to online video condensation [11,22], our approach is more robust to different scenes. Example frames of the condensation videos are shown in Fig. 11.

8 Conclusion

A robust condensation approach based on piecewise condensation framework has been proposed in this paper. The piecewise condensation framework can condense results with high visual quality in different scenes, even the sequence captured by intermittent moving camera. We divide the input video into several clips. Then we present a self-adaptive background modeling approach with a feedback scheme and a selective diffusion strategy to keep the integrity of foreground objects, followed by a sticky trajectory strategy to remove noisy fragments and reduce blinking effect. The process of condensation is with high speed and low memory cost. Besides, an adaptive truncation is designed to refine the low condensation ratio brought by online tube filling. Experimental results show the superiority of the proposed approach.

Acknowledgment. This work was supported by 863 Program 2014AA015104, and National Natural Science Foundation of China 61273034, and 61332016.

References

1. Petrovic, N., Jojic, N., Huang, T.S.: Adaptive video fast forward. Multimed. Tools Appl. **26**, 327–344 (2005)
2. Smith, M.A., Kanade, T.: Video skimming and characterization through the combination of image and language understanding. In: 1998 IEEE International Workshop on Content-Based Access of Image and Video Database, Proceedings, pp. 61–70. IEEE (1998)
3. Höferlin, B., Höferlin, M., Weiskopf, D., Heidemann, G.: Information-based adaptive fast-forward for visual surveillance. Multimed. Tools Appl. **55**, 127–150 (2011)
4. Kim, C., Hwang, J.N.: An integrated scheme for object-based video abstraction. In: Proceedings of the Eighth ACM International Conference on Multimedia, pp. 303–311. ACM (2000)
5. Kang, H.W., Chen, X.Q., Matsushita, Y., Tang, X.: Space-time video montage. In: 2006 IEEE Computer Society Conference on Computer Vision and Pattern Recognition, vol. 2, pp. 1331–1338. IEEE (2006)
6. Pritch, Y., Rav-Acha, A., Gutman, A., Peleg, S.: Webcam synopsis: peeking around the world. In: IEEE 11th International Conference on Computer Vision, ICCV 2007, pp. 1–8. IEEE (2007)

7. Pritch, Y., Rav-Acha, A., Peleg, S.: Nonchronological video synopsis and indexing. IEEE Trans. Pattern Anal. Mach. Intell. **30**, 1971–1984 (2008)
8. Rav-Acha, A., Pritch, Y., Peleg, S.: Making a long video short: dynamic video synopsis. In: 2006 IEEE Computer Society Conference on Computer Vision and Pattern Recognition, vol. 1, pp. 435–441. IEEE (2006)
9. Li, Z., Ishwar, P., Konrad, J.: Video condensation by ribbon carving. IEEE Trans. Image Process. **18**, 2572–2583 (2009)
10. Feng, S., Lei, Z., Yi, D., Li, S.Z.: Online content-aware video condensation. In: 2012 IEEE Conference on Computer Vision and Pattern Recognition (CVPR), pp. 2082–2087. IEEE (2012)
11. Zhu, J., Feng, S., Yi, D., Liao, S., Lei, Z., Li, S.: High performance video condensation system. IEEE Trans. Circuits Syst. Video Technol. **25**, 1113–1124 (2015)
12. Sun, L., Xing, J., Ai, H., Lao, S.: A tracking based fast online complete video synopsis approach. In: 2012 21st International Conference on Pattern Recognition (ICPR), pp. 1956–1959. IEEE (2012)
13. Irani, M., Anandan, P.: Video indexing based on mosaic representations. Proc. IEEE **86**, 905–921 (1998)
14. Liu, X., Mei, T., Hua, X.S., Yang, B., Zhou, H.Q.: Video collage. In: International Conference on Multimedia 2007, Augsburg, Germany, September 2007, pp. 461–462 (2007)
15. Fu, W., Wang, J., Zhao, C., Lu, H.: Object-centered narratives for video surveillance. In: 19th IEEE International Conference on Image Processing (ICIP) 2012, vol. 8556, pp. 29–32 (2012)
16. Goldman, D.B., Curless, B., Salesin, D., Seitz, S.M.: Schematic storyboarding for video visualization and editing. ACM Trans. Graph. **25**, 862–871 (2006)
17. Correa, C.D., Ma, K.L.: Dynamic video narratives. ACM Trans. Graph. **29**, 88 (2010)
18. Huang, C.R., Chung, P.C.J., Yang, D.K., Chen, H.C., Huang, G.J.: Maximum a posteriori probability estimation for online surveillance video synopsis. IEEE Trans. Circuits Syst. Video Technol. **24**, 1417–1429 (2014)
19. Barnich, O., Van Droogenbroeck, M.: ViBe: a universal background subtraction algorithm for video sequences. IEEE Trans. Image Process. **20**, 1709–1724 (2011)
20. Olivier, B., Marc, V.D.: ViBe: a powerful random technique to estimate the background in video sequences. In: ICASSP, pp. 945–948 (2009)
21. Fu, W., Wang, J., Gui, L., Lu, H., Ma, S.: Online video synopsis of structured motion. Neurocomputing **135**, 155–162 (2014)
22. Van Droogenbroeck, M., Paquot, O.: Background subtraction: experiments and improvements for ViBe. In: 2012 IEEE Computer Society Conference on Computer Vision and Pattern Recognition Workshops (CVPRW), pp. 32–37. IEEE (2012)

Unsupervised Person Re-identification via Graph-Structured Image Matching

Bolei Xu[⊠] and Guoping Qiu

The University of Nottingham, Ningbo, China
bolei.xu@nottingham.edu.cn

Abstract. This paper presents a novel unsupervised framework to solve both person re-identification and partial person re-identification problems. For each pedestrian image, we first use an established image segmentation method to generate superpixels to construct an Attributed Region Adjacency Graph (ARAG) in which nodes corresponding with superpixels and edges representing correlations between superpixels. We then apply region-based Normalized Cut to the graph to merge similar neighbouring superpixels to form natural image regions corresponding to various body parts and backgrounds. To tackle the occlusion problem often encountered in these applications, we apply Denoising Autoencoder with nonnegativity constrains to learn robust and part-based representation of image patches in each node of the graph. Finally, the similarity of an image pair is measured by the Earth Mover's Distance (EMD) between the robust image signatures of the nodes in the corresponding ARAGs. We evaluate our methods on both person re-id and partial person re-id datasets and the results show that our new framework outperforms state of the art methods.

1 Introduction

Person re-identification (person re-id) aims to match the images of a pedestrian captured by non-overlapping camera views. It has received increasing attentions not only due to its wide application in video surveillance including multi-camera tracking, human retrieval and event detection, but also for its ability to save huge human labor in searching for a target person in large video datasets.

In this paper, we study both person re-id problem and a more difficult problem called partial person re-identification (partial re-id) [1]. While typical person re-id task (Fig. 1(a)) only decide whether two input images containing full people bodies are taken from the same person, partial re-id aims to match a full body image to an occluded partial probe image which are taken from non-overlapping cameras. Partial re-id task is more closer to the real world scenario than the general re-id task, since the full body image is usually difficult to capture in the public scene. For example, the body of a particular person can be occluded by the static obstacles as shown in Fig. 1(b). To solve the partial re-id problem, Zheng *et al.* [1] propose a sparse representation framework to achieve local patch-level matching and global part-based matching. Although their method

© Springer International Publishing AG 2017
C.-S. Chen et al. (Eds.): ACCV 2016 Workshops, Part III, LNCS 10118, pp. 301–314, 2017.
DOI: 10.1007/978-3-319-54526-4_23

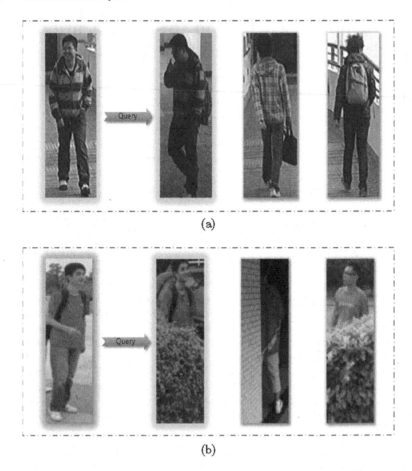

(a)

(b)

Fig. 1. Comparison of two different tasks (a) Person re-id task only aims to match between human body images. (b) Partial re-id task that considers the occlusion problem in the real life.

achieved superior performance than the previous re-id models, their framework relies on the person label information in the learning process, which means that the scalability of their model is poor and it requires large amount of time to manually label each image when applying the model to large datasets. Thus, an unsupervised model is more desirable in order to save the cost of data labelling.

In this work, we have developed a novel unsupervised framework to address the person re-id and partial person re-id problem. We first segment each image by constructing an Attributed Region Adjacency Graph (ARAG) based on super-pixels generated through Quick-Shift and further optimized by region-based Normalized Cut to re-cluster image regions. The segmented sub-images are represented as the nodes in the graph. As occlusion is usually a severe problem especially in partial person re-id task, we train a Denoising Autoencoder with Nonnegativity Constrains to learn robust and part-based images representation from the segmented image patches. The extracted feature for each image region is then able to form signatures to represent the whole image. Finally, we calculate image distance by means of Earth Mover Distance [2] to match between different graph-represented image pairs. The main contributions of this study are three-fold:

- To the best of our knowledge, it is the first paper to apply unsupervised model for solving both person re-identification and partial person re-identification problem without any requirement on label information.
- We propose an efficient framework to construct graph-structured image representation and measure similarity between images.
- Experimental results on publicly available testing datasets show that the new method outperforms existing unsupervised methods.

2 Related Work

The methods addressing the person re-identification problem can be divided into two categories: supervised and unsupervised.

Supervised Re-id Model. Previous work on solving this problem is dominated by supervised learning methods. These methods aim to identify an optimal matching function or distance metric to map the original image to a manually-designed feature space where the distance of the same image pair is closer than those of different image pairs. However, the features employed by these methods are usually handcrafted [3–5]. In recent work, deep learning has shown its ability to achieve state-of-the-art performance. Li et al. [6] and Ahmed et al. [7] proposed to apply deep learning methods to simultaneously learn discriminative features and the corresponding similarity metric for person re-identification, which has been shown to significantly outperform methods employing handcrafted feature descriptors. The success of deep learning framework on addressing the person re-identification problem is due to its ability to learn effective image representations. However, it is difficult to train a large deep learning network on small datasets as described in [6], which means that the generalization ability of deep learning model is poor when applying to small size datasets. Also, these supervised learning methods heavily rely on labeled training data which is always in short supply. A number of transfer learning-based methods have been developed to solve the scalability limitation of supervised methods, however, when lighting and view conditions have changed, it is still necessary to label new image pairs and to update the model [8].

Unsupervised Re-id Model. Previous unsupervised learning methods were mainly focused on feature design. Liu et al. [9] applied feature importance mining scheme to find the optimal weights for global feature types. Farenzena et al. [10] tried to obtain good view invariance in pedestrian images by the Symmetry-Driven Accumulation of Local Features (SDALF). Other unsupervised methods aimed to detect the saliency of person's appearance. Zhao et al. [11] constructed different saliency models based on patch-based representation.

Wang *et al.* proposed a generative probabilistic topic modeling to discover localized foreground saliency. Recently, Kodirov *et al.* [12] apply an unsupervised sparse coding method to learn from unlabelled data to generate features that are more discriminative than hand-craft features. Lisanti *et al.* [13] propose an iterative re-weighted sparse ranking method to ensure the appropriate person candidate are selected at each iteration. Peng *et al.* [14] develop a cross-dataset transfer learning approach to transfer feature representation from labelled source datasets to the unlabelled target datasets. However, these methods usually assume full-body images are available and are not designed for partial body matching.

Partial Re-id Model. The first pioneer work on solving partial re-id is proposed by [1]. They decompose image into regular grid patches, matching is performed at the local patch level as well as global spatial layout level. However, their classification model relies on manually labelled information as done in most previous work on person re-id problem, which is not required in our model. On the other hand, in our work, the image is segmented by the pixel clustering techniques which is more capable of describing the images natural structures than regular grid partitions as applied in [1].

3   Methodology

3.1   Image Segmentation

In this section, we propose an efficient unsupervised segmentation method to decompose image into several image patches in order to separate human body parts and background, and also to produce signatures to represent each image. Some previous work segment the image with pre-defined grid structure and apply it to the whole datasets [1,10,15]. We argue that such method is not able to always distinguish human body part from the noisy background especially when applying it to solving the partial person re-id task, since it is unknown which part of human body would show up in the images that is captured by non-overlapping cameras. Thus, it is essential to preprocess each image by means of dividing it into local patches according to its own image structures.

Attributed Region Adjacency Graph. The image pixels are initially clustered by quick shift [16] to generate superpixels. In previous related work, Boltz *et al.* [17] propose to directly produce image signatures based on superpixels and then apply the EMD framework to calculate distances between different image pairs. However, such kind of method would incur heavy computational cost as there would be too numerous superpixels and EMD is intrinsically a model with high computational cost [18]. On the other hand, superpixelization simply performs image region segmentation without taking into consideration the correlations among the segmented image regions. Shen *et al.* [19] suggested that considering the patches correlation could prevent mismatching patches with similar appearance in the person re-id task.

To overcome these problems, we propose an Attributed Region Adjacency Graph (ARAG) that is able to model the correlation of superpixels and can also be applied to re-cluster superpixels by means of region-based Normalized Cut. If we first consider each superpixel generated by Quick-Shift as a node in a graph, we can construct an ARAG as $\mathbf{G} = (\mathbf{V}, \mathbf{E}, \mathbf{S})$, where $\mathbf{V}$ is the set of nodes that represents the superpixels, $\mathbf{E}$ is the set of edges representing the spatial neighborhood information between nodes i and j, where $i, j \in \{1, ..., N\}$ and N is the total number of superpixels, and $\mathbf{S}$ is the edge affinity matrix and its element $s(i, j)$ measure the dissimilarity between node i and j. To achieve efficiency to preprocess each image, the node attribute is represented by the mean value $\mathbf{F}_{(i)}$ of the 8 color channels (RGB, YCbCr, HS).

Region Re-clustering. Although the number of superpixel is much fewer than the number of pixels in the image, it is still computational expensive to match images using EMD based on superpixels. Thus, we intend to re-cluster superpixels and produce a more simplified graph to represent image structure as shown in Fig. 2.

To re-cluster nodes in the graph is equivalent to removing the edges that connecting two parts. The dissimilarity of two parts can be measured as the total weight of the removed edges:

$$cut(\mathbf{A}, \mathbf{B}) = \sum_{i \in \mathbf{A}, j \in \mathbf{B}} s(i, j) \tag{1}$$

where $\mathbf{A}$ and $\mathbf{B}$ are the two disjoint sets: $\mathbf{A} \cup \mathbf{B} = \emptyset, \mathbf{A} \cap \mathbf{B} = \mathbf{V}$. This can be formulated as finding the minimum cut problem, which are well studied [20,21]. However, as Eq. 1 does not consider any intragroup information, minimum cut always favors grouping small set of nodes, which means to recursively apply minimum cut would lead to unbalanced and overclustered result [22].

Normalized Cut (NCut) [22] is thus motivated to produce balanced clusters which is more robust than other spectral graph partitioning methods and has been widely applied in computer vision work [23–25]. Ncut performs $k-$way cut by minimizing:

$$NCut_k(V) = \frac{cut(P_1, V - P_1)}{vol(P_1)} + \frac{cut(P_2, V - P_2)}{vol(P_2)} + ... + \frac{cut(P_k, V - P_k)}{vol(P_k)} \tag{2}$$

where $vol(P_i) = \sum_{u \in P_i, v \in V} s(u, v)$, and $\{P_1, P_2, ..., P_k\}$ are the segmented region of the graph. The rationale of NCut is to find an optimal image partitions that the intragroup has the most similarity while the intergroup has the most dissimilarity. While minimizing normalized cut is a NP-complete problem, Shi and Malik [22] show that Eq. 2 can be efficiently solved by relaxing a continuous underestimator.

In this work, we re-cluster image regions based on the pre-segmented superpixels instead of the original image pixels as done in the conventional NCut

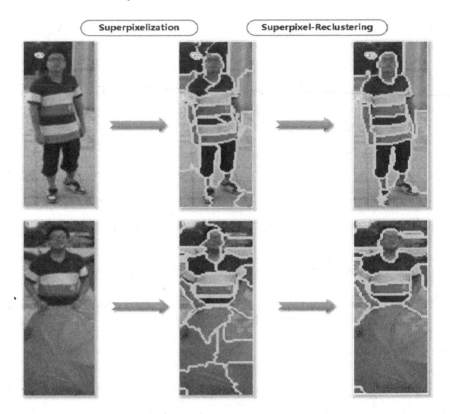

Fig. 2. To re-cluster superpixels by means of region-based Normalized Cut is able to reduce the number of segmented regions in the image while preserve the salient features of the overall image.

framework. Following the work of [26], we design a region-based weight matrix that is different from the conventional NCut framework:

$$s(i,j) = \begin{cases} e^{\frac{-\|\mathbf{F}_{(i)}-\mathbf{F}_{(j)}\|_2^2}{\sigma_I}}, & \text{if } i \text{ and } j \text{ are adjacent} \\ 0, & \text{otherwise} \end{cases} \tag{3}$$

where σ_I is the scaling factor referring to the sensitivity of the difference between node i and j, and $\|\cdot\|_2$ is the vector norm operator. As the number of superpixels is much smaller than the image pixels, one tremendous advantage of applying region-based NCut on ARAG rather than conventional NCut is that it is able to reduce computational complexity with a more simplified graph structure.

3.2   Learning Robust and Part-Based Representation

In this section, we apply Denoising Autoencoder (DA) to extract feature representation from image patches. The rationales of using the Denoising Autoencoder to solve the person re-identification problem are as follows. From a high-level perspective, humans often only need to have a partial image of a person to recognize the subject. From a practical point of view, we often only have partial image of the subject available. By corrupting the images we are simulating the situations where only a partial image of the subjects is a available or the images are very noisy. The DA can reconstruct the original clean image from a corrupted version and should therefore be robust against noise and occlusion, which is a severe problem especially on solving partial re-id problem.

Denosing Autoencoder. A Denosing Autoencdoer [27] first corrupts an input image $x \in \mathbb{R}^d$ to construct a partially destroyed version $\tilde{x}$ through a stochastic mapping $\tilde{x} \sim q_D(\tilde{x}|x)$. This is usual done by applying *salt-and-peper noise*: randomly choose a fraction v elements of the input x and set them to 0 while the rest are left untouched. The corrupted input $\tilde{x}$ is then mapped to the hidden representation by means of the basic autoencoder $y = f_\theta(\tilde{x}) = s(W\tilde{x} + b)$ where s is usually a non-linear activation function, W is the weights and b is the bias to be learned, from which we are able to reconstruct $z = g_{\theta'}(y) = s(W'y + b')$ with parameter set $\theta' = \{W', b'\}$. The training goal is to minimize the average reconstruction error, i.e., to make z as close as possible to the original uncorrupted input image x, by modifying the parameters θ and θ'. A typical loss function often used in the literature is the squared error loss:

$$Loss = \frac{1}{N}\sum_{k=1}^{N}\frac{1}{2}\|x^k - z^k\|^2 \tag{4}$$

where N is the total training data set size. Here the input x refers to the partitioned image patch which is first segmented by the method as described in Sect. 3.1 and cropped by a bounding box from the original person image.

Denoising Autoencoder with Nonnegativity Constraint (DANC). In order to improve the performance of denoising autoencoder, we regularize it by introducing nonnegativity constraint to produce part-based representation. It is shown in the previous related work [28,29] that training a multilayer network with nonnegative weights is able to learn part-based additive representation of data, which improves the ability of network to disentangle the hidden structure of the data. We then modify Eq. 4 by further regularize the weights to train DANC:

$$Loss = \frac{1}{N}\sum_{k=1}^{N}\frac{1}{2}\|x^k - z^k\|_2^2 + \lambda_1\|\mathbf{W}\|_1 + \frac{1}{2}\lambda_2\|\mathbf{W}\|_2^2 \tag{5}$$

where $\mathbf{W} = W\bigcup W'$, and we constrain weight matrix W and W' to have only non-negative elements while the bias is remained unconstrained.

3.3   Image Matching by Earth Mover's Distance

We then are able to measure the distance between different image pairs by means of Earth Mover's Distance (EMD) [2]. We choose EMD as the metric to address partial re-id as it is able to achieve many-to-many matching, which is a desired property for addressing partial re-id problem. In the partial re-id task, an image pair from one person might only have a few regions with similar image features because of the severe occlusion. EMD is shown to have great performance when matching only parts of an images [2].

EMD is based on the solution to the Monge-Kantorovich transportation problem [30] that measuring the minimum cost of transforming one image signature into another one. Given two images P and Q, we can partition images by the process as described in Sect. 3.1. After the image structures are further refined by region-based NCut, it can be easily converted into the EMD signature. Thus the two images after segmentation can be represented as $P(m) = \{(p_1, l_{p_1}), ..., (p_m, l_{p_m})\}$ and $Q(n) = \{(q_1, l_{q_1}), ..., (q_n, l_{q_n})\}$, where p_i and q_j refer to the feature vector of segmented parts in image P and Q respectively, l_{p_i} and l_{q_j} are their weights that represent the fraction of image pixel in that region and m, n are the numbers of segmented clusters in image P and Q. Here we use learned feature representation from DANC for creating image signature which is different from that used in conducting region-based NCut on images. Apart from the visual feature, the centroid position of each cluster $(\bar{x}, \bar{y})$ is added to the feature vector. Then the EMD is calculated as:

$$EMD(P(m), Q(n)) = \min WORK(P(m), Q(n), \mathbf{O}) \tag{6}$$

$$WORK(P(m), Q(n), \mathbf{O}) = \sum_{i=1}^{m} \sum_{j=1}^{n} d_{ij} f_{ij} \tag{7}$$

subject to the constraints as shown in [2]. Here $\mathbf{O} = [f_{ij}]$ is the flow from node i in image graph P to node j in image graph Q, and d_{ij} is the ground distance between node i and node j. The ground distance d_{ij} is defined as follows:

$$d_{ij} = [L(y^P, y^Q) + \alpha L(\bar{x}, \bar{y})]^{\frac{1}{2}} \tag{8}$$

where L represents the Euclidean distance and α refers to the importance of spatial information.

4   Experiments

4.1   Datasets and Settings

CUHK01 Dataset. CUHK01 [31] is a dataset that has 971 individuals, and each person has two views captured by two different camera view. Following its protocol, the dataset is randomly divided into non-overlapping training set (485 persons) and test set (486 persons) (Fig. 3).

VIPER Dataset. VIPeR is a more challenge dataset with relatively lower image resolution. The dataset contains 632 pedestrian pairs that captured by two different cameras. Following the evaluation protocol, we randomly split the data set into half (i.e. $p = 316$) for training and the rest 316 for testing [34].

Samples from CUHK01 data set

Samples from VIPeR data set

Samples from Partial RE-ID data set

Fig. 3. An overview of three datasets.

Partial REID Dataset. We also evaluate our framework on the partial re-id task. The only publicly available partial re-id dataset is the Partial REID dataset [1]. It contains 600 images of 60 people, and each person in the dataset has 5 different viewpoints of both full-body and partial-body images. We randomly select 70% of individuals from dataset (i.e. $p = 42$) for experiment. Following the evaluation protocol of [1], all the partial images from every person form the probe set and the holistic person images is regarded as the gallery set.

Settings. For each image in both datasets, we first smooth the image by Gaussian filter with $\sigma_{Gaussian} = 1$. The scaling factor in Eq. 3 is set to 6, and the α in Eq. 8 is set to 0.5. In the process of segmentation, we remove the image patch with relatively lower pixel number (less than 1% of total image pixel number) and resize each image patch to 20×20 to be the input for DANC and the number of hidden neurons in DANC is set to be 200. The experiments were conducted following 20 random splits and the reported average Cumulative Matching Characteristic (CMC) curves are single-shot results.

Compared Models. We compared our method with four state-of-art unsupervised person re-identification methods (eSDC [11], SDALF [10]), ISR [13] and CDLT [14]. We also compare our method with two well-known metric learning method including LMNN [32] and ITML [33] on the three datasets.

4.2   Experiment Result

Experiment on CUHK01 Dataset. From Table 1, it can be seen that our method achieves the best rank-1 accuracy among all the methods. The improvement of our methods include two aspects: (1) Although the image resolution is low, the feature learned by DANC is more robust than the handcrafted feature that applied other methods. (2) When matching image pairs, EMD is able to find the most similar parts from image patch of human body regardless of any variant in the background.

Experiment on VIPeR Dataset. On the VIPeR dataset, CDLT shows better performance than our method. However, it should be noted that the method proposed by them is based on transfer learning, which means a number of source dataset has to be labelled while our method does not require any label information.

Experiment on Partial REID Dataset. In Table 1, the experiment results on Partial REID dataset shows that our framework is also able to outperform the eSDC and SDALF by a large margin. Under the partial re-id setting, these re-id methods are not able to handle the misalignment of human body with the existence of obstruction. Apart from comparing with re-id models, our framework also show superior performance than the metric learning methods and SRC. This is because these methods mainly focus on the matching with entire input image

Table 1. Performance of rank-1 result on three datasets. '-' refers to no reported result or no public code on the corresponding datasets.

	CUHK01	VIPeR	Partial REID
eSDC	22.83	26.7	28.5
SDALF	9.9	19.9	20.5
ISR	-	27	-
CDLT	27.1	**31.5**	-
LMNN	13.45	21.92	26.52
ITML	15.98	14.36	27.43
Ours	**28.4**	30.2	**32.6**

while our framework is based on the patch-level matching that is more appropriate for the partial re-id setting. It should also be noted that these supervised methods use labeled information to learn discriminative functions than thus are expected to achieve better performance. However, our model does not require any labeled information while the performance of our framework is better than these supervised methods. It suggests that our model is scalable to large scale application when the labels of identity are not available.

4.3   Comparison with Other Segmentation-Based Models

We also compare our framework with other segmentation-based model. The first one is to directly calculate EMD based on superpixel (Quick-Shift + EMD) [17], and the second one is to directly segment image using NCut and then calculate image distance by EMD (NCut + EMD). These two methods apply feature representation as described in Sect. 3.1. In Table 2, it shows that our framework is able to achieve much better performance than other two models. Comparing with Quick-Shift + EMD, our framework applies ARAG to consider the internal relationship between superpixels, and achieves superior performance by re-clustering them through the region-based NCut to produce a simplified graph structure for matching. Then comparing with NCut + EMD, our approach uses superpixelization as pre-processing method to remove image noise. Also, the created region has much large feature difference than that between pixels, which makes it more accurate to perform NCut on regions rather than on pixels.

Table 2. Comparing with other segmentation-based methods

	CUHK01	VIPeR	Partial REID
Ours	**28.40**	**30.2**	**32.6**
Quick-Shift + EMD	17.52	19.88	23.84
NCut + EMD	13.98	18.27	21.52

4.4   Evaluating Segmentation Result

In this section, we aim to evaluate the necessity of applying NCut as second segmentation to reduce the segmented patch number of each image. In Fig. 4, it is shown that applying NCut based on the segment result of quick-shift can significantly reduce the patch number of images, which is able to lower down computational expense to further applying EMD to match image pairs.

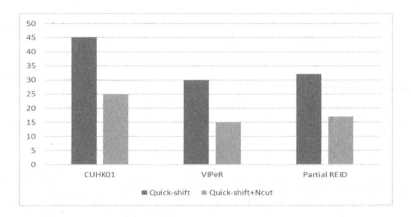

Fig. 4. Comparison of average patch number on three datasets by applying different segmentation methods.

5   Conclusions

In this paper, we have presented a novel framework to address the partial re-id task. In the first step, we segment image region based on the Region Adjacency Graph and refine the segment result by means of Normalized Cut to model the relationship between different image regions. We built a Denoising Autoencoder with Nonnegativity Constraint to learn robust and part-based feature representation from image patches. Then, each region of image is able to be converted to EMD signature, and the EMD is employed to achieve many-to-many matching between different image pairs. We evaluated our framework on several publicly available datasets, in which our method showed superior performance to many previous supervised and unsupervised methods on both person re-id and partial person re-id task.

Acknowledgments. This work is supported by Ningbo Science and Technology Bureau (Project No. 2012B10055 and 2013D10008) and by the International Doctoral Innovation Centre (IDIC) at the University of Nottingham Ningbo China.

References

1. Zheng, W.S., Li, X., Xiang, T., Liao, S., Lai, J., Gong, S.: Partial person re-identification. In: Proceedings of the IEEE International Conference on Computer Vision, pp. 4678–4686 (2015)
2. Rubner, Y., Tomasi, C., Guibas, L.J.: The earth mover's distance as a metric for image retrieval. Int. J. Comput. Vis. **40**, 99–121 (2000)
3. Zheng, W.S., Gong, S., Xiang, T.: Person re-identification by probabilistic relative distance comparison. In: 2011 IEEE Conference on Computer Vision and Pattern Recognition (CVPR), pp. 649–656. IEEE (2011)

4. Gray, D., Tao, H.: Viewpoint invariant pedestrian recognition with an ensemble of localized features. In: Forsyth, D., Torr, P., Zisserman, A. (eds.) ECCV 2008. LNCS, vol. 5302, pp. 262–275. Springer, Heidelberg (2008). doi:10.1007/978-3-540-88682-2_21

5. Li, W., Zhao, R., Wang, X.: Human reidentification with transferred metric learning. In: Lee, K.M., Matsushita, Y., Rehg, J.M., Hu, Z. (eds.) ACCV 2012. LNCS, vol. 7724, pp. 31–44. Springer, Heidelberg (2013). doi:10.1007/978-3-642-37331-2_3

6. Li, W., Zhao, R., Xiao, T., Wang, X.: Deepreid: deep filter pairing neural network for person re-identification. In: Proceedings of the IEEE Conference on Computer Vision and Pattern Recognition, pp. 152–159 (2014)

7. Ahmed, E., Jones, M., Marks, T.K.: An improved deep learning architecture for person re-identification. In: Proceedings of the IEEE Conference on Computer Vision and Pattern Recognition, pp. 3908–3916 (2015)

8. Wang, H., Gong, S., Xiang, T.: Unsupervised learning of generative topic saliency for person re-identification (2014)

9. Liu, C., Gong, S., Loy, C.C.: On-the-fly feature importance mining for person re-identification. Pattern Recogn. **47**, 1602–1615 (2014)

10. Farenzena, M., Bazzani, L., Perina, A., Murino, V., Cristani, M.: Person re-identification by symmetry-driven accumulation of local features. In: 2010 IEEE Conference on Computer Vision and Pattern Recognition (CVPR), pp. 2360–2367. IEEE (2010)

11. Zhao, R., Ouyang, W., Wang, X.: Unsupervised salience learning for person re-identification. In: Proceedings of the IEEE Conference on Computer Vision and Pattern Recognition, pp. 3586–3593 (2013)

12. Kodirov, E., Xiang, T., Gong, S.: Dictionary learning with iterative laplacian regularisation for unsupervised person re-identification. In: BMVC 2015, vol. 3, p. 8, September 2015

13. Lisanti, G., Masi, I., Bagdanov, A.D., Del Bimbo, A.: Person re-identification by iterative re-weighted sparse ranking. IEEE Trans. Pattern Anal. Mach. Intell. **37**, 1629–1642 (2015)

14. Peng, P., Xiang, T., Wang, Y., Pontil, M., Gong, S., Huang, T., Tian, Y.: Unsupervised cross-dataset transfer learning for person re-identification. In: Proceedings of the IEEE Conference on Computer Vision and Pattern Recognition 2016, pp. 1306–1315 (2016)

15. Chang, Y.C., Chiang, C.K., Lai, S.H.: Single-shot person re-identification based on improved random-walk pedestrian segmentation. In: 2012 International Symposium on Intelligent Signal Processing and Communications Systems (ISPACS), pp. 1–6. IEEE (2012)

16. Vedaldi, A., Soatto, S.: Quick shift and kernel methods for mode seeking. In: Forsyth, D., Torr, P., Zisserman, A. (eds.) ECCV 2008. LNCS, vol. 5305, pp. 705–718. Springer, Heidelberg (2008). doi:10.1007/978-3-540-88693-8_52

17. Boltz, S., Nielsen, F., Soatto, S.: Earth mover distance on superpixels. In: 2010 17th IEEE International Conference on Image Processing (ICIP), pp. 4597–4600. IEEE (2010)

18. Ling, H., Okada, K.: An efficient earth mover's distance algorithm for robust histogram comparison. IEEE Trans. Pattern Anal. Mach. Intell. **29**, 840–853 (2007)

19. Shen, Y., Lin, W., Yan, J., Xu, M., Wu, J., Wang, J.: Person re-identification with correspondence structure learning. In: Proceedings of the IEEE International Conference on Computer Vision, pp. 3200–3208 (2015)

20. Wu, Z., Leahy, R.: An optimal graph theoretic approach to data clustering: theory and its application to image segmentation. IEEE Trans. Pattern Anal. Mach. Intell. **15**, 1101–1113 (1993)
21. Weiss, Y.: Segmentation using eigenvectors: a unifying view. In: The Proceedings of the Seventh IEEE International Conference on Computer Vision, vol. 2, pp. 975–982. IEEE (1999)
22. Shi, J., Malik, J.: Normalized cuts and image segmentation. IEEE Trans. Pattern Anal. Mach. Intell. **22**, 888–905 (2000)
23. Ngo, C.W., Ma, Y.F., Zhang, H.J.: Video summarization and scene detection by graph modeling. IEEE Trans. Circuits Syst. Video Technol. **15**, 296–305 (2005)
24. Costeira, J., Kanade, T.: A multi-body factorization method for motion analysis. In: Giralt, G., Hirzinger, G. (eds.) Robotics Research, pp. 368–377. Springer, London (1996)
25. Sarkar, S., Soundararajan, P.: Supervised learning of large perceptual organization: graph spectral partitioning and learning automata. IEEE Trans. Pattern Anal. Mach. Intell. **22**, 504–525 (2000)
26. Tao, W., Jin, H., Zhang, Y.: Color image segmentation based on mean shift and normalized cuts. IEEE Trans. Syst. Man Cybern. Part B: Cybern. **37**, 1382–1389 (2007)
27. Vincent, P., Larochelle, H., Bengio, Y., Manzagol, P.A.: Extracting and composing robust features with denoising autoencoders. In: Proceedings of the 25th International Conference on Machine learning, pp. 1096–1103. ACM (2008)
28. Chorowski, J., Zurada, J.M.: Learning understandable neural networks with non-negative weight constraints. IEEE Trans. Neural Netw. Learn. Syst. **26**, 62–69 (2015)
29. Hosseini-Asl, E., Zurada, J.M., Nasraoui, O.: Deep learning of part-based representation of data using sparse autoencoders with nonnegativity constraints. IEEE Trans. Neural Netw. Learn. Syst. **27**(12), 2486–2498 (2016)
30. Hitchcock, F.L.: The distribution of a product from several sources to numerous localities. J. Math. Phys. **20**, 224–230 (1941)
31. Li, W., Wang, X.: Locally aligned feature transforms across views. In: Proceedings of the IEEE Conference on Computer Vision and Pattern Recognition, pp. 3594–3601 (2013)
32. Weinberger, K.Q., Saul, L.K.: Distance metric learning for large margin nearest neighbor classification. J. Mach. Learn. Res. **10**, 207–244 (2009)
33. Davis, J.V., Kulis, B., Jain, P., Sra, S., Dhillon, I.S.: Information-theoretic metric learning. In: Proceedings of the 24th international conference on Machine learning, pp. 209–216. ACM (2007)
34. Gray, D., Brennan, S., Tao, H.: Evaluating appearance models for recognition, reacquisition, and tracking. In: Proceedings of the IEEE International Workshop on Performance Evaluation for Tracking and Surveillance (PETS), vol. 3, no. 5, October 2007

Saliency-Based Person Re-identification by Probability Histogram

Zongyan Zhang[1], Cairong Zhao[1], Duoqian Miao[1(✉)], Xuekuan Wang[1], Zhihui Lai[2], and Jian Yang[3]

[1] Department of Computer Science and Technology,
Tongji University, Shanghai, China
dqmiao@tongji.edu.cn
[2] College of Computer Science and Software Engineering,
Shenzhen University, Shenzhen, China
[3] School of Computer Science and Engineering,
Nanjing University of Science and Technology, Nanjing, China

Abstract. Person re-identification has attracted increasing interests due to its broad application in automatic search and video surveillance. It is easy for humans to recognize person identities but difficult for computers. Thus, knowing how humans to recognize person identities is helpful to improve the performance of the computers person re-identification. In this paper, we propose an effective feature representation based on salient regions and a pool of multiple metric learning for person re-identification. The proposed feature representation extracts local details (salience regions) and global distribution of pedestrian images. To reduce the effects of illumination changes, we apply probability histogram in four kinds of color spaces where similar color can be characterized by the similar histogram distribution to different color spaces. Moreover, a pool of multiple metric learning is applied to all features captured from different spaces and models. The proposed method has been evaluated on two public datasets. Experimental results show that the proposed method outperforms others.

1 Introduction

Person re-identification is a task to find a person from a series of non-overlapping camera views in which a person has the same identity. It is a challenging task because the person seems quite different under the change of viewpoints, poses, appearance and illumination. Besides, non-overlapping camera views make background information useless, thus the used pedestrian images contain lots of noises.

To find the correct person, an effective feature representation which should be robust to environment, poses and viewpoints changes is necessary. Color and texture are the most pervasive and based features used in person re-identification (e.g. [1–5]). These descriptors have been successfully applied to solve the problem

---

Z. Zhang and C. Zhao—Authors contributed equally.

© Springer International Publishing AG 2017
C.-S. Chen et al. (Eds.): ACCV 2016 Workshops, Part III, LNCS 10118, pp. 315–329, 2017.
DOI: 10.1007/978-3-319-54526-4_24

Fig. 1. Samples of pedestrian images in different camera views of person re-identification.

of person re-identification. In addition, misalignments caused by variations of viewpoints and poses are commonly exist in person re-identification as Fig. 1. The remarkable spatial difference should be considered in feature representation.

Several effective approaches greatly improve the robustness of feature representation and advance the person re-identification, such as LOMO [5], SDALF [6], salience match [7], SCNCD [8]. These feature describe pedestrian image from local details or the whole image which neglect the complementarity between local details with global geometric correspondence. Moreover, because of the variations of pedestrian images, an effective feature representation is still a challenging problem in person re-identification.

The other aspect of person re-identification is learning a distance or similarity function to divide persons into similar pairs and dissimilar pairs. Many metric learning algorithms have been proposed to address this problem, for example KISSME [1], PCCA [9], LFDA [10], MFA [11]. However, existing approaches have not considered that different feature extracted by various model may not be effectively handled by the single metric.

In this paper, we propose a novel person re-identification model which fuses a hybrid feature representation with a multiple metric learning. The hybrid feature representation describes a pedestrian image from the local detail and global geometric distribution, which fusion probability histogram feature extracted from each overlapping sliding windows as global description and some small salient regions as local detail. The probability histogram based on different color spaces is robust to variations of illumination. Salient regions have great invariance against viewpoints and poses changes. To learning a discriminant metric, we propose a novel model where different features from different space are handled by different metric respectively.

The rest of the paper is organized as follow. In Sect. 2, a brief review of related works in the person re-identification field is given. The proposed person re-identification approach is described in Sect. 3. The experiment and comparison between our method with exist ones are showed in Sect. 4. Finally, conclusion are drawn in Sect. 5.

2   Related Work

In the past few years, many researchers have proposed different approaches to tackle the challenge of person re-identification. These approaches mainly follow two different aspects: feature representation and metric learning.

Feature Representation. To stably describe pedestrians appearance in disjoint camera views, various approaches have been proposed. Farenzena et al. [6] proposed the Symmetry-Driven Accumulation of Local Features (SDALF) method which exploits the symmetry and dissymmetry in pedestrian image to select the major part of the body figure. The application of the symmetry and dissymmetry is considered to handle viewpoint variations. Zheng et al. [12] proposed to segment pedestrian image into six horizontal stripes equally and compare pedestrian image in each stripe. This feature is robust to the viewpoint changes in horizontal direction. Yang et al. [8] proposed a salient color names based color descriptor (SCNCD) which designs a novel method to describe colors. SCNCD utilizes the probability distribution to represent the distance of color with color name. A higher probability means the color name is nearer to the color. SCNCD is robust to the variations of illumination because similar colors have similar probability distributions. Liao et al. [5] proposed an efficient Local Maximal Occurrence (LOMO) approach which considers the spatial details within a horizontal stripe. To handle viewpoint changes, LOMO maximizes the local occurrence of the same histogram bin among sub-windows which at the same horizontal location. These feature representations describe pedestrian images from global distribution, consider little about local details and none of them combines the local details with global geometric correspondence to represent pedestrian images.

Moreover, Zhao et al. [7] proposed the salience match where salience is used to handle the change of poses and viewpoints. They considered the saliency information in pedestrian images and the salience patches match each other by adjacency constrained search. These methods use salience region to describe pedestrian images, but the time complexity is very high.

Metric Learning. On the other hand, discriminative learning models have been extensively studies for person re-identification. Prosser et al. [13] proposed a RankSVM method which reformulates the person re-identification problem as ranking problem. RankSVM learns a subspace where potential true match is given higher ranking. Pedagadi et al. [10] apply local fisher discriminant analysis (LFDA) to the person re-identification. Koestinger et al. [1] utilize KISSME to learn a distance metric from equivalence constraints. KISSME learns a Mahalanobis distance to minimize the intra-class distance and maximize the inter-class distance to compute the match in different feature space. Liao et al. [5] extended KISSME approach and proposed a cross-view quadratic discriminant analysis (XQDA). XQDA further learns a discriminant subspace and a metric together.

3 Saliency-Based Person Re-identification by Probability Histogram

A person from a series of non-overlapping camera views usually has quite different appearance. For example, some parts of a pedestrian on different camera view can be misaligned due to the diverse poses and viewpoints. To handle such problems, some small salient regions, that can represent the certain pedestrian as specific as possible, would be detected as local features from pedestrian images, and these features describe the major color information of body or legs as well as the characteristic of clothing. Moreover inspired by Spatial Pyramid Matching [14], we utilize a multi-scale sliding windows to describe global geometric correspondence of a person image. As show in Fig. 2, the proposed person re-identification approach consists of five phases: (1) feature extraction, (2) background suppression, (3) multiple metric learning, (4) salient region detection, and (5) distance fusion.

3.1 Feature Extraction

Before feature extraction, each pedestrian image has been segmented into a grid of multi-scale local patches. In this paper, we use a patch size of $10 * 10/15 * 15$, with overlapping step of 5 pixels. Within each patch, we extract color probability histogram in four color space and a texture feature: LBP.

Since there is no single color space and descriptor that could be against the variations of illumination, we project the original RGB images into each color space $S \in \{\text{HSV}, \text{normalizedrgb}, \text{YUV}, \text{rgs}\}$ and compute color features in each color space respectively. Since color value in each channel is smooth and successive, we divide the channel into 8 sections and the color space into $8 * 8 * 8 = 512$ subspaces equally. Throughout the paper, each channel in all color space is normalized to the range $[0, 1]$. $S_{c_j}^i$ denotes the i-th section of color channel c_j (j = 1, 2, 3). $m_{c_j}^i$ is the center of section that can be computes by:

$$m_{c_j}^i = \frac{\sum v}{n} (\forall v \in s_{c_j}^i, n = 32) \tag{1}$$

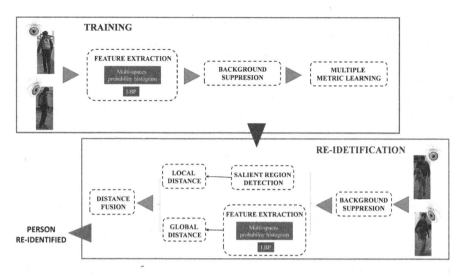

Fig. 2. The system architecture based on five main stages: (1) feature extraction, (2) background suppression, (3) multiple metric learning, (4) salient region detection, and (5) distance fusion.

where n is the number of values in $S_{c_j}^i$. The collection of all space center can be denoted as $z = [m_{c_1}^1 m_{c_2}^1 m_{c_3}^1, m_{c_1}^1 m_{c_2}^1 m_{c_3}^2, m_{c_1}^1 m_{c_2}^1 m_{c_3}^3, ..., m_{c_1}^8 m_{c_2}^8 m_{c_3}^7, m_{c_1}^8 m_{c_2}^8 m_{c_3}^8]$.

The distance between color c and k-nearest neighbor color space centers is computed by:

$$dist(c, z_p) = \sum_{i=1}^{3} \|c_i - z_{p_i}\|^2 \tag{2}$$

where i means the i-th channel of color c and z_p.

The probability distribution based the above distance is used to describe color c, defined as:

$$d_p(c) = \begin{cases} \frac{exp(-dist(c,z_p))}{\sum_{l=1}^{k} exp(-dist(c,z_l))} & , z_p \in KNN(C) \\ 0 & , z_p \notin KNN(C) \end{cases} \tag{3}$$

where K means the number of nearest neighbors (we set k = 8 in this paper), p = [1,512] refers to index of z. z_p and $z_l(p, l = 1, 2, .., K)$ belong to K nearest color space center of color c. To calculate the probability distribution of color c, we first use KNN algorithm to find K nearest space centers in Euclidean space. Then the distance from color c to each KNN color space center is utilized to embody the similarity between color c and each KNN color space center. After normalization, the probability distribution of color c over 512 space is defined as d_p. Moreover, it is easy to prove that the sum of the probability distribution of color c over all color space is 1. Finally, the color probability histogram feature of patch Xb in p-th part of z is the sum of all pixels in this patch, which can be defined as:

$$Xb^p = \sum d_p(c_i) \qquad (4)$$

where c_i is the color of a pixel in patch Xb. So the color feature dimension of patch Xb is $4 * 512 = 2048$.

In addition, Local Binary Pattern (LBP) [15] is extracted from each patch. A feature which fuses color and texture feature is effective to represent the pedestrian image.

3.2   Salient Region Evaluate

In Sect. 3.1, we proposed a global pedestrian image representation model based on a continuity and effective color representation method. Thus, local characteristics also embedded in the image has been lost, but these features has rich discriminative information.

In real life, human could recognize person identities based on some small salient regions. These small salient regions represent pedestrian main characteristic which is distinctive and reliable in person re-identification. Besides, pedestrian body usually contain more distinctive information than legs. Thus, we choose these specific regions as our salient regions:

(a) Body salient region which represents major color information of body;
(b) Legs salient region which represents major color information of legs;
(c) Characteristic salient region which contains characteristics of body, as show in Fig. 3;

The example of salient region detected by our method is shown in Fig. 4.

Fig. 3. Examples of salience region which contain characteristic of body. These regions has the invariable ability to against the changes of poses and viewpoints.

To detect these small salient regions, we propose a detect method called similarrank to compute the saliency of patch. Similarrank is improved from the pagerank [16] which is first applied to person re-identification to our knowledge. Serval suppositions should be made in similarrank:

Fig. 4. Examples of salient regions detected by our method. The body salient regions are shown in the red boxes. The legs salient regions are in the green boxes. The Characteristic salient regions are shown in the blue boxes. (Color figure online)

For body/legs salient region:

(a) the more patches similar to patch A, the more salient/important patch A is;
(b) the more salient/important patches similar to patch A, the more salient/important patch A is;

For characteristic salient region:

(a) the more patches dissimilar to patch A, the more salient/important patch A is;
(b) the more salient/important patches dissimilar to patch A, the more salient/important patch A is;

The patch b_i is similar to patch b_j means that similar score for patch b_i to b_j is much large than zero. The similar score for patch b_i to b_j is computed as:

$$S(b_i, b_j) = \begin{cases} exp(-dist(f_i, f_j)) &, exp(-dist(f_i, f_j)) \geq \theta \\ 0 &, exp(-dist(f_i, f_j)) < \theta \end{cases} \tag{5}$$

where θ is a similarity threshold (we choose $\theta = 0.9$ in this paper), f_i and f_j denotes the feature of patch b_i and b_j respectively, $dist(f_i, f_j)$ is the Euclidean distance between f_i and f_j. Furthermore, we divide each pedestrian images into M horizontal stripes. The similar score vector of patch b_i in stripe m can be denoted as $ds_{im} = \{S(b_i, b_1), ..., S(b_i, b_j), ..., S(b_i, b_n)\}$ where $j \neq i$ and n is the number of patches in stripe m, patch b_j represents all other patches in the horizontal stripe positioned at similar height. In addition, the $dist(f_i, f_j)$ in ds_{im} has been normalized by L2-norm, thus $S(b_i, b_j)$ and $S(b_j, b_i)$ are not equal. The set of similar score vectors of all patches in a stripe m can be written as $ds_m = \{ds_{1m}, ..., ds_{im}, ..., ds_{nm}\}$, where ds_{im} has been normalized by L1-norm. The set of similar score vectors of all stripes can be denoted as $ds = \{ds_1, ..., ds_m, ..., ds_M\}$, where M is the number of stripes. A set of similarRank score can be denoted as $SR = \{SR_1, ..., SR_m, ..., SR_M\}$, where $SR_m = \{SR_{11}, ..., SR_{im}, ..., SR_{nM}\}$ is a set of similarRank score in stripe m. Finally, the similarRank score of each patch b_i can be compute as Algorithm 1. The details are shown in Fig. 5:

Algorithm 1 SimilarRank algorithm

Input: The set of all stripes' similar vector: *ds;* The max Iterations Times: *max;*

   Output: A set of similar Rank score *SR;*

   Initialize: SR$_{im}$=1, ∀ SR$_{im}$∈sR;

        Iterations Times *Iter*=1;

Repeat

   For each SR$_{im}$ in SR

$$SR_{im} = \frac{1-q}{|SR|} + q \sum_{\forall S(b_j, b_i) \in ds} \frac{S(b_j, b_i)}{n_j}, \; n_j = \left| \{s \mid s \in ds_{jm} \, \& \, s \neq 0\} \right| ;$$

   End

   Iter++;

Until Iter=max or |ΔSR|<γ

Fig. 5. Algorithm 1 to compute the similarRank score of each patch b_i.

Body/Legs Salient Region. For stripe m, the patch which has the maximum of SR_m is denoted as $bmax_m$. The final score is computed as

$$SI(bmax_m) = \sum_{\forall m' \neq m} S(bmax_m, bmax_{m'}) \tag{6}$$

Top half of stripes are selected as body region, and the patch b_m which displays the maximum of $SI(b_m)$ in body region is chosen as the body salient region. Bottom half of stripes are selected as legs region, and the patch b_m which shows the maximum of $SI(b_m)$ in legs region is chosen as the legs salient region.

Characteristic Salient Region. For stripe m, the patch which has the minimum of SR_m is denoted as $bmin_m$. The final score is computed as

$$ST(bmin_m) = \sum_{\forall m' \neq m} S(bmin_m, bmin_{m'}) \tag{7}$$

The patch b_m which is the minimum of $ST(b_m)$ in body region is chosen as the characteristic salient region.

To handle misalignment caused by large viewpoints and poses variations, we compute Euclidean distance between query salient patch b and all patches in gallery images and choose the minimum as the final salient distance $dist_{salient}$.

3.3 Salient Region Evaluate

In Sects. 3.1 and 3.2, we proposed a model to represent the global distribution and local details of pedestrian image. But some patches in background hide interferential information, which have a great negative impact on pedestrian representation. Because person re-identification is a task to find a person from a series of non-overlapping camera views which inevitably make the background of the pedestrian image inconstant. Such negative impact of background will reduce the accuracy of person re-identification. To handle this problem, we set different weight for each patch according to their column. The weight w_c of c-th patch bc in horizontal is defined as:

$$w_c = exp(-\frac{(\mu_c - \mu)^2}{2l^2}) \tag{8}$$

where l is the half of patch size, μ is the half of image width and μ_c is the center column of b_c in pixel. Therefore, the representation near the edge of the image has been suppressed by weight w.

3.4 Metric Learning and Distance Fusion

In person re-identification, the feature of a pedestrian usually input into metric learning as a vector. But as there is no single color space could be against to the variations of illumination, there is no single transformation function suit for a vector which contain different kinds of feature. The combined feature space may also be too complex to be robustly handle by single metric. Therefore, we propose metric each feature space separately and train a pool of multiple metrics.

In [1], Kostinger et al. proposed a metric method called KISSME based on the log likelihood ratio test of two Gaussian distributions. In [5] Liao et al. extend the KISSME approaches to a cross-view metric learning called XQDA. XQDA first learn a subspace $W \in R^{d*n}$ with cross-view data where d means the original dimensional and n means the dimensional subspace. And then it learns a kernel matrix M in the n dimensional subspace. The distance between probe X_p with gallery X_g is computed as

$$distM(X_p, X_g) = (X_p - X_g)^T WMW^T (X_p - X_g) \tag{9}$$

where X_p, X_g mean the feature representation of pedestrian image I_q, I_g.

We assume the result of metric learning in feature space k is defined as W_k, M_k, for k = 1, ..., n (n is the number of space). The result of a probe person p and a gallery person g being the same in feature space k can be defined as the credibility P_k which transform from distance

$$P_k(p = g|M_k) = \sigma((X_p - X_g)^T W_k M_k W_k^T (X_p - X_g)) \tag{10}$$

where $\sigma(z) = exp(-Z)$. The final global credibility can be denoted as:

$$P_k(p = g|M_1, ..., M_k) = \sum_{k=1}^{4} \beta_k P_k(p = g|M_k) \tag{11}$$

where β_k is parameter of feature space weight.

In real life, saliency various in pedestrian images. Only salient regions in salient pedestrian image has the higher credibility. Differences among salient regions can be considered as the saliency of pedestrian image which computed as:

$$\lambda = (1 - \frac{f_b f_l}{\sqrt{|f_b|}\sqrt{|f_l|}})(1 - \frac{f_c f_l}{\sqrt{|f_c|}\sqrt{|f_l|}})(1 - \frac{f_c f_b}{\sqrt{|f_c|}\sqrt{|f_b|}}) \qquad (12)$$

where f_b, f_l, f_c represent the feature of body salient region, legs salient region, characteristic salient region respectively. The more different among salient regions, the more salient this pedestrian image is.

In order to output a final decision, all result must be pooled. The pooled result can be obtained by computing the integrated credibility considering local and global result. The salient value λ of pedestrian image can be considered as the weight of local result. Thus, the final decision is computed as

$$P_k(p = g|M_1, ..., M_k) = (1 - 0.7\lambda)\sum_{k=1}^{4}\beta_k P_k(p = g|M_k) + 0.7\lambda dist_{salient} \qquad (13)$$

Formula (13) is used to compute the final ranking for re-identification.

4   Experiments

In this section, we evaluate our method on two publicly available datasets, i.e. the VIPeR dataset [17], and the CUHK01 dataset [18]. These datasets have been selected because they both are very challenge datasets for person re-identification and contain many real scenarios, i.e. viewpoints, poses and illumination changes, different backgrounds, image resolutions, occlusions, etc. In our experiments on both datasets, we randomly partition the dataset into two equal part, one for training and another for testing, without overlap on every single person. For each dataset, evaluation procedure is repeated 10 times. We set $\beta = [0.5, 0.3, 0.3, 0.1]$ in each color space i.e. HSV, normal rgb, YUV, rgs.

All the results are shown on the basis of recognition rate by the Cumulative Matching Characteristic (CMC) curve. Images from one camera are used as probe and images from other camera are used as gallery. Each probe image is matched with every gallery image. The rank of correct match is expressed in the CMC curve. In our experiments, our approach is compared with the state-of-the-art methods, i.e. LOMO+XQDA, MFA, kLFDA, KISSME, LFDA.

4.1   Experiment on VIPeR

The VIPeR dataset is captured by two cameras in outdoor environment. It contains 632 low spatial resolution image pairs which relate to 632 persons and are captured by two different cameras. Images from camera A are mostly captured from 0° to 90° while others from camera B mostly from 90° to 180°. All images in VIPeR dataset are 128 * 48 pixels. Some examples from VIPeR dataset show in Fig. 6.

Fig. 6. Examples from VIPeR dataset. The two rows show the different appearances of the same person in different camera views.

We compare our method with the state-of-the-art methods including LOMO+XQDA, MFA, kLFDA, KISSME, LFDA. The experiment results show in Table 1 and Fig. 7. Each methods are following the same evaluation protocol as ours. Results demonstrate that our method achieves better result than the state-of-art methods on higher ranks. As shown, our method achieves the highest rank 1 score by reaching a recognition rate of 44.3% which more 4.1% than the second. From rank 1 to 20, our method performs better than all other methods.

Table 1. Comparison with the state-of-the-art methods on VIPeR dataset.

Method	Rank = 1	Rank = 5	Rank = 10	Rank = 15	Rank = 20
Ours	**44.3**	**74.6**	**85.8**	**91.8**	**94.9**
LOMO+XQDA	40.2	68.3	80.8	87.1	91.1
MFA	20.5	49.0	63.4	73.1	78.2
LFDA	18.3	44.6	57.3	66.7	73.0
KISSME	22.5	49.6	64.1	71.8	83.5
kLFDA	22.2	47.2	60.3	69.0	76.0

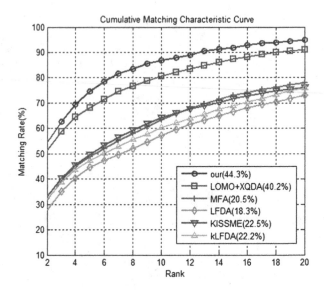

Fig. 7. The CMC curve of ours method and state-of-the-art method on VIPeR dataset

4.2   Experiment on CUHK-01

The CUHK-01 dataset has images captured by non-overlapping camera views in a campus environment. Images in this dataset are of higher resolution. The dataset contain 971 persons images captured from two different cameras. Each person has two images in each camera view. Camera A is from a frontal view and camera B is from a side view. All images are normalized to 160 * 60 pixel for evaluations. It also contain significant variations on viewpoints, poses, illumination, and their images are with occlusions and background clutters. Some example from CUHK-01 dataset show in Fig. 8.

We split the dataset into a training set which contain 485 person and a testing set which contain 486 person. We compare our method with five state-of-the-art approaches include LOMO+XQDA, MFA, kLFDA, KISSME, LFDA. The experimental results are shown in Fig. 9 and Table 2. From the experimental results, we can see our method performs better than other ones by reaching a correct recognition rate of 65.3% at rank 1. As shown, in the most rank, our method performs always better recognition rate than all other existing methods. From the analysis of all the results, we can conclude that, in general, our method is outperform the state-of-the-art approaches and more robust to the viewpoint and illumination variations.

The intrinsic reasons for significant improvement is as follows. Firstly, we analysis person images from different scales. It describes pedestrian image from local details and global distribution. Secondly, we extract probability histogram from different color space to represent pedestrian images. It is much more stable and effective for the obvious change of illumination. In addition, we model each feature space separately. A pool of multiple metric learning is built to metric an optimal distance.

Fig. 8. Examples from CUHK-01 dataset. Each region show the different appearances of the same person in different camera views.

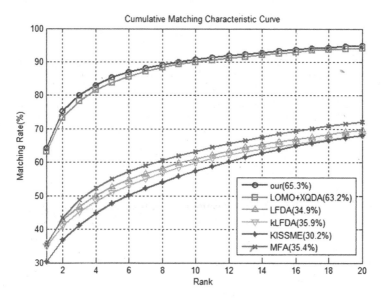

Fig. 9. The CMC curve of ours method and state-of-the-art method on CUHK-01 dataset

Table 2. Comparison with the state-of-the-art methods on CUHK-01 dataset.

Method	Rank = 1	Rank = 5	Rank = 10	Rank = 15	Rank = 20
Ours	**65.3**	**85.6**	**91.15**	**94.0**	**94.6**
LOMO+XQDA	63.2	83.9	90.0	92.6	94.2
MFA	35.4	55.1	63.3	68.5	72.1
LFDA	34.9	50.9	59.9	64.8	68.0
KISSME	30.2	47.7	57.5	63.9	68.2
kLFDA	35.9	52.7	61.1	66.2	69.8

5   Conclusion

In this paper, we have proposed a novel feature representation which consider the local detail and global geometric correspondence of pedestrian image. Some small salient regions detected to describe the local detail of pedestrian image. Probability histograms is computed in different color space to address the illumination problem. Besides, we proposed a pool of multiple metric learns. All the distance metric results are translated into reliability credibility and pooled as final decision. The evaluated on two public datasets VIPeR, CUHK-01, showing outperforms the state-of-the-art on both datasets. In the future work, we are considering how to learn a color transformation between different camera views and how salient region can be used to weight patches.

References

1. Koestinger, M., Hirzer, M., Wohlhart, P., et al.: Large scale metric learning from equivalence constraints. In: 2012 IEEE Conference on Computer Vision, Pattern Recognition (CVPR), pp. 2288–2295. IEEE (2012)
2. Prates, R., Dutra, C.R.S., Schwartz, W.R.: Predominant color name indexing structure for person re-identification. In: 2016 IEEE International Conference on Image Processing (ICIP), pp. 779–783. IEEE (2016)
3. Lisanti, G., Masi, I., Bagdanov, A.D., et al.: Person re-identification by iterative re-weighted sparse ranking. IEEE Trans. Pattern Anal. Mach. Intell. **37**(8), 1629–1642 (2015)
4. Chen, D., Yuan, Z., Hua, G., et al.: Similarity learning on an explicit polynomial kernel feature map for person re-identification. In: Proceedings of the IEEE Conference on Computer Vision and Pattern Recognition, pp. 1565–1573 (2015)
5. Liao, S., Hu, Y., Zhu, X., et al.: Person re-identification by local maximal occurrence representation, metric learning. In: Proceedings of the IEEE Conference on Computer Vision and Pattern Recognition, pp. 2197–2206 (2015)
6. Farenzena, M., Bazzani, L., Perina, A., Murino, V., Cristani, M.: Person re-identification by symmetry-driven accumulation of local features. In: CVPR (2010)
7. Zhao, R., Ouyang, W., Wang, X.: Unsupervised salience learning for person re-identification. In: Proceedings of the CVPR (2013)

8. Yang, Y., Yang, J., Yan, J., Liao, S., Yi, D., Li, S.Z.: Salient color names for person re-identification. In: Fleet, D., Pajdla, T., Schiele, B., Tuytelaars, T. (eds.) ECCV 2014. LNCS, vol. 8689, pp. 536–551. Springer, Heidelberg (2014). doi:10.1007/978-3-319-10590-1_35

9. Mignon, A., Jurie, F.: PCCA: a new approach for distance learning from sparse pairwise constraints. In: IEEE Conference on Computer Vision and Pattern Recognition (CVPR), pp. 2666–2672. IEEE (2012)

10. Pedagadi, S., Orwell, J., Velastin, S., Boghossian, B.: Local fisher discriminant analysis for pedestrian re-identification. In: 2013 IEEE Conference on Computer Vision and Pattern Recognition (CVPR), pp. 3318–3325. IEEE (2013)

11. Yan, S., Xu, D., Zhang, B., Zhang, H.J., Yang, Q., Lin, S.: Graph embedding and extensions: a general framework for dimensionality reduction. IEEE Trans. Pattern Anal. Mach. Intell. 29(1), 40–51 (2007)

12. Zheng, W.S., Gong, S., Xiang, T.: Person re-identification by probabilistic relative distance comparison. In: IEEE Conference on Computer Vision and Pattern Recognition (CVPR), pp. 649–656. IEEE (2011)

13. Prosser, B., Zheng, W.S., Gong, S., Xiang, T., Mary, Q.: Person re-identification by support vector ranking. In: BMVC (2010)

14. Lazebnik, S., Schmid, C., Ponce, J.: Beyond bags of features: spatial pyramid matching for recognizing natural scene categories. In: IEEE Computer Society Conference on Computer Vision and Pattern Recognition, vol. 2, pp. 2169–2178. IEEE (2006)

15. Ojala, T., Pietikainen, M., Maenpaa, T.: Multiresolution gray-scale and rotation invariant texture classification with local binary patterns. IEEE Trans. Pattern Anal. Mach. Intell. 24(7), 971–987 (2002)

16. Page, L., Brin, S., Motwani, R., et al.: The PageRank citation ranking: bringing order to the web (1999)

17. Gray, D., Brennan, S., Tao, H.: Evaluating appearance models for recognition, reacquisition, and tracking. In: PETS (2007)

18. Li, W., Zhao, R., Wang, X.: Human reidentification with transferred metric learning. In: Lee, K.M., Matsushita, Y., Rehg, J.M., Hu, Z. (eds.) ACCV 2012. LNCS, vol. 7724, pp. 31–44. Springer, Heidelberg (2013). doi:10.1007/978-3-642-37331-2_3

Gait Gate: An Online Walk-Through Multimodal Biometric Verification System Using a Single RGB-D Sensor

Mohamed Hasan[1,2]([✉]), Yasushi Makihara[1], Daigo Muramatsu[1], and Yasushi Yagi[1]

[1] Institute of Scientific and Industrial Research, Osaka University, Osaka, Japan
{hasan,makihara,muramatsu,yagi}@am.sanken.osaka-u.ac.jp
[2] Shoubra Faculty of Engineering, Benha University, Cairo, Egypt

Abstract. This paper introduces the Gait Gate as the first online walk-through access control system based on multimodal biometric person verification. Face, gait and height modalities are simultaneously captured by a single RGB-D sensor and fused at the matching-score level. To achieve the real-time requirements, mutual subspace method has been used for the face matcher. An acceptance threshold has been learned beforehand using data of a set of subjects disjoint from the targets. The Gait Gate has been evaluated through experiments in actual online situation. In experiments, 1324 walking sequences have resulted from the verification of 26 targets. The verification results show an average computation time of less than 13 ms and an accuracy of 6.08% FAR and 7.21% FRR.

1 Introduction

Biometric authentication systems have been widely deployed in various applications like security, forensics, surveillance, and entrance, access and boarder control. These systems rely on the evidence of fingerprints, iris, face, gait, hand vein, signature, voice, etc. for people verification or identification. Current systems may be unimodal or multimodal depending on the adopted number of biometric traits.

From a usability viewpoint, current authentication systems suffer from two main limitations. First, targets of verification have to pay a dedicated effort to provide the required biometrics, for example, touching screens or looking at a camera. Moreover, the time consumed by the overall process increases proportionally with the number of targets because most of the current systems can only authenticate a single subject at a time. Imagine the queue of a number of employees waiting for authentication at an entrance-controlled area.

Therefore, there is a need for a biometric authentication system with an increased throughput. Although a contact-less fingerprint scanner [1] and a high-throughput walk through iris verification system [2] have been recently introduced, the biometric readers in such systems still require a dedicated target's

© Springer International Publishing AG 2017
C.-S. Chen et al. (Eds.): ACCV 2016 Workshops, Part III, LNCS 10118, pp. 330–344, 2017.
DOI: 10.1007/978-3-319-54526-4_25

effort and can not simultaneously authenticate multiple subjects. Hence, there is a high demand for a walk-through access control system that perform authentication without requiring an excessive effort from the targets.

Face and gait are promising modalities for such a walk-through system due to their effectiveness even if the subject is at a relatively large distance. Although face recognition has been extensively studied [3,4], its performance degrades in uncontrolled situations due to the following reasons: PIE (pose, illumination and expression) variations [5], significantly low image resolution and occlusion. Compared to face modality, gait recognition [6] is a relatively new method having the advantage of being effective even with significantly low-resolution images. However, it suffers from the variation in views [7,8], walking speeds [9] and clothing [10].

To mitigate the limitations of individual face and gait recognition, the two modalities have been fused [11–15] in multimodal biometric systems. Generally, fusion of heterogeneous modalities [16–19] can take place at the feature level, score level, or decision level. In particular, face and gait fusion is a promising approach because such modalities can be captured by a single sensor (camera).

Almost all the existing multimodal biometric systems are tested offline without addressing the online requirements of a walk-through system like computation time and real-environment evaluation. For example, offline systems may adopt time-consuming approaches for silhouette extraction [15] or face matching [20]. In addition, assuming simple backgrounds [20], such systems are not evaluated in online real environments.

In this paper, we introduce the first online walk-through access control system based on multimodal biometric verification named Gait Gate. The proposed system uses a single RGB-D sensor and combines face, gait and height modalities at the matching-score level. The contributions of this paper are twofold:

(1) Online multimodal biometric verification system using a single RGB-D sensor. The proposed system captures the biometrics during the targets walking, i.e., no dedicated effort is needed. The system meets the real-time requirements of a walk-through access control system.
(2) Experimental evaluation in an actual online situation. The experiments have been performed in an actual environment with no assumption of the background. Although a gait recognition system aiming at working in an online situation has been introduced [21], it was not actually tested online.

2   Online Walk-Through Verification System

2.1   Overview

An overview of the proposed system is depicted in Fig. 1. The online walk-through access control system is based on multimodal biometric verification using a single RGB-D sensor, i.e., Microsoft Kinect v2. The verification target inputs a claimed ID to the system. Afterwards, the face, gait and height modalities are captured as the target walks in front of the sensor. No excessive

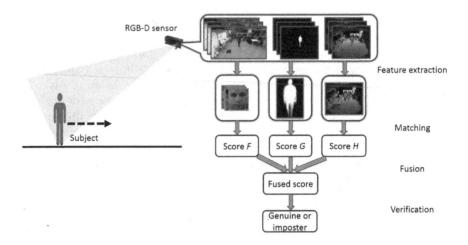

Fig. 1. Overview of the proposed wall through access control system. A single RGB-D sensor is used to extract face, gait and height modalities.

effort is required by the target to provide the biometric traits. The captured modalities are matched against that gallery templates associated with the input claimed ID. Modality fusion is performed at the matching-score level.

2.2 Pre-processing

RGB color, depth, body index, face, and skeleton frames are processed to extract the data required for modalities description. The face modality F is described through the following steps. First, the face region and facial landmark points are localized in the RGB frame. The facial landmarks are alignment points at the left eye, right eye, nose, right and left mouth corners. Second, the average depth of the facial landmarks is computed and assumed to be the depth of the face template d_f. Third, the face image is size-normalized to a standard depth d_s using a scaling ratio of (d_s/d_f). Fourth, the size-normalized image is converted from RGB color to gray scale. Sequence of the gray-scale size-normalized face images is regarded as the set of the face modality.

The body index frame indicates which depth pixels belong to tracked subjects. Hence, this frame is processed to extract the 2D silhouette of each subject and the extracted silhouette is size-normalized into a 88×128 pixel-sized frame. A sequence of the size-normalized silhouette frames is used to describe the gait modality G by the gait energy image (GEI) [22].

The skeleton frame provides the ground plane parameters. Height is computed as the distance between the top point of the 2D body silhouette in camera space and the ground plane. A sequence of height measurements are then averaged to compute the average height which describes the height modality H.

2.3 Face Matcher

Given a pair of face sequences F_p and F_g for the probe and gallery respectively, we use the mutual subspace method (MSM) [23] to calculate the dissimilarity score between them. MSM is adopted here as it is a powerful and fast set-to-set matching technique suitable for our target of an online system. In this technique, face images of the probe set and those of the gallery set are modeled as subspaces and the similarity measure is calculated as a canonical angle between these two subspaces.

Stated formally, each face image in the probe/gallery set is unfolded into a d – dimensional vector in a raster scan way. The set of n face images $(x_i \in \mathbb{R}^d)$ in a probe/gallery set is modeled as:

$$X = \begin{bmatrix} x_1 \dots x_n \end{bmatrix}, \quad X \in \mathbb{R}^{d \times n}. \tag{1}$$

The X matrix is factorized using singular value decomposition as:

$$X = U \Sigma V^T, \tag{2}$$

where U and Σ represent the left singular vectors and the singular values respectively. The first k columns of U are then selected by thresholding on the cumulative sum of the singular values (σ_i), where k is the minimum value satisfying the following equation:

$$\frac{\sum_{i=0}^{k} \sigma_i}{\sum_{i=0}^{n} \sigma_i} \geq threshold. \tag{3}$$

Let U_p and U_g denote the selected subsets of the left singular vector for the probe and gallery subspaces respectively. The canonical angle between the two subspaces is calculated as the maximum eigen value σ_{max} of the following matrix Z:

$$Z = U_g{}^T U_p U_p{}^T U_g. \tag{4}$$

Finally, the dissimilarity score between the probe and gallery face images is given by:

$$S_F = 1 - \sigma_{max}. \tag{5}$$

2.4 Gait Matcher

The GEI of the probe G_p and gallery G_g are represented by:

$$\begin{aligned} G_p &= \{G_{p,i}\}, \quad i = 1 \dots N_p \\ G_g &= \{G_{g,j}\}, \quad j = 1 \dots N_g, \end{aligned} \tag{6}$$

where N_p and N_g represent the number of periods of the probe and gallery GEIs.

The dissimilarity score between G_p and G_g is calculated by Euclidean distance as:

$$S_G = \min_{i,j} \| G_{p,i} - G_{g,j} \|_2. \tag{7}$$

2.5 Height Matcher

Let the top point of the body silhouette at the i-th frame be represented by $(x_{t,i} \ \ y_{t,i} \ \ z_{t,i})^T$ in camera space (which is obtained from a depth map) and the ground plane by $(Ax + By + Cz + D = 0)$. The height modality is described as:

$$H = \frac{1}{N_f} \sum_{i=1}^{N_f} \frac{|Ax_{t,i} + By_{t,i} + Cz_{t,i} + D|}{\sqrt{A^2 + B^2 + C^2}}, \tag{8}$$

where N_f is the number of silhouette frames for which the ground plane parameters have been successfully acquired.

The dissimilarity score between the probe H_p and gallery H_g heights is calculated as the absolute difference:

$$S_H = |H_p - H_g|. \tag{9}$$

2.6 Fusion Rule

To eliminate subject dependency, matching scores are normalized using Z-score normalization method before fusion. Let $S_x(i,j)$ be the scores of feature $x \in$ {face (F), gait (G), height (H)} between the i-th probe and the j-th gallery. If the identity claimed by the probe is c, then the normalized scores are given by:

$$\bar{S}_x(i,j) = \frac{S_x(i,j) - \mu_x(c)}{\sigma_x(c)}, \tag{10}$$

where $\mu_x(c)$ and $\sigma_x(c)$ stand for mean and standard deviation that will be described later. These are the subject's specific normalization parameters associated with feature x and the the c-th subject.

In this study, N_t gallery templates are registered for each target. Hence, the minimum score is calculated for each modality as:

$$\bar{S}_{x_{min}}(i) = \min_{j} \bar{S}_x(i,j), \quad j = 1, ..., N_t \quad and \quad x \in \{F, G, H\}. \tag{11}$$

Given the normalized minimum scores, the fusion score S_{fusion} is calculated by:

$$S_{fusion} = f(\bar{S}_{F_{min}}, \bar{S}_{G_{min}}, \bar{S}_{H_{min}}), \tag{12}$$

where $f(.)$ is a fusion rule that combines the input scores. The SUM fusion rule is considered in this work due to its reported robustness [17,20]. Verification is performed by comparing the fusion score with an acceptance threshold T.

2.7   Parameters Learning

The subject's specific normalization parameters in Eq. (10) are learned prior to the verification process. Upon registering N_t gallery templates for all targets, each template of the c-th subject is matched against all the gallery templates. The mean $(\mu_F(c), \mu_G(c), \mu_H(c))$ and standard deviation $(\sigma_F(c), \sigma_G(c), \sigma_H(c))$ of the resulting matching scores for each modality represent the c-th subject's specific normalization parameters.

The acceptance threshold T in Sect. 2.6 is also learned beforehand using data of a set of subjects (validation set) disjoint from the targets. For each subject in the validation set, a number of templates are registered and matched against each other. The modalities description, matching scores computation, normalization and fusion are performed using the methods mentioned earlier in Sect. 2. Distribution of the resulting fusion scores are used to learn the threshold T which nearly achieves equal error rate (EER) of false acceptance rate (FAR) and false rejection rate (FRR).

2.8   Implementation

The proposed system can simultaneously verify up to N_s subjects in real-time through multi-threading. A single main thread is assigned for simultaneous modality extraction of up to N_s tracked subjects. A number of N_s concurrent threads are assigned for the verification of each subject. If the modality sequences of the i-th subject ($i = 1, ..., N_s$) are extracted, the i-th thread is called to complete the verification process.

The main thread (Algorithm 1) starts by loading the targets' gallery templates, the targets' normalization parameters and the acceptance threshold. Afterwards, the Kinect frames are acquired and processed to extract the face, silhouette and height sequences for each tracked subject. A silhouette frame counter C is incremented upon updating the silhouette sequence. Upon collecting N_{sf} silhouette frames of a subject, the associated verification thread is called, C is cleared and the main thread is continued.

At the beginning of the verification thread (Algorithm 2), face, gait and height features of the probe are extracted using the sequences obtained in Algorithm 1. The extracted features are matched against the associated gallery templates and the verification process is completed as mentioned earlier.

3   Experiments

3.1   Experimental Setup

The walking scenario is depicted in Fig. 2. A subject starts walking 2 m before the place where we start feature extraction to account for walking acceleration.

Algorithm 1. Main thread of multi-users multi-modal biometric verification system.

Input: Gallery templates, normalization parameters and acceptance threshold (T).
Output: Action (A).

Initialize the Kinect sensor.
for *Subject* **do**
  $C[Subject] \leftarrow 0$
end for
for *Frame* **do**
  Acquire the latest color, depth, body index, skeleton and face frames.
  Extract the 2D body silhouette for each subject.
  for *Subject* **do**
    Get ground plane parameters.
    Get face bounding box in color space.
    Generate a size-normalized face template.
    Add the current face image to the face sequence.
    Add the current body silhouette to the silhouette sequence.
    Compute the height.
    Add the current height measurement to the height sequence.
    $C[subject] \leftarrow C[subject] + 1$.
    if $C[subject] = N_{sf}$ **then**
      $R \leftarrow$ verification thread.
      $C[subject] \leftarrow 0$.
      if $R = Reject$ **then**
        $A =$ turn on an alarm.
      end if
    end if
  end for
end for

Algorithm 2. Verification thread of the multi-users multi-modal biometric verification system.

Input: Face, silhouette and height sequences of the probe, gallery data (F_g, G_g, H_g), normalization parameters ($\mu_F(c), \mu_G(c), \mu_H(c), \sigma_F(c), \sigma_G(c), \sigma_H(c)$) of the claimed identity (c) and the acceptance threshold (T).
Output: Verification result (R).

for $X \in \{ F, G, H \}$ **do**
  $X_p \leftarrow$ probe's X-modality sequence.
  $S_X \leftarrow X$-modality matching (X_p, X_g).
  $\bar{S}_X \leftarrow$ score normalization ($S_X, \mu_X(c), \sigma_X(c)$).
  $\bar{S}_{X_{min}} \leftarrow$ MIN($\bar{S}_X$) over the N_t gallery templates.
end for
$S_{fusion} \leftarrow$ score fusion ($\bar{S}_{F_{min}}, \bar{S}_{G_{min}}, \bar{S}_{H_{min}}$).
if $S_{fusion} < T$ **then**
  $R =$ Accept
else
  $R =$ Reject
end if
return R

Features extraction starts if the subject's depth is below a threshold which was experimentally set to 4 m to meet the requirements of the Kinect body index tracking. We implemented our algorithm using C++ and the Kinect SDK and run on a computer with 4 GHz CPU and 32 GB RAM. The hyper-parameters are experimentally decided as: $N_{sf} = 30, N_s = 6, N_t = 10$ and $d_s = 1$ m. The acceptance threshold T is set to -3.3 according to the description in Sect. 2.7.

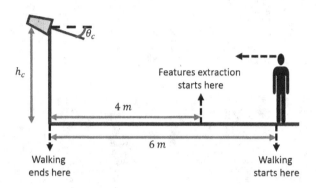

Fig. 2. Experimental setup of the walk-through access control system. $h_c = 2.5$ m is the camera height and $\theta_c = 25°$ is the camera tilt.

To collect a target's gallery template, each subject is asked to walk from a specified starting point to an ending point as shown in Fig. 2. This walking scenario is called here a waking sequence. We collected ten gallery templates for each subject. Samples from the gallery templates of one walking sequence are shown in Fig. 3.

To evaluate the proposed system, gallery templates are first collected for each of the 26 targets (evaluation set). After gallery collection, 1324 sequences have been collected for the 26 targets across three weeks. Walking sequences have been collected at different times per day for each subject and the claimed identity has been selected to equate the number of genuine (accept) and imposter (reject) claims.

To learn the acceptance threshold in Sect. 2.7, 340 walking sequences have been collected for the disjoint 10 subjects in the validation set. These sequences have been collected with a time lapse of 2–3 days/subject to account for the daily variation in the gait modality.

3.2   Results

The breakdown of the average computation time for the main (Algorithm 1) and verification (Algorithm 2) threads is reported in Tables 1 and 2, respectively.

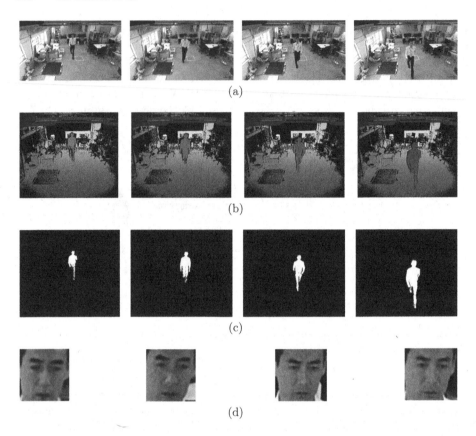

(a)

(b)

(c)

(d)

Fig. 3. Sample frames from a gallery walking sequence: (a) color RGB frames (b) depth frames (c) body index frames (note that the Kinect RGB, depth and body index frames are not synchronized) (d) size-normalized face images at $d_s = 1\,\text{m}$. (Color figure online)

The reported results confirms that thanks to the multi-threading strategy, the proposed system satisfies the real-time requirements of an online system due to two reasons. First, the average computation time of the main thread is 12.21 ms which is far less than the Kinect frame rate (30 fps). Second, the average time consumed by one verification thread (393.91 ms) is less than the time between two calls for the same verification thread which is given by the time required to capture a sequence of 30 silhouette frames (around 1000 ms).

The verification accuracy result is represented by 6.08% FAR and 7.21% FRR. These values are higher than that reported by offline systems (e.g. [20]) due to the difficulties of an actual online situation in a real environment. Such difficulties are discussed in the following section.

Table 1. Breakdown of the computation time of the main thread of Algorithm 1.

Component	Time (ms)
Kinect frames acquirement	**0.07**
RGB and depth frames processing	**1.42**
Height measurement	0.87
Body index frame processing to get 2D silhouette	0.81
Updating the silhouette sequence	0.73
Total Silhouette frame processing	**2.41**
Face region and facial landmarks detection in RGB space	0.12
Facial landmarks mapping from RGB to depth space	6.54
Average depth computation of the facial landmarks	1.65
Total time for face frame processing	**8.31**
Total time of a single iteration	**12.21**

Table 2. Breakdown of the computation time of the verification thread of Algorithm 2.

Component	Time (ms)
Probe sequences retrieving	**0.02**
Average height computation and height matching	**0.02**
Face cropping	0.11
Computing X vector of fae frames (Eq. 1)	0.61
Computing SVD of X	50.46
Selecting subset of eigen vectors	0.16
Computing the canonical angle	156.27
Total time for face matching	**207.61**
GEI computation	171.78
GEI matching	0.33
Total time for gait matching	**172.11**
Score fusion and saving matching scores to HDD	**14.15**
Total matching time	**393.91**

3.3 Discussion

We calculated the FAR and FRR for the results of the evaluation data. The detection error tradeoff (DET) curve for such data is shown in Fig. 4.

The FAR and FRR values for all subjects in the evaluation set are shown in Fig. 5. The most obvious failure cases are for subjects with IDs 17, 18 and 19 who are all female subjects. Sample GEI frames for these failure cases from the falsely rejected results are shown in Fig. 6. On the other hand, results for subject

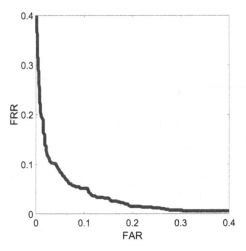

Fig. 4. DET curve for the evaluation data.

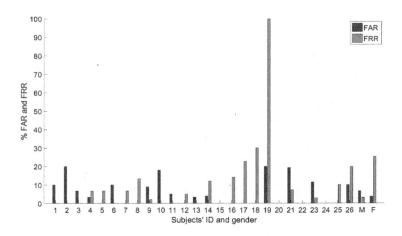

Fig. 5. FAR and FRR for all subjects in the evaluation group. The average FAR and FRR based on subjects' gender (male M and female F) is also shown.

IDs 15, 22 and 24 represent the most successful cases. Sample GEI frames for such successful cases from the truly accepted results are shown in Fig. 7.

We hypothesized that the clothes variation associated with female subjects degrades the accuracy of gait matching. To further confirm our hypothesis, we plot the average FAR and FRR based on the subjects' gender in Fig. 5.

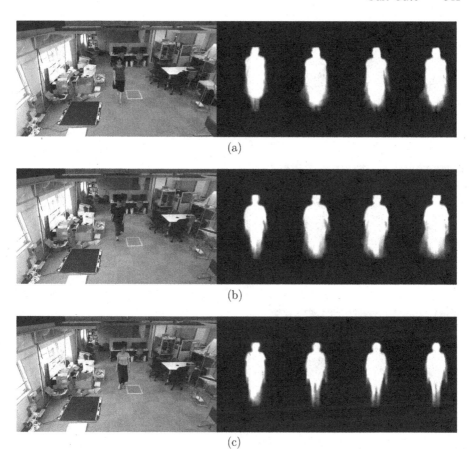

Fig. 6. Sample RGB and GEI frames for the most obvious failure cases: (a) Sample color RGB and GEI frames from the gallery templates of the subject ID 17 are shown in the first two columns while the last three columns show sample GEI frames from the falsely rejected results for the same subject. (b) Sample color RGB and GEI frames from the gallery templates of the subject ID 18 are shown in the first two columns while the last three columns show sample GEI frames from the falsely rejected results for the same subject. And (c) Sample color RGB and GEI frames from the gallery templates of the subject ID 19 are shown in the first two columns while the last three columns show sample GEI frames from the falsely rejected results for the same subject. (Color figure online)

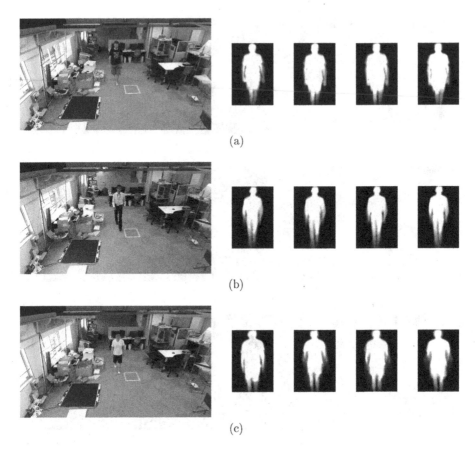

Fig. 7. Sample RGB and GEI frames for the most successful cases: (a) Sample color RGB and GEI frames from the gallery templates of the subject ID 15 are shown in the first two columns while the last three columns show sample GEI frames from the truly accepted results for the same subject. (b) Sample color RGB and GEI frames from the gallery templates of the subject ID 22 are shown in the first two columns while the last three columns show sample GEI frames from the truly accepted results for the same subject. And (c) Sample color RGB and GEI frames from the gallery templates of the subject ID 24 are shown in the first two columns while the last three columns show sample GEI frames from the truly accepted results for the same subject. (Color figure online)

4   Conclusions

This paper introduces the Gait Gate as the first online walk-through access control system based on multimodal biometric verification. Face, gait and height modalities have been captured using a single RGB-D sensor and fused at the matching-score level. The proposed system achieved an average frame rate of 80 fps confirming the real-time requirements. The proposed system has been evaluated in actual online situation through different days/times. The resulting

FAR (6.08%) and FRR (7.21%) are greatly affected by variation in gait modality due to the clothing changes. Future work will address the problem of clothing variation effect on gait matching and the extension to the identification scenario.

Acknowledgement. This work was supported by JSPS Grants-in-Aid for Scientific Research (A) JP15H01693, the JST CREST "Behavior Understanding based on Intention-Gait Model" project and Nanjing University of Science and Technology.

References

1. Safran Identity and Security: The worlds fastest contactless fingerprint scanner (2014). http://www.morpho.com
2. SRI International: Identity on the move walk-through identity reader (2016). https://www.sri.com/engage/products-solutions/iom-portal-system
3. Abate, A.F., Nappi, M., Riccio, D., Sabatino, G.: 2D and 3D face recognition: a survey. Pattern Recogn. Lett. **28**, 1885–1906 (2007)
4. Jafri, R., Arabnia, H.R.: A survey of face recognition techniques. JIPS **5**, 41–68 (2009)
5. Gross, R., Matthews, I., Cohn, J., Kanade, T., Baker, S.: Multi-PIE. Image Vis. Comput. **28**, 807–813 (2010)
6. Nixon, M.S., Tan, T., Chellappa, R.: Human Identification Based on Gait, vol. 4. Springer Science & Business Media, Heidelberg (2010)
7. Kusakunniran, W., Wu, Q., Zhang, J., Li, H.: Gait recognition under various viewing angles based on correlated motion regression. IEEE Trans. Circuits Syst. Video Technol. **22**, 966–980 (2012)
8. Makihara, Y., Sagawa, R., Mukaigawa, Y., Echigo, T., Yagi, Y.: Gait recognition using a view transformation model in the frequency domain. In: Leonardis, A., Bischof, H., Pinz, A. (eds.) ECCV 2006. LNCS, vol. 3953, pp. 151–163. Springer, Heidelberg (2006). doi:10.1007/11744078_12
9. Makihara, Y., Tsuji, A., Yagi, Y.: Silhouette transformation based on walking speed for gait identification. In: Proceedings of the 23rd IEEE Conference on Computer Vision and Pattern Recognition, pp. 717–722 (2010)
10. Hossain, M.A., Makihara, Y., Wang, J., Yagi, Y.: Clothing-invariant gait identification using part-based clothing categorization and adaptive weight control. Pattern Recogn. **43**, 2281–2291 (2010)
11. Shakhnarovich, G., Darrell, T.: On probabilistic combination of face and gait cues for identification. In: Proceedings of the Fifth IEEE International Conference on Automatic Face and Gesture Recognition, pp. 169–174. IEEE (2002)
12. Kale, A., RoyChowdhury, A.K., Chellappa, R.: Fusion of gait and face for human identification. In: Proceedings of the IEEE International Conference on Acoustics, Speech, and Signal Processing (ICASSP 2004), vol. 5, p. V-901. IEEE (2004)
13. Zhang, T., Li, X., Tao, D., Yang, J.: Multimodal biometrics using geometry preserving projections. Pattern Recogn. **41**, 805–813 (2008)
14. Zhou, X., Bhanu, B.: Feature fusion of side face and gait for video-based human identification. Pattern Recogn. **41**, 778–795 (2008)
15. Hofmann, M., Schmidt, S.M., Rajagopalan, A.N., Rigoll, G.: Combined face and gait recognition using alpha matte preprocessing. In: 2012 5th IAPR International Conference on Biometrics (ICB), pp. 390–395. IEEE (2012)

16. Jain, A., Nandakumar, K., Ross, A.: Score normalization in multimodal biometric systems. Pattern Recogn. **38**, 2270–2285 (2005)
17. Fierrez-Aguilar, J., Ortega-Garcia, J., Garcia-Romero, D., Gonzalez-Rodriguez, J.: A comparative evaluation of fusion strategies for multimodal biometric verification. In: Kittler, J., Nixon, M.S. (eds.) AVBPA 2003. LNCS, vol. 2688, pp. 830–837. Springer, Heidelberg (2003). doi:10.1007/3-540-44887-X_96
18. Ross, A., Jain, A.: Information fusion in biometrics. Pattern Recogn. Lett. **24**, 2115–2125 (2003)
19. He, M., Horng, S.J., Fan, P., Run, R.S., Chen, R.J., Lai, J.L., Khan, M.K., Sentosa, K.O.: Performance evaluation of score level fusion in multimodal biometric systems. Pattern Recogn. **43**, 1789–1800 (2010)
20. Muramatsu, D., Iwama, H., Makihara, Y., Yagi, Y.: Multi-view multi-modal person authentication from a single walking image sequence. In: 2013 International Conference on Biometrics (ICB), pp. 1–8. IEEE (2013)
21. Choudhary, A., Chaudhury, S.: Gait recognition based online person identification in a camera network. In: Jawahar, C.V., Shan, S. (eds.) ACCV 2014. LNCS, vol. 9008, pp. 145–156. Springer, Heidelberg (2015). doi:10.1007/978-3-319-16628-5_11
22. Man, J., Bhanu, B.: Individual recognition using gait energy image. IEEE Trans. Pattern Anal. Mach. Intell. **28**, 316–322 (2006)
23. Iwashita, Y., Sakano, H., Kurazume, R.: Gait recognition robust to speed transition using mutual subspace method. In: Murino, V., Puppo, E. (eds.) ICIAP 2015. LNCS, vol. 9279, pp. 141–149. Springer, Heidelberg (2015). doi:10.1007/978-3-319-23231-7_13

3D Object Recognition with Enhanced Grassmann Discriminant Analysis

Lincon Sales de Souza$^{(\boxtimes)}$, Hideitsu Hino, and Kazuhiro Fukui

Graduate School of Systems and Information Engineering,
University of Tsukuba, Tsukuba, Japan
lincons@cvlab.cs.tsukuba.ac.jp, {hinohide,kfukui}@cs.tsukuba.ac.jp

Abstract. Subspace representation has become a promising choice in the classification of 3D objects such as face and hand shape, as it can model compactly the appearance of an object, represent effectively the variations such as the change in pose and illumination condition. Subspace based methods tend to require complicated formulation, though, we can utilize the notion of Grassmann manifold to cast the complicated formulation into a simple one in a unified manner. Each subspace is represented by a point on the manifold. Thank to this useful correspondence, various types of conventional methods have been constructed on a manifold by the kernel trick using a Grassmann kernel. In particular, discriminant analysis on Grassmann manifold (GDA) have been known as one of the useful tools for image set classification. GDA can work as a powerful feature extraction method on the manifold. However, there remains room to improve its ability in that the discriminative space is determined depending on the set of data points on the manifold. This suggests that if the data on a manifold are not so discriminative, the ability of GDA may be limited. To overcome this limitation, we construct a set of more discriminative class subspaces as the input for GDA. For this purpose, we propose to project class subspaces onto a generalized difference subspace (GDS), before mapping class subspaces onto the manifold. The GDS projection can magnify the angles between class subspaces. As a result, the separability of data points between different classes is improved and the ability of GDA is enhanced. The effectiveness of our enhanced GDA is demonstrated through classification experiments with CMU face database and hand shape database.

1 Introduction

In this paper, we discuss a framework for characterizing 3D objects with image sets, which are obtained from a video or multiple camera system, focusing on classification tasks of face and hand shape. Subspace representation is very effective for image set based classification, since a set of images of an object can be effectively modeled by a low-dimensional subspace of a high-dimensional vector space [1–5]. For example, the appearance of a face under varying lighting conditions can be compactly represented with a low (from 4 to 9)-dimensional subspace [5]. In subspace-based classification, an input subspace is classified by

© Springer International Publishing AG 2017
C.-S. Chen et al. (Eds.): ACCV 2016 Workshops, Part III, LNCS 10118, pp. 345–359, 2017.
DOI: 10.1007/978-3-319-54526-4_26

using the canonical angles between it and reference subspace, where the input
and reference subspace are generated from the sets of input and reference images
of each class, respectively. The concept of canonical angle is a natural extension of
angle between two vectors [6,7]. We can effectively measure the closeness between
two m-dimensional subspaces by using m canonical angles between them. Mutual
subspace method (MSM) [8] is well known as a fundamental classification method
using canonical angles.

Subspace based method using canonical angles is often formulated as a complicated procedure. To avoid this issue, it is useful to introduce the Grassmann
manifold $\mathcal{G}(m, D)$, which is defined as a set of m-dimensional linear subspaces
of $\mathbb{R}^D$ [9]. In this framework, a subspace-based method is regarded as a simple
classification method on a Grassmann manifold, where each single subspace is
treated as a point. For example, MSM corresponds to the simplest method using
the distance between two points on a Grassmann manifold. According to this
correspondence, various types of classification methods have been constructed on
a Grassmann manifold [10,11]. In particular, discriminant analysis on a Grassmann manifold (GDA) has been known as one of the useful tools for image set
classification. GDA can be easily conducted as a kernel discriminant analysis
through the kernel trick with a Grassmann kernel, as will be described in details
later.

GDA can work as a powerful feature extraction method on a Grassmann
manifold. However, GDA cannot touch and operate data points on the manifold
directly, because it uses the framework of the kernel trick. GDA is capable of
finding out the optimal discriminant space from the given data set. Hence, if
class subspaces were not separable in a vector space, the corresponding data

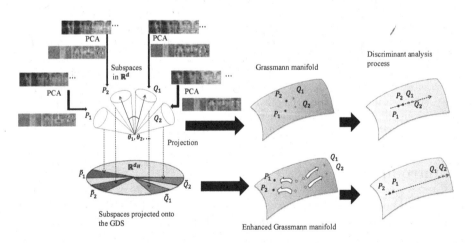

Fig. 1. Conceptual diagram of Grassmann manifold enhanced by GDS projection. The
upper process shows the case without GDS projection and the lower process shows the
case with GDS projection. There are four subspaces of 2 classes in a high dimensional
vector space. P_1 and P_2 belong to one class, and Q_1 and Q_2 to another class.

points on the manifold are not also separable. In this case, the best performance that GDA can achieve is limited. This implies that there still remains room to improve the discriminative ability of GDA by using more discriminative class subspaces. From this viewpoint, we propose to project class subspaces onto a generalized difference subspace (GDS) [12], before mapping each class subspace on a Grassmann manifold. We expect the GDS projection can enhance the discriminative ability of GDA, as it can magnify the angles between different class subspaces to provide more discriminative sample for GDA, as shown in Fig. 1. The validity of our enhanced GDA is demonstrated through experiments with the CMU Multi PIE face database [13] and hand shape database [14].

The paper is organized as follows. In Sect. 2, we introduce the basic idea of this paper. In Sects. 3 and 4, we review GDS, Grassmann manifold and related concepts. In Sect. 5, we define the concept of the enhanced GDA and describe the proposed framework. In Sect. 6, we evaluate the framework through experiments with CMU face database and hand shape database. Section 7 concludes the paper.

2   Basic Idea on Enhanced Grassmann Discriminant Analysis

Our basis idea is to utilize more discriminative class subspaces as the inputs for GDA, as mentioned previously. There are several approaches for obtaining desirable class subspaces, instead of naive class subspaces, which is generated from a set of raw appearance images of an object by using the principal component analysis (PCA). We can generate such desirable class subspace from certain kind of discriminative feature vectors, which are extracted from raw appearance images of an object. However, the subspace representation may be not valid in the vector space of the extracted feature unlike in that of the original raw appearance images.

In this paper, we propose to project class subspaces onto a generalized difference subspace (GDS), before mapping them onto a Grassmann manifold. The advantage of the proposed approach against the approach based on feature vectors is that more discriminative class subspaces can be obtained directly and smartly from a set of given class subspaces. This approach works even when the subspace representation is not valid. In addition, it has high scalability for further extensions.

Generalized difference subspace (GDS) is defined as a subspace, which represents a "difference component" among multiple class subspaces [12]. GDS is a further extension of difference subspace (DS) for two class subspaces, which is a natural generalization of a difference vector of two vectors. GDS projection can magnify the canonical angles between different class subspaces by removing a common subspace of them. The details will be described later. As a result, the data points mapped onto the Grassmann manifold become more separable as shown in Fig. 1.

Although GDA and GDS projection seem to aim at the similar effect, their mechanisms are quite different in that GDA works on a Grassmann manifold, while GDS projection works in a high dimensional vector space before mapping onto the manifold. From this difference, we expect a kind of synergistic effect from both to enhance the ability of GDA. In the following, we refer the GDA with GDS projection as enhanced GDA (eGDA). Further, GDS has been kernelized by using the kernel PCA to kernel GDS (KGDS) to deal with nonlinear class subspaces [12]. We refer the GDA with KGDS projection as enhanced kernel GDA (eKGDA).

3   Conventional Grassmann Discriminant Analysis

In this section we define canonical angles, outline the concept of Grassmann manifold and the most used kernel with it, and explain the algorithm of GDA.

3.1   Definition of Canonical Angles

Suppose we have two subspaces that we want to compare their similarity: a d_p-dimensional subspace $\mathcal{P}$ and a d_q-dimensional subspace $\mathcal{Q}$, and both lay in a d-dimensional vector space. For convenience, we suppose $d_p \leq d_q$. The canonical angles $\{0 \leq \theta_1, \ldots, \theta_{d_p} \leq \frac{\pi}{2}\}$ between $\mathcal{P}$ and $\mathcal{Q}$ are recursively defined as follows [7,15]:

$$\cos \theta_i = \max_{\boldsymbol{u}_i \in \mathcal{P}} \max_{\boldsymbol{v}_i \in \mathcal{Q}} \boldsymbol{u}_i^\top \boldsymbol{v}_i$$

$$\text{s.t. } \|\boldsymbol{u}_i\| = \|\boldsymbol{v}_i\| = 1, \ \boldsymbol{u}_i^\top \boldsymbol{u}_j = \boldsymbol{v}_i^\top \boldsymbol{v}_j = 0, \ j = 1, \ldots, i-1, \qquad (1)$$

where $\boldsymbol{u}_i$ and $\boldsymbol{v}_i$ are the canonical vectors that form the i-th smallest canonical angle, θ_i. The first canonical angle θ_1 is the smallest angle between $\mathcal{P}$ and $\mathcal{Q}$. The second canonical angle θ_2 is the smallest angle in a direction orthogonal to that of θ_1. The remaining θ_i for $i = 3, \ldots, d_p$ are calculated analogously, in a direction orthogonal to all smaller canonical angles.

There are several methods to calculate canonical angles [7,15,16]. The simplest and most practical method is singular value decomposition (SVD). Let the subspaces be represented as matrices of unitary orthogonal bases, $P = [\boldsymbol{\Phi}_1 \ldots \boldsymbol{\Phi}_{d_p}] \in \mathbb{R}^{d \times d_p}$ and $Q = [\boldsymbol{\Psi}_1 \ldots \boldsymbol{\Psi}_{d_q}] \in \mathbb{R}^{d \times d_q}$, where $\boldsymbol{\Phi}_i$ are the bases for $\mathcal{P}$ and $\boldsymbol{\Psi}_i$ are the bases for $\mathcal{Q}$. Let the SVD of $P^\top Q \in \mathbb{R}^{d_p \times d_q}$ be $P^\top Q = U \Sigma V^\top$, s.t. $\Sigma = \text{diag}(\kappa_1, \ldots, \kappa_{d_p})$, where $\{\kappa_i\}_{i=1}^{d_p}$ represents the set of singular values. The canonical angles $\{\theta_i\}_{i=1}^{d_p}$ can be obtained as $\{\cos^{-1}(\kappa_1), \ldots, \cos^{-1}(\kappa_{d_p})\}$ ($\kappa_1 \geq \ldots \geq \kappa_{d_p}$). The corresponding canonical vectors, $\boldsymbol{u}_i, \boldsymbol{v}_i (i = 1, \ldots, d_p)$ are obtained by the equations $[\boldsymbol{u}_1 \boldsymbol{u}_2 \ldots \boldsymbol{u}_{d_p}] = AU$ and $[\boldsymbol{v}_1 \boldsymbol{v}_2 \ldots \boldsymbol{v}_{d_p}] = BV$. The similarity between the two subspaces $\mathcal{P}$ and $\mathcal{Q}$ is measured by t angles as follows:

$$S[t] = \frac{1}{t} \sum_{i=1}^{t} \cos^2 \theta_i, \ 1 \leq t \leq d_p. \qquad (2)$$

3.2 Grassmann Manifold

The Grassmann manifold $\mathcal{G}(m, d)$ is defined as the set of m-dimensional linear subspaces of $\mathbb{R}^d$. It is an $m(d - m)$-dimensional compact Riemannian manifold and can be derived as a quotient space of orthogonal groups $\mathcal{G}(m, d) = \mathcal{O}(d)/\mathcal{O}(m) \times \mathcal{O}(d - m)$, where $\mathcal{O}(m)$ is the group of $m \times m$ orthonormal matrices. The Grassmann manifold can be embedded in a reproducing kernel Hilbert space by the use of a Grassmann kernel. In this case, the most popular kernel is the projection kernel k_p, which can be defined as $k_p(\mathcal{Y}_1, \mathcal{Y}_2) = \sum_{i=1}^{m} \cos^2 \theta_i$. We can measure the distance between two points on a Grassmann manifold by using this projection kernel.

3.3 Algorithm of Grassmann Discriminant Analysis

Discriminant analysis on Grassmann manifold (GDA) [10] is conducted as kernel LDA with the Grassmann kernels. Its predecessor, linear discriminant analysis (LDA) [17], followed by a K-NN classifier, is well known and has been successfully used for classification. Let $\boldsymbol{x}_1, \ldots, \boldsymbol{x}_N$ be the data vectors and $y_1, \ldots, y_N$ ($y_i \in 1, \ldots, C$) be the class labels. Each class c has N_c number of samples. Let $\boldsymbol{\mu}_c = \frac{1}{N_c} \sum_{i|y_i=c} \boldsymbol{x}_i$ be the mean of class c, and $\boldsymbol{\mu} = \frac{1}{N} \sum_i \boldsymbol{x}_i$ be the overall mean. LDA searches for the discriminant direction $\boldsymbol{w}$ which maximizes the Rayleigh quotient $R(\boldsymbol{w}) = \boldsymbol{w}' S_b \boldsymbol{w} / \boldsymbol{w}' S_w \boldsymbol{w}$ where S_b and S_w are the between-class and within-class covariance matrices respectively:

$$S_b = \frac{1}{N} \sum_{c=1}^{C} N_c (\boldsymbol{\mu}_c - \boldsymbol{\mu})(\boldsymbol{\mu}_c - \boldsymbol{\mu})^\top, \tag{3}$$

$$S_w = \frac{1}{N} \sum_{c=1}^{C} \sum_{i|y_i=c} (\boldsymbol{x}_i - \boldsymbol{\mu}_c)(\boldsymbol{x}_i - \boldsymbol{\mu}_c)^\top. \tag{4}$$

The optimal $\boldsymbol{w}$ is obtained from the largest eigenvector of $S_w^{-1} S_b$. Since $S_w^{-1} S_b$ has rank $C-1$, there are $C-1$ optima $W = \{\boldsymbol{w}_1, \ldots, \boldsymbol{w}_{C-1}\}$. By projecting data onto the space spanned by W, we achieve dimensionality reduction and feature extraction of data onto the most discriminant subspace.

Kernel LDA [18–20] can be formulated by using the kernel trick as follows. Let $\Gamma : \mathbb{R}^d \to \mathcal{F}$ be a non-linear map from the input space $\mathbb{R}^d$ to a feature space $\mathcal{F}$, and $\Gamma = [\boldsymbol{\gamma}_1, \ldots, \boldsymbol{\gamma}_N]$ be the feature matrix of the mapped training points $\boldsymbol{\gamma}_i$. Assuming $\boldsymbol{w}$ is a linear combination of those feature vectors, $\boldsymbol{w} = \Gamma \boldsymbol{\alpha}$, we can use the kernel trick and rewrite the Rayleigh quotient in terms of $\boldsymbol{\alpha}$ as:

$$Ra(\boldsymbol{\alpha}) = \frac{\boldsymbol{\alpha}^\top \Gamma^\top S_b \Gamma \boldsymbol{\alpha}}{\boldsymbol{\alpha}^\top \Gamma^\top S_w \Gamma \boldsymbol{\alpha}} = \frac{\boldsymbol{\alpha}^\top K(V - e_N e_N^\top/N)K\boldsymbol{\alpha}}{\boldsymbol{\alpha}^\top (K(I_N - V)K + \sigma^2 I_N)\boldsymbol{\alpha}} = \frac{\boldsymbol{\alpha}^\top \Sigma_b \boldsymbol{\alpha}}{\boldsymbol{\alpha}^\top (\Sigma_w + \sigma^2 I_N)\boldsymbol{\alpha}}, \tag{5}$$

where K is the kernel matrix, e_N is a vector of ones that has length N, V is a block-diagonal matrix whose c-th block is the matrix $e_{N_c} e_{N_c}^\top/N_c$, and

$\Sigma_b = K(V - e_N e_N^\top / N)K$. The term $\sigma^2 I_N$ is used for making the computation stable, and for regularizing the covariance matrix $\Sigma_w = K(I_N - V)K$. It is composed of the covariance shrinkage factor $\sigma^2 > 0$, and the identity matrix I_N of size N. The set of optimal vectors α are computed from the eigenvectors of $(\Sigma_w + \sigma^2 I_N)^{-1} \Sigma_b$.

4   Projection onto Generalized Difference Subspace

In this section, we outline the concept of generalized difference subspace (GDS) and explain how to generate a GDS $\mathcal{H}$ from multiple class subspaces. As mentioned previously, a GDS represents "difference components" among multiple class subspaces [12], as an extension of difference subspace (DS) between two class subspaces.

We firstly describe the concept of DS, which is a natural extension of the difference vector between two vectors, u and $v \in \mathbb{R}^d$. To formally define DS, let us consider two d_m-dimensional subspaces $\mathcal{P}$ and $\mathcal{Q}$ in $\mathbb{R}^d$. When there is no intersection between the two subspaces, d_m canonical angles $\{\theta_i\}_{i=1}^{d_m}$ are obtained between the subspaces. Let d_i be the difference vector, $u_i - v_i$, between the canonical vectors $u_i \in \mathcal{P}$ and $v_i \in \mathcal{Q}$, which form the i-th canonical angle θ_i. A DS is defined by the normalized $\{d_i\}_{i=1}^{d_m}$ [12].

In addition to the above geometrical definition, a DS can be also analytically defined by using the sum matrix, $S = P + Q$, of the two orthogonal projection matrices P and Q, which correspond to the orthogonal projection operators onto the subspaces $\mathcal{P}$ and $\mathcal{Q}$, respectively. The projection matrices P and Q are defined as $\sum_{i=1}^{d_m} \boldsymbol{\Phi}_i \boldsymbol{\Phi}_i^\top$ and $\sum_{i=1}^{d_m} \boldsymbol{\Psi}_i \boldsymbol{\Psi}_i^\top$, respectively, where $\boldsymbol{\Phi}_i$ and $\boldsymbol{\Psi}_i$ represent the bases of $\mathcal{P}$ and $\mathcal{Q}$. The DS is spanned by the d_m eigenvectors of matrix S that correspond to eigenvalues smaller than 1.

According to the analytical definition, the concept of DS has been extended to the generalized difference subspace (GDS) for multiple class subspaces [12]. Given $C(\geq 2)$ m-dimensional class subspaces, $\{\mathcal{P}_c\}_{c=1}^C$, a generalized difference subspace (GDS), $\mathcal{H}$, can be defined as the subspace produced by removing the principal component subspace (PCS) of all the class subspaces from the sum subspace, $\mathcal{S}$, of those subspaces. From this definition, the GDS is defined as the subspace spanned by d_h eigenvectors, $\{d_i\}_{i=1}^{d_h}$ corresponding to the d_h smallest eigenvalues of the sum matrix $(S = \sum_{c=1}^C P_c)$ of orthogonal projection matrices P_c of the class c. The optimal d_m is experimentally determined according to the orthogonal degree between the class subspaces projected onto the GDS.

Further, the concepts of DS and GDS have been extended to nonlinear kernel DS (KDS) and kernel GDS (KGDS), $\mathcal{H}^\Phi$, by using the kernel trick [12]. In these methods, each image set is represented by a nonlinear subspace.

5   Enhancement of the Discriminative Ability of GDA

In this section, we explain the algorithm of our enhanced GDA, referring to the flow chart shown in Fig. 2.

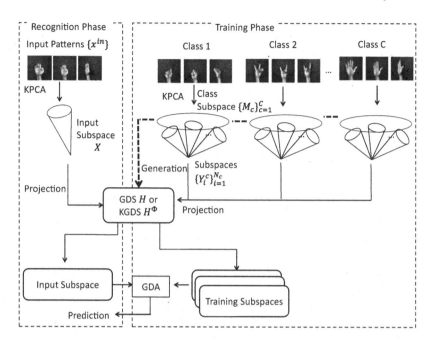

Fig. 2. Scheme of the eGDA/ eKGDA framework. In the training phase, each i-th image set of c-th class is modeled by a subspace matrix Y_i^c. Each c-th class is modeled by a larger class subspace M_c, which is generated from a set of Y_i^c. A GDS $\mathcal{H}$ or KGDS, $\mathcal{H}^\Phi$ is generated from a set of $\{M_c\}_{c=1}^C$. The subspaces $\{Y_i^c\}_{i_1}^{N_c}$ are projected onto the GDS/KGDS, and further mapped onto the Grassmann manifold. Finally, the discriminant analysis is applied to the set of mapped subspaces. In the recognition phase, a set input patterns $\{\boldsymbol{x}^{in}\}$ is also modeled as a subspace matrix X. It is projected onto the GDS/ KGDS and then mapped onto the manifold. The prediction of an input set is done by using 1-NN in the obtained discriminant space on the manifold.

Given N_c training sets $\{\boldsymbol{x}_l^{i,c}\}_{l=1}^{L_i^c}$ for each c-th class ($c = 1, \ldots, C$) and a set of L_{in} input images $\{\boldsymbol{x}_l^{in}\}_{l=1}^{L_{in}}$. Each of these sets contains images in different illumination conditions, or express different angles of a face.

In our framework, an image with the size $w \times h$ is represented by a $d(= w \times h)$-dimensional vector and each image set is represented by a subspace; $\{\boldsymbol{x}_l^{i,c}\}_{l=1}^{L_i^c}$ and $\{\boldsymbol{x}^{in}\}$ are represented by $\mathcal{Y}_i^c$ and $\mathcal{X}$, respectively. The orthogonal basis of each subspace is obtained as the eigenvectors corresponding to the m largest eigenvalues of the image set auto-correlation matrix, $R_i^c = \frac{1}{L_i^c} \sum_{l=1}^{L_i^c} \boldsymbol{x}_l^{i,c} \boldsymbol{x}_l^{i,c^\top}$. In the following, each subspace m-dimensional $\mathcal{Y}$ is represented by the $d \times m$ matrix, Y, which has the corresponding orthogonal basis as its column vectors.

In order to utilize effectively the feature extraction function of GDS, we introduce the global class subspaces $\mathcal{M}_c$, which is denoted by a matrix $M_c \in \mathbb{R}^{d \times d_m}$, which represents compactly all the subspaces belonging to the same class c. The orthogonal basis of $\mathcal{M}_c$ can be obtained as the eigenvectors corresponding to the d_m largest eigenvalues of the auto-correlation matrix:

$$R^c = \frac{1}{N_c} \sum_{i=1}^{N_c} R_i^c = \frac{1}{L_i^c N_c} \sum_{i=1}^{N_c} \sum_{l=1}^{L_i^c} x_l^{i,c} x_l^{i,c\top}. \tag{6}$$

Next, to generate a GDS, we calculate the total sum matrix, S, which is defined previously as:

$$S = \sum_{c=1}^{C} \sum_{j=1}^{d_m} \Phi_j^c \Phi_j^{c\top}, \tag{7}$$

where Φ_j^c is a basis of the d_m-dimensional $\mathcal{M}_c$. As seen in Sect. 4, the orthogonal basis of the GDS can be obtained d_h eigenvectors, $\{d_i\}_{i=1}^{d_h}$ corresponding to the d_h smallest eigenvalues of the sum matrix S. The subspaces Y_i^c are projected onto the GDS and their projections are denoted by $\{\widetilde{Y}_i^c\}_{i=1}^{N_c} \in \mathbb{R}^{d_h \times m}$. The input subspace of X is also projected onto the GDS and its projection is denoted by $\widetilde{X}$.

We apply the GDA to these projected subspaces through the procedure in Sect. 3.3. For example, the kernel matrix, K, is calculated as the similarity matrix between class subspaces $\widetilde{Y}_q$ and $\widetilde{Y}_w$. We call the GDA/KGDA, which is constructed from these more discriminant subspaces on GDS/KGDS, as enhanced GDA/KGDA, eGDA/eKGDA. The step-by-step training and testing algorithms of eGDA and eKGDA are shown in Algorithms 1 and 2, respectively.

6 Evaluation Experiments

In this section, we discuss the validity of the proposed method through face and hand shape recognition tasks. For this purpose, we compared the proposed method with known state-of-the-art subspace-based methods, namely the mutual subspace method (MSM) [21], RBF kernel constrained mutual subspace method (KCMSM) [12], Grassmann discriminant analysis (GDA) [10], and its RBF kernel extension (KGDA) [22].

6.1 Experiments on Hand Shape Recognition

We conducted two types of experiments with the Tsukuba hand shape dataset as shown in Fig. 3. This database contains 30 (hand classes) × 100 (subjects) image sets, each of which contains 28 hand shape images, consisting of 4 frames × 7 different view-points. In the experiments, all the images were resized to 24 × 24 pixels.

In the first preliminary experiment, to aid the understanding of the mechanism of our enhanced GDA, we performed visualizations of three different hand classes, using three image sets of 15 subjects. Once the class subspaces are mapped onto a $C - 1$-dimensional discriminant space on the manifold, they can be treated as points on the discriminant space. Thus, we can easily visualize the distribution of the points to check the metric structure. For example,

---

Algorithm 1. Training algorithm of eGDA/ eKGDA

---

input: pattern sets $\{x_l^{i,c}\}_{l=1}^{L_i^c}$, with class label c

if *eGDA* **then**

 for $c = 1, \ldots, C$ **do**

  for $i = 1, \ldots, N_c$ **do**

   $R_i^c \leftarrow \frac{1}{L_i^c} \sum_{l=1}^{L_i^c} x_l^{i,c} x_l^{i,c^\top}$ // calculate set covariance matrix

   $Y_i^c \leftarrow \text{EVD}(R_i^c)$ // apply eigendecomposition

  end

  $R^c \leftarrow \frac{1}{N_c} \sum_{i=1}^{N_c} R_i^c$ // calculate class covariance matrix

  $M_c \leftarrow \text{EVD}(R^c)$ // apply eigendecomposition

 end

 $P, H \leftarrow \text{EVD}(\sum_{c=1}^{C} M_c M_c^\top)$ // obtain GDS and principal subspace

 foreach Y_i^c **do** $\tilde{Y}_i^c \leftarrow H^\top Y_i^c$ // project all subspaces onto the GDS

else if *eKGDA* **then**

 for $c = 1, \ldots, C$ **do**

  for $i = 1, \ldots, N_c$ **do**

   $K_i^c \leftarrow \frac{1}{L_i^c} \sum_{l=1}^{L_i^c} k(x_l^{i,c}, x_l^{i,c})$ // calculate set kernel matrix

   $Y_i^c \leftarrow \text{EVD}(K_i^c)$ // apply eigendecomposition

  end

  $K^c \leftarrow \frac{1}{N_c} \sum_{i=1}^{N_c} K_i^c$ // calculate class kernel matrix

  $M_c \leftarrow \text{EVD}(K^c)$ // apply eigendecomposition

 end

 $P^\Phi, H^\Phi \leftarrow \text{EVD}(\sum_{c=1}^{C} M_c M_c^\top)$ // obtain KGDS and principal subspace

 foreach Y_i^c **do** $\tilde{Y}_i^c \leftarrow H^{\Phi^\top} Y_i^c$ // project all subspaces onto the KGDS

end

for $q = 1, \ldots, N$ **do**

 for $w = 1, \ldots, N$ **do**

  $[S_{train}]_q^w \leftarrow k_p(\tilde{Y}_q, \tilde{Y}_w)$ // generate similarity matrix

 end

end

$\alpha^* \leftarrow \max_\alpha Ra(\alpha)$ // solve LDA problem

$F_{train} = \alpha^{*\top} S_{train}$ // compute training coefficients

return F_{train}

---

for three classes, $C = 3$, points are on 2-dimensional discriminant space. Figure 4 shows the scatter points plotted on the discriminant space in two different combinations of three classes. We can see that in both cases, the proposed eGDA is able to generate a more discriminative space on the manifold in comparison with the conventional GDA. Although the conventional GDA can generate the discriminative space, its class separability looks lower than that obtained by the eGDA. This may be because the original GDA can only select the most discriminant directions as expected and cannot further adjust the layout of the points

Algorithm 2. Input evaluation algorithm of eGDA/ eKGDA

---

input: pattern set with L' input images $\{x^{in}\}$

if $eGDA$ then

  $R_{in} \leftarrow \frac{1}{L'} \sum_{l'=1}^{L'} x_{l'}^{in} x_{l'}^{in\top}$ // calculate set covariance matrix

  $X \leftarrow \text{EVD}(R_{in})$ // apply eigendecomposition

  $\widetilde{X} \leftarrow H^\top X$ // project subspace onto the GDS

else if $eKGDA$ then

  $K_{in} \leftarrow \frac{1}{L'} \sum_{l'=1}^{L'} k(x_{l'}^{in} x_{l'}^{in})$ // calculate set kernel matrix

  $X \leftarrow \text{EVD}(K_{in})$ // apply eigendecomposition

  $\widetilde{X} \leftarrow H^{\Phi^\top} X$ // project subspace onto the KGDS

end

for $q = 1, \ldots, N$ do

  $[S_{test}]_q \leftarrow k_p(\widetilde{Y}_q, \widetilde{X})$ // generate similarity matrix

end

$F_{test} = \boldsymbol{\alpha}^{*\top} S_{test}$ // compute test coefficients

$\text{pred}(x^{in}) \leftarrow \text{NN}(F_{train}, F_{test})$ // perform 1-NN classification

return $\text{pred}(x^{in})$ // return a class prediction

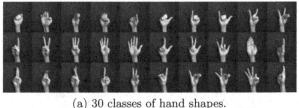

(a) 30 classes of hand shapes.

(b) Seven views of hand images.

Fig. 3. Sample images from the Tsukuba hand shape dataset. It shows the (a) 30 classes of hand shapes, and (b) seven views of a hand shape belonging to the 30-th class.

on the manifold. In our eGDA, thanks to the GDS projection, we can obtain more discriminant space.

In the second experiment, we evaluated the performances of the proposed eGDA and eKGDA, in the classification problem of 30 kinds of hand shapes. We used 15×30 image sets from 15 subjects as training sets, 15×30 image sets from 15 subjects as validation sets, and 70×30 image sets from the remaining 70 subjects as testing sets, where each image set contains 28 hand images. Table 1 shows the results. The accuracy refers to the performance on the test sets by using the shown parameters, which are the optimal parameters found by using the validation sets. m_{train} and m_{test} refer to the dimension of training and test

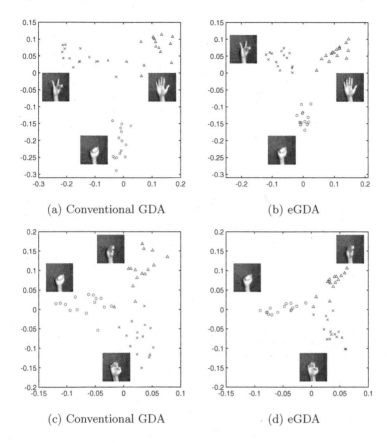

Fig. 4. Scatter points of three hand shape classes that are somewhat different by using (a) Conventional GDA; (b) eGDA. In addition, scatter points of three hand shape classes that are somewhat similar: (c) Conventional GDA; (d) eGDA.

subspaces in MSM and KCMSM. Plain m is the dimension for both test and training in manifold frameworks. d_m is the dimension of class subspaces used to generate a GDS in the proposed methods. σ refers to the RBF kernel parameter, and d_p is the dimension of the principal subspace. Where there is a "-", it means that parameter is not applicable.

We can see that the proposed eKGDA obtained the best performance among the methods, followed by KCMSM, while GDA and KGDA did not perform very well. These results suggest the usefulness of the feature extraction by the GDS projection in the GDA formulation as expected.

6.2 Face Recognition Experiments on CMU Multi PIE Dataset

We conducted experiments to check the validity of the enhanced GDA/KGDA on two kinds of tasks on face recognition.

Table 1. Results of the experiment using the Tsukuba hand shape dataset. The accuracy refers to the performance on the test sets by using the shown parameters, which are the optimal parameters found by using the validation sets. m_{train} and m_{test} refer to the dimension of training and test subspaces in MSM and KCMSM. Plain m is the dimension for both test and training in manifold frameworks. d_m is the dimension of class subspaces used to generate a GDS in the proposed methods. σ refers to the RBF kernel parameter, and d_p is the dimension of the principal subspace. Where there is a "-", it means that parameter is not applicable.

	Accuracy	m_{train}	m_{test}	σ	d_p
MSM	62.30	30	6	-	-
KCMSM	69.77	100	16	0.5	20
			m	σ	d_p
GDA	61.13		16	-	-
KGDA	67.09		14	1	-
		d_m	m	σ	d_p
eGDA	69.20	150	16	-	5
eKGDA	71.69	150	14	1	20

(a) Illumination conditions.

(b) Facial expressions.

Fig. 5. Sample images from CMU Multi PIE dataset (a) from several illumination conditions and (b) facial expressions.

In the first experiment, we performed face recognition using only the subset of front faces with neutral expression from CMU Multi PIE dataset [13].

We used face images of 128 subjects, whose data were collected in 4 sessions. In this experiment, 20 face images of the same person, which were captured under different 20 kinds of illumination conditions is treated as one image set. Figure 5(a) shows examples of face images with neutral expression. The images were cropped and resized to 16×16 pixels. We used 12-fold cross-validation to evaluate the performance of each method. For each fold, two sessions were used for testing, one for training and one for validation. The parameters used on each fold's test sets were those that minimize error on the respective fold's validation sets. Figure 6 shows average recognition accuracies for six different methods, along with the standard deviations.

The experimental result in Fig. 6(a) shows that eGDA has the highest average value. We also conducted a t-test between eGDA and KCMSM with 12 samples

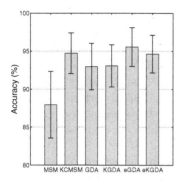

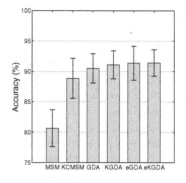

(a) First CMU experiment.   (b) Second CMU experiment.

Fig. 6. The results of the experiments. Averages of accuracies are shown with standard deviations. (a) First CMU experiment and (b) second CMU experiment.

and significance level $\alpha = 0.05$. From the test results, we can conclude with more than 95% confidence ($p = 0.0184$) that the proposed eGDA performed better than KCMSM.

In the second experiment, we conducted face classification under more difficult conditions, including image sets with other types of facial expressions, such as smile, surprise and disgust of the 128 subjects, as shown in Fig. 5(b). The types of facial expressions of each session are listed in Table 2. We executed a 10-fold cross-validation where two sessions were selected for testing, one for training and one for validation. Parameter optimization uses the validation sets as explained previously.

Figure 6(b) shows average recognition accuracies for six different methods in the second experiment, along with the standard deviations. The experimental result shows the advantage of our eGDA and eKGDA against the conventional methods. To confirm the validity of them, we conducted a t-test between KGDA

Table 2. Facial expressions present in the sessions of CMU Multi PIE dataset. The number within each cell indicates how many sets with that expressions exist for one person, in that session.

Expression	Session 1	Session 2	Session 3	Session 4
Neutral	1	1	1	2
Smile	1		1	
Surprise		1		
Squint		1		
Disgust			1	
Scream				1
Total	2	3	3	3

and eKGDA, with 10 samples and significance level $\alpha = 0.05$. From the results, we can conclude with more than 95% confidence ($p = 0.0423$) that the proposed method, eKGDA, performed better than the conventional KGDA.

In the first experiment, the challenge was that there were few training data for the methods, just one subspace per person class. In such cases, the Grassmann manifold formulation like GDA usually presents a drop in performance, as they need more subspaces to estimate the structure of the Grassmann manifold. The proposed method could alleviate this issue by enhancing discriminative ability of each class subspace even in such situation.

In the second experiment, the addition of five other expressions largely increased the difficulty of the face classification task, because of the fact that in most cases, the learned expressions may be a little different from those used to optimize parameters, and also different from the ones that show up during test phase. The addition of expressions also caused large inner class variations. We can see a drop in performance of the conventional KCMSM, as it does not have the manifold mechanism for collapsing the inner-class variations. In contrast, the proposed methods could still perform well in this case.

7 Conclusions

In this paper we have proposed an enhanced Grassmann discriminant analysis and its kernel version to address more effectively the classification of 3D object with image sets, focusing on the applications of face and hand shapes classification. The key idea of our enhanced Grassmann manifold is to project class subspaces onto a generalized difference subspace before mapping them on a Grassmann manifold. The GDS projection can extract the differences between classes and generate data points with optimized between-class separability on the manifold, which are more desirable for GDA. The validity of our enhanced Grassmann discriminant analysis was evaluated through classification experiments with CMU face dataset and hand shape dataset, where it outperformed state-of-the-art methods such as the kernel Grassmann discriminant analysis and the kernel constrained mutual subspace method. As a future work, we seek to comprehend the relationship between the two types of mapping in GDS projection and Grassmann manifold more clearly.

Acknowledgement. This work is supported by JSPS KAKENHI Grant Number 16H02842.

References

1. Shashua, A.: On photometric issues in 3d visual recognition from a single 2d image. Int. J. Comput. Vis. **21**, 99–122 (1997)
2. Belhumeur, P.N., Kriegman, D.J.: What is the set of images of an object under all possible lighting conditions? In: Proceedings of 1996 IEEE Computer Society Conference on Computer Vision and Pattern Recognition, CVPR 1996, pp. 270–277. IEEE (1996)

3. Georghiades, A.S., Belhumeur, P.N., Kriegman, D.J.: From few to many: illumination cone models for face recognition under variable lighting and pose. IEEE Trans. Pattern Anal. Mach. Intell. **23**, 643–660 (2001)
4. Lee, K.C., Ho, J., Kriegman, D.J.: Acquiring linear subspaces for face recognition under variable lighting. IEEE Trans. Pattern Anal. Mach. Intell. **27**, 684–698 (2005)
5. Basri, R., Jacobs, D.W.: Lambertian reflectance and linear subspaces. IEEE Trans. Pattern Anal. Mach. Intell. **25**, 218–233 (2003)
6. Hotelling, H.: Relation between two sets of variables. Biometrica **28**, 322–377 (1936)
7. Afriat, S.: Orthogonal and oblique projectors and the characteristics of pairs of vector spaces. Proc. Camb. Philos. Soc. **53**, 800–816 (1957)
8. Yamaguchi, O., Fukui, K., Maeda, K.: Face recognition using temporal image sequence. In: Proceedings of International Conference on Automatic Face and Gesture Recognition, pp. 318–323 (1998)
9. Chikuse, Y.: Statistics on Special Manifolds. Lecture Notes in Statistics, vol. 174. Springer, Heidelberg (2013)
10. Hamm, J., Lee, D.D.: Grassmann discriminant analysis: a unifying view on subspace-based learning. In: Proceedings of the 25th International Conference on Machine Learning, pp. 376–383. ACM (2008)
11. Turaga, P., Veeraraghavan, A., Srivastava, A., Chellappa, R.: Statistical computations on Grassmann and Stiefel manifolds for image and video-based recognition. IEEE Trans. Pattern Anal. Mach. Intell. **33**, 2273–2286 (2011)
12. Fukui, K., Maki, A.: Difference subspace and its generalization for subspace-based methods. IEEE Trans. Pattern Anal. Mach. Intell. **37**, 2164–2177 (2015)
13. Gross, R., Matthews, I., Cohn, J., Kanade, T., Baker, S.: Multi-PIE. Image Vis. Comput. **28**, 807–813 (2010)
14. Ohkawa, Y., Fukui, K.: Hand shape recognition using the distributions of multi-viewpoint image sets. IEICE Trans. Inf. Syst. **E95-D**, 1619–1627 (2012)
15. Hotelling, H.: Relations between two sets of variates. Biometrika **28**, 321–377 (1936)
16. Maeda, K., Watanabe, S.: A pattern matching method with local structure. Trans. IEICE **68**, 345–352 (1985)
17. Fukunaga, R.: Statistical Pattern Recognition. Wiley, New York (1990)
18. Scholkopft, B., Mullert, K.R.: Fisher discriminant analysis with kernels. Neural Netw. Signal Process. IX **1**, 1 (1999)
19. Baudat, G., Anouar, F.: Generalized discriminant analysis using a kernel approach. Neural Comput. **12**, 2385–2404 (2000)
20. Li, Y., Gong, S., Liddell, H.: Constructing structures of facial identities using kernel discriminant analysis. In: The 2nd International Workshop on Statistical and Computational Theories of Vision (2001)
21. Yamaguchi, O., Fukui, K., Maeda, K.: Face recognition using temporal image sequence. In: 1998 Proceedings of Third IEEE International Conference on Automatic Face and Gesture Recognition, pp. 318–323. IEEE (1998)
22. Harandi, M.T., Salzmann, M., Jayasumana, S., Hartley, R., Li, H.: Expanding the family of grassmannian kernels: an embedding perspective. In: Fleet, D., Pajdla, T., Schiele, B., Tuytelaars, T. (eds.) ECCV 2014. LNCS, vol. 8695, pp. 408–423. Springer, Heidelberg (2014). doi:10.1007/978-3-319-10584-0_27

An Extended Sparse Classification Framework for Domain Adaptation in Video Surveillance

Farshad Nourbakhsh[1,2]($\boxtimes$), Eric Granger[1,2], and Giorgio Fumera[1,2]

[1] Laboratoire d'imagerie de vision et d'intelligence artificielle
École de technologie supérieure, Université du Québec, Montréal, Canada
fnourbakhsh@livia.etsmtl.ca
[2] Department of Electrical and Electronic Engineering,
University of Cagliari Piazza d'Armi, 09123 Cagliari, Italy

Abstract. Still-to-video face recognition (FR) systems used in video surveillance applications capture facial trajectories across a network of distributed video cameras and compare them against stored distributed facial models. Currently, the performance of state-of-the-art systems is severely affected by changes in facial appearance caused by variations in, e.g., pose, illumination and scale in different camera viewpoints. Moreover, since an individual is typically enrolled using one or few reference stills captured during enrolment, face models are not robust to intra-class variation. In this paper, the Extended Sparse Representation Classification through Domain Adaptation (ESRC-DA) algorithm is proposed to improve performance of still-to-video FR. The system's facial models are thereby enhanced by integrating variational information from its operational domain. In particular, robustness to intra-class variations is improved by exploiting: (1) an under-sampled dictionary from target reference facial stills captured under controlled conditions; and (2) an auxiliary dictionary from an abundance of unlabelled facial trajectories captured under different conditions, from each camera viewpoint in the surveillance network. Accuracy and efficiency of the proposed technique is compared to state-of-the-art still-to-video FR techniques using videos from the Chokepoint and COX-S2V databases. Results indicate that ESRC-DA with dictionary learning of unlabelled trajectories provides the highest level of accuracy, while maintaining a low complexity.

1 Introduction

With the availability of low-cost video cameras and high capacity memory, technologies for video surveillance (VS) have become more prevalent in recent years. VS networks are increasingly deployed by public security organizations in e.g., airports, train stations and border crossings. Accurate and robust systems are required to recognize individuals and their actions from video feeds.

In VS, decision support systems can rely on facial information (along with other sources, like soft biometrics) to alert an analyst as to the presence of individuals of interest. The ability to automatically recognize faces in videos recorded

© Springer International Publishing AG 2017
C.-S. Chen et al. (Eds.): ACCV 2016 Workshops, Part III, LNCS 10118, pp. 360–376, 2017.
DOI: 10.1007/978-3-319-54526-4_27

across a distributed network of surveillance cameras can greatly enhance security and situational awareness.

Watch-list screening is among the most challenging applications in VS [1,2]. During enrolment of a target individual, facial regions of interests (ROIs) are isolated from one or few reference still images that were captured under controlled conditions. Then, during operations, each ROI captured in videos are matched against the facial models of each individual enrolled to the system. Robust spatio-temporal FR is typically performed using a person tracker (based on head, face and other information). This allows to accumulate the matching scores for each enrolled individual over a trajectory, e.g., a set of facial ROIs corresponding to a same person tracked in the scene. Thus, a spatio-temporal fusion module compares these accumulated scores with decisions thresholds in order to detect target individuals associated with each trajectory [3].

Currently, the performance of state-of-the-art systems for still-to-video FR is severely affected by variations in, e.g., pose, scale, blur, illumination and camera viewpoint [4]. Systems for FR in VS are typically implemented with individual-specific face detectors (e.g., two-class classification systems) [5]. During enrolment, it is assumed that a detector is designed to encode a facial model, using labelled ROIs extracted from target reference stills versus cohort and other non-target ROIs, all of which are captured under the same controlled conditions (in the enrolment domain, ED). In watchlist screening each individual is typically enrolled using one or few reference stills captured during enrolment, face models are often poor representatives of the faces to be recognized during operations. Moreover, during operations, video ROIs are captured in a camera field of view (FoV) (in an operational domain, OD) under uncontrolled conditions. Capture conditions may vary dynamically within an OD according to environmental conditions and individual behaviours. Therefore, the face model of target individuals are not robust to the intra-class variations of ROIs in an OD, and many yield poor FR performance.

Still-to-video FR can be addressed using techniques proposed in literature for single sample per person (SSPP) problems [6,7]. In order to improve robustness to intra-class variability, FR techniques specialized for SSPP problems must often rely on adaptation, multiple face representations, synthetic face generation, and enlarged auxiliary reference datasets. However, multiple representation and synthetic generation techniques alone are only effective to the extent where reference target ROIs captured in the environmental domain (ED) are representative of an OD [8].

An important issue in still-to-video FR is that probe ROIs are captured over multiple distributed surveillance cameras, and each one represents a non-stationary OD. Their data distribution differs significantly from ROIs captured with a still camera in the ED [9]. Any distributional change (either domain shift or concept drift) can degrade system performance [10]. Context-aware systems could efficiently adapt to different and changing capture conditions [11]. However, in the most common approach, prior expert knowledge of the expected OD is employed to define typical contexts and to design specialized individual detectors. Then, a suitable detector is selected dynamically among the pool for a given OD. In practice, however, this approach would only provide coarse adaptation because still-to-video FR systems are deployed in diverse and unknown capture condition. It is difficult to predict, and to collect adequate labelled data for each context. Given an adequate number of labelled reference samples, numerous adaptive classifiers [12] and ensemble methods could also be employed for adaptation in non-stationary environments, some of which are specialized for video-to video FR.

Several transfer learning methods have recently been proposed to design accurate recognition systems that will perform well on OD data given knowledge from the ED. Since the learning tasks and feature spaces between ED and OD are the same, but their data probability distributions are different, our transfer learning scenario is related to domain adaptation (DA). According to the information transferred between an ED and an OD, two unsupervised DA approaches from literature are relevant for still-to-video FR [9,13,14]. Instance transfer methods attempt to exploit parts of the OD data for learning in the ED. In contrast, feature representation transfer methods exploit OD data to find a good common feature representation space that reduces the difference between ED and OD spaces and the classification error. Note that most methods in literature initially require an adequate number of labelled reference samples from the ED.

In this paper, we focus on sparse modelling techniques that are suitable for SSPP problems, and allow for instance-based DA. In particular, a framework based on the Extended Sparse Representation Classification (ESRC) [15] algorithm is proposed for the design of still-to-video FR systems, as needed in watchlist screening applications that is called ESRC-DA. Assume that each individual of interest is enrolled to the system using one reference still image capture under controlled condition. Facial models based on one reference still conditions and that facial diverge from the faces captured with video surveillance cameras. Therefore with the ESRC-DA algorithm, unsupervised DA is exploited for accurate recognition of individuals. A large auxiliary dictionary of unlabelled ROIs is extracted from an abundance of facial trajectories captured under different OD conditions, from each camera FoV in the surveillance network. This is combined with an under-sampled dictionary extracted from few reference labelled facial ROIs captured from reference target stills under controlled ED conditions. Apart from improving robustness to intra-class variations in ODs, an advantage of the new ESRC-DA algorithms is the ease for managing information from multi-source environments by organizing and combining different domains. Indeed, a learned dictionary that combines ROIs from ED and OD data for SRC may not be optimal if they have different data distributions. Finally, dictionary learning methods are recommended to compactly represent the variational information from potentially large sets of unlabelled video ROIs.

The rest of the manuscript is organized as follows. Section 2 introduces some notation of sparse modelling in still-to-video FR. Section 3 introduces the proposed method based on ESRC with domain adaptation. Section 4 is devoted to showing the effectiveness of ESRC-DA on two public video-surveillance datasets

(Chokepoint and COX-S2V), and to its comparison with state of the art reference methods. Finally, some concluding discussion are provided.

2 Sparse Modelling in Still-to-Video FR

Watch-list screening is among the most challenging applications in of video surveillance. In VS, FR attempts systems to detect the presence of individuals of interest enrolled to the system. During enrolment some target individual, facial regions of interests (ROIs) are isolated in one or few reference still images that were captured under controlled conditions in the ED. It is assumed that discriminant features are extracted into ROI patterns using a state-of-the-art face descriptor to design a facial model. During operations, ROIs isolated from videos are matched against the facial models where these ROIs are captured under uncontrolled conditions in the OD. Therefore, the performance of state-of-the-art FR systems is severely affected by occlusion, variations in capture conditions (pose, scale, expression, illumination, blur, etc.) [8]. In addition, watchlist screening systems must employ a limited reference samples in order to design the facial models. To overcome the aforementioned challenges of SSPP problems and enhance robustness to intra-class variability, FR systems may exploit auxiliary data captured from unknown individuals or actors during some calibration process.

SRC recently has received much attention in FR literature due to its potential to handle intra class variability. The main assumption in sparse modelling is that as any test sample (ROI pattern) from a class (individual) enrolled with a sufficient number of training samples will be approximately represented in the linear span of training set of corresponding class. The sparse representation of a sample can be formulated as

$$y = Dx_0, \tag{1}$$

where matrix D is defined as over-complete dictionary with N distinct classes that has enough samples for each class. The column of matrix D contains training samples, and x_0 is a sparse coefficient vector. Ideally, the entries of x_0 are always zero, except for the same class as y. By increasing the number of classes that increases the sparsity of coefficient vector, Eq. (1) is able to handle some degree of noise, and provides a unique solution. The Eq. (1) can be solved with a l_1 minimization as follows:

$$x_0 = \min_x \|y - Dx\|_2^2 + \lambda \|x\|_1 \tag{2}$$

where $\lambda > 0$, is the scalar regularization parameter that controls reconstruction error and sparsity. According to the Eq. (2) that is the main part of SRC algorithm, the test sample y can be reconstructed linearly with a relevant basis of dictionary D respect into the sparse coefficient with non zero value for the corresponding class and almost zero for non classes. [16] has proposed a sparsity concentration index (SCI) that is defined as below

$$SCI(x) = \frac{N. \max_i |\delta_i(x)|_1 / |x|_1 - 1}{N - 1} \tag{3}$$

For the solution found by Eq. (2), SCI is 1 when the test sample belongs to the one of the N classes and it is zero when the sparse coefficient are spread over all classes. SRC method has been applied successfully on FR application [16] with a large amount of training data that has a direct effect on classification result. It is clear that this technique is not fit with an application that provides few number of training samples that is called under-sampled dictionary, like still-to-video video surveillance.

A few FR systems based on the SRC algorithm have been proposed in the literature for VS applications. Naseem et. al. [17] extended the SRC for video-based FR framework and compared the performance of proposed systems with state of art SIFT methods. They have improved the video-to-video FR system by fusing the scores obtained with SRC and SIFT. Nagendra et al. [18] applied their method on video-based FR to identify a video trajectory of a person while rejecting unknown individuals. Optimized combination of Gabor and HOG features are obtained to produce a robust descriptor for video-to-video FR system that is followed by regularized SRC algorithm. Cui et al. [19] proposed a video-to-video FR system that measured the similarity between two image sets that uses joint sparse representation.

There are very few systems specialized for still-to-video FR. Jianquan et al. [20] have extended the problem of FR with SSPP to patch based dictionary learning to improve robustness to variations. The literature mainly focuses on closed-set classification in video-to-video FR or still-to-still FR for a fixed number of individuals. Moreover, the problem of still-to-video FR surveillance become a very challenging when ROIs captured in the operational domain video are different from ROIs captured during enrolment. Transfer learning methods provide techniques to cope with the domain difference between ED and OD.

Transductive transfer learning requires the source and target task be the same with different domain. DA is defined as a relaxed notion of transductive transfer learning by availability of part of unlabelled data in target domains that divide to supervised and unsupervised categories and sparse coding and dictionary based methods have been proposed for DA. Fore example, [21] modelled dictionaries across different domains with a parametric mapping function. Ni et al. [22,23] proposed an unsupervised DA dictionary learning framework by generating a set of intermediate dictionaries, which smoothly connect the ED and OD. It allows the synthesis of data associated with the intermediate domains that can then be used to build classifiers for domain shifts. A semi-supervised DA dictionary learning framework was proposed for learning a dictionary to optimally represent both ED and OD data [24]. Both the domains are projected into a common low-dimensional space, allowing disregarding irrelevant information. Finally, a view independent representation is achieved by storing all the intermediate dictionaries. Another technique is introduced by [25,26] with learning a target classifier from classifiers trained on the source domain(s). Zheng et al. [27] propose a dictionary learning approach based on DA between videos from two domains by forcing the two videos from a same frame in different domain have the same sparse representation to learn two separate dictionaries. In this way, the source view video can be directly applied on the target view video. Shekhar et al. [24] proposed a DA method by mapping source and target domain to a low dimensional subspace and learn a common dictionary.

ESRC [15] is sparse representation modelling method that is adapted into SSPP where external data can be used to learn intra-class variabilities. Given a limited set of labelled reference data for target individuals during enrolment, and an abundance of unlabelled non target videos that can captured from different camera domains, ESRC can perform DA by representing the variations of faces appearing during operations.

The issue of undersampled data is addressed in [15] with ESRC that extends Eq. (2) to the following

$$\mathbf{x_0} = \min_{\mathbf{x}} \|\mathbf{y} - [\mathbf{D}, \mathbf{E}] \left[\begin{smallmatrix} \mathbf{x_d} \\ \mathbf{x_e} \end{smallmatrix} \right] \|_2^2 + \lambda \|\mathbf{x}\|_1 \tag{4}$$

Where $\mathbf{D}$ is an under-sampled dictionary populated with target data, $\mathbf{E}$ is external dictionary with non target data and $\mathbf{x_e}$ is corresponds to the variant bases. After obtaining sparse coefficient $\mathbf{x}$, a test image $\mathbf{y}$ will be assigned to the class with the minimum class-wise reconstruction error $r(\mathbf{y})$, defined as:

$$r(\mathbf{y}) = \|\mathbf{y} - [\mathbf{D}, \mathbf{E}] \left[\begin{smallmatrix} \delta(\mathbf{x_d}) \\ \mathbf{x_e} \end{smallmatrix} \right] \|_2^2 \tag{5}$$

where $r_k(\mathbf{y})$ and $\delta_\mathbf{k}(\mathbf{x_d})$ refer to the k-th class, $k = 1, \ldots, N$ that is a vector whose only nonzero entries are the entries in $\mathbf{x_0}$ that are associated with class k and the test image $\mathbf{y}$ will be assigned to the class with the minimum class-wise reconstruction error that is $r_k(\mathbf{y})$.

One of the solutions to handle large amount of external data with different domains is to select data randomly from external data to reduce the time complexity. The drawback of this method is that the performance of ESRC affected by amount of data exploited to cover all the common variations in all domains. The second option is to apply DL methods allow for a compressed representation of all domains from external data. For instance, Shafiee et al. [28] have investigated the impact on performance of three different DL methods for SRC. They used Metaface DL, Fisher Discriminative DL (FDDL), Sparse Modelling Representative Selection (SMRS) to obtain compact representation of training data. They showed that the FDDL method provides a high recognition accuracy compare to other methods. K-Means Singular Value Decomposition (KSVD) [29] and Method of Optimal Directions (MOD) [30] are two popular unsupervised DL techniques which have been used in the literature. These EM style methods alternate between dictionary and sparse coding. The difference between these two methods are in dictionary updating – KSVD updates atom by atom, and MOD updates all atoms simultaneously. Graph compression is another option to learn a compact representation of different domains. Depending on the type of encoding, these methods produce a lossy or lossless compression. Data can be presented as a collection of feature vectors or representation of the similarity/dissimilarity relations among data samples. Nourbakhsh et al. [31] have proposed a graph compression method based on matrix factorization that focuses on structural information of the input data and the extension of the proposed method on DL is presented on [32]. Finally, several methods that combine DL and classification have recently been proposed like SVDL [33,34] to produce a compact auxiliary dictionary and classification in the same time.

3   ESRC with Domain Adaptation

In this paper, a still-to-video FR system for watchlist screening is proposed. It can efficiently adapt to multiple non-stationary sources of ROIs based on contextual information from the specific operational VS environment. Despite the very limited number of labelled still ROIs from enrolment (ED), the framework relies on a bank of unlabelled non-target video ROIs captured (during a prior calibration phase) from each different camera FoVs (OD) in a distributed surveillance network. The proposed system is robust to variations of an individual's facial appearance caused by changes in, e.g., pose, illumination, scale and camera viewpoint in the VS environment by extracting a compact auxiliary dictionary from different ODs. ESRC-DA benefits from the abundance of external facial data that is readily available in VS applications, followed by DL to compress the external (variational) dictionary and improve accuracy. Figure 1 illustrates the processing pipeline of the proposed ESRC-DA method for still-to-video FR. It is divided into design and operational phases.

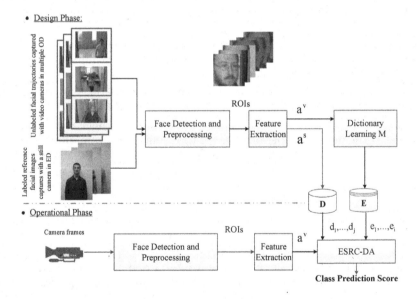

Fig. 1. Block diagram of ESRC-DA method that generates matching scores for an input ROI captured with one surveillance camera.

3.1 Design Phase

The system designed consist in learning external dictionary $\mathbf{E}$ from an abundance of facial trajectories captured under difference conditions and camera FoVs during the calibration process and under-sampled dictionary $\mathbf{D}$ of labelled reference ROIs captured with a high quality still camera in the $\mathbf{ED}$ (see Algorithm 1). Video trajectories are gathered from several cameras with different ODs. The Faces of unknown people or actors are captured during data collection therefore data are not labelled. For each frame, (**line 1**) face detection and preprocessing module performs a segmentation on each frame I_i^v to isolate a region of interest (ROI) to each different face (**line 2**). Then features are extracted from each normalized ROI_r and stored as a pattern $\mathbf{a}_{r,i}^v$ (**line 4**) to produce an overcomple dictionary $\mathbf{M} = [\mathbf{m}_1, \ldots, \mathbf{m}_r]$ that passes to dictionary learning module which finds a compact representation of ROI patterns from video calibration (**line 5**). DL reduces the size of overcomplete dictionary $\mathbf{M}$ to a compressed auxiliary dictionary $\mathbf{E} = [\mathbf{e}_1, \ldots, \mathbf{e}_i]$ (**line 7**). The still ROIs captured during the enrolment process are stored in dictionary $\mathbf{D}$ as a gallery faces model by extracting faces from each ROI of still image $\mathbf{I}_j^s$ and preprocessing each ROI (**line 9**). In the next step, feature extraction is used on each ROI_k to produce a pattern $\mathbf{a}_{k,j}^s$ (**line 11**) stored in the undersampled dictionary $\mathbf{D}$, (**line 12**).

DL is presented as a method to reduce the size of overcomplete dictionary to produce a compressed dictionary that obtain almost the same performance as uncompressed data. The large amount of external data makes inefficient to process it as a dictionary that covers intra class variability. DL provides a compact representation of data by partitioning the input data into coherent cluster. A cluster can be seen as a set of elements being mutually similar and dissimilar to elements belonging to other cluster that the compressed graph is equal to identity matrix in those cases. All gathered data from different domains have similarity to other elements so they can be grouped as cluster to reduce the size of original data. Therefore DL, clusters data from different ODs to produce a compact dictionary that is representative of domain variability.

3.2 Operational Phase

During operation, the proposed system is designed to obtain a score that depicts the similarity of a given input probe ROI to each to a watchlist individual enrolled to the system (see Algorithm 2). Faces are captured using a face detector that extracts ROIs in each frame I_i^v (**line 1 & 2**). Pattern is $\mathbf{a}_{n,i}^v$ calculated by applying feature extraction module (**line 4**). For ESRC, the main assumption is intraclass variation of any watchlist face can be approximated by a sparse linear combination of the intraclass differences from sufficient number of generic faces. ESRC assign $\mathbf{a}_{n,i}^v$ to the corresponding gallery watchlist dictionary $\mathbf{D}$ with respect to the external dictionary $\mathbf{E}$ so column of matrix $\mathbf{D}$ and $\mathbf{E}$ are normalized to have unit l_2-norm (**line 5**) and the coefficients $\mathbf{x}_0$ is obtained by

minimizing the following equation $\min_{x} \|a_{n,i}^{v} - [\mathbf{D}, \mathbf{E}]\left[\begin{smallmatrix} x_d \\ x_e \end{smallmatrix}\right]\|_2^2 + \lambda\|x\|_1$ (line 6).
Finally, $r_{k,i,n}(\mathbf{y}) = \|a_{n,i}^{v} - [\mathbf{d_k}, \mathbf{E}]\left[\begin{smallmatrix} \delta(x_d) \\ x_e \end{smallmatrix}\right]\|_2^2$ computes the similarity of the probe
input ROI to each facial model (line 8).

Algorithm 1. Design Phase for all Cameras FoVs

---

* **Input:** Unlabeled facial trajectories captured with video cameras in multiple OD
and labelled reference facial images captures with a still camera in ED
* **Output:** Undersampled dictionary $\mathbf{D} = [\mathbf{d_1}, ..., \mathbf{d_K}]$ of still images, $\mathbf{E} = [\mathbf{e_1}, ..., \mathbf{e_i}]$
as an external dictionary

// **Unlabelled facial trajectories captured**
with video cameras in multiple OD
1: For each frame I_i in the video sequence, $i = 1, ..., \infty$
2: ROI ← face detector and preprocessing on frame I_i^v
3: **for** $r=1$ to number of ROIs in frames **do**
4:      $a_{r,i}^v$ ← feature extraction on ROI$_r$
5:      $\mathbf{M} \leftarrow [\mathbf{M}; a_{r,i}^v]$
6: **end for**
7: $\mathbf{E} = [\mathbf{e_1}, ..., \mathbf{e_i}]$ ← dictionary learning (See Sect. 3 on DL)

// **Labelled reference facial images captures**
with a still camera in Ed
8: For each still images I_j in the sequence, $j = 1, ..., \infty$
9: ROI ← face detector and preprocessing on image I_j^s
10: **for** $k=1$ to number of ROIs in image **do**
11:      $a_{k,j}^s$ ← feature extraction on ROI$_j$
12:      $\mathbf{D} = [\mathbf{d_1}, ..., \mathbf{d_{k,j}}] \leftarrow [\mathbf{D}; a_{k,j}^s]$
13: **end for**

---

4   Experimental Results

4.1   Methodology for Validation

The proposed system is validated and compared experimentally using videos
from two challenging real-world data sets: Chokepoint and COX-S2V. They are
video surveillance datasets that emulate watchlist screening applications. The
main characteristics of these two datasets with respect to others is that they
contain a high-quality still face image and surveillance videos for each subject. Videos are captured over a distributed networks of cameras that covers
a range of variations changes in, e.g., pose, illumination, blur, scale. These are
presently among the most representative public data sets for watchlist screening
applications.

Chokepoint dataset [35] contains high quality frontal still image of each individuals and videos captured from those same people passing through different

Algorithm 2. Operational Phase for Frames Captured with one Camera

* **Input:** Undersampled dictionary $\mathbf{D} = [\mathbf{d_1}, ..., \mathbf{d_K}]$ of still images, external dictionary $\mathbf{E} = [\mathbf{e_1}, ..., \mathbf{e_i}]$ of videos from different domains,

* **Output:** Matching score for a probe transaction (ROI) for each facial model in dictionary $\mathbf{D}$, $r_{[k=1,...,K]}(y)$,

// **Transaction Level Processing**

1: **for** each frame I_i in the video sequence, $i = 1, ..., \infty$ **do**
2:     ROI $\leftarrow$ Face Detector and Preprocessing on frame I_i^v
3:     **for** $n=1$ to number of ROIs in frame **do**
4:         $\mathbf{a_{n,i}^v} \leftarrow$ Feature Extraction on ROI$_n$
5:         Normalize the columns of $\mathbf{D}$ and $\mathbf{E}$ to have unit l_2-norm
6:         Solve the l_1-minimization problem
$$\min_{x} \|\mathbf{a_{n,i}^v} - [\mathbf{D}, \mathbf{E}]\left[\begin{smallmatrix} \mathbf{x_d} \\ \mathbf{x_e} \end{smallmatrix}\right]\|_2^2 + \lambda\|\mathbf{x}\|_1$$
7:         **for** $k=1$ to number of facial model in Dictionary $\mathbf{D}$ **do**
8:             Compute the residuals
$$S = 1 - r_{k,i,n}(\mathbf{y}) = \|\mathbf{a_{n,i}^v} - [\mathbf{d_k}, \mathbf{E}]\left[\begin{smallmatrix} \delta(\mathbf{x_d}) \\ \mathbf{x_e} \end{smallmatrix}\right]\|_2^2 \text{ decision}$$
9:         **end for**
10:     **end for**
13: **end for**

portal with 3 surveillance cameras. It is composed of videos of 25 individual with 19 subjects are male and 6 are female. The 48 out of 54 video sequences are one person at a time and the rest remainder contain a mixture of people. Videos are captured from 3 cameras positioned over two portals during 4 sessions (1 month intervals), at 30 fps, and with an image resolution is 800×600 pixels. Each ROI or face images are scaled to a common size of 96×96 pixels. This dataset features variations in pose illumination, lighting, scale, and blur that makes it challenging for still-to-video FR. Figure 2(a) shows ROIs of 5 selected target individuals and their test video correspondence is recorded with 3 cameras above different portals.

COX-S2V [36] consists of 1000 subjects, where each subject has a high quality still images under controlled conditions, and four lower-quality facial trajectories captured under uncontrolled conditions. Each trajectory has 25 faces, where ROIs taken from these videos encounter changes in illumination, expression, scale, viewpoint, and blur.

In all experiments with Chokepoint dataset, 5 target individuals among 25 are selected randomly to design a watch-list that includes a high quality frontal captured images of selected people. The background model for the operational phase is designed based on selecting 10 unknown individual video sequences along 5 video sequences of already selected people in the watch-list. Moreover, the remaining 10 individual's video sequences are used as external data to build auxiliary dictionary. For the experiment with COX-S2V, three times randomly 20 individuals among 1000 subjects are selected to build a watch-list from high quality data and their corresponding low quality video trajectories are applied

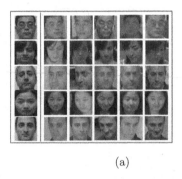   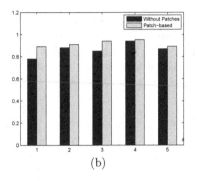

(a)   (b)

Fig. 2. (a) An example of ROIs extracted from mugshots of 5 redundancy selected target individuals of interest and some of the ROIs from corresponding operational video; (b) Comparison of ESRC-DA w/DL MFA with and without patch-based extraction from ROIs.

for testing. Background model is made of 100 video sequence of individuals to be applied on operational phase.

As described in Sect. 2, there are many methods in literature for DL like KSVD, MOD and SMRS. Nourbakhsh et. al [31] proposed a method to compress a graph based on matrix factorization approach (MFA) that needs an efficient time to construct the reduced graph and it is parameter free. Let M is a $N \times N$ similarity matrix and $K \leq N$ a constant and the compression rate is K/N. The goal of DL is to produce a reduced matrix R with order $K \times K$ and a many to one mapping function $\psi : [n] \rightarrow [k]$ between vertices of the original graph and reduced graph. The mapping function is expressed in terms of a left stochastic matrix. A least squares approximation is applied on following minimizer by dropping a left-stochastic constrain to a real matrix X to calculate the reduced matrix and mapping function.

The optimization can be addressed as a EM method which alternates updates of the variable R and updates of the variable X. The minimization approach converges to a stationary point by updating a decrease of the objective function in every iteration.

$$min \quad f(X,R) = ||M - X^T R X||_2^2 \tag{6}$$

where $s.t.$ $X \in S,$ $R \in R^{k \times k}$
and

$$f(X,R) = \sum_{(i,j)\in\{1,2,...,N\}} \sum_{(k,h)\in\{1,2,3,...K\}} \delta_{(k,i)\neq(h,j)} X_{ki} X_{hj}$$

$$\times (M_{ij} - R_{kh})^2 + \sum_{i\in\{1,2,...,N\}} \sum_{k\in\{1,2,3,...K\}} X_{ki} (M_{ii} - R_{kk})^2$$

It has been shown that DL algorithm reduces the complexity of many algorithms from n^2 to $(k^2 + n)$ by replacing the original data with its corresponding factorization.

Receiver Operating Characteristic (ROC) is applied to evaluate the performance of the proposed system that is defined as true positive rate (TPR) versus false positive rate (FPR). TPR is the proportion of correctly detected as individual of interest over the total number of target ROIs and FPR is the proportion of non-target detected as individual of interest over the total number of non-target ROIs. Area Under the ROC (AUC) is a global measure of performance which is defined as the probability of classification over the range of TPR and FPR. Finally Precision-Recall Operating Characteristic (PROC)curve constitutes a graphical representation of detector performance where the impact of data imbalance is considered. The precision between positive predictions (precision $PR = \frac{TP}{TP+FP}$) is combined with the TPR (or recall) to draw a PROC curve. To measure the efficiency of different algorithms, this paper also shows the time complexity. It is estimated as the total number of dot products (DPs) needed by an algorithms to produce a matching score in response to one probe ROI captured in operational videos.

4.2   Results and Discussion

Several experiments were conducted on the Chokepoint dataset to characterize the effectiveness of the proposed method. AUROC and AUPR curves were produced using ESRC-DA *w/*DL MFA on the test videos of five randomly selected watchlist individuals. These curves show the impact of exploiting DA. ESRC-DA can achieve a higher level of performance, even when the proportion of target to non-target is imbalanced (as seen in PR curves). Figure 3 shows the ROC and P-R curves obtained with the proposed system (ESRC-DA w/DL MFA) on Chokepoint data when each ROI is represented with HOG features. The faces captured in video and stills were scaled and normalized to 48 × 48 pixels ROIs and represented using HOG descriptors. In the next experiment, these ROIs were divided to 9 fixed size patches (forming 9 different external dictionaries), where the scores computed from local matching were combined using the average score-level fusion rule.

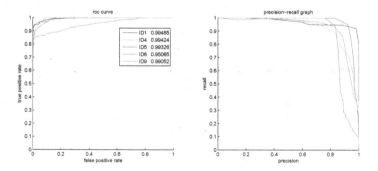

Fig. 3. ROC and P-R curves for the ESRC-DA methods on Chokepoint videos. Area under the curve of ROC is reported for five randomly selected individuals in the legend.

Table 1. Average performance of ESRCA-DA and references methods on Chokepoint dataset.

Classifiers	AUC	AUPR	Time comp (DPs)
NN (Template matching)	0.6991 ± 0.0411	0.2635 ± 0.0663	1,890
SRC [17]	0.8866 ± 0.0523	0.5447 ± 0.1427	3,572,100
SRC with Reject Criteria (SCI)	0.9041 ± 0.0509	0.5576 ± 0.1539	3,572,100
RSC [37]	0.8007 ± 0.0600	0.2478 ± 0.0759	3,572,100
ESRC w/o DL (Random) [15]	0.9360 ± 0.0311	0.6317 ± 0.1418	228,614,400
RADL w/o DL (Random) [34]	0.8313 ± 0.0701	0.3000 ± 0.1164	228,614,400
RADL w/DL [34]	0.8211 ± 0.0601	0.2106 ± 0.0762	230,748,921
SVDL w/DL [33]	0.7123 ± 0.0141	0.3112 ± 0.0543	230,748,921
ESRC-DA w/DL KSVD [29]	0.9450 ± 0.0455	0.6432 ± 0.1783	228,614,400
ESRC-DA w/DL MOD [30]	0.9408 ± 0.0404	0.6553 ± 0.1655	228,614,400
ESRC-DA w/DL SMRS [38]	0.9342 ± 0.0399	0.6316 ± 0.1719	228,614,400
ESRC-DA w/DL MFA [31]	$\mathbf{0.9716 \pm 0.0128}$	$\mathbf{0.7697 \pm 0.0673}$	228,614,400

Figure 2(b) shows the AUC performance with and without patch and patch-based extraction using ESR-DA w/DL MFA. It measure the effectiveness of local patch-based matching to handle variations in pixel features. Results show that that patch based method improve the performance processing ROIs distorted by, e.g., pose variations but are computationally slower.

Table 1 shows the average AUC and P-R accuracy, and time complexity of ESRC-DA w/DL versus state of the art methods which are NN, SRC, SRC with SCI, RSC, ESRC w/o DL, RADL w/o DL, RADL w/DL and SVDL w/DL for still-to-video FR. The time complexity $O(N_d \times N_v)$ is calculated in terms of the overall number of dot products (DPs) to process a ROI in operational phase, where N_d is dimensionality and N_v is number of vector. The worse case scenario is considered for each method. Fore example the time complexity of SRC and ESRC is considered quadratic to the number of training samples [39]. The table shows the average performance over 5 replications with random selection of individuals. The reference methods can be divided to three main categories: methods without DL that are NN, SRC, RCS, ESRC w/o DL and RADL w/o DL where SRC and NN are used as a base line methods, algorithms with different DL and classification that are RADL w/DL and SVDL. The last category is DA methods that are ESRC followed with KSVD, MOD, SMRS and MFA as DL. The ESR-DA w/DL MFA shows a significantly higher level of AUC and AUPR accuracy, with lower time complexity to reconstruct DL $O(k^2 + n)$ compare to state of the art methods and it is parameter free. The proposed system (ESRC-DA w/DL) is also compared using COX-S2V data with the state-of the art methods in Table 2.

The proposed system (ESRC-DA w/DL) is also compared using Chokepoint dataset with still-to-video state-of the art methods in Table 3. The result shows

Table 2. Average performance of ESRCA-DA and references methods on COXS2V dataset.

Classifiers	AUC	AUPR	Time comp (DPs)
SVDL w/DL [33]	0.6993 ± 0.5671	0.4409 ± 0.6291	432,364,725
ESRC-DA w/DL KSVD [29]	0.9729 ± 0.0313	0.5280 ± 0.2813	432,224,100
ESRC-DA w/DL MOD [30]	0.9781 ± 0.0312	0.5612 ± 0.2213	432,224,100
ESRC-DA w/DL SMRS [38]	0.9743 ± 0.0405	0.5712 ± 0.1581	432,224,100
ESRC-DA w/DL MFA [31]	0.9900 ± 0.0113	0.6321 ± 0.0456	432,224,100

that ESRC-DA/w DL MFA provide a high level of performance with a lower time complexity. For the time complexity, the codes are implemented in MATLAB, using a 3.40 GHz and 8 GB RAM computer.

Table 3. Average performance of ESRCA-DA and references methods on Chokepoint.

Classifiers	AUC	AUPR	Time complexity
AAMT-FR [3]	0.6490 ± 0.070	0.793 ± 0.03	0.217 ± 0.06 s
Ensemble of e-SVMs (1 block) [8]	0.9228 ± 0.54	0.909 ± 0.284	0.435 ± 0.07 s
ESRC-DA w/DL MFA [31]	0.9643 ± 0.051	0.7151 ± 0.022	0.119 ± 0.04 s

5   Conclusion

In this paper, an Extended SRC framework for still-to-video FR is proposed to accurately recognize individuals of interest across a distributed network of video surveillance cameras. To overcome the limitations of labelled reference still ROIs that are captured during enrolment for face modelling, this algorithm exploits are abundance of external video ROIs that can typically be captured with different cameras in the operational domain. The Extended SRC through Domain Adaptation (ESRC-DA) algorithm enhanced the robustness of facial models by integrating information from its operational an under-sampled dictionary from target reference stills with an auxiliary dictionary learned using facial trajectories captured under different operational capture conditions. The proposed matrix factorization approach is well adapted for this context, and it is shown that ESRC-DA w/DL MFA outperforms state-of-the-art methods. Results were obtained on two real-world video surveillance data sets (Chokepoint and COX-S2V) and using various face representations and matching schemes.

Exploiting unlabelled facial ROIs captured in videos of the operational environment allow for the ESRC-DA algorithm to achieve a higher level of FR accuracy than state-of-the-art systems. Furthermore, it represents a cost-effective solution for still-to-video FR, as required in several VS applications.

References

1. Chen, S., Mau, S., Harandi, M.T., Sanderson, C., Bigdeli, A., Lovell, B.C.: Face recognition from still images to video sequences: a local-feature-based framework. EURASIP J. Image Video Process. **2011**, 790598 (2011)
2. Kamgar-Parsi, B., Lawson, W., Kamgar-Parsi, B.: Toward development of a face recognition system for watchlist surveillance. IEEE Trans. PAMI **33**, 1925–1937 (2011)
3. Dewan, M.A.A., Granger, E., Marcialis, G.L., Sabourin, R., Roli, F.: Adaptive appearance model tracking for still-to-video face recognition. Pattern Recogn. **49**, 129–151 (2016)
4. Pato, J., Millett, L.: Biometric recognition: challenges and opportunities. Whither Biometrics Committee, National Research Council of the NSA (2010)
5. Pagano, C.C., Granger, E., Sabourin, R., Gorodnichy, D.O.: Detector ensembles for face recognition in video surveillance. In: IJCNN, pp. 1–8 (2012)
6. Tan, X., Chen, S., Zhou, Z.H., Zhang, F.: Face recognition from a single image per person: a survey. Pattern Recogn. **39**, 1725–1745 (2006)
7. Kan, M., Shan, S., Su, Y., Xu, D., Chen, X.: Adaptive discriminant learning for face recognition. Pattern Recog. **46**, 2497–2509 (2013)
8. Bashbaghi, S., Granger, E., Sabourin, R., Bilodeau, G.: Ensembles of exemplar-SVMs for video face recognition from a single sample per person. In: IEEE International Conference on Advanced Video and Signal Based Surveillance, pp. 1–6 (2015)
9. Patel, V.M., Gopalan, R., Li, R., Chellappa, R.: Visual domain adaptation: a survey of recent advances. IEEE Sig. Process. Mag. **32**, 53–69 (2015)
10. Minku, L.L., White, A.P., Yao, X.: The impact of diversity on online ensemble learning in the presence of concept drift. IEEE Trans. Knowl. Data Eng. **22**, 730–742 (2010)
11. Snidaro, L., Garca, J., Llinas, J.: Context-based information fusion: a survey and discussion. Inf. Fusion **25**, 16–31 (2015)
12. Huang, Z., Wang, R., Shan, S., Li, X., Chen, X.: Log-euclidean metric learning on symmetric positive definite manifold with application to image set classification. In: ICML. JMLR Workshop and Conference Proceedings (2015)
13. Margolis, A.: Automatic annotation of spoken language using out-of-domain resources and domain adaptation. IEEE Trans. Knowl. Data Eng. (2011)
14. Pan, S.J., Yang, Q.: A survey on transfer learning. IEEE Trans. Knowl. Data Eng. **22**, 1345–1359 (2010)
15. Deng, W., Hu, J., Guo, J.: Extended SRC: undersampled face recognition via intraclass variant dictionary. IEEE Trans. PAMI **34**, 1864–1870 (2012)
16. Wright, J., Yang, A.Y., Ganesh, A., Sastry, S.S., Ma, Y.: Robust face recognition via sparse representation. IEEE Trans. Pattern Anal. Mach. Intell. **31**, 210–227 (2009)
17. Naseem, I., Togneri, R., Bennamoun, M.: Sparse representation for video-based face recognition. In: Tistarelli, M., Nixon, M.S. (eds.) ICB 2009. LNCS, vol. 5558, pp. 219–228. Springer, Heidelberg (2009). doi:10.1007/978-3-642-01793-3_23
18. Nagendra, S., Baskaran, R., Abirami, S.: Video-based face recognition and face-tracking using sparse representation based categorization. Procedia Comput. Sci. **54**, 746–755 (2015). International Conference on Data Mining and Warehousing
19. Cui, Z., Chang, H., Shan, S., Ma, B., Chen, X.: Joint sparse representation for video-based face recognition. Neurocomputing **135**, 306–312 (2014)

20. Gu, J., Liu, L., Hu, H.: Patch-based sparse dictionary representation for face recognition with single sample per person. In: Yang, J., Yang, J., Sun, Z., Shan, S., Zheng, W., Feng, J. (eds.) Biometric Recognition. LNCS, vol. 9428, pp. 120–126. Springer, Heidelberg (2015). doi:10.1007/978-3-319-25417-3_15

21. Qiu, Q., Patel, V.M., Turaga, P., Chellappa, R.: Domain adaptive dictionary learning. In: Fitzgibbon, A., Lazebnik, S., Perona, P., Sato, Y., Schmid, C. (eds.) ECCV 2012. LNCS, vol. 7575, pp. 631–645. Springer, Heidelberg (2012). doi:10.1007/978-3-642-33765-9_45

22. Ni, J., Qiu, Q., Chellappa, R.: Subspace interpolation via dictionary learning for unsupervised domain adaptation. In: IEEE Conference on Computer Vision and Pattern Recognition (2013)

23. Qiu, Q., Ni, J., Chellappa, R.: Dictionary-based domain adaptation methods for the re-identification of faces. In: Gong, S., Cristani, M., Yan, S., Loy, C.C. (eds.) Person Re-Identification. ACVPR, pp. 269–285. Springer, Heidelberg (2014). doi:10.1007/978-1-4471-6296-4_13

24. Shekhar, S., Patel, V.M., Nguyen, H.V., Chellappa, R.: Generalized domain-adaptive dictionaries. In: IEEE Conference on Computer Vision and Pattern Recognition (2013)

25. Duan, L., Tsang, I.W., Xu, D., Chua, T.: Domain adaptation from multiple sources via auxiliary classifiers. In: Proceedins of International Conference on Machine Learning (2009)

26. Duan, L., Tsang, I.W., Xu, D.: Domain transfer multiple kernel learning. IEEE Trans. Pattern Anal. Mach. Intell. **34**, 465–479 (2012)

27. Guo, H., Jiang, Z., Davis, L.S.: Discriminative dictionary learning with pairwise constraints. In: Lee, K.M., Matsushita, Y., Rehg, J.M., Hu, Z. (eds.) ACCV 2012. LNCS, vol. 7724, pp. 328–342. Springer, Heidelberg (2013). doi:10.1007/978-3-642-37331-2_25

28. Shafiee, S., Kamangar, F., Athitsos, V., Huang, J.: The role of dictionary learning on sparse representation-based classification. In: International Conference on PErvasive Technologies Related to Assistive Environments (2013)

29. Aharon, M., Elad, M., Bruckstein, A.: K-SVD: an algorithm for designing overcomplete dictionaries for sparse representation. Trans. Sig. Proc. **54**, 4311–4322 (2006)

30. Engan, K., Aase, S.O., Hakon Husoy, J.: Method of optimal directions for frame design. In: International Conference on Acoustics, Speech, and Signal Processing (1999)

31. Nourbakhsh, F., Bulò, S.R., Pelillo, M.: A matrix factorization approach to graph compression with partial information. Int. J. Mach. Learn. Cybern. **6**, 523–536 (2015)

32. Nourbakhsh, F., Granger, E.: Learning of graph compressed dictionaries for sparse representation classification. In: International Conference on Pattern Recognition Applications and Methods, ICPRAM, pp. 309–316 (2016)

33. Yang, M., Van Gool, L., Zhang, L.: Sparse variation dictionary learning for face recognition with a single training sample per person. In: The IEEE International Conference on Computer Vision (ICCV) (2013)

34. Wei, C., Wang, Y.F.: Undersampled face recognition via robust auxiliary dictionary learning. IEEE Trans. Image Process. **24**, 1722–1734 (2015)

35. Wong, Y., Chen, S., Mau, S., Sanderson, C., Lovell, B.C.: Patch-based probabilistic image quality assessment for face selection and improved video-based face recognition. In: (CVPR) Workshops on Biometrics. IEEE (2011)

36. Huang, Z., Shan, S., Wang, R., Zhang, H., Lao, S., Kuerban, A., Chen, X.: A benchmark and comparative study of video-based face recognition on COX face database. IEEE Trans. Image Proccess. **24**, 5967–5981 (2015)

37. Yang, M., Zhang, L., Yang, J., Zhang, D.: Robust sparse coding for face recognition. In: International Conference on Computer Vision and Pattern Recognition (2011)

38. Elhamifar, E., Sapiro, G., Vidal, R.: See all by looking at a few: sparse modeling for finding representative objects. In: IEEE Conference on Computer Vision and Pattern Recognition (2012)

39. Donoho, D.L., Tsaig, Y.: Fast solution of l_1-norm minimization problems when the solution may be sparse. IEEE Trans. Inf. Theory **54**, 4789–4812 (2008)

BCP-BCS: Best-Fit Cascaded Matching Paradigm with Cohort Selection Using Bezier Curve for Individual Recognition

Jogendra Garain[1(✉)], Adarsh Shah[2], Ravi Kant Kumar[1],
Dakshina Ranjan Kisku[1], and Goutam Sanyal[1]

[1] Department of Computer Science and Engineering,
National Institute of Technology Durgapur, Durgapur, West Bengal, India
jogs.cse@gmail.com
[2] Microsoft India (R&D) Private Limited, Hyderabad, India

Abstract. The concept of cohort selection has been emerged as a very interesting and potential topic for ongoing research in biometrics. It has the capability to provide the traditional biometric systems to having a higher performance rate with lesser complexity and cost. This paper describes a novel matching technique incorporated with Bezier curve cohort selection. The Best-Fit matching with dynamic threshold has been proposed here to reduce the number of false match. This algorithm is applied for matching of Speeded Up Robust Feature (SURF) points detected on face images to find out the matching score between two faces. After that, Bezier curve is applied as a cohort selection technique. All the cohort scores are plotted in a 2D plane as if these are the control points of a Bezier curve and then a Bezier curve of degree n is plotted on the same plane using De Casteljau algorithm where number of control point is $n + 1$. A template contains more discriminative features more it is having distance from the curve. All the templates having score point far from the curve are included into the account of cohort subset. For each enrolled user a specific cohort subset is determined. As long as the subset is formed, T-norm cohort score normalization technique is applied to obtain the normalized scores which are further used for person identification and verification. Experiments are conducted on FEI face database and results are showing dominance over the non-cohort system.

1 Introduction

The challenge, to overcome the difficulties with a face biometric system and to achieve higher accuracy, is continuously knocking to the researcher's mind. The reason why face is considered the most challenging biometric trait in computer-based identity verification of individuals is its dynamicity. Another difficulty with face biometric is insufficient information available with face images to be used for matching. Although traditional face biometric systems give a decent accuracy but only for neutral, frontal and full profile face images of good quality and captured in controlled environment with good illumination condition. But this

© Springer International Publishing AG 2017
C.-S. Chen et al. (Eds.): ACCV 2016 Workshops, Part III, LNCS 10118, pp. 377–390, 2017.
DOI: 10.1007/978-3-319-54526-4_28

kind of favor in face images is not always available in real scenario. So when there is a change in facial expressions and/or pose the performance starts to degrade. It drastically degrades when rotated, occluded and half profile faces are given as input. In order to minimize the effects of such face variations, researchers are continuously trying to find out a solution of it. As of now, cohort selection can solve a certain extent of these problem [1–4] and further, this effort would help to build a high-end robust system with desired accuracy. In this work, SURF points [5] are used as features and the proposed matching algorithm is applied to obtain the matching scores between all face image pairs. Provided the matching scores set, most similar and dissimilar cohorts scores are selected to form the cohort subset. The work is tested on four different sets (100 each) of Brazilian face collected from FEI face database [6] and the recognition accuracy achieved is remarkable. The paper organization is as follows. Section 2 describes the related works. Section 3 contains a brief description about cohort set and cohort score. Section 4 discusses about the matching algorithms and the mechanics, a Bezier curve is used for cohort selection. Database setup and Experimental outputs are described in Sect. 5 and the last section concludes the proposed methodology.

2   Related Work

It is evident from recent literatures on biometrics that identity verification through cohort selection has the ability to minimize the shortcomings which occur due to abrupt variations and inadequate information found on face images. Initially, the model of cohort selection was assembled with automated speaker verification system. Auckenthaler et al. [7] applied Bayes' theorem based cohort and score normalization to upgrade the performance of a speaker verification system. Later, it was applied to face and fingerprint biometrics. Aggarwal et al. [8] utilized a set of additional template as reference to represent a fingerprint template as a composite one to improve the performance of a biometric authentication system. The work reported in [4] uses library images for superior performance and similarity measure of a face pair is found to be invariant to facial expression, pose and illumination. Here, authors have represented a face image by a signature. They have used a Library of face images of different identities among which a list of identities is considered as the signature of each particular face. The list is ordered as per the similarity scores of a probe face and the Library images. Wolf et al. [9] proposed a method, Two-Shot Similarity (TSS) scores extending their previous method, One-Shot Similarity (OSS) [10], to increase the performance of a recognition system by using background samples?. Sparse Representation-Based Classification (SRC) proposed by Wagner et al. in [11] is a method to align a test image to a set of frontal training images. However, it needs a rich set of training images to perform well. Deng et al. [12] proposed an Extended Sparse Representation-Based Classification (ESRC) method where they showed that the variation among the images of same person can be predicted by a sparse linear combination of some additional subjects which may be from the gallery or from the outside gallery. The authors are basically keeping the information into a dictionary about the intraclass variation of each subject. The probable deviations of test images to training images can now be predicted with the help of this dictionary containing prior knowledge. It works well with a few training samples even if with a single sample per subject. Merati et al. [2] have shown that a traditional biometric system can give better recognition accuracy as well as reduce computational cost up to 50% if suitable cohort selection and cohort score normalization techniques are applied. Here they have proposed a method to extract the cohort coefficients using a polynomial regression. With the help of sorted cohort, the authors in [2,13] proved that the closer and outlier cohorts are more helpful to distinguish the identities from each other. Garain et al. [14] proposed a method (MMCC) as cohort selection technique which reduces 60% of the total computational cost. Tulyakov et al. [15] stated that combining more than one score normalization techniques enhance the performance of a biometric system better than a single normalization technique. Sun et al. [1] used an additional cohort set, different from the set of face images being compared, for face pair matching. They have used the cohort coefficients as complementary information due to insufficient information embedded with these two faces to match. The authors in [16] proposed a picture-specific cohort selection method to match two pictures of a face. They determine two cohort set for two input images from a set of fixed number of cohort samples. Thereafter the images are compared based on the raw matching scores between them and the similarity scores with the cohort subsets. Garain et al. [3] describes how the matching scores are used as control points of a Bezier curve to select cohort subset. The distance of all control points from the curve is calculated first then the mean of theses distances is set as a threshold of cohort selection. We have incorporated this cohort selection technique with a novel matching algorithm and have used SURF points as features whereas authors in [3] used SIFT points.

3   Concept of Cohort Score

Comparison or matching can be conducted in two different ways when a query or a probe face is given to a face recognition system as input referring a gallery face as its identity, such as (a) comparison between input face and reference face, and (b) comparison between input face and gallery faces. The former provides genuine score and the later gives a set of scores called cohort scores. The set of gallery images other than claimed one is called cohort set or Library images or background samples. Cohort set not necessarily be formed with gallery images, it may be a different set of images which are not being used as enrolled or test images. Figure 1 shows that the score comes from a pair of query template and a claimed template is called genuine score and when it comes from a pair of query template and a cohort template is called cohort score. The selection of Cohort subset is not all time requirement, the whole database can be used as cohort subset for normalizing the matching scores when the number of enrolled templates in the database is very less. But for a large databases if the entire database is used for score normalization then accuracy will not increase in proportion to

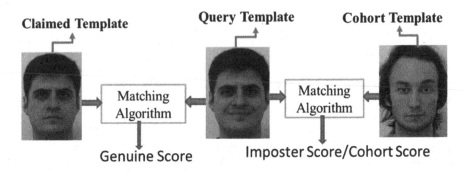

Fig. 1. Genuine score vs. Cohort score.

hike of execution time. So rather using the complete dataset as cohort, a subset is chosen and it reduces the computational cost remarkably. A query image not having sufficient information to represent itself uses cohort set as a counterpart of it.

4   Cohort Selection Mechanics

4.1   SURF Point Detection

Speeded-Up Robust Features (SURF) [5] is a faster key point detector than Scale Invariant Feature Transform (SIFT) [17]. Each key point is described here by a vector of length 64 unlike 128 in SIFT. It uses Integral image which trim down the execution time. The determinants of Hessian matrix H(x; σ) at any point on an image I at scale σ calculated using Eq. 1 are used to find the key points. To construct the key points scale invariant, rather scaling up or down the image size, Box filters of different sizes are applied on the original image directly for approximation of second order Gaussian derivatives. It speeds up the execution too.

$$H(x;\sigma) = \begin{bmatrix} L_{xx}(x,\sigma) & L_{xy}(x,\sigma) \\ L_{xy}(x,\sigma) & L_{yy}(x,\sigma) \end{bmatrix} \quad (1)$$

where, $L_{xx}(x,\sigma)$, $L_{xx}(x,\sigma)$, $L_{xx}(x,\sigma)$ are the second order Gaussian derivatives of an image I at a point x. Figure 2 shows SURF points detection from a sample face image. SURF points are detected in both query and claimed face images then matching is performed to calculate the matching score.

4.2   Calculate Matching Scores

A matching score $M_s(f_q, f_c)$ is a function of a face pair which counts the total number of matched points found between them. Let f_q is a query face and f_c is a claimed face and P_q, P_c are two sets of SURF points on the query and claimed

Input   Output

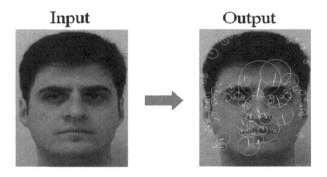

Fig. 2. SURF points detection.

faces respectively, then the function of matching score can be expressed by the Eq. 2 as follows.

$$M_s(f_q, f_c) = \{count(x)|x \implies M_T(p_x, p_y)\} \tag{2}$$

where $(p_x \in P_q, p_y \in P_c)$ is a matched pair for a particular matching technique M_T. Our proposed work uses Best-Fit matching as M_T discussed below.

Best-Fit Matching: Each SURF point detected on the query face is compared with the SURF points detected on the claimed face one after another. Let p_q and p_c are a pair of SURF point on a query face and a claimed face respectively, then (p_q, p_c) is considered a match pair if the distance between p_q and p_c is less than the threshold value and smallest among the distances between p_q and p_i where p_i is any SURF point on the claimed face image. Matching score between a face pair is the total number of match pair. Here, $(p_q \in f_q, p_c \in f_c)$ is considered as a matched pair iff $(D(p_q, p_c) \leq \theta$ and $D(p_q, p_c) = min(D(p_q, p_i)|\forall p_i \in f_c)$. Here $min(X)$ is a function which finds the minimum element from the input set X and D is a function which calculates how much dissimilar two SURF points are. In Algorithm 1, the threshold value (θ) is dynamic. For a particular SURF point on the query face image (f_q), the value of θ depends on the mean (μ) of the distances from that particular SURF point to all other SURF points detected on the claimed face (f_c). Since this mean value changes for almost all the SURF points on a query face, the θ value also changes. SURF point matching between two faces of same person i.e. genuine match is shown in Fig. 3. The left face is a neutral face and the right face is with smiling expression. Total number of SURF points on left face matched with the SURF points on the right face is measured as matching score (M_s).

Algorithm 1. Best-Fit Matching

INPUT: Claimed face f_c and Query face f_q

OUTPUT: Matching Score M_s

Let $\{P_c | c = 1 \text{ to } m\}$ is a set of SURF points on f_c and $\{P_q | q = 1 \text{ to } n\}$ is a set of SURF points on f_q

Initially $M_s \leftarrow 0$

for $i = 1$ to n **do**

   Set $D \leftarrow$ **NULL**

   for $j = 1$ to m **do**

      $D_{ij} \rightarrow d_i - d_j$ where d_i and d_j are the ith and j^{th} SURF points

                              descriptor of query face and claimed face respectively.

      $D = D \cup d_{ij}$

   end for

   Find $\mu = mean(D)$

   Find $\theta = \mu/2$

   Find $s = min(D)$

   if $s \leq \theta$ **then**

      $M_s = M_s + 1$

   end if

end for

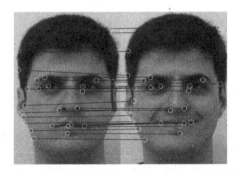

Fig. 3. Surf point matching using Best-Fit.

4.3 Bezier Curve Plotting

Number of imposter score/cohort score is one less than the total number of enrolled subject (S) in the database if only one image per enrolled subject are there and cohort set is formed from the enrolled images. But for every query image there is only one genuine score (g), the matching score between query face and claimed face. With the help of De Casteljau algorithm [18] plot an nth degree Bezier curve as if all the $n + 1$ cohort scores (δ) are control points. How to plot an nth degree Bezier curve controlled by $n + 1$ cohort scores (δ), is mathematically shown below in Eq. 3.

$$B(t) = \sum_{i=0}^{n} b_i^n(t).\delta_{i+1},\ 0 \le t \le 1 \tag{3}$$

where $b_i^n = \binom{n}{i}.t^i.(1-t)^{n-i}$ is the Bernstein basis polynomial of degree n.

4.4 Selection of Cohorts Using Bezier Curve

Given a Bezier curve, it has been estimated that it passes through as many control points as possible. Moreover, a Bezier curve always passes through the dense area. A set of control point is closed to one another, can be said in other way that the corresponding user's templates contain almost same characteristics. We are interested on those points which are far from the curve because those points are having discriminative features [13]. So those points are chosen to be selected in the cohort subset as depicted in Fig. 4. This subset is formed for each and every enrolled user and when an enrolled face is claimed as its identity by any query image, their matching score is normalized with the help of this cohort subset and a cohort score normalization technique like T-norm given in Eq. 4.

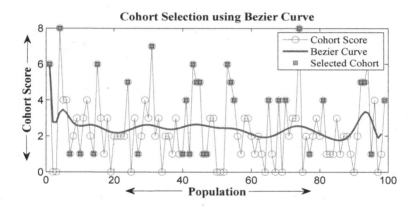

Fig. 4. Cohort selection using Bezier curve.

$$S_{(T-norm)} = \frac{g - \mu}{\sigma} \tag{4}$$

where $S_{(T-norm)}$ is the normalized score, g is genuine score, μ is the mean of all cohort scores(δ) and σ is standard deviation.

Since the zero valued points don't affect the normalized score, these are ignored even if they are far from the curve. The following flowchart, shown in Fig. 5, gives a pictorial overview of the proposed method of cascading Best-Fit with Bezier curve cohort selection and the Algorithm 2 states about the Bezier curve cohort selection step by step.

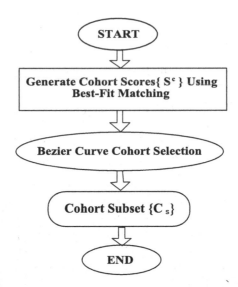

Fig. 5. Flowchart of cascading Best-Fit with Bezier curve cohort selection.

Algorithm 2. Bezier Cohort [3]

INPUT:Set of cohort scores $S = \{\delta_1, \delta_2, \delta_3...\delta_{n-1}\}$,$n \rightarrow$ no of enrolled template and $\delta_i \rightarrow$ cohort score of ith Subject.

OUTPUT: Cohort Subset $\{C_s\}$.

Step1: Plot a Bezier curve using De Casteljau algorithm considering all δ_i as control points.

Step2: Find the Euclidean distances (d_i) of each control point δ_i from the curve.

Step3: Calculate the mean (μ) of all distances(d_i),$i = 1, 2, ..., n - 1$.

Step4:

if $\delta_1 > 0$ then

$C_s = \{ \delta_1\}$ $otherwise$ $C_s = \{\phi\}$, $\phi \rightarrow NULL$

end if

Step5: $C_s = \{C_s \cup d_i\}$ $where$ $d_i > \mu$, $i = 2, 3, 4...n - 1$

---

5   Database and Experimental Setup

Total 2800 color face images of 200 individuals of Brazilian male and female are captured at the Artificial Intelligence Laboratory of FEI in So Bernardo do Campo, So Paulo, Brazil from June 2005 to March 2006 build FEI face Database [6]. There are 14 different instances of each individual in size of 640×480 pixels and with variation of rotation permitted up to $180°$. For enrolled identity and cohort selection purpose we have used 200 neutral (11th instance) and 100 smiling (12th instance) face images respectively which are available in a subset additionally provided in the database. This subset contains one neutral and one smiling face image of each 200 individuals, already cropped to

360×260 pixels. First 100 smiling faces are used as cohort set and other 100 are used for testing. In addition three more instances, $10°$ rotated (5th instance), $20°$–$30°$ rotated (4th instance) and poor lighting (14th instance), are collected from the main database and manually cropped to size 360×260 pixels. This set of 600 face images is made in two subsets each containing 300 face images (100 of each variation). The first subset is used as training dataset and remaining is for testing purpose along with the remaining 100 smiling faces. Some sample face images from the dataset used in this experiment, are shown in Fig. 6.

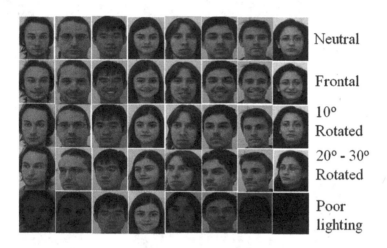

Fig. 6. FEI face database samples.

Detection of SURF points on a face from each dataset (neutral, smiling, rotated1 ($10°$), rotated2 ($20°$–$30°$) and poor lighting) as well as matching with enrolled face (neutral) are shown as an example in Fig. 7(a) and (b). All the faces shown in Fig. 7(a) and (b) belong to same person with different variations. Applying the Bezier-cohort method, first the set of smiling faces are used to determine the cohort subset of each enrolled face. Thereafter another set of 100 smiling face images are used for testing along with other three sets mentioned above. Figure 8 shows some sample face images of cohort subset obtained during experiment. The faces in the first column are enrolled face and all other columns hold corresponding cohort faces. As the proposed cohort selection method is person dependant, the number of cohort varies from person to person. Figure 9 plots the number of cohort for each of the 200 enrolled persons. In Fig. 10, the frequencies of "number of cohort" are displayed. It is observed that the "number of cohort" belongs in between 12 to 36 which is not very less to provide complementary information to represent each enrolled face image and not very large to accomplish in a moderate computational cost. It is also noticeable in this Figure that more than 50% subjects have 22 to 26 number of cohort which is the favourable aspect of this cohort selection algorithm. The performance

(a)

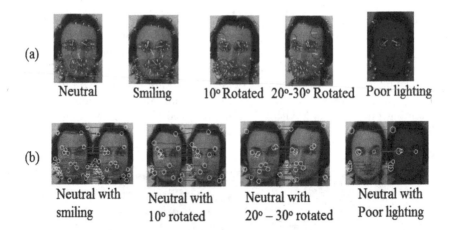

Neutral   Smiling   10° Rotated 20°-30° Rotated Poor lighting

(b)

Neutral with   Neutral with   Neutral with   Neutral with
smiling   10° rotated   20° – 30° rotated Poor lighting

Fig. 7. (a) SURF points on five different faces (neutral, smiling, 10° rotated, 20°–30° rotated and captured in poor lighting) of a same person (b) SURF points matching of 4 different faces with the neutral face.

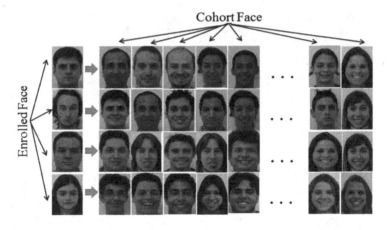

Fig. 8. Enrolled face (First column) and their corresponding cohort faces.

of this authentication system after incorporating Bezier curve cohort selection method is found superior to the systems which are not using cohort selection as well as the literature with cohort selection [3,13] shown in Table 2. The trade-off between false acceptance rate (FAR) and genuine acceptance rate (GAR) plotted as ROC curve shown in Figs. 11 and 12. The ROC curves for different test data set (smiling, rotated and poor lighting face) are plotted in Figs. 11(a)–(d) separately to highlight the improvement of the system after applying cohort selection. Figure 12 is showing the same for combination of all four dataset i.e. for total 400 test images. Table 1 depicts about the effect of cohort selection in respect to FAR, FRR, EER, and accuracy. It gives a clear view that the highest accuracy is achieved for smiling dataset and the accuracy is falling down with

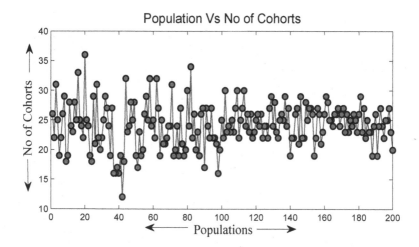

Fig. 9. No of cohorts per enrolled person.

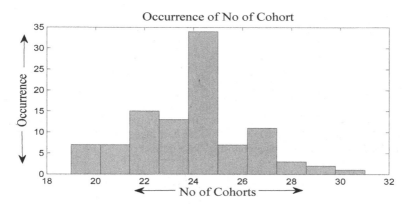

Fig. 10. Frequency of "No of cohorts"

increase in rotation angle. The analysis in Table 1 displays higher false rejection rate (FRR) for poor lighting conditions with the face dataset in comparison to other because without enhancing the quality, the SURF points detected on those faces are not sufficient. However the improvement of the system due to application of cohort faces is quite interesting. During experiment the proposed matching algorithm is tested by setting different values $(5\,\mu, 4\,\mu, 3\,\mu, 2\,\mu, \mu, \mu/2$ and $\mu/4)$ as threshold (θ). It is observed that at $\theta = \mu/2$, the algorithm gives best result by reducing maximum number of false matches. As the value is increased, the matching score increases up to 4 but it adds more number of false match relatively and reduce the systems accuracy by increasing false acceptance rate (FAR). Table 2 compares the performance of the proposed work with MMCC [14] and Bezier Curve cohort selection with SIFT point [3].

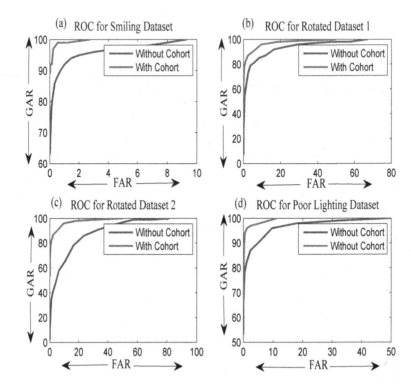

Fig. 11. Receiver Operating Characteristic curve of applying cohort vs. without apply-ing cohort for - (a) smiling dataset (b) 10° rotated face dataset (c) 20°–30° rotated face dataset and (d) poor lighting condition face dataset.

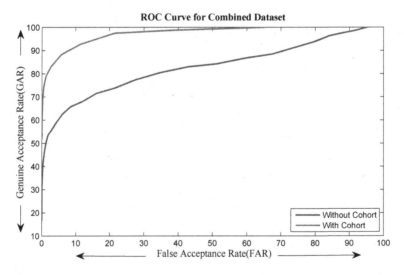

Fig. 12. Receiver Operating Characteristic curve for the dataset of 400 test faces mix-ing up all four dataset (smiling, rotated1, rotated2 and poor lighting).

Table 1. Performance Comparison between before and after use of cohort selection.

Dataset used	Evaluation parameters							
	Before cohort selection				After cohort selection			
	FAR	FRR	EER	Accuracy	FAR	FRR	EER	Accuracy
Smiling	3.02	3	3.01	96.99	0.53	1	0.77	**99.24**
10° Rotated	8.28	15	11.64	88.36	9.6	4	6.8	**93.20**
20°–30° Rotated	23.47	14	18.74	81.27	12.48	9	10.74	**89.26**
Poor lighting	3.88	26	14.94	85.06	1.60	1	1.30	**98.70**
Overall	8.51	34.3	21.38	78.60	5.75	11.75	8.75	**91.25**

Table 2. Performance Comparison with previous method.

Cohort selection technique	No. of test images	Dataset used (FEI)	EER (%)	GAR (%)	Accuracy (%)
Method 1 [14]	100	Poor lighting	5.67	96.53	94.33
Method 2 [3]	100	Poor lighting	1.48	98	98.55
Proposed	100	Poor lighting	1.30	99	**98.70**

6 Conclusion

This paper has presented a human recognition system which makes use of Best-Fit cascaded matching paradigm with Bezier curve based cohort selection method. The Best-Fit algorithm is cascaded with dynamic threshold and Bezier curve method in order to achieve successful cohort selection from a set of face images and further, this crucial selection of cohort scores is then used to reduce the FAR and increase genuine acceptance rate. Overall efficacy is found to be convincing. As part of recognition system, since any image enhancement technique is not used for pre-processing of face images, the number of SURF points detected on the faces captured in poor lighting condition is very less shown in Fig. 7(a) even zero for few faces. So the performance of this proposed method can be increased by enhancing the image quality. It is observed during experiments that with the increase of rotation angle the accuracy is also decreasing. And some false match pair are also coming due to hair, moustache and beard. Some suitable image pre-processing and outlier removal techniques can resolve this problem up to some extent. It can be a future scope of this work. Therefore after refurbishing this work with image pre-processing and enhancement techniques, we can achieve a more robust face biometric system.

References

1. Sun, Y., Nasrollahi, K., Sun, Z., Tan, T.: Complementary cohort strategy for multimodal face pair matching. IEEE Trans. Inf. Forensics Secur. **11**, 937–950 (2016)
2. Merati, A., Poh, N., Kittler, J.: User-specific cohort selection and score normalization for biometric systems. IEEE Trans. Inf. Forensics Secur. **7**, 1270–1277 (2012)

3. Garain, J., Kumar, R.K., Kisku, D.R., Sanyal, G.: Selection of user-dependent cohorts using bezier curve for person identification. In: Campilho, A., Karray, F. (eds.) ICIAR 2016. LNCS, vol. 9730, pp. 566–572. Springer, Heidelberg (2016). doi:10.1007/978-3-319-41501-7_63
4. Schroff, F., Treibitz, T., Kriegman, D., Belongie, S.: Pose, illumination and expression invariant pairwise face-similarity measure via doppelgänger list comparison. In: 2011 International Conference on Computer Vision, pp. 2494–2501. IEEE (2011)
5. Bay, H., Tuytelaars, T., Gool, L.: SURF: speeded up robust features. In: Leonardis, A., Bischof, H., Pinz, A. (eds.) ECCV 2006. LNCS, vol. 3951, pp. 404–417. Springer, Heidelberg (2006). doi:10.1007/11744023_32
6. Thomaz, C.E., Giraldi, G.A.: A new ranking method for principal components analysis and its application to face image analysis. Image Vis. Comput. **28**, 902–913 (2010)
7. Auckenthaler, R., Carey, M., Lloyd-Thomas, H.: Score normalization for text-independent speaker verification systems. Digit. Sig. Proc. **10**, 42–54 (2000)
8. Aggarwal, G., Ratha, N.K., Bolle, R.M.: Biometric verification: looking beyond raw similarity scores. In: 2006 Conference on Computer Vision and Pattern Recognition Workshop (CVPRW 2006), p. 31. IEEE (2006)
9. Wolf, L., Hassner, T., Taigman, Y.: The one-shot similarity kernel. In: 2009 IEEE 12th International Conference on Computer Vision, pp. 897–902. IEEE (2009)
10. Wolf, L., Hassner, T., Taigman, Y.: Similarity scores based on background samples. In: Zha, H., Taniguchi, R., Maybank, S. (eds.) ACCV 2009. LNCS, vol. 5995, pp. 88–97. Springer, Heidelberg (2010). doi:10.1007/978-3-642-12304-7_9
11. Wagner, A., Wright, J., Ganesh, A., Zhou, Z., Ma, Y.: Towards a practical face recognition system: robust registration and illumination by sparse representation. In: IEEE Computer Society Conference on Computer Vision and Pattern Recognition, vol. 2, p. 3 (2009)
12. Deng, W., Hu, J., Guo, J.: Extended SRC: undersampled face recognition via intraclass variant dictionary. IEEE Trans. Pattern Anal. Mach. Intell. **34**, 1864–1870 (2012)
13. Merati, A., Poh, N., Kittler, J.: Extracting discriminative information from cohort models. In: 2010 Fourth IEEE International Conference on Biometrics: Theory Applications and Systems (BTAS), pp. 1–6. IEEE (2010)
14. Garain, J., Kumar, R.K., Sanyal, G., Kisku, D.R.: Cohort selection of specific user using max-min-centroid-cluster (MMCC) method to enhance the performance of a biometric system. Int. J. Secur. Appl. **9**, 263–270 (2015)
15. Tulyakov, S., Zhang, Z., Govindaraju, V.: Comparison of combination methods utilizing T-normalization and second best score model. In: IEEE Computer Society Conference on Computer Vision and Pattern Recognition Workshops, CVPRW 2008, pp. 1–5. IEEE (2008)
16. Sun, Y., Tistarelli, M., Poh, N.: Picture-specific cohort score normalization for face pair matching. In: 2013 IEEE Sixth International Conference on Biometrics: Theory, Applications and Systems (BTAS), pp. 1–8. IEEE (2013)
17. Lowe, D.G.: Distinctive image features from scale-invariant keypoints. Int. J. Comput. Vis. **60**, 91–110 (2004)
18. Casteljau, P.d.F.d.: Outillage méthodes calcul. Enveloppe Soleau P **2108** (1959)

Benchmark and Evaluation of Surveillance Task (BEST)

BEST: Benchmark and Evaluation of Surveillance Task

Chongyang Zhang[✉], Bingbing Ni, Li Song, Guangtao Zhai,
Xiaokang Yang, and Wenjun Zhang

Institute of Image Communication and Network Engineering,
Shanghai Jiao Tong University, Shanghai 200240, China
sunny_zhang@sjtu.edu.cn

Abstract. Smart/Intelligent video surveillance technology plays the central role in the emerging smart city systems. Most intelligent visual algorithms require large-scale image/video datasets to train classifiers or acquire discriminative features using machine learning. However, most existing datasets are collected from non-surveillance conditions, which have significant differences as compared to the practical surveillance data. As a consequence, many existing intelligent visual algorithms trained on traditional datasets perform not so well in the real world surveillance applications. We believe the lack of high quality surveillance datasets has greatly limited the application of the computer vision algorithms in practical surveillance scenarios. To solve this problem, one large-scale and comprehensive surveillance image and video database and test platform, called Benchmark and Evaluation of Surveillance Task (abbreviated as BEST), is developed in this work. The original images and videos in BEST were all collected from on-using surveillance cameras, and have been carefully selected to cover a wide and balanced range of outdoor surveillance scenarios. Compared with the existing surveillance/non-surveillance datasets, the proposed BEST dataset provides a realistic, extensive and diversified testbed for a more comprehensive performance evaluation. Our experimental results show that, performance of seven pedestrian detection algorithms on BEST is worse than that on the existing datasets. This highlights the difference between non-surveillance data and real surveillance data, which is the major cause of the performance decreases. The dataset is open to the public and can be downloaded at: http://ivlab.sjtu.edu.cn/best/Data/List/Datasets.

1 Introduction

Video surveillance technology plays more and more important roles in the emerging public security management systems. Clearly todays video surveillance systems need to change the security paradigm from "investigation to preemption" [1]. Automatic visual analysis technologies can move today's video surveillance systems from the investigative to preventive paradigm. The aim of developing smart/intelligent visual surveillance is to replace the traditional passive video surveillance, which is proving ineffective as the number of cameras exceeds the

C.-S. Chen et al. (Eds.): ACCV 2016 Workshops, Part III, LNCS 10118, pp. 393–407, 2017.
DOI: 10.1007/978-3-319-54526-4_29

capability of human operators to monitor them. In short, the goal of visual surveillance is not only to put cameras in the place of human eyes, but also to accomplish the entire surveillance task as automatically as possible [2,3], such as awareness of location, identity and activity of objects in the monitored space, preempt incidents or detect abnormal events real-timely, enhance forensic capabilities through content based video/image retrieval. It also has a wide spectrum of promising applications, including access control in special areas, human identification at a distance, crowd flux statistics and congestion analysis, detection of anomalous behaviors, and interactive surveillance using multiple cameras, etc. [3].

Smart/Intelligent Surveillance is the use of computer vision and pattern recognition technologies to analyze information from situated sensors [1–3]. As an active research topic in computer vision, smart visual surveillance attempts to detect, recognize and track certain objects from image sequences, and more generally to understand and describe object behaviors. Recent advances in computer vision, such as deep learning, multi-modal analysis, large-scale spatio-temporal analysis, have shown great potential for some high level understanding tasks in smart video surveillance application. These include high performance human/object detection and tracking, cross camera human identification and re-identification, and action/activity/event detection. These novel techniques require large scale surveillance datasets to model of various visual understanding tasks as well as evaluation of algorithmic performances. Most intelligent visual algorithms require large-scale image/video datasets to train classifiers or learn discriminative features using machine learning. The training dataset is a strong dependency of the machine learning algorithms. However, most existing datasets are collected from non-surveillance videos, which have significant data differences compared to the practical surveillance data. In this way, many existing intelligent visual algorithms are trained on traditional datasets, and thus many of them perform not so well in the real surveillance application systems.

To this end, one large-scale and comprehensive surveillance image and video database platform, called Benchmark and Evaluation of Surveillance Task (with BEST being short), is developed to aim to highlight vision related surveillance tasks. The original images and videos in BEST were all collected using on-using surveillance cameras, and images were captured with significant scenarios, background clutter, occlusions, and viewpoint/illumination variations, which makes the dataset very challenging. Based on this newly collected multi-task multi-camera on-using surveillance databases, this benchmark aims at bringing together cutting-edge researches in the field of surveillance task aware based intelligent surveillance algorithms and applications. The datasets have been released in our website for download and usage: http://ivlab.sjtu.edu.cn/best/Data/List/Datasets.

2   Related Works

In the past decades, an increasing number of benchmarks have been proposed to push forward the performance of computer vision, e.g., ImageNet [4],

PASCAL VOC [5] for visual object classification, INRIA [6], ETH [7], Caltech [9] and KITTI [10] datasets for pedestrian detection, KTH [11] and Weizmann [12] datasets that consist of people action videos, PETS [13] and TRECVID [15] used for the object tracking, event detection, or retrievals. Two datasets are available in [16] to provide a realistic, camera-captured, diverse set of videos used for change detection. As for the Person ReID benchmark, the iLIDS-VID dataset [17] contains of 600 image sequences for 300 people in two non-overlapping camera views, and the PRID2011 dataset [18] includes 400 image sequences for 200 people from two cameras. However, most existing datasets are collected from non-surveillance videos. The comparison (Table 1) of the self-collected data and practical surveillance data show that there exist significant data differences. The properties, like object appearance, image resolution, illumination, view point, and occlusion of objects differ greatly. Despite an extensive set of ideas has been explored for intelligent surveillance applications, most existing algorithms are trained on traditional datasets. Consequently the accuracy and robustness of the most existing algorithms may perform not so well in the real surveillance systems.

Table 1. Comparison of BEST and existing datasets.

	Existing non-surveillance datasets [4,6,9,11,12]	Existing surveillance datasets [13,15–17]	BEST
Data resource (Surveillance or not)	Mostly Not (Self-sampling or Internet)	Partly Yes	Yes
Scenario (Surveillance or not)	Mostly Not	Partly Yes (Simulated)	Yes (Real)
Task (Surveillance or not)	Not	Partly Yes (Partly are simulated)	Yes (Partly are simulated)
Diversity	Yes	Not	Yes

We believe the lack of high quality surveillance-oriented datasets greatly limits the application of the computer vision in the practical surveillance domain. To this end, we collect and organize a large-scale and comprehensive surveillance image and video database platform, called Benchmark and Evaluation of Surveillance Task, with BEST being short. The original images and videos in BEST were all collected using on-using monitoring cameras, and they have been selected to cover a wide range of surveillance scenarios and are representative of typical outdoor visual data. Compared with the existing surveillance or non-surveillance datasets, the BEST dataset provide a realistic, camera-captured, diverse set of surveillance images or videos, which is much larger in scale and diversity:

Fig. 1. Different scenarios in BEST

Resource. The original images and videos in BEST were all collected using on-using surveillance cameras.

Scale. The BEST benchmark contains over 10 million original surveillance images and more than 10k surveillance video clips; and the well-labeled images have reached a scale of more than 100 thousands samples.

Diversity.

– Scenarios: The datasets contain more than 20 surveillance scenarios, such as streets, roads, highways, campus, entrances, squares, and so on (see Fig. 1).
– Resolution: Spatial resolutions of the images/videos vary from 320 × 240, 720 × 576, 1280 × 720, to 1920 × 1080.
– Illumination: We divide the illumination into several different situations: normal, foggy, rainy, cloudy, dusk, night and others.
– View: Due to the variation of camera positions, various view angles can be got in the datasets, such as front side, left side, back side and right side. The division and label of this property may help with the object detection and recognition task.
– ...

There are three datasets have been constructed and released in the BEST benchmark:

– SPID: Surveillance Pedestrian Image Dataset
– SHAD: Surveillance Human Action Dataset
– SPRD: Surveillance Person Re-Identification Dataset

These datasets will be revised/expanded from time to time based on feedback from the academia and the industry.

3   SPID: Surveillance Pedestrian Image Dataset

3.1   Description

As a subset of BEST2016 dataset, this dataset was constructed for the pedestrian detection task in surveillance images. The images in Surveillance Pedestrian

Image Dataset (SPID) are extracted from the videos recorded by daily used surveillance cameras. All the pedestrians are well labeled in several different properties. Each image contains at least one pedestrian. The datas resources satisfy the diversification benchmark including various intensities, different time periods, multiple scenarios and so on. SPID contains about 30k well-labeled images and 110k pedestrian objects. We split the dataset into training and testing sets roughly in half. The testing set contains 14550 images and the training set contains 15439 images as well. For detailed statistics about the data, see the bottom row of Table 2.

Table 2. Comparisons between proposed SPID and existing pedestrian detection datasets

Dataset	imaging setup	Testing			Training			Properties								
		# pedestrians	# pos. images	# neg. images	# pedestrians	# pos. images	# neg. images	color images	occlusion labels	view labels	illumi. labels	pose labels	video seqs.	no select. bias	view point	publication
INRIA [6]	photo	1208	614	1218	566	288	453	√							hori.	2005
ETH [7]	mobile cam.	2388	499	-	12k	1804	-	√					√	√	-	2007
Daimler [8]	mobile cam.	15.6k	-	6.7k	56.6k	21.8k	-						√		-	2009
Caltech [9]	mobile cam.	192k	67k	61k	155k	65k	56k	√	√				√	√	hori.	2009
KITTI [10]	mobile cam.	-	7518	-	4445	7481	-	√					√	√	hori.	2012
SPID	fixed cam.	53.7k	15.4k	-	56.2k	14.5k	-	√	√	√	√	√	√	√	bird	2016

The images in SPID are all collected from on-using surveillance systems. Some of them are collected from the monitoring cameras used in companies, university campus; some of them are got from public security cameras. The surveillance cameras are set along the streets in the cities and campus, or beside the squares and highways. Most cameras in cities are placed 6–10 m above the ground, the other cameras like those set at campus entrance are placed 1 m high. We gather 297 pieces of videos, totally about 61 h long. Basically each piece of the videos is about 20 min long, we extract the videos to images per second and pick up the images with at least one pedestrian to label. The resolution of the images are various from 320×240, 720×576, 1280×720, to 1920×1080. Our dataset contains both low resolution and HD images. Figure 2 shows multiple diverse examples of pedestrians in SPID.

3.2   Ground Truth Annotation

The SPID images are well-labeled and one .xml format ground truth annotation file can be got for each labeled image and bounding box pedestrian sample. The xml file contains the total visible pedestrian number of a single image and the corresponding bounding box (BB) coordinate of each pedestrian. If a pedestrian is occluded, the BB only contains the visible part. If an image contains multiple

Fig. 2. Examples of diverse pedestrians in SPID. The first four rows show the various view of pedestrians (right side, back side, left side and front side). The fifth row shows special pose pedestrians (stoop). Next row shows the pedestrians captured during night and illumination is bad. The following row shows pedestrians with various attachments. The last row shows the low resolution pedestrians.

pedestrian objects, we label them respectively, i.e. this dataset only has 'person' label and no 'group' label.

Each BB describes a specific pedestrian object. For each pedestrian sample, we also provide one annotation file with several useful properties: the name of its source image, the specific size and level, scene, view, pose, intensity condition, occlusion level, and attachment information. In addition, the top colors, bottom color, top style, bottom style of this pedestrians cloth are also labeled. The main properties are descripted as follows.

Scene. The cameras are settled at different areas in cities and campuses. The scenes of the video are diverse, which contain roads, city highways, campus roads and school entrances (see Fig. 1). Therefore the scenes of our dataset represent the typical surveillance scenes in daily life.

Illumination. The videos we collected are recorded during continuous time periods; the basic illumination of one image or sample in different time and weathers are always different: its is high during the sunny days and maybe low during the night or cloudy weather, as shown in Fig. 1. We divide the illumination into several different situations: normal, foggy, rainy, cloudy, dusk, night and other. The distribution statistics of the illumination in SPID can be seen in Fig. 3(a).

View. Due to the variation of camera positions, pedestrians show various angles of view, such as front side, left side, back side and right side. The property of view is more useful for the granular pedestrian detection due to the different appearance features with the different views: When a pedestrian is standing as the front side view, his/her face is one key appearance feature to detect the person (as many algorithms do in PASCAL). However, as the other three sides, face is occluded. Totally 9 type views are designed in SPID: front/back side, left/right side, left-front/right-front side, left-back/right-back side, and others. The distribution statistics of the view in SPID can be seen in Fig. 3(b). The division and label of this property may help to design some other features for this condition.

Occlusion. Occlusion influences the accuracy of pedestrian detection significantly [8]. Caltech dataset is the only existing pedestrian dataset that labeled this property. Unlike Caltech, we divide the occlusion levels by the ratio of occluded parts to the whole pedestrian body. Occluded ratio is grouped into 4 levels in SPID: Occ = none, Occ < 33%, Occ in [33%–66%], Occ > 66%, others.

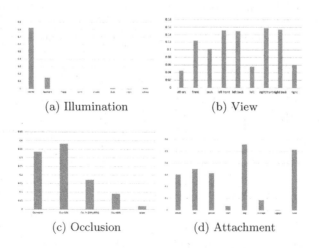

(a) Illumination   (b) View

(c) Occlusion   (d) Attachment

Fig. 3. Statistics of different properties in SPID data set.

The levels are measured by human subjectively when labeling. The distribution statistics of the occlusion in SPID can be seen in Fig. 3(c).

Attachment. In many times, one pedestrian will hold one attachment (like bag, bike, etc.) with him, and the attachment may change the contour of one pedestrian. This will result the decrease of detection performance for the algorithms trained using dataset without any pedestrian-with-attachment. In SPID, the property of with or without attachment is labeled to help for the training of pedestrian detection methods. The common attachments of pedestrians contain hat, glasses, scarf, bag, dunnage, luggage and others. The distribution statistics of the attachment in SPID can be seen in Fig. 3(d).

Appearance. The color and style of clothes are important attributes for person identification. We choose some fundamental color and style to make a rough division. Colors include 11 types such as black, white, red, yellow, and so on. The styles contain purity, horizontal stripes, vertical stripes, checks and pattern.

4 SPRD: Surveillance Person Re-Identification Dataset

4.1 Description

The SPRD (Surveillance Person Re-Identification Dataset) contains 9700 pedestrian images taken from arbitrary viewpoints from 24 real surveillance cameras, also under varying illumination conditions. Each sequence contains about 200 to 300 images from the same person. For each camera view, we record continuous human sequences to facilitate multi-shot person Re-ID. The naming rule of this dataset is as follows. Each folder contains images from a unique person. They might include different camera views. The number of images of the same person varies. The naming rule for each image is as follows. For example, image file "057_02_0001_00562.jpg" denotes camera ID 57, person ID 2, sequence number 1, and frame ID 562. We follow a multi-shot person Re-ID setting. Namely, instead of using only a single image, sequence of images can be used together to represent a person. Basically, a pair of person image sequences of the used for training should be from different viewing cameras (this can be read from the camera ID).

Compared with existing datasets for the purpose of person re-identification, such as iLIDS-VID [17] and PRID2011 [18], the proposed SPRD dataset possesses the following advantages. First, it contains more view variations than other datasets. In particular, the number of viewpoints (cameras) used for capturing one person is usually more than ten. In contrast, previous dataset only contains person images from a limited number of viewpoints, e.g., fewer than three. Second, the proposed SPRD dataset is annotated with part bounding boxes, which makes part-matching based person re-identification algorithms accessible. Last, the proposed dataset has longer sequences than most existing person re-identification datasets, which enables multi-shot based person re-identification algorithms. Detailed comparisons are given in Table 3.

Table 3. Comparisons of various person re-identification datasets.

Dataset	# Camera views	Has part annotation
iLIDS-VID [17]	2	No
PRID [18]	2	No
SPRD	10+	Yes

5 SHAD: Surveillance Human Action Dataset

5.1 Description

The task of surveillance action recognition is to recognize human actions in surveillance videos. Monitoring and understanding surveillance human actions play an important role in smart video surveillance applications. Many surveillance action recognition algorithms have been proposed in recent years; however, most of existing human action datasets used to evaluate proposed algorithms are constructed using self-collected video, not the real on-using surveillance data. Thus, in the paper we introduce the Surveillance Human Action Dataset (SHAD) which is collected from real surveillance videos. The dataset contains 290 high definition video clips recorded by 25 surveillance cameras and selected from about 47 h of video data. Six human action classes that occur frequently in real surveillance are included: walking, sitting, bending, squat, falling and cycling (Table 4). Each action is performed by large number of people and recorded with both cluttered background and varying illumination. Additionally, we provide annotations in which distinct people ID are assigned and bounding boxes are annotated to localize each people in the frame. Meanwhile, for some people the visible human key points are annotated. To the best of our knowledge, SHAD is currently the most challenging dataset for surveillance human action recognition and we hope the release of the dataset will help accelerate the research progress of the field. Comparisons between proposed SHAD and existing human action datasets is given in Table 5.

Table 4. Summarizations of labeled actions for multi-action recognition tasks.

Clip label	Clip number	Number of labeled actions						Action number
		Walking	Sitting	Bending	Squat	Falling	Cycling	
Walking	50	51	1	0	0	1	0	53
Sitting	40	57	40	5	0	0	12	114
Bending	50	49	0	50	0	0	1	100
Squat	50	96	0	1	50	0	3	150
Falling	50	35	0	8	1	50	0	94
Cycling	50	19	0	0	0	0	52	71
Total	290	307	41	64	51	51	68	582

Table 5. Comparisons between proposed SHAD and existing human action datasets.

	KTH [7]	Weizmann [11]	TRECVID [13]	VIRAT [14]	SHAD
Number of action types	6	10	10	23	6
Avg. number of samples per class	100	9	3–1670	10–1500	40–50
Resolution	160 × 120	180 × 144	720 × 576	1920 × 1080	1920 × 1080
Human height	80–100	60–70	20–200	20–180	7–536
Human to video height ration	65–85%	42–50%	4–36%	2–20%	0.65–49.6%
Number of scene	N/A	N/A	5	7	20
Natural background clutter	No	No	Yes	Yes	Yes
Bounding boxes	Cropped	Cropped	No	Yes	Yes
Labeled human ID	No	No	No	No	Yes
Keypoint annotations	No	No	No	No	Yes

5.2 Dataset Properties

Scene. The dataset contains video clips coming from 20 surveillance cameras. For each camera, it captures different scenes. The backgrounds are cluttered with campus buildings, trees, bicycles and so on. As mentioned above, the illumination also varies due to the large record duration.

Viewpoints. Due to the fact that people are not required to perform actions in restricted areas, action performers show various viewpoints, such as front side, back side, left side or right side, towards the cameras as shown in Fig. 5.

Fig. 4. Examples of pedestrians in SPRD.

Meanwhile, some surveillance cameras are close to each other, which makes the same action recorded by different cameras and captured in different viewpoints.

Scale. Because the surveillance cameras can cover wide range of areas, the heights of people in our dataset vary dramatically, which can be better visualized by Fig. 6 which depicts the distribution of people heights in our dataset (Fig. 6).

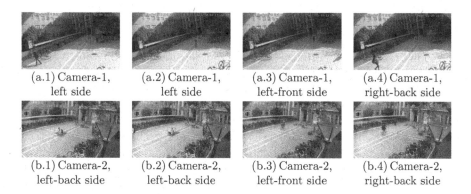

| (a.1) Camera-1, left side | (a.2) Camera-1, left side | (a.3) Camera-1, left-front side | (a.4) Camera-1, right-back side |

| (b.1) Camera-2, left-back side | (b.2) Camera-2, left-back side | (b.3) Camera-2, left-front side | (b.4) Camera-2, right-back side |

Fig. 5. An example of various viewpoints shown in two different cameras. The target human is bounded by green rectangle box. The rest two columns are examples that the same action is captured by these two cameras.

5.3   Annotations

To give benefits for SHAD dataset users, we also provide some annotations for this dataset. However, annotating a large video dataset presents a challenge on its own. For one video clip, two types of annotations are included. The first annotation is frame-level: for certain frame in one clip, bounding boxes are applied to localize all the visible people in the current frame and each people is assigned with a unique people ID. The people ID for the same people throughout the whole clip will be kept the same which make it possible to keep the trajectory of each people. The second annotation is human-level, which aims at providing a more detailed description on people himself or herself. For example, the locations of visible human key-points (like shoulder, elbow, wrist and so on) are annotated because the human pose can be well represented by the configuration of human key-points. To be specific, there are 13 key-points included: head, left shoulder, right shoulder, left elbow, right elbow, left wrist, right wrist, left hip, right hip, left knee, right knee, left ankle and right ankle.

6   Performance Evaluation of Pedestrian Detection Using SPID

With the remarkable data difference between the surveillance images and non-surveillance datasets, the performance of intelligent algorithms can vary significantly when applying them into the BEST directly. In this section, performance evaluation of 7 popular pedestrian detection methods using existing non-surveillance dataset and the surveillance pedestrian datasetSPID, is given to show the variance of detection performance.

ROC curves are the most used evaluation methodology to use compare the performance of different PD algorithms [9]. The ROC curves show the relationship between miss rate and FPPI (False Positive Per Image). In INRIA dataset only the testing positive 288 images are considered. The test set (set06–set10) of Caltech is used, we extracted images with 30 frames interval and obtained totally 4024 images. The SPID test set contains 15439 images. For each image, the ground truth bounding boxes (BBgt) and detected bounding boxes (BBdt) are loaded. The aspect ratio of all the boxes is preprocessed to 0.41 and the overlap threshold of bounding boxes is set to 0.5. Specific accuracy calculation method is the same as that in [19].

The performance comparison of six existing pedestrian detection methods using hand-crafted features: k-poselets [20], ACFCaltech+ (ACF [21] retrain using Caltech dataset), LDCFCaltech (LDCF [22] retrain using Caltech dataset), ACFInria, LDCFInria and DPMInria (DPM [23] retrain using Inria dataset) and one deep-learned-feature based method (Faster R-CNN [24].), is shown in Fig. 7. From Fig. 7, its easy to find that the detection performance decreasing significantly when applying the existing methods on SPID directly. For the performance of these method retrained using SPID dataset (DPMSpid, Faster R-CNNSpid), the retrained methods using SPID or Caltech cant work very well on surveillance images (see the DPMSpid and Faster R-CNNSpid lines in Fig. 7). In other words, due to the remarkable discrepancy between the surveillance images and non-surveillance datasets, developing of specific pedestrian detection algorithms are needed for the surveillance applications. The detail of this evaluation can be seen in [25] (Fig. 7).

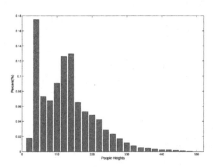

Fig. 6. The distribution of human heights SHAD. A wide range of heights is covered and the majority of heights are less than 220 pixels.

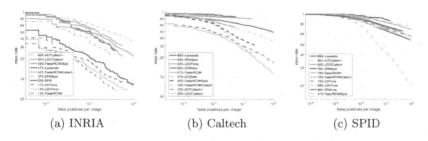

(a) INRIA   (b) Caltech   (c) SPID

Fig. 7. Evaluation results under reasonable condition on three test datasets. (a), (b) and (c) shows the pedestrian detection results on INRIA, Caltech, and SPID, respectively.

7   Conclusion

This paper introduces the BEST platform: Benchmark and Evaluation of Surveillance Task, which contains multiple task-driven datasets using videos and images captured by on-using surveillance systems. The images in BEST are well labeled, and some popular intelligence algorithms, such as pedestrian detection methods, including hand-crafted detectors and deep learning methods, are evaluated using the existing datasets (such as INRIA, Caltech) and the proposed BEST dataset SPID. Evaluation results show that the data differences, such as scale, view, illumination and occlusion between existing public datasets and SPID have large impact on the detection performances.

Throwing new light on existing datasets, we hope that the proposed benchmarks will complement the gap. This benchmark encapsulates a rigorous and comprehensive academic benchmarking effort for testing and evaluation existing and new algorithms for surveillance tasks. It will be revised/expanded from time to time based on received feedback, and will maintain a comprehensive ranking of submitted methods for years to come.

Acknowledgement. This work was partly funded by NSFC (No. 61571297, No. 61371146, No. 61527804, 61521062), 111 Program (B07022), and China National Key Technology R&D Program (No. 2012BAH07B01). The authors also thank the following organizations for their surveillance data supports: SEIEE of Shanghai Jiao Tong University, The Third Research Institute of Ministry of Public Security, Tianjin Tiandy Digital Technology Co., Shanghai Jian Qiao University, and Qingpu Branch of Shanghai Public Security Bureau.

References

1. Shu, C.F., Hampapur, A., Lu, M., Brown, L., Connell, J., Senior, A., Tian, Y.: IBM smart surveillance system (s3): a open and extensible framework for event based surveillance. In: IEEE Conference on Advanced Video and Signal Based Surveillance, pp. 318–323 (2005)

2. Hampapur, A., Brown, L., Connell, J., Pankanti, S.: Smart surveillance: applications, technologies and implications. In: Joint Conference of the Fourth International Conference on Information, Communications and Signal Processing, 2003 and the Fourth Pacific Rim Conference on Multimedia, pp. 1133–1138 (2004)
3. Hu, W., Tieniu, T., Wang, L., Maybank, S.: A survey on visual surveillance of object motion and behaviors. IEEE Trans. Syst. Man Cybern. Part C: Appl. Rev. **34**(3), 334–352 (2004)
4. Deng, J., Dong, W., Socher, R., Li, L.J., Li, K., Li, F.F.: ImageNet: a large-scale hierarchical image database, pp. 248–255 (2009)
5. Everingham, M., Van Gool, L., Williams, C.K.I., Winn, J., Zisserman, A.: The PASCAL visual object classes (VOC) challenge. Int. J. Comput. Vis. **88**(2), 303–338 (2010)
6. Dalal, N., Triggs, B.: Histograms of oriented gradients for human detection. In: IEEE Conference on Computer Vision and Pattern Recognition, pp. 886–893 (2005)
7. Ess, A., Leibe, B., Van Gool, L.: Depth and appearance for mobile scene analysis. In: IEEE International Conference on Computer Vision, pp. 1–8 (2007)
8. Enzweiler, M., Gavrila, D.M.: Monocular pedestrian detection: survey and experiments. IEEE Trans. Pattern Anal. Mach. Intell. **31**(12), 2179–2195 (2009)
9. Dollar, P., Wojek, C., Schiele, B., Perona, P.: Pedestrian detection: a benchmark. In: IEEE Conference on Computer Vision and Pattern Recognition, pp. 304–311 (2009)
10. Geiger, A., Lenz, P., Urtasun, R.: Are we ready for autonomous driving? The kitti vision benchmark suite. In: IEEE Conference on Computer Vision and Pattern Recognition, pp. 3354–3361 (2012)
11. Schuldt, C., Laptev, I., Caputo, B.: Recognizing human actions: a local SVM approach. In: 17th International Conference on Proceedings of the Pattern Recognition, (ICPR 2004), vol. 3, pp. 32–36 (2004)
12. Blank, M., Gorelick, L., Shechtman, E., Irani, M., Basri, R.: Action as space-time shapes. IEEE Trans. Pattern Anal. Mach. Intell. **29**(12), 1395–1402 (2005)
13. http://www.cvg.reading.ac.uk/PETS2016/a.html/
14. Oh, S., Hoogs, A., Perera, A., Cuntoor, N.: A large-scale benchmark dataset for event recognition in surveillance video. In: Proceedings of IEEE Computer Vision and Pattern Recognition, pp. 3153–3160 (2011)
15. Over, P., Awad, G.M., Fiscus, J.G., Antonishek, B., Michel, M., Kraaij, W., Smeaton, A.F., Qunot, G.: TRECVID 2015 an overview of the goals, tasks, data, evaluation mechanisms and metrics. In: Proceedings of TRECVID 2015. NIST, USA (2015)
16. http://www.changedetection.net/
17. Wang, T., Gong, S., Zhu, X., Wang, S.: Person re-identification by video ranking. In: Fleet, D., Pajdla, T., Schiele, B., Tuytelaars, T. (eds.) ECCV 2014. LNCS, vol. 8692, pp. 688–703. Springer, Heidelberg (2014). doi:10.1007/978-3-319-10593-2_45
18. Hirzer, M., Beleznai, C., Roth, P.M., Bischof, H.: Person re-identification by descriptive and discriminative classification. In: Scandinavian Conference on Image Analysis, pp. 91–102 (2011)
19. Dollr, P., Wojek, C., Schiele, B., Perona, P.: Pedestrian detection: an evaluation of the state of the art. IEEE Trans. Pattern Anal. Mach. Intell. **34**(4), 743–761 (2012)
20. Gkioxari, G., Hariharan, B., Girshick, R., Malik, J.: Using k-poselets for detecting people and localizing their keypoints. In: IEEE Conference on Computer Vision and Pattern Recognition, pp. 3582–3589 (2014)

21. Dollar, P., Appel, R., Belongie, S., Perona, P.: Fast feature pyramids for object detection. IEEE Trans. Pattern Anal. Mach. Intell. **36**(8), 1532–1545 (2014)
22. Nam, W., Dollr, P., Han, J.H.: Local decorrelation for improved detection. Adv. Neural Inf. Process. Syst. **1**, 424–432 (2014)
23. Felzenszwalb, P.F., Girshick, R.B., Mcallester, D., Ramanan, D.: Object detection with discriminatively trained part-based models. IEEE Trans. Softw. Eng. **32**(9), 1627–1645 (2014)
24. Ren, S., He, K., Girshick, R., Sun, J.: Faster R-CNN: towards real-time object detection with region proposal networks. In: Advances in Neural Information Processing Systems (2015)
25. Wang, D., Zhang, C., Cheng, H., Shang, Y., Mei, L.: SPID: surveillance pedestrian image dataset and performance evaluation for pedestrian detection. In: 13th Asian Conference on Computer Vision Workshop on Benchmark and Evaluation of Surveillance Task (2016)

Multiple-Shot Person Re-identification via Riemannian Discriminative Learning

Yuheng Lu[1,2], Ruiping Wang[1,2,3(✉)], Shiguang Shan[1,2,3], and Xilin Chen[1,2,3]

[1] Key Laboratory of Intelligent Information Processing of Chinese
Academy of Sciences (CAS), Institute of Computing Technology, CAS,
Beijing 100190, China
yuheng.lu@vipl.ict.ac.cn, {wangruiping,sgshan,xlchen}@ict.ac.cn
[2] University of Chinese Academy of Sciences, Beijing 100049, China
[3] Cooperative Medianet Innovation Center, Beijing, China

Abstract. This paper presents a Riemannian discriminative learning framework for multiple-shot person re-identification. Firstly, image regions are encoded into covariance matrices or a Gaussian extension as robust feature descriptors. Since these matrices lie on some specific Riemannian manifolds, we introduce a manifold averaging strategy to fuse the feature descriptors from multiple images for a holistic representation, and exploit Riemannian kernels to implicitly map the averaged matrices to a Reproducing Kernel Hilbert Space (RKHS), where conventional discriminative learning algorithms can be conducted. In particular, we apply kernel variants of two typical methods, i.e., the Linear Discriminant Analysis (LDA) and Metric Learning to Rank (MLR), to demonstrate the flexibility of the framework. Extensive experiments on five public datasets exhibit impressive improvements over existing multiple-shot re-identification methods as well as representative single-shot approaches.

1 Introduction

Person re-identification, a task of recognising pedestrian appearance in different time and locations captured with a multi-camera network without field of view overlap, has attracted wide interest in the field of surveillance. Applications in security, medical guardianship, tracking, and even online image retrieval on clothes [1] demonstrate the great while growing significance of the problem.

Current practices on person re-identification have mainly concentrated on two stages, extracting distinctive while stable features [2–9], and/or learning discriminative cross-view metrics [10–16]. Due to the enormous challenges in the task including vast variations in (i) pedestrian viewpoint and pose, (ii) environment illumination and occlusion, and (iii) camera position, configuration and resolution, person re-identification still remains an open problem.

Among all the literatures endeavouring to tackle this problem, most focus on the single-shot scenario [10,12,14–30], which refers to querying one single image at a time in a gallery constituted by single images, where one identity

© Springer International Publishing AG 2017
C.-S. Chen et al. (Eds.): ACCV 2016 Workshops, Part III, LNCS 10118, pp. 408–425, 2017.
DOI: 10.1007/978-3-319-54526-4_30

is presented by usually one image (*Single-vs-Single*, or *SvS*), or *independent* multiple images (*Single-vs-Multiple*, or *SvM*), as shown in Fig. 1(a,b). Recently, approaches based on the multi-shot scenario [2–9,11,13,31–39], where multiple images of the same person captured in the same camera are grouped as the probe to match the gallery formed also by multiple-image *groups* (*Multiple-vs-Multiple*, or *MvM*), have started to blossom (Fig. 1(c)). Owing to the practical accessibility and wider variation coverage of the extra information, which usually brings higher and more robust performance, the MvM case is addressed in this paper. According to different matching schemes, we categorise multi-shot person re-identification methods into four classes: (i) closest point-based approaches [3,4] where a pair of closest points are exhaustively searched for to calculate the distance between each group pair; (ii) score voting-based methods [31,32] that average all similarity scores of all image pairs of two groups as the group similarity; (iii) set structure-based approaches [11,34] which model the distribution structure of each image group (or set) and measure the similarity of the set models for matching; and (iv) signature-based ones [2,6,9] that generate a signature for each group to facilitate subsequent matching, converting the problem into the single-shot case.

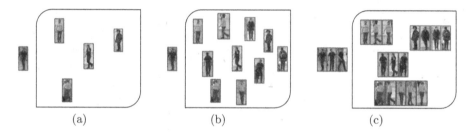

(a)   (b)   (c)

Fig. 1. Three cases of person re-identification matching scenario: (a) Single-vs-Single (SvS), (b) Single-vs-Multi (SvM) and (c) Multi-vs-Multi (MvM). Images outside and inside the boxes are the probe and gallery respectively. Images corresponding to different identities are bordered with different colors. (Color figure online)

Considering that signature generating is not only flexible in feature modeling and concise in multi-frame encapsulation, but also naturally compatible with a wide variety of subsequent processes applicable in the well-studied single-shot cases, we propose a signature-based approach, following the majority [2,5–7,9,13,35–39] of the multi-shot community. Specifically, having noticed the encouraging performance of region covariance matrix (a.k.a. Symmetric Positive Definite (SPD) matrix) [6,7,28,30,40], we adopt it as our feature representation of images and further extend it into a pixel-oriented local Gaussian descriptor, which is also written in the form of SPD matrix under the framework of information geometry [41,42]. Considering the fact that both the region covariance matrix and the Gaussian extension lie on the SPD Riemannian manifold, conventional operations that work in Euclidean space are not directly applicable.

Hence, we utilize Riemannian metrics to integrate the feature descriptors (i.e. SPD matrices) of multiple images and measure their similarities, and further exploit the corresponding Riemannian kernel functions to map points on Riemannian manifold to a high-dimensional Hilbert space for performing discriminative learning. We name our approach Multi-shot Riemannian Discriminative Learning (MRDL), and show the flowchart in Fig. 2.

MRDL has the following three main advantages. (i) As a multi-shot approach, it models signatures to fuse rich information from multiple images based on robust feature descriptors, i.e., the region covariance and Gaussian matrix which characterize correlations between feature dimensions. (ii) MRDL performs discriminative learning on the manifold, strengthening the discrimination power of the approach compared to other methods that conduct unsupervised matching, and retaining the geometric structure of the manifold space by operating on the SPD Riemannian manifold. (iii) What presented in this paper is a framework with remarkable extensibility. Along with abundant types of features applicable in the front end and various manifold distance metrics replaceable in the middle part, the rear end is also an arena for the performance of quantities of discriminative learning algorithms.

The rest of this paper is organised as follows. Related works of both the overall person re-identification field and the multi-shot branch are introduced in Sect. 2. Section 3 presents the region covariance and Gaussian descriptors we employed. Then in Sect. 4, we deliver the proposed method MRDL, including the kernel derivation of two typical kinds of discriminative learning algorithms: linear discriminant analysis and metric learning. Subsequently, extensive experiments are conducted on five benchmark datasets in Sect. 5. Lastly, we summarise our work and discuss possible future research directions in Sect. 6.

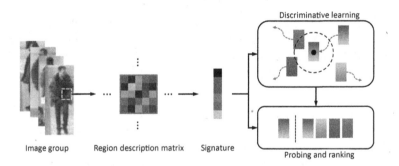

Fig. 2. Workflow of the proposed method. We firstly compute the region description matrices (covariance or Gaussian) of each image, and those from the same region of different images are fused to form the signature for the image group. Afterwards, group descriptors of training samples are processed in the learning phase to find a discriminative subspace for classification. In the testing phase, signature of a probe image group is projected into the learned space and matched with the gallery samples.

2   Related Works

In the early years, discriminative and robust feature extraction [3,4,8,9,18–20,22] was a dominating research topic in the area of person re-identification. Quite a few methods sought for representative and robust features with sophisticated designs, such as spatio shape and appearance context modeling [3], human body symmetry and asymmetry based local feature accumulation [4], fisher vector encoding [8], and local maximal occurrence representation [18]. Meanwhile, supervised learning was also adopted to guide extracting more discriminative features, including support vector machine (SVM) based ranking [19], weak classifiers boosting [20], deep neural network training [22], and bag-of-words based patch description [9]. On the other hand, learning discriminative metrics [10,11,13] has rapidly grown to share the person re-identification empire in recent years. Zheng $et\ al.$ [10] use triplets to learn distance metric. Wu $et\ al.$ [11] adopt a structural SVM based algorithm to rank image sets. Pedagadi $et\ al.$ [13] apply locality preserving strategy to Fisher discriminant analysis.

Practically, one can usually obtain more than one frame from a video track for each person. Owing to the additional information compared with single images, multiple frames alleviate the disadvantages brought by variations in viewpoint, pose and occlusion. As mentioned in the previous section, we categorise approaches specifically designed for or applicable to multi-shot scenario into four classes: closest point-based, score voting-based, set structure-based and signature-based approaches.

The first class of **closest point**-based methods [3,4,8] select the minimum distance of all sample pairs between two image sets as the set distance. On the contrary, works on **score voting** usually average similarities of all possible sample pairs between the sets [31,32], or reconstruction residuals of each probe sample from the gallery group [33], to get the set distance. For **set structure** modeling, Wu $et\ al.$ [11,34] formulate image sets as affine hulls, and then conduct discriminative metric learning [11] or locality constrained collaborative representation with l-2 regularization [34] on the hulls. Finally, most of the large number of **signature**-based approaches compute the mean in the vectorized feature space [5,9,13,35–37] or on the SPD manifold of region covariance matrices [6,7], while the others model spatio-temporal appearance [2,38] or body action [39] to generate a holistic representation for each image group. Among them, mean Riemannian covariance grid (MRCG) [6] is the most related one to our proposed method. However, MRCG is unsupervised, while ours performs discriminative learning on the Riemannian manifold, introducing great discrimination power to the learned space, which will be detailed in the next section.

3   Region Covariance and Gaussian Descriptor

Suppose we have a dataset with I identities, the i-th of which includes T_i image groups. The t-th group is composed of $F_{i,t}$ image frames, and each image is spatially partitioned into S local regions for more precise modeling since different body parts may show sharply contrasted appearances. Tuzel *et al.* [40] proposed the region covariance matrix as a fast and robust descriptor, which aggregates the covariances of all pixel-level feature pairs inside a spatial region. Let $\mathbf{P} = [\mathbf{p}_1, \mathbf{p}_2, ..., \mathbf{p}_n]$ be the data matrix of an image region with n pixels, where $\mathbf{p}_k \in \mathbb{R}^d$ denotes the d-dimensional feature descriptor of the k-th pixel, the region is encoded by the $d \times d$ covariance matrix:

$$\mathbf{C} = \frac{1}{n-1} \sum_{k=1}^{n} (\mathbf{p}_k - \bar{\mathbf{p}})(\mathbf{p}_k - \bar{\mathbf{p}})^T, \tag{1}$$

where $\bar{\mathbf{p}}$ is the mean of $\mathbf{p}_k (k = 1, ..., n)$.

Since the SPD covariance matrix $\mathbf{C}$ captures the second-order statistics of the features, the first-order statistics, feature mean, is lost. Considering that $\mathbf{C}$, which encodes the pixel feature variation pattern, and $\bar{\mathbf{p}}$ that captures the general position of the features in the original space are complementary, we embed them in a more informative Gaussian representation: $(\bar{\mathbf{p}}, \mathbf{C})$. Specifically, the mean vector $\bar{\mathbf{p}}$ and $d \times d$ covariance matrix $\mathbf{C}$ are mapped together into the space of $(d + 1) \times (d + 1)$ SPD matrices. Under the framework of information geometry [41,42], the embedding is fulfilled through two mappings, from affine transformation $(\bar{\mathbf{p}}, \mathbf{C}^{1/2})$ to a simple Lie group, and subsequently to the SPD matrix space:

$$\mathbf{G} = |\mathbf{C}|^{-\frac{1}{d+1}} \begin{pmatrix} \mathbf{C} + \bar{\mathbf{p}}\bar{\mathbf{p}}^T & \bar{\mathbf{p}} \\ \bar{\mathbf{p}}^T & 1 \end{pmatrix}. \tag{2}$$

We name $\mathbf{G}$ as the region Gaussian matrix. Considering local appearances of the same person usually shift horizontally in different cameras, each pedestrian image is divided into six horizontal *stripes* equally, as in [10,37]. For the s-th $(s = 1, ..., 6)$ stripe, a region Gaussian matrix $\mathbf{G}^s$ is calculated through Eq. 2.

Then, for the t-th image group of the i-th identity, the region descriptor of the s-th stripe of the f-th frame is denoted as $\mathbf{G}_{i,t}^{f,s}$, as shown in Fig. 3. Considering averaging is the most widely applied strategy [5–7,9,13,35–37] in the multi-shot community, the SPD matrices of multiple images in the same area are averaged to generate the image group signature $S_{i,t} = \{\overline{\mathbf{G}}_{i,t}^s, s = 1, ..., 6\}$, which will be explained in detail later in Sect. 4.1.

Similarly, MRCG computes region covariance matrices for densely sampled patches on each image. Afterwards, a covariance grid is produced for each image group as signature for direct patch-level matching. The main difference between MRCG, an unsupervised robust feature extraction method, and our MRDL is that we design a way to perform *discriminative learning* on the Riemannian manifold with each object represented by *multiple* covariance matrices. Besides, instead of on a dense patch grid, we extract region covariances on stripes, which not only takes severe horizontal appearance shifts caused by viewpoint and pose changes into consideration, but also greatly reduces model complexity. Furthermore, we merge feature mean with the covariance matrix, resulting in a more powerful local feature representation, the region Gaussian matrix. Extensive experiments in Sect. 5 exhibit impressive improvements in our approach.

4 Multi-shot Riemannian Discriminative Learning

4.1 Symmetric Positive Definite Matrix Manifold

Residing on the symmetric positive definite matrix manifold (SPD matrix manifold), which is a special Riemannian manifold, the covariance/Gaussian matrix distinguishes itself from the common feature descriptors in Euclidean space with the manifold characterising non-linear data distributions.

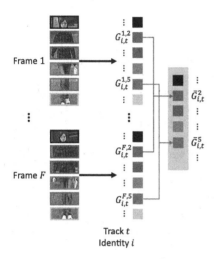

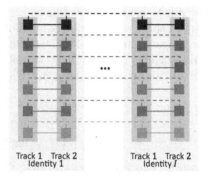

Fig. 3. Image group signature generation. Firstly, the SPD matrices (colored square) of each image stripe are computed, then those of the same region from all images in a group are averaged to generate the signature. Best viewed in color.

Fig. 4. Graph embedding technique in the learning phase. Positive pairs are constructed with the *same-position* matrices of *same*-identity image tracks (solid lines); negative pairs are formed also with the *same-position* matrices but of *different*-identities tracks (dash lines connected *only*). Best viewed in color.

Two different distance metrics are widely applied in Riemannian geometry: the log-Euclidean distance (LED) [43] and affine-invariant distance (AID) [44]:

$$d_{LED}(\mathbf{C}_1, \mathbf{C}_2) = \left\| \log(\mathbf{C}_1) - \log(\mathbf{C}_2) \right\|_F \tag{3}$$

$$d_{AID}(\mathbf{C}_1, \mathbf{C}_2) = \sqrt{\sum_{k=1}^{d} \ln^2 \lambda_k(\mathbf{C}_1, \mathbf{C}_2)}, \tag{4}$$

where $\log(\mathbf{C})$ denotes the matrix logarithm operation, $\|\boldsymbol{\cdot}\|_F$ denotes the Frobenius norm, and $\lambda_k(\mathbf{C}_1, \mathbf{C}_2)$ $(k = 1, ..., d)$ are the generalized eigenvalues of $\mathbf{C}_1, \mathbf{C}_2$. Note that in this section, to keep simplicity, $\mathbf{C}$ can denote either covariance matrix (Eq. 1) or Gaussian matrix (Eq. 2).

With the metrics above, one can find the mean SPD matrix on a manifold $\mathcal{M}$ with m samples by using the Karcher or Fréchet mean, as in [6]:

$$\overline{\mathbf{C}} = \arg\min_{\mathbf{C} \in \mathcal{M}} \sum_{k=1}^{m} d^2(\mathbf{C}, \mathbf{C}^k), \tag{5}$$

where d can be either Eq. 3 or Eq. 4. We utilize this averaging method to generate signatures in the previous section.

4.2   Kernel Linear Discriminant Analysis

Although the SPD matrices lie on the manifold, to fully activate the discriminative power beneath the samples, we can also extend the kernel algorithms in Euclidean space to SPD matrix manifold with appropriate kernel functions. Here we investigate three Riemannian kernel functions derived from LED and AID, the log-Euclidean trace (LET) kernel [45], the log-Euclidean Gaussian (LEG) kernel [46] and the affine-invariant Gaussian (AIG) kernel [46]:

$$k_{LET}(\mathbf{C}_1, \mathbf{C}_2) \quad = \mathrm{tr}[\log(\mathbf{C}_1) \cdot \log(\mathbf{C}_2)] \tag{6}$$

$$k_{LEG}(\mathbf{C}_1, \mathbf{C}_2) = \exp(-d_{LED}^2(\mathbf{C}_1, \mathbf{C}_2)/2\sigma^2) \tag{7}$$

$$k_{AIG}(\mathbf{C}_1, \mathbf{C}_2) = \exp(-d_{AID}^2(\mathbf{C}_1, \mathbf{C}_2)/2\sigma^2). \tag{8}$$

where σ is the Gaussian bandwidth. Note that AIG is not strictly positive definite. But since AID is a true geodesic distance and AIG could yield good results, it is compared in our experiments.

Let $k(\mathbf{C}_1, \mathbf{C}_2) = \langle \phi(\mathbf{C}_1), \phi(\mathbf{C}_2) \rangle$ denote a kernel function where $\phi(\cdot)$ maps points on $\mathcal{M}$ to a high dimensional Reproducing Kernel Hilbert Space (RKHS) $\mathcal{H}$ for richer representations and inner products. For the three kernels mentioned above, LET performs explicit mapping while LEG and AIG map implicitly. Suppose we have N training samples on $\mathcal{M}$. We first perform kernel linear discriminant analysis (KLDA) by solving the following optimization similar to [47]:

$$\alpha^* = \arg\max_{\alpha} \frac{\alpha^T \mathbf{K} \mathbf{L}_B \mathbf{K} \alpha}{\alpha^T \mathbf{K} \mathbf{L}_W \mathbf{K} \alpha}, \tag{9}$$

where $\mathbf{K}$ is the kernel Gram matrix: $\mathbf{K}_{jl} = k(\mathbf{C}_j, \mathbf{C}_l)$ $(j, l = 1, ..., N)$, and $\mathbf{L}_B$ and $\mathbf{L}_W$ are the Laplacian matrices of the between-class and within-class graph matrices $\mathbf{E}_B$, $\mathbf{E}_W$ respectively:

$$\mathbf{E}_{B,jl} = \begin{cases} 1/m_{jl}, & \text{if } \mathbf{C}_j \text{ and } \mathbf{C}_l \text{ are in the negative sets of each other} \\ 0, & \text{else} \end{cases} \tag{10}$$

$$\mathbf{E}_{W,jl} = \begin{cases} 1/n_{jl}, & \text{if } \mathbf{C}_j \text{ and } \mathbf{C}_l \text{ are in the positive sets of each other} \\ 0, & \text{else} \end{cases} \tag{11}$$

and m_{jl}, n_{jl} here indicate the corresponding set sizes. For each sample, a positive set and a negative set are formed by all same-class samples and different-class samples respectively. The optimization objective $\boldsymbol{\alpha}$ defines a discriminative projection direction in the RKHS space: $\mathbf{w} = \sum_{k=1}^{N} \boldsymbol{\alpha}_k \phi(\mathbf{C}_k)$.

In order to focus on close sample pairs that should be paid more attention to during the classification, inspired by [13], we take advantage of the locality structure [48] of samples by utilizing the affinity matrix $\mathbf{A}$, which is obtained through a local scaling method [49]: assign the distance with the q-th nearest neighbor as the distance scaling factor θ_k for each point k:

$$\mathbf{A}_{jl} = \exp\left(-\frac{d^2(\mathbf{C}_j, \mathbf{C}_l)}{\theta_j \theta_l} \right). \tag{12}$$

Here q can be set proportional to N. Instead of applying the affinity matrix on both $\mathbf{E}_B, \mathbf{E}_W$, we perform neighbor emphasizing by only penalizing close negative pairs since positive sample pairs are way less in number and all expected to be drawn near: $\mathbf{E}'_B = \mathbf{E}_B \cdot \mathbf{A}$. The affinity $\mathbf{A}$ functions as graph weight here.

While, as described above, a signature of an image group contains 6 independent SPD matrices. To consider matrices semantically irrelevant in body parts separately, we relate only the ones of the same stripe by introducing a graph embedding technique [50] to generate the positive/negative sets, as illustrated in Fig. 4. For any local matrix, a positive pair is only constructed with the *same-position* matrix from other image groups (tracks) with the *same* identity (solid lines connected), while negative pairs are formed with also the *same-position* matrices of other tracks with *different* identities (dash lines connected only). As another way to look at the embedding technique, the 6 stripes are treated as 6 different classes, making a training task with I identities has $c = 6I$ classes.

The problem in Eq. 9 can be tackled by solving the generalized eigenvalue problem: $\mathbf{K}\mathbf{L}_B\mathbf{K}\boldsymbol{\alpha} = \lambda\mathbf{K}\mathbf{L}_W\mathbf{K}\boldsymbol{\alpha}$. Once the $(c-1)$ leading eigenvectors $\boldsymbol{\alpha}_1, ..., \boldsymbol{\alpha}_{c-1}$ are obtained, the coefficient matrix of $(c-1)$ projection directions is naturally settled: $\mathcal{A} = [\boldsymbol{\alpha}_1, ..., \boldsymbol{\alpha}_{c-1}] \in \mathbb{R}^{N \times (c-1)}$.

In the testing phase, the SPD matrices in a signature $S_{test} = \{\mathbf{C}_{test}^s, s = 1, ..., 6\}$ are projected into the same $(c-1)$-dimensional subspace through $\mathcal{A}^T\mathbf{K}_s$, where $\mathbf{K}_s = [k(\mathbf{C}_{train}^1, \mathbf{C}_{test}^s), ..., k(\mathbf{C}_{train}^N, \mathbf{C}_{test}^s)]^T$, and matched with points of *corresponding* spatial position. In the end, the distance between two signatures $S_{i1,t1}, S_{i2,t2}$ are obtained by averaging the 6 SPD matrix pair distances:

$$d(S_{i1,t1}, S_{i2,t2}) = \frac{1}{6} \sum_{s=1}^{6} d(\mathbf{C}_{i1,t1}^s, \mathbf{C}_{i2,t2}^s). \tag{13}$$

4.3   Kernel Metric Learning to Rank

To further exhibit the flexibility of our framework, we reformulate a metric learning algorithm to learn another discriminative subspace. Considering that person re-identification is usually formulated as a ranking problem [9,11,16,19,38], the metric learning to rank [51] algorithm, which is specifically designed for ranking and also applied in [11], is adopted with the kernel variant. We refer to it as kernel metric learning to rank (KMLR). It learns a metric matrix with respect to which, the ranking list of gallery samples for each probe resembles the corresponding ground truth list as much as possible.

Given two points in the RKHS space ϕ_j, ϕ_l and any metric matrix $\mathbf{M} = \boldsymbol{\Phi}\mathcal{A}\mathcal{A}^T\boldsymbol{\Phi}^T = \boldsymbol{\Phi}\mathbf{W}\boldsymbol{\Phi}^T$ where $\boldsymbol{\Phi} = [\phi_1, ..., \phi_N]$ and $\mathbf{W} = \mathcal{A}\mathcal{A}^T$, distance between them is represented with Frobenius inner products:

$$\begin{aligned}
\|\phi_j - \phi_l\|_{\mathbf{M}}^2 &= (\phi_j - \phi_l)^T\mathbf{M}(\phi_j - \phi_l) = (\mathbf{K}_j - \mathbf{K}_l)^T\mathbf{W}(\mathbf{K}_j - \mathbf{K}_l) \\
&= \mathrm{tr}(\mathbf{W}(\mathbf{K}_j - \mathbf{K}_l)(\mathbf{K}_j - \mathbf{K}_l)^T) = \langle\mathbf{W}, (\mathbf{K}_j - \mathbf{K}_l)(\mathbf{K}_j - \mathbf{K}_l)^T\rangle_F,
\end{aligned} \tag{14}$$

where $\mathbf{K}_. = [\langle\phi_1, \phi_.\rangle, ..., \langle\phi_N, \phi_.\rangle]^T$.

By defining the opposite of the latter part as a kernel feature map for sample pair (ϕ_j, ϕ_l): $\varphi_{j,l} = -(\mathbf{K}_j - \mathbf{K}_l)(\mathbf{K}_j - \mathbf{K}_l)^T$, we use the partial order feature [52] to present a ranking list $\mathbf{y}_k$ with respect to a sample ϕ_k:

$$\psi_k(\mathbf{y}_k) = \sum_{j\in\mathcal{X}_k^+}\sum_{l\in\mathcal{X}_k^-} y_k^{jl}\left(\frac{\varphi_{k,j} - \varphi_{k,l}}{|\mathcal{X}_k^+|\cdot|\mathcal{X}_k^-|}\right), \tag{15}$$

where $\mathcal{X}_k^+$ and $\mathcal{X}_k^-$ are the positive and negative sample set w.r.t. ϕ_k, $|\cdot|$ denotes the set size, and

$$y_k^{jl} = \begin{cases} +1 \text{ if } \phi_j \text{ ranks prior to } \phi_l \\ -1 \text{ if } \phi_j \text{ ranks posterior to } \phi_l \end{cases}. \tag{16}$$

Then with a ground truth ranking $\mathbf{y}_k^*$ for ϕ_k, the optimization objective is formed by introducing the structural SVM framework:

$$\mathbf{W}^* = \arg\min_{\mathbf{W}}\{\mathrm{tr}(\mathbf{W}) + \beta\cdot\xi\},$$

$$\text{s.t. } \frac{1}{N}\sum_{k=1}^{N}(\underbrace{\langle\mathbf{W}, \psi_k(\mathbf{y}_k^*)\rangle_F}_{(i)} - \underbrace{\langle\mathbf{W}, \psi_k(\mathbf{y}_k)\rangle_F}_{(ii)}) \geq \frac{1}{N}\sum_{k=1}^{N}\underbrace{\Delta(\mathbf{y}_k^*, \mathbf{y}_k)}_{(iii)} - \underbrace{\xi}_{(iv)}, \tag{17}$$

$$\forall\mathbf{y}_k \neq \mathbf{y}_k^*, \ \mathbf{W} \succeq 0, \ \xi \geq 0$$

where ξ is the slack variable and β is the trade-off constant. $\Delta(\mathbf{y}_k^*, \mathbf{y}_k)$, the loss function of $\mathbf{y}_k$ w.r.t. $\mathbf{y}_k^*$, is set to the difference of AUC (area under the ROC curve) scores and serves as the SVM margin. Intuitively, the expression Eq. 17 finds an optimal coefficient matrix $\mathbf{W}$ by which after projected from the original RKHS space, the actual rankings of every sample to all relevant samples obtained simply by Euclidean distances resemble the ground truth rankings close enough (term (i)), closer than all the other possible rankings (term (ii)) with margins (term (iii)), relaxable with a slack variable (term (iv)). It can be solved by the cutting-plane algorithm [53] in an iterative manner.

The same graph embedding and matching scheme in Sect. 4.2 are applied here.

5 Experiments

5.1 Feature Representation

As introduced before, our MRDL is flexible with pixel-level feature choices. However, to make better comparisons, we applied the most widely used [10,11,19,20,31,33–35,37] color and texture mixture, RGB+YUV+HS with Schmid and Gabor filters (RGB+YUV+HS+SG). As the same configurations in the literature, the 13-channel Schmid filters have parameters τ and σ set to (2,1), (4,1), (4,2), (6,1), (6,2), (6,3), (8,1), (8,2), (8,3), (10,1), (10,2), (10,3) and (10,4) respectively. Also, the 16-channel Gabor filters use parameters γ, λ, θ and σ^2 set to (0.3,4,0,2), (0.3,8,0,2), (0.4,4,0,1), (0.4,8,0,1), (0.3,4,$\pi/2$,2), (0.3,8,$\pi/2$,2), (0.4,4,$\pi/2$,1), (0.4,8,$\pi/2$,1) respectively, each producing a magnitude and phase as responses. In addition, the pixel spatial position (x, y) is appended. Together with the 8 color channels, each pixel is presented by a 39-dimensional feature vector. Thus, the region covariance matrix is of size 39×39 and the region Gaussian matrix 40×40. Meanwhile, we compared the feature with two other ones, RGB+Gradient used in [6] and RGB+YUV+HS+LBP [14], in the experiments.

5.2 Datasets and Evaluation Protocols

We conducted experiments on five benchmark person re-identification datasets: i-LIDS [27], CAVIAR (CAVIAR4REID) [23], PRID2011 [24], CUHK01 [25] and SPRD [54]. The famous **i-LIDS** is constructed with 119 identities with 476 image frames from 2 cameras in an airport arrival hall. Each identity is represented by 2 to 8 (mostly 4) frames of normalized size 64×128 pixels. This dataset is characterized by large illumination changes and severe occlusions, as shown in Fig. 5(a). **CAVIAR** (Fig. 5(b)), a classic multi-shot dataset, consists of 72 identities (50 overlapping ones) in 2 cameras in a shopping center, 10 frames for each person in each view of size varying from 17×39 to 72×144. The images were selected manually to maximize variations in resolution changes, light conditions, occlusions and pose changes, making CAVIAR much more difficult to conquer. As a video-based dataset, **PRID2011** is recorded by 2 cameras outside a building with 385 identities in view A and 749 in view B, and we only make use of the 200 identities appear in both views. The images in each video track vary from 5 frames to hundreds with a unified size of 64×128. Figure 5(c) demonstrates the viewpoint change and stark difference in illumination, background and camera characteristics. Despite of the frame number, **CUHK01** is a pretty large set of 971 identities, 2 views and 2 frames each (60×160) captured in a campus. Images in camera A are all of front and back view, and those in camera B are of lateral view (Fig. 5(d)). As a recently released multi-shot dataset, **SPRD** (Fig. 5(e)) contains image sequences of 37 identities taken from 24 real surveillance cameras. This dataset undergoes huge variations in track and frame number, image size, pose, view, illumination, occlusion, background, even within

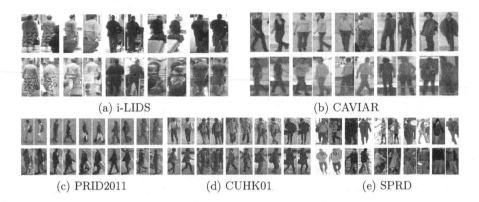

(a) i-LIDS   (b) CAVIAR

(c) PRID2011   (d) CUHK01   (e) SPRD

Fig. 5. Sample images of five introduced person re-identification datasets: (a) i-LIDS, (b) CAVIAR, (c) PRID2011, (d) CUHK01, (e) SPRD. In each dataset except SPRD, the first row presents images from camera A and the second from camera B. For SPRD, images from two random cameras are shown for each person. Each identity in each view is exampled by two frames.

the same sequence. The reason why VIPeR [26], the most well-known dataset, is not adopted is that it is single-shot-based with only 1 frame per view.

As to evaluation protocols, we split the identities of all datasets into equally sized training and testing sets, except for SPRD which had been divided into 3 sessions and was evaluated in 3-fold cross validation[1] Specifically, the training identities of i-LIDS, CAVIAR, PRID2011 and CUHK01 were 59, 25, 100 and 486 respectively. All of the datasets except i-LIDS and SPRD are captured by two views, thus naturally forms two image groups for each identity. For i-LIDS, we kept at most 4 frames for each person, and equally divided the frames of each identity into two image groups randomly, as in [4,6]. For SPRD, all image tracks were used. Also, considering the huge variation in frame number of the PRID2011 tracks, we selected at most 10 frames randomly in each track, making it a moderate size to generate signatures. In the testing phase, image groups from camera A were used as probes, and those from camera B constituted the gallery. For SPRD, a random group (not necessarily from the same camera view, making the task more difficult on SPRD) of each identity was selected as the gallery, and the others as probes. All random splits were performed 10 times. The performance is measured using the Cumulative Matching Characteristic (CMC) curve, which shows the probability of finding the correct match in top r ranks.

Several image pre-processing steps were applied before feature extraction. First, except for CUHK01, all images were resized into a uniform resolution of 64×128. Besides, all color channels of each image were normalized by histogram equalization to handle global illumination changes.

---

[1] http://ivlab.sjtu.edu.cn/best.

5.3   Module Validation

In this subsection, we validate the effectiveness of the proposed approach[2] by step-by-step verification. For space limits, all results are listed in one table. As demonstrated in Table 1, we firstly analysed each individual component of the method with KLDA, as well as a brief investigation in the case with KMLR. The experiments were conducted with the aforementioned RGB+YUV+HS+SG as features and LEG (Eq. 7) as kernel function. The Gaussian bandwidth σ in kernel function was derived from the mean distance of training data, and the scaling factor assigner q for Eq. 12 was set to $0.1N$. Here we exhibit CMC accuracies of $r = 1, 5, 10$ for quantitative comparisons.

Table 1. Component validation, multi-shot strategy comparison, feature comparison and kernel function comparison of the proposed method. Here *Cov.* and *Gau.* stand for region covariance and Gaussian matrix as local representations. *Unsup.*, *GE* and *NE* refer to unsupervised matching, supervised learning with graph embedding and neighbor emphasizing respectively. Default item selections are bolded. CMC accuracies of $r = 1, 5, 10$ are exhibited, with the highest ones in each validation group highlighted.

Validation group	Validation item	i-LIDS			CAVIAR		
		$r=1$	$r=5$	$r=10$	$r=1$	$r=5$	$r=10$
Component validation	Gau.(Unsup.)	0.4500	0.6333	0.7333	0.1600	0.5600	0.7400
	Gau.+KMLR	0.5083	0.7167	0.8417	0.4200	0.7200	**0.9200**
	Gau.+KLDA(GE)	0.5750	0.7833	0.8500	**0.5000**	**0.7600**	0.8800
	Gau.+KLDA(GE+NE)	**0.5917**	**0.8167**	**0.8917**	**0.5000**	**0.7600**	**0.9200**
	Cov.+KLDA(GE+NE)	0.5833	0.7917	0.8667	0.4600	**0.7600**	**0.9200**
Multi-shot strategy (Gau.(Unsup.))	Closest-point	0.3167	0.5667	0.7250	0.0800	0.3600	0.6000
	Score-voting	0.2500	0.5667	**0.7500**	**0.1800**	0.3600	0.6400
	Signature(mean)	**0.4500**	**0.6333**	0.7333	0.1600	**0.5600**	**0.7400**
Feature	RGB+Gradient	0.4917	0.7167	0.8167	0.3600	0.7600	**0.9600**
	RGB+YUV+HS+LBP	0.5250	0.7667	0.8750	0.4200	**0.7800**	0.8800
	RGB+YUV+HS+SG	**0.5917**	**0.8167**	**0.8917**	**0.5000**	0.7600	0.9200

Validation group	Validation item	SPRD			i-LIDS			CAVIAR		
		$r=1$	$r=5$	$r=10$	$r=1$	$r=5$	$r=10$	$r=1$	$r=5$	$r=10$
Kernel function	LET	0.2000	0.5286	0.9351	0.3667	0.5083	0.5917	0.2600	0.5600	0.7200
	LEG	0.3429	0.7714	0.9675	**0.5917**	**0.8167**	**0.8917**	**0.5000**	0.7600	0.9200
	AIG	**0.5429**	**0.8871**	**1.0000**	0.5750	0.7917	0.8417	**0.5000**	**0.8200**	**0.9400**

It can be observed that compared with the unsupervised version, the proposed discriminative learning approach greatly improves the recognition accuracies, which is extremely obvious on the difficult CAVIAR dataset. Meanwhile, MRDL with KLDA performs generally better than with KMLR, thus the rest validations in this subsection will be held on MRDL(KLDA). Neighbor emphasizing technique boosts the re-identification rate in different degrees on i-LIDS and CAVIAR, mainly after rank-5, thus it is beneficial for practical monitoring systems where usually top-10 matches are displayed to the user. On the other hand, the precisions of region Gaussian matrix are higher than those of region

---

[2] The source code is released on our website: http://vipl.ict.ac.cn/resources/codes.

covariance matrix mainly on top ranks, especially in the strict multi-shot case CAVIAR, proving it is better to consider mean information in local descriptors.

Afterwards, we compared the applied multi-shot strategy (signature generating by averaging) with the previously discussed closest-point and score-voting ones. It is obvious that averaging yields much better results in most cases, showing that though averaging may lose frame-specific information, it also filters out noise which is particularly common in reID datasets.

Subsequently, we applied three kinds of pixel-level features in the framework: RGB+YUV+HS+SG, RGB+Gradient [6,7] and RGB+YUV+HS+LBP [14]. We used region Gaussian matrix as descriptor and KLDA (with neighbor emphasizing) as kernel learning method. The results exhibit that RGB+YUV+HS+SG has generally stronger description ability. We attribute it to richer color channels of RGB+YUV+HS+SG, compared to RGB+Gradient, and diverse image filters that spatially spread over neighbor pixels compared to the gradients and LBPs.

Last but not least, we tested the three kernel functions, LET (Eq. 6), LEG (Eq. 7) and AIG (Eq. 8), on three datasets with RGB+YUV+HS+SG and region Gaussian matrix. It is obvious that different kernel functions vary dramatically in the final performance, and LEG and AIG produce higher accuracies than LET, verifying that discriminative learning on the original manifold preserves Riemannian geometry better compared to in the explicitly mapped Euclidean space. While, considering AIG lacks positive definiteness, we prefer to use LEG in our work. In addition, we are optimistic that the result would be even better if more appropriate Riemannian metrics or kernel functions are applied.

5.4   Comparison Results and Analysis

We compared MRDL with both the multi-shot community and some representative single-shot methods on all datasets except SPRD (Table 2), since different methods reported results on different datasets and none reported on SPRD.

Table 2. Approach comparison chart on i-LIDS (IL), CAVIAR (CA), PRID2011 (PR) and CUHK01 (CU). A check mark denotes a certain method reported results on a corresponding dataset. These works are categorized into single-shot (Sin.) ones and multi-shot ones, which are further divided into closest point-based (CP), score voting-based (SV), set structure-based (SS) and signature-based (SG) groups.

Category		Approach	IL	CA	PR	CU	Category		Approach	IL	CA	PR	CU
Sin.	Feature extraction	SCR [28]	✓				CP		SDALF [4]	✓			
		MLF [21]				✓			HPE [5]	✓			
		DeepReID [22]				✓			LFDA [13]		✓		
	Metric learning	LADF [29]		✓				Mean	MRCG [6]	✓			
		EnsembleReID[14]	✓	✓					COSMATI [7]	✓			
		RMLLC(R) [16]	✓	✓			SG		DVDL [35]		✓	✓	
SV		ICT [32]		✓					Saliency [36]				✓
		SRID [33]			✓				AFDA [37]			✓	
SS		SBDR [11]	✓	✓				Model	STFV3D+LFDA[39]			✓	
		LCRNP [34]		✓					DVR[38]			✓	

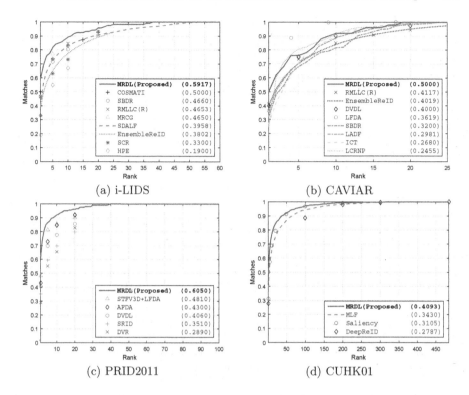

Fig. 6. CMC curves of the proposed method and comparison methods on four public person re-identification datasets: (a) i-LIDS, (b) CAVIAR, (c) PRID2011, (d) CUHK01. Particularly, the rank-1 accuracies are listed behind the name of each method. Those of which the CMC curves are unavailable are represented as markers. Here the MRDL is with KLDA and region Gaussian matrix descriptor.

The CMC curves are shown in Fig. 6. With only half identities in the supervised methods, the horizontal axis of CMC curves of the unsupervised SDALF, LCRNP, HPE, MRCG and SCR are compressed by 50% for fair comparison. Also, results of methods (LFDA, LADF, DeepReID) with different testing identity numbers are rescaled in the same way.

The proposed MRDL achieves the highest rank-1 accuracy on all evaluated datasets with improvements of 9.17%, 8.83%, 12.40% and 6.63% over the state-of-the-art methods, exhibiting impressive effectiveness of the framework. It can be observed that MRDL is not only discriminative in datasets with small between-class variations, but also robust on the ones with large within-class variations in illumination, resolution, occlusion, pose and view. In addition, even the unsupervised version of our MRDL with region Gaussian matrices only is better than or comparable with most of the works on i-LIDS, demonstrating the representation ability of our local descriptors. Besides, we should note that on CAVIAR, LFDA took different frames in a group as independent training samples, which, actually as a single-shot protocol, introduces multi-modality and is easier to produce higher accuracies in large within-group variance datasets as CAVIAR. Thus, it's inappropriate to be directly compared with the multi-shot MRDL where one image group is treated as one sample. While, DVDL, STFV3D+LFDA and DVR utilized all frames in each track of the PRID2011 dataset, holding richer information since the very beginning, but are still inferior to our method in performance. Also, DeepReID on CUHK01 was trained on 871 identities, naturally possessing an edge in the learning phase. But the accuracies of our MRDL transcend those of it with a large margin.

6 Conclusion

We proposed an effective discriminative learning framework in Riemannian manifold for multiple-shot person re-identification. Different from other multi-shot approaches, our method represents local stripes as SPD matrices, averages on manifold to generate signatures, and perform kernelized learning algorithms also on the Riemannian manifold. Experiments demonstrated the impressive effectiveness, especially the improvements brought by the Riemannian discriminative learning phase, even for the simple LDA, and superiority of the method over the state of the arts on five benchmark datasets. We will further explore more suitable kernel functions and metric learning algorithms, such as those based on fixed boundary sample pairs or relative comparison triplets.

Acknowledgement. This work is partially supported by 973 Program under contract No. 2015CB351802, Natural Science Foundation of China under contracts Nos. 61390511, 61379083, 61272321, 61271445, and Youth Innovation Promotion Association CAS No. 2015085.

References

1. Gong, S., Cristani, M., Yan, S., Loy, C.C.: Person Re-identification, vol. 1. Springer, Heidelberg (2014)
2. Gheissari, N., Sebastian, T.B., Hartley, R.: Person reidentification using spatiotemporal appearance. In: IEEE Conference on Computer Vision and Pattern Recognition, vol. 2, pp. 1528–1535. IEEE (2006)
3. Wang, X., Doretto, G., Sebastian, T., Rittscher, J., Tu, P.: Shape and appearance context modeling. In: IEEE International Conference on Computer Vision, pp. 1–8. IEEE (2007)
4. Farenzena, M., Bazzani, L., Perina, A., Murino, V., Cristani, M.: Person re-identification by symmetry-driven accumulation of local features. In: IEEE Conference on Computer Vision and Pattern Recognition, pp. 2360–2367. IEEE (2010)
5. Bazzani, L., Cristani, M., Perina, A., Farenzena, M., Murino, V.: Multiple-shot person re-identification by HPE signature. In: International Conference on Pattern Recognition, pp. 1413–1416. IEEE (2010)
6. Bak, S., Corvee, E., Bremond, F., Thonnat, M.: Multiple-shot human re-identification by mean Riemannian covariance grid. In: IEEE International Conference on Advanced Video and Signal-Based Surveillance, pp. 179–184. IEEE (2011)

7. Bąk, S., Charpiat, G., Corvée, E., Brémond, F., Thonnat, M.: Learning to match appearances by correlations in a covariance metric space. In: Fitzgibbon, A., Lazebnik, S., Perona, P., Sato, Y., Schmid, C. (eds.) ECCV 2012. LNCS, vol. 7574, pp. 806–820. Springer, Heidelberg (2012). doi:10.1007/978-3-642-33712-3_58
8. Ma, B., Su, Y., Jurie, F.: Local descriptors encoded by fisher vectors for person re-identification. In: Fusiello, A., Murino, V., Cucchiara, R. (eds.) ECCV 2012. LNCS, vol. 7583, pp. 413–422. Springer, Heidelberg (2012). doi:10.1007/978-3-642-33863-2_41
9. Zheng, L., Shen, L., Tian, L., Wang, S., Wang, J., Tian, Q.: Scalable person re-identification: a benchmark. In: IEEE International Conference on Computer Vision, pp. 1116–1124. IEEE (2015)
10. Zheng, W.S., Gong, S., Xiang, T.: Person re-identification by probabilistic relative distance comparison. In: IEEE Conference on Computer Vision and Pattern Recognition, pp. 649–656. IEEE (2011)
11. Wu, Y., Minoh, M., Mukunoki, M., Lao, S.: Set based discriminative ranking for recognition. In: Fitzgibbon, A., Lazebnik, S., Perona, P., Sato, Y., Schmid, C. (eds.) ECCV 2012. LNCS, vol. 7574, pp. 497–510. Springer, Heidelberg (2012). doi:10.1007/978-3-642-33712-3_36
12. Hirzer, M., Roth, P.M., Köstinger, M., Bischof, H.: Relaxed pairwise learned metric for person re-identification. In: Fitzgibbon, A., Lazebnik, S., Perona, P., Sato, Y., Schmid, C. (eds.) ECCV 2012. LNCS, vol. 7577, pp. 780–793. Springer, Heidelberg (2012). doi:10.1007/978-3-642-33783-3_56
13. Pedagadi, S., Orwell, J., Velastin, S., Boghossian, B.: Local fisher discriminant analysis for pedestrian re-identification. In: IEEE Conference on Computer Vision and Pattern Recognition, pp. 3318–3325. IEEE (2013)
14. Xiong, F., Gou, M., Camps, O., Sznaier, M.: Person re-identification using kernel-based metric learning methods. In: Fleet, D., Pajdla, T., Schiele, B., Tuytelaars, T. (eds.) ECCV 2014. LNCS, vol. 8695, pp. 1–16. Springer, Heidelberg (2014). doi:10.1007/978-3-319-10584-0_1
15. Liao, S., Li, S.Z.: Efficient PSD constrained asymmetric metric learning for person re-identification. In: IEEE International Conference on Computer Vision, pp. 3685–3693. IEEE (2015)
16. Chen, J., Zhang, Z., Wang, Y.: Relevance metric learning for person re-identification by exploiting listwise similarities. IEEE Trans. Image Process. 24, 4741–4755 (2015)
17. Zhao, R., Ouyang, W., Wang, X.: Unsupervised salience learning for person re-identification. In: IEEE Conference on Computer Vision and Pattern Recognition, pp. 3586–3593. IEEE (2013)
18. Liao, S., Hu, Y., Zhu, X., Li, S.Z.: Person re-identification by local maximal occurrence representation and metric learning. In: IEEE Conference on Computer Vision and Pattern Recognition, pp. 2197–2206. IEEE (2015)
19. Prosser, B., Zheng, W.S., Gong, S., Xiang, T.: Person re-identification by support vector ranking. In: British Machine Vision Conference, pp. 21.1–21.11. BMVA Press (2010)
20. Gray, D., Tao, H.: Viewpoint invariant pedestrian recognition with an ensemble of localized features. In: Forsyth, D., Torr, P., Zisserman, A. (eds.) ECCV 2008. LNCS, vol. 5302, pp. 262–275. Springer, Heidelberg (2008). doi:10.1007/978-3-540-88682-2_21
21. Zhao, R., Ouyang, W., Wang, X.: Learning mid-level filters for person re-identification. In: IEEE Conference on Computer Vision and Pattern Recognition, pp. 144–151. IEEE (2014)

22. Li, W., Zhao, R., Xiao, T., Wang, X.: Deepreid: deep filter pairing neural network for person re-identification. In: IEEE Conference on Computer Vision and Pattern Recognition, pp. 152–159. IEEE (2014)

23. Cheng, D.S., Cristani, M., Stoppa, M., Bazzani, L., Murino, V.: Custom pictorial structures for re-identification. In: British Machine Vision Conference, vol. 1, pp. 68.1–68.11. BMVA Press (2011)

24. Hirzer, M., Beleznai, C., Roth, P.M., Bischof, H.: Person re-identification by descriptive and discriminative classification. In: Heyden, A., Kahl, F. (eds.) SCIA 2011. LNCS, vol. 6688, pp. 91–102. Springer, Heidelberg (2011). doi:10.1007/978-3-642-21227-7_9

25. Li, W., Zhao, R., Wang, X.: Human reidentification with transferred metric learning. In: Lee, K.M., Matsushita, Y., Rehg, J.M., Hu, Z. (eds.) ACCV 2012. LNCS, vol. 7724, pp. 31–44. Springer, Heidelberg (2013). doi:10.1007/978-3-642-37331-2_3

26. Gray, D., Brennan, S., Tao, H.: Evaluating appearance models for recognition, reacquisition, and tracking. In: IEEE International Workshop on Performance Evaluation for Tracking and Surveillance, vol. 3. IEEE (2007)

27. Zheng, W.S., Gong, S., Xiang, T.: Associating groups of people. In: British Machine Vision Conference, pp. 23.1–23.11. BMVA Press (2009)

28. Bak, S., Corvee, E., Brémond, F., Thonnat, M.: Person re-identification using spatial covariance regions of human body parts. In: IEEE International Conference on Advanced Video and Signal Based Surveillance, pp. 435–440. IEEE (2010)

29. Li, Z., Chang, S., Liang, F., Huang, T.S., Cao, L., Smith, J.R.: Learning locally-adaptive decision functions for person verification. In: IEEE Conference on Computer Vision and Pattern Recognition, pp. 3610–3617. IEEE (2013)

30. Ma, B., Su, Y., Jurie, F.: BiCov: a novel image representation for person re-identification and face verification. In: British Machine Vision Conference, pp. 57.1–57.11. BMVA Press (2012)

31. Su, C., Yang, F., Zhang, S., Tian, Q., Davis, L.S., Gao, W.: Multi-task learning with low rank attribute embedding for person re-identification. In: IEEE International Conference on Computer Vision, pp. 3739–3747. IEEE (2015)

32. Avraham, T., Gurvich, I., Lindenbaum, M., Markovitch, S.: Learning implicit transfer for person re-identification. In: Fusiello, A., Murino, V., Cucchiara, R. (eds.) ECCV 2012. LNCS, vol. 7583, pp. 381–390. Springer, Heidelberg (2012). doi:10.1007/978-3-642-33863-2_38

33. Karanam, S., Li, Y., Radke, R.: Sparse re-id: block sparsity for person re-identification. In: IEEE Conference on Computer Vision and Pattern Recognition Workshops, pp. 33–40. IEEE (2015)

34. Wu, Y., Mukunoki, M., Minoh, M.: Locality-constrained collaboratively regularized nearest points for multiple-shot person re-identification. In: Korea-Japan Joint Workshop on Frontiers of Computer Vision. CiteSeer (2014)

35. Karanam, S., Li, Y., Radke, R.J.: Person re-identification with discriminatively trained viewpoint invariant dictionaries. In: IEEE International Conference on Computer Vision, pp. 4516–4524. IEEE (2015)

36. Martinel, N., Micheloni, C., Foresti, G.L.: Saliency weighted features for person re-identification. In: Agapito, L., Bronstein, M.M., Rother, C. (eds.) ECCV 2014. LNCS, vol. 8927, pp. 191–208. Springer, Heidelberg (2015). doi:10.1007/978-3-319-16199-0_14

37. Li, Y., Wu, Z., Karanam, S., Radke, R.: Multi-shot human re-identification using adaptive fisher discriminant analysis. In: British Machine Vision Conference. BMVA Press (2015)

38. Wang, T., Gong, S., Zhu, X., Wang, S.: Person re-identification by video ranking. In: Fleet, D., Pajdla, T., Schiele, B., Tuytelaars, T. (eds.) ECCV 2014. LNCS, vol. 8692, pp. 688–703. Springer, Heidelberg (2014). doi:10.1007/978-3-319-10593-2_45

39. Liu, K., Ma, B., Zhang, W., Huang, R.: A spatio-temporal appearance representation for video-based pedestrian re-identification. In: IEEE International Conference on Computer Vision, pp. 3810–3818. IEEE (2015)

40. Tuzel, O., Porikli, F., Meer, P.: Region covariance: a fast descriptor for detection and classification. In: Leonardis, A., Bischof, H., Pinz, A. (eds.) ECCV 2006. LNCS, vol. 3952, pp. 589–600. Springer, Heidelberg (2006). doi:10.1007/11744047_45

41. Amari, S.I., Nagaoka, H.: Methods of Information Geometry, vol. 191. American Mathematical Society, Providence (2007)

42. Lovrić, M., Min-Oo, M., Ruh, E.A.: Multivariate normal distributions parametrized as a Riemannian symmetric space. J. Multivar. Anal. **74**, 36–48 (2000)

43. Arsigny, V., Fillard, P., Pennec, X., Ayache, N.: Geometric means in a novel vector space structure on symmetric positive-definite matrices. SIAM J. Matrix Anal. Appl. **29**, 328–347 (2007)

44. Pennec, X., Fillard, P., Ayache, N.: A Riemannian framework for tensor computing. Int. J. Comput. Vis. **66**, 41–66 (2006)

45. Wang, R., Guo, H., Davis, L.S., Dai, Q.: Covariance discriminative learning: a natural and efficient approach to image set classification. In: IEEE Conference on Computer Vision and Pattern Recognition, pp. 2496–2503. IEEE (2012)

46. Jayasumana, S., Hartley, R., Salzmann, M., Li, H., Harandi, M.: Kernel methods on the Riemannian manifold of symmetric positive definite matrices. In: IEEE Conference on Computer Vision and Pattern Recognition, pp. 73–80. IEEE (2013)

47. Baudat, G., Anouar, F.: Generalized discriminant analysis using a kernel approach. Neural Comput. **12**, 2385–2404 (2000)

48. He, X., Niyogi, P.: Locality preserving projections. In: Neural Information Processing Systems, vol. 16, pp. 153–160. MIT (2004)

49. Zelnik-Manor, L., Perona, P.: Self-tuning spectral clustering. In: Advances in Neural Information Processing Systems, pp. 1601–1608 (2004)

50. Yan, S., Xu, D., Zhang, B., Zhang, H.J., Yang, Q., Lin, S.: Graph embedding and extensions: a general framework for dimensionality reduction. IEEE Trans. Pattern Anal. Mach. Intell. **29**, 40–51 (2007)

51. McFee, B., Lanckriet, G.R.: Metric learning to rank. In: International Conference on Machine Learning, pp. 775–782. ACM (2010)

52. Joachims, T.: A support vector method for multivariate performance measures. In: International Conference on Machine Learning, pp. 377–384. ACM (2005)

53. Joachims, T., Finley, T., Yu, C.N.J.: Cutting-plane training of structural svms. Mach. Learn. **77**, 27–59 (2009)

54. Zhang, C., Ni, B., Song, L., Yang, X., Zhang, W.: BEST: benchmark and evaluation of surveillance task. In: Chen, C.-S., Lu, J., Ma, K.-K. (eds.) ACCV 2016 Workshops. LNCS, vol. 10118, pp. 393–407. Springer, Heidelberg (2016)

Visually Similar K-poselets Based Human Pose Recognition

Shoucheng Ni, Weiwei Liu, Hao Cheng, and Chongyang Zhang$^{(\boxtimes)}$

Institute of Image Communication and Network Engineering,
Shanghai Jiao Tong University, Shanghai, China
sunny_zhang@sjtu.edu.cn

Abstract. In the paper, we propose the visually similar k-poselets to recognize human poses (e.g., stoop, squat) in still images. Compared with the original k-poselets that are collected according to similar keypoints configurations, we further introduce appearance similarity constraints to generate visually similar k-poselets. The number of selected visually similar k-poselets for each pose category is iteratively decreased based on discriminative criterion. The pose dictionary, constructed with learned visually similar and discriminative k-poselets of different poses, is applied in pose recognition. The experimental results on our released human pose database verify the effectiveness of the proposed visually similar k-poselets based pose recognition method.

1 Introduction

Human pose recognition from still images serves as an important pre-requisite step to many computer vision applications, such as human action understanding, human-machine interaction, surveillance, etc. Though continuous attention has been attracted and considerable progress has been made in recent years, human pose recognition remains a challenging task due to factors such as human pose variation, viewpoint change, partial occlusion and so on.

Previous work on human pose recognition can be divided into holistic-based and part-based approaches. Holistic-based approaches are motivated by the fact that body appearances of different poses vary dramatically and thus the descriptor of global appearance can serve as a discriminative representation of human poses: [6] applies a distance transform of the binary silhouette image to extract the feature; [1] takes the combination of geometry parameters (e.g., first principal axis, mean point, width and height of bounding box) as the input of a fuzzy logic system (FLS); [12] applies nonnegative matrix factorization (NMF) to both clean human poses and background images to calculate NMF coefficients for pose classification. Precise silhouette subtraction of human body is the essential step to guarantee accurate acquisitions of body silhouette and extractions of global appearance feature; however, foreground extraction or segmentation of human body from the cluttered background in single images nowadays can not be so precise and thus it is the bottleneck in holistic-based approaches. On the contrary, part-based approaches model the human pose as a representation of

© Springer International Publishing AG 2017
C.-S. Chen et al. (Eds.): ACCV 2016 Workshops, Part III, LNCS 10118, pp. 426–440, 2017.
DOI: 10.1007/978-3-319-54526-4_31

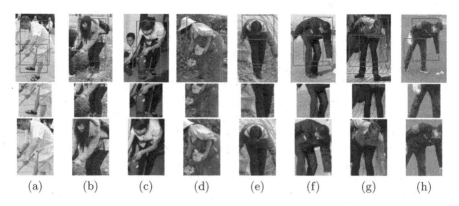

(a)   (b)   (c)   (d)   (e)   (f)   (g)   (h)

Fig. 1. Examples of k-poselets ($k = 2$). Original images are shown in the first row. Red depicts examples of the first poselet as shown in the second column, while blue depicts examples of the second poselet as shown in the third row. (Color figure online)

body parts or keypoints configurations. [15] proposes an approach for efficiently localizing a pose by generating a Pose-Specific Part Model (PSPM). Meanwhile, human poses are also well characterized by the configuration of keypoints (e.g., head, shoulders, knees); however, models [4,11,13] with high complexity are indispensable to accurate pose estimations and keypoint predictions.

As argued in [17], human poses are articulated configurations of human body parts. To avoid the problem of silhouette subtraction existing in holistic-based approaches, we introduce a part-based model called k-poselets [5] into human pose recognition framework. A k-poselet is a deformable part model with k parts and each part is aligned to a specific keypoints configuration. As depicted in Fig. 1, k-poselets ($k = 2$) enable each part to move with respect to each other with a deformation cost that is learned jointly with corresponding HOG [2] templates. Inspired by k-poselets' capability of accurate localizations of human keypoints and rich representations of body parts, we attempt to construct a k-poselets based pose dictionary to recognize human poses in images.

In this work we propose visually similar k-poselets, which solves the problem of weak and overly-generic detectors trained from visually incoherent/dissimilar categories [9]. As depicted in Fig. 1, the first four poselets (a)–(d) are visually dissimilar to the last four ones (e)–(h), although they are generated as one 2-poselets. The observation that poselets with similar keypoints configurations may not necessarily be visually similar motivates us to introduce visually similar k-poselets so as to decrease intra-class variability between collected poselets that are used to train the corresponding k-poselets. The concept "visually similar" has been widely applied in solving computer vision problems, such as scene classification [14], object recognition [3], action recognition [7], person re-identification [16] and so on. The key insight of our work is to seek visually similar instances to train detectors performing well on these visually similar ones.

After the visually similar k-poselets are generated, a compact pose dictionary is learned by selecting the most discriminative ones of each pose category so that it can discriminate different pose categories as well as possible. To favour more discriminative k-poselets, weights are assigned according to the self-defined discrimination. The pose recognition is conducted with the learned pose dictionary by ranking and keeping the top L k-poselets detections that have the highest response scores concerning the input image. Similarly, different weights are assigned to the L scores specific to sorting indexes. Parameters mentioned above are iteratively inferred and updated to optimize overall recognition performances.

The main contributions of this paper are three-fold: (i) the visually similar k-poselet is developed with appearance similarity constraints and two distinct criteria are presented to select discriminative k-poselets; (ii) the human pose recognition algorithm is proposed based on the learned pose dictionary consisting of visually similar and discriminative k-poselets; (iii) the human pose database containing 1080 images with keypoints annotated is released, which includes 3 pose categories and covers a wide variety of both viewpoints and backgrounds.

The remaining of the paper is organized as follows. Section 2 gives detail explanations on how to generate and select discriminative visually similar k-poselets. We present the visually similar k-poselets based pose recognition algorithm in Sect. 3. The released human pose database and experimental results are introduced and presented in Sect. 4. Finally, conclusions are drawn in Sect. 5.

2   Visually Similar K-poselets Learning

2.1   Visual Similarity

The original k-poselet is collected according to similar keypoints configurations. However, as mentioned above, image patches sharing similar keypoints configurations may vary dramatically in appearance and the k-poselet trained from a set of visually dissimilar patches is ambiguous. Thus, the appearance similarity constraint is introduced to seek visually similar k-poselets.

Given the set of one k-poselet $kpl = (p_1, \cdots, p_k)$ collected by original algorithm in [5], where p_i is the i-th poselet containing N_I patches $p_{i,1}, \cdots, p_{i,N_I}$ that share similar keypoints configurations, the j-th instance kpl_j consists of every j-th patch in k poselets: $p_{1,j}, \cdots, p_{k,j}$ $(1 \leq j \leq N_I)$. Hierarchical clustering is applied to further cluster N_I instances in k-poselet kpl into several subsets.

Patches in the same poselet may differ in sizes; however, the ratio between width and height of all these patches is a constant, thus we directly resize all patches in the same poselet with a fixed width and height and the resized patches are denoted as $p'_{i,1}, \cdots, p'_{i,N_I}$ $(1 \leq i \leq k)$. The visual dissimilarity between patches is measured by the distance between corresponding HOG features. Because resized patches in different poselets is not the same, the correlation distance $dist_{cor}$ is applied to ensure fair comparisons. The distance $dist(i, j)$ between the i-th and j-th instance is calculated as

$$dist(i,j) = \sum_{m=1}^{k} dist_{cor}\big(HOG(p'_{m,i}), HOG(p'_{m,j})\big). \tag{1}$$

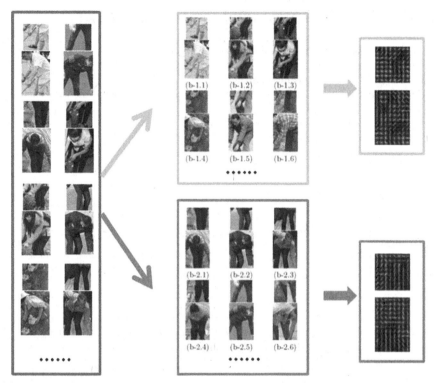

(a) visually dissimilar k-poselets (b) visually similar k-poselets (c) HOG templates

Fig. 2. Examples of visually similar k-poselets ($k = 2$). (b-1) and (b-2) represent two clusters of visually similar 2-poselets generated by applying hierarchical clustering on (a). Compared with k-poselets collected by original algorithm, proposed k-poselets exhibit visual similarity within the same cluster.

The hierarchical tree is built based on the complete distance between two clusters as the same scenario described in [10]. The split of the tree is conducted according to a predefined distance threshold $dist_{thre}$ so as to ensure the maximum correlation distance between any two instances in the same cluster is no larger than the threshold (the threshold is set to 1.85 for the case of 2-poselet). At the same time, clusters with too few patches (less than 10) are rejected. Figure 2 is an example of generated clusters of visually similar k-poselets by applying hierarchical clustering on the original k-poselet set shown in Fig. 1.

Similar to [5], each visually similar k-poselet is described by a weight vector and the detection score of k-poselet kpl given image I is calculated by scoring all the k-poselets placement hypothesis to find the maximum score:

$$score_{det}(kpl, I) = \max_{\forall l \in L_h} score(l \mid kpl, I), \qquad (2)$$

where L_h represents the set of all the possible hypothesis.

Algorithm 1. Hierarchical clustering based visually similar k-poselets generation

---

Require: I_{train}: Training images that belong to certain pose category; N_K: Number of visually similar k-poselets expected to be generated; $dist_{thre}$: Threshold of the correlation distance;

Ensure: kpl: Generated visually similar k-poselets;

1: **Initialize:** The number of visually similar k-poselets to be trained for each pose category: $\overline{N_K} = N_K, kpl = \emptyset$;
2: **while** $\overline{N_K} \neq 0$ **do**
3:     $kpl' \leftarrow hierarchical\_cluster(I_{train}, dist_{thre})$;
4:     N'_K: the number of generated visually similar k-poselets kpl';
5:     **if** $N'_K \leq \overline{N_K}$ **then**
6:         $kpl \leftarrow \left(kpl \cup kpl'\right)$;
7:     **end if**
8:     **if** $N'_K > \overline{N_K}$ **then**
9:         $kpl \leftarrow \left(kpl \cup \{kpl'_i\}_{i=1}^{\overline{N_K}}\right)$;
10:     **end if**
11:     $\overline{N_K} \leftarrow \overline{N_K} - N'_K$;
12: **end while**
13: **return** kpl;

For each pose category, N_K visually similar k-poselets are expected to be generated, which is described in Algorithm 1.

2.2   Discriminative Learning

Considering some k-poselets may appear in more than two pose categories, discriminative k-poselets that are capable of discriminating between the corresponding pose and other poses, are supposed to be selected.

K-poselets Detection Tuple. Given N_P different pose categories and pre-trained N_K visually similar k-poselets of each pose category, the k-poselet dictionary $\boldsymbol{D_{kpl}}$ is initialized as $\{kpl(i)\}_{i=1}^{N_D}$, where $kpl(i)$ is the i-th visually similar k-poselet with pose label $p_{K,i} \in [1, N_P]$ and N_D is the number of k-poselets in $\boldsymbol{D_{kpl}}$ which is equal to $N_K \times N_P$ initially.

Given one image I and the k-poselets dictionary $\boldsymbol{D_{kpl}}$, N_D detection scores are calculated: $score(I, \boldsymbol{D_{kpl}}) = \{score_{det}(kpl(i), I)\}_{i=1}^{N_D}$. The k-poselets detection tuple $D_L(I, \boldsymbol{D_{kpl}})$ is defined as

$$D_L(I, \boldsymbol{D_{kpl}}) = \{S_L, K_L, P_L\}, \tag{3}$$

where L is the number of the largest detection scores to be evaluated, $S_L = (s_L + 1) \in \mathbb{R}^{1 \times L}$ and s_L^i is the i-th largest detection score in $score(I, \boldsymbol{D_{kpl}})$, $K_L = \{K_L^i\}_{i=1}^L$ and $K_L^i \in [1, N_D]$ is the index of k-poselet that achieves the i-th largest detection score, $P_L = \{P_L^i\}_{i=1}^L$ and $P_L^i \in [1, N_P]$ is the index of pose that the K_L^i-th k-poselet belongs to.

Assume that there are N_{train} training images and the pose label of the m-th training image I_m is $p_{I,m}(1 \leq m \leq N_{train}, p_{I,m} \in [1, N_P])$, the corresponding k-poselets detection tuple is expressed as $D_{L,m}(I_m, \boldsymbol{D_{kpl}}) = \{S_{L,m}, K_{L,m}, P_{L,m}\}$.

We further introduce the detection matrix $M_{det} \in \mathbb{R}^{N_D \times L \times N_P}$ to summarize the detection performance of $\boldsymbol{D_{kpl}}$ given N_{train} training images:

$$M_{det}(i,j,h) =$$

$$\begin{cases} \dfrac{\sum\limits_{m=1}^{N_{train}} f(i,K^j_{L,m};h,P^j_{L,m})(S^j_{L,m}-S^m_{max\_diff})}{\sum\limits_{m=1}^{N_{train}} f(i,K^j_{L,m};h,P^j_{L,m})} & \sum\limits_{m=1}^{N_{train}} f(i,K^j_{L,m};h,P^j_{L,m}) \neq 0 \\ 0 & otherwise \end{cases}, \quad (4)$$

where

$$\Delta(condition) = \begin{cases} 1 & condition \; is \; true \\ 0 & otherwise \end{cases}, \quad (5)$$

$$f(i,K^j_{L,m};h,P^j_{L,m}) = \Delta((i == K^j_{L,m}) \wedge (h == P^j_{L,m})), \quad (6)$$

$$S^m_{max\_diff} =$$

$$\begin{cases} min(S_{L,m}) & \begin{aligned}&((h = p_{I,m}) \wedge (\forall P_{L,m} = p_{I,m})) \vee \\ &((h \neq p_{I,m}) \wedge (\forall P_{L,m} \neq p_{I,m}))\end{aligned} \\ max(S^{Index}_{L,m}) & \begin{aligned}&((h = p_{I,m}) \wedge (P^{Index}_{L,m} \neq p_{I,m}) \wedge (S^{Index}_{L,m} < S^j_{L,m})) \vee \\ &((h \neq p_{I,m}) \wedge (P^{Index}_{L,m} = p_{I,m}) \wedge (S^{Index}_{L,m} < S^j_{L,m}))\end{aligned} \end{cases} \quad (7)$$

$M_{det}(i,j,h)$ is the average detection score difference between the qualified score $S^j_{L,m}$ and the maximum score $S^m_{max\_diff}$ that is smaller than $S^j_{L,m}$ and achieved by k-poselets belonging to pose categories different from $P^j_{L,m}$.

Discriminative K-poselets Selection. The number of visually similar k-poselets in $\boldsymbol{D_{kpl}}$ is iteratively decreased so that a compact dictionary consisting of discriminative and visually similar k-poselets can be generated. Based on different criteria of evaluating the discrimination of one k-poselet, two selection modes (Max-Average selection or Average-Max selection) are proposed.

Different weights are assigned to L detection scores. $W_O \in \mathbb{R}^{1 \times L}$, initialize as w_O, is the weight vector and W^l_O represents the weight of the l-th largest score. Given the detection matrix M_{det} and W_O, the weighted detection matrix concerning the i-th k-poselets and the j-th pose category is calculated as

$$w^{i,j}_K = \langle M_{det}(i,:,j), W_O \rangle \; (1 \leq i \leq N_D, 1 \leq j \leq N_P), \quad (8)$$

where $\langle \mathbf{a}, \mathbf{b} \rangle$ represents the dot production between vector $\mathbf{a}$ and vector $\mathbf{b}$.

In Max-Average selection (**MA-selection**), the discrimination of one k-poselet is measured by the average discrimination of the corresponding pose category against the other categories. Assume the i-th k-poselet $kpl(i)$ in $\boldsymbol{D_{kpl}}$ belongs to the p_i-th pose category, the discrimination of $kpl(i)$ is calculated as

$$w^i = \frac{\sum\limits_{j=1}^{N_P}(w^{i,p_i}_K - w^{i,j}_K)}{N_P - 1}. \quad (9)$$

Algorithm 2. Average-Max based visually similar k-poselets selection

---

Require: w_K: Weighted detection matrix; N_K: Initial number of visually similar k-poselets of each pose category; N_K': Number of visually similar k-poselets to be selected for each pose category; N_P: Number of pose categories; D_{kpl}: Original k-poselets dictionary;

Ensure: D_{kpl}': Updated k-poselets dictionary; w: discrimination of k-poselets;

1: **Initialize:** Number of selected subsets containing $\left\lceil \frac{N_K'}{N_P-1} \right\rceil$ k-poselets: $N_{ceil} = 0$,

   number of selected subsets containing $\left\lfloor \frac{N_K'}{N_P-1} \right\rfloor$ k-poselets: $N_{floor} = 0$, $D_{kpl}' = \emptyset$;

2: **for** $j = 1$ to N_P **do**

3:      $\{I_{j,n}\}_{n=1}^{N_P-1}, I_{j,n} = \emptyset$: k-poselets indexes in the n-th subset;

4:      $\{k_{j,m}\}_{m=1}^{N_K}$: indexes in D_{kpl} of N_K k-poselets belonging to j-th pose category;

5:      $w_{k,j}^{m,n} = \begin{cases} w_K^{k_{j,m},j} - w_K^{k_{j,m},n+1} & 1 \le m \le N_K, 1 \le n \le N_P - 1, j = 1 \\ w_K^{k_{j,m},j} - w_K^{k_{j,m},n} & 1 \le m \le N_K, 1 \le n \le j-1, j \ne 1, j \ne N_P \\ w_K^{k_{j,m},j} - w_K^{k_{j,m},n+1} & 1 \le m \le N_K, j \le n \le N_P - 1, j \ne 1, j \ne N_P \\ w_K^{k_{j,m},j} - w_K^{k_{j,m},n} & 1 \le m \le N_K, 1 \le n \le N_P - 1, j = N_P \end{cases}$;

6:      $w_{k,j} = \{w_{k,j}^{m,n}\}, 1 \le m \le N_K, 1 \le n \le N_P - 1$;

7:      **while** $N_{floor} \ne \left(N_P - 1 - \left(N_K' - \left\lfloor \frac{N_K'}{N_P-1} \right\rfloor \times (N_P - 1) \right) \right)$ **do**

8:          Sort $w_{k,j}$ and $w_{k,j}^{m',n'}$ is the largest in $w_{k,j}$;

9:          $w^{k_{j,m'}} \leftarrow w_{k,j}^{m',n'}, I_{j,n'} \leftarrow (I_{j,n'} \cup k_{j,m'})$;

10:          **if** $|I_{j,n'}| = \left\lfloor \frac{N_K'}{N_P-1} \right\rfloor$ **then**

11:              $w_{k,j} \leftarrow \left(w_{k,j} - \left(\{w_{k,j}^{m',n}\}_{n=1}^{N_P-1} \vee \{w_{k,j}^{m,n'}\}_{m=1}^{N_K} \right) \right)$;

12:              $N_{floor} \leftarrow N_{floor} + 1$;

13:          **end if**

14:          **if** $|I_{j,n'}| < \left\lfloor \frac{N_K'}{N_P-1} \right\rfloor$ **then**

15:              $w_{k,j} \leftarrow \left(w_{k,j} - \{w_{k,j}^{m',n}\}_{n=1}^{N_P-1} \right)$;

16:          **end if**

17:      **end while**

18:      **while** $N_{ceil} \ne \left(N_K' - \left\lfloor \frac{N_K'}{N_P-1} \right\rfloor \times (N_P - 1) \right)$ **do**

19:          Sort $w_{k,j}$ and $w_{k,j}^{m',n'}$ is the largest in $w_{k,j}$;

20:          $w^{k_{j,m'}} \leftarrow w_{k,j}^{m',n'}, I_{j,n'} \leftarrow (I_{j,n'} \cup k_{j,m'})$;

21:          **if** $|I_{j,n'}| = \left\lceil \frac{N_K'}{N_P-1} \right\rceil$ **then**

22:              $w_{k,j} \leftarrow \left(w_{k,j} - \left(\{w_{k,j}^{m',n}\}_{n=1}^{N_P-1} \vee \{w_{k,j}^{m,n'}\}_{m=1}^{N_K} \right) \right)$;

23:              $N_{ceil} \leftarrow N_{ceil} + 1$;

24:          **end if**

25:          **if** $|I_{j,n'}| < \left\lceil \frac{N_K'}{N_P-1} \right\rceil$ **then**

26:              $w_{k,j} \leftarrow \left(w_{k,j} - \{w_{k,j}^{m',n}\}_{n=1}^{N_P-1} \right)$;

27:          **end if**

28:      **end while**

29:      $\{k_{j,m}\}_{m=1}^{N_K'} \leftarrow \{I_{j,n}\}_{n=1}^{N_P-1}, D_{kpl}' \leftarrow \left(D_{kpl}' \cup \{D_{kpl}(k_{j,m})\}_{m=1}^{N_K'} \right)$;

30: **end for**

31: **return** D_{kpl}', w;

For the j-th pose category, N'_K k-poselets $\{kpl(k_{j,m})\}_{m=1}^{N'_K}$ with the N'_K maximum discrimination $\{w^{k_{j,m}}\}_{m=1}^{N'_K}$ are to be selected: $w^{k_{j,1}} \geq \cdots w^{k_{j,m}} \geq \cdots \geq w^{k_{j,N'_K}}$, $p_{k_{j,m}} = j$.

In Average-Max selection (**AM-selection**), the discrimination of one k-poselet is measured by the maximum discrimination of the corresponding pose category against any of the other categories. For the j-th pose category, N'_K k-poselets are averagely assigned to $(N_P - 1)$ subsets, each of which is selected to discriminate the j-th pose category from one of the rest $(N_P - 1)$ pose categories: to be specific, $\left(N'_K - \left\lfloor \frac{N'_K}{N_P-1} \right\rfloor \times (N_P - 1)\right)$ subsets are assigned with each containing $\left\lceil \frac{N'_K}{N_P-1} \right\rceil$ k-poselets, meanwhile each of the rest $\left(N_P - 1 - \left(N'_K - \left\lfloor \frac{N'_K}{N_P-1} \right\rfloor \times (N_P - 1)\right)\right)$ subsets contains $\left\lfloor \frac{N'_K}{N_P-1} \right\rfloor$ k-poselets. The detailed procedure of Average-Max selection is illustrated in Algorithm 2.

To ensure the weights of k-poselets W_K is within 0 to 1, the sigmoid function is applied and α is set to 10 in the experiment:

$$W_K^i = \frac{1}{1 + \exp^{-\alpha w^i}}. \tag{10}$$

3 K-poselets Based Pose Recognition

3.1   Pose Recognition

Given the generated dictionary $\boldsymbol{D_{kpl}}$ and one single image I, firstly the k-poselets detection tuple $D_L(I, \boldsymbol{D_{kpl}})$ is calculated as described in Sect. 2.2. The pose score of I is defined as a mapping Φ from $D_L(I, \boldsymbol{D_{kpl}})$: $score_{pose} = \Phi(D_L(I, \boldsymbol{D_{kpl}})) \in \mathbb{R}^{1 \times N_P}$, where the score of the i-th pose category is calculated as $score_{pose}^i = \sum_{j=1}^{L} W_O^j S_L^j W_K^{Ki} C_i^j$, where $C_i = \{C_i^j\}_{j=1}^L \in \mathbb{N}^{1 \times L}$ and $C_i^j = \Delta(P_L^j == i)$.

$score_{pose}^i$, representing the likelihood of image I belonging to the i-th pose category, can be further expressed as

$$score_{pose}^i = F(D_L, i; W_O) = [W_O^1, \cdots, W_O^L] \begin{bmatrix} S_L^1 W_K^{K_K^1} C_i^1 \\ \vdots \\ S_L^L W_K^{K_K^L} C_i^L \end{bmatrix} = \langle W_O^T, \phi(D_L, i) \rangle. \tag{11}$$

The estimated pose label is supposed to maximize the corresponding pose score; however, some images may not be detected by any k-poselets in dictionary $\boldsymbol{D_{kpl}}$, which results in $S_L = \{\underbrace{0, 0, \cdots, 0, 0}_{L}\}$. In this case, it is not recognized as any pose label in N_P categories:

$$\hat{y} = \begin{cases} \arg\max\limits_{y \in [1, N_P]} F(D_L, y; W_O) & \exists i \in [1, L], S_L^i \neq 0 \\ 0 & \forall i \in [1, L], S_L^i = 0 \end{cases}. \tag{12}$$

3.2   Parameter Inference

Given N_{val} validation images, suppose the ground truth pose label of i-th image is y_i and the label inferred from Eq. 12 is $\hat{y}_i$, the objective function is defined as

$$\min_{W_O, \xi \geq 0} \frac{1}{2}(W_O - w_O)(W_O - w_O)^{\mathrm{T}} + \frac{C}{N_{val}} \sum_{i=1}^{N_{val}} \xi_i, \tag{13}$$
$$s.t. \, \forall \hat{y}_i, \, W_O \phi(D_L(i), y_i) - W_O \phi(D_L(i), \hat{y}_i) \geq \Delta(y_i == \hat{y}_i) - \xi_i,$$

which can be solved by a cutting plane procedure [8] and we use the package provided by A. Vedaldi[1].

3.3   Dictionary Learning

The k-poselets selection is conducted in an iterative manner: after each iteration, the number of k-poselets for each pose category is reduced to discard less discriminative k-poselets so that the k-poselets dictionary $\boldsymbol{D_{kpl}}$ can be updated. In next iteration, the detection matrix M_{det} and weighted detection matrix w_K are changed due to the updated $\boldsymbol{D_{kpl}}$, which further affects the selection of k-poselets. The proposed iterative dictionary learning algorithm is capable of easing the influence of undesired or discarded k-poselets in dictionary generation, as summarized in Algorithm 3.

Algorithm 3. Iterative k-poselets dictionary learning

Require: $\boldsymbol{D_{kpl}}$: Initial dictionary of visually similar k-poselets; N_K: Number of k-poselets for each pose category in initial dictionary; T_{iter}: Iteration times; N_{iter}: Number of k-poselets to discard for each pose category in one iteration;
Ensure: $\boldsymbol{D'_{kpl}}$: Iteratively generated dictionary of visually similar k-poselets;
1: **Initialize:** Number of visually similar k-poselets after each iteration: $N'_K = N_K, \boldsymbol{D'_{kpl}} = \boldsymbol{D_{kpl}}$;
2: **for** $t = 1$ to T_{iter} **do**
3:      Calculate M_{det} and w_K according to Eq. 4 and Eq. 8 respectively;
4:      $N'_K \leftarrow N'_K - N_{iter}, \boldsymbol{D_{kpl}} \leftarrow \boldsymbol{D'_{kpl}}$;
5:      **if MA-selection then**
6:          $[\boldsymbol{D'_{kpl}}, w] \leftarrow$MA-selection$(\boldsymbol{D_{kpl}}, N'_K)$;
7:      **end if**
8:      **if AM-selection then**
9:          $[\boldsymbol{D'_{kpl}}, w] \leftarrow$AM-selection$(\boldsymbol{D_{kpl}}, N'_K)$;
10:     **end if**
11:     Update W_K and W_O according to Eq. 10 and Eq. 13 respectively;
12: **end for**
13: **return** $\boldsymbol{D'_{kpl}}$;

---

[1] http://www.robots.ox.ac.uk/~vedaldi//svmstruct.html.

4   Experimental Results

4.1   Released Database

Our human pose database are collected from the Internet. After images that contain one person are cropped, the locations of 13 human keypoints are annotated manually. To be specific, the annotated keypoints are head, left shoulder, right shoulder, left elbow, right elbow, left wrist, right wrist, left hip, right hip, left knee, right knee, left ankle and right ankle. If one keypoint can be clearly observed, its location will be annotated and recorded.

The human poses in the database consist of 3 categories: stoop, running, squat. After selection, there are 1080 annotated human pose images and the number of each category is 390, 220 and 470 respectively. Figure 3 shows some examples of selected images for each category. As we can see, the images in the released database have cluttered background and images belonging to the same pose category include wide visual variations.

(a) stoop

(b) running

(c) squat

Fig. 3. Examples of different human pose images in our database.

4.2   Experiments on Pose Recognition

Experiment Details. In experiments, images for each category is split into training, validation and testing images with the ratio of 6:1:3. For each pose, 50 2-poselets are generated initially and 5 2-poselets are left out in each iteration. The descending weights $W_{O\_de} \in \mathbb{R}^{1 \times L}$ assigned to L detection scores are initialized as $w^l_{O\_de} = \frac{L+1-l}{\sum_{i=1}^{L}(L+1-i)}$ $(1 \le l \le L)$, which gives more credits to higher ranked k-poselets detection score.

Table 1. Human pose recognition performance on our released database. **para** indicates the combination of $T_{iter} - Sel_{mode} - L$, where $Sel_{mode} = 0/1$ means AM/MA selection is applied.

para	1-0-10	1-1-10	1-0-15	1-1-15	1-0-20	1-1-20	2-0-10	2-1-10
Stoop	59.83%	62.39%	61.54%	59.83%	57.26%	59.83%	57.26%	62.39%
Running	48.48%	48.48%	50.00%	46.97%	43.94%	42.42%	46.97%	46.97%
Squat	26.24%	24.11%	32.62%	24.11%	26.95%	25.53%	29.08%	26.24%
Acc.	42.90%	42.90%	**46.60%**	41.67%	41.36%	41.36%	42.90%	43.52%
mAP	44.85%	45.00%	**48.05%**	43.64%	42.72%	42.60%	44.44%	45.20%
para	2-0-15	2-1-15	2-0-20	2-1-20	3-0-10	3-1-10	3-0-15	3-1-15
Stoop	55.56%	61.54%	61.54%	61.54%	55.56%	56.41%	59.83%	62.39%
Running	50.00%	42.42%	50.00%	42.42%	56.06%	50.00%	53.03%	45.45%
Squat	33.33%	25.53%	27.66%	26.95%	31.21%	29.08%	26.24%	28.37%
Acc.	44.75%	41.98%	44.44%	42.59%	45.06%	43.21%	43.83%	44.14%
mAP	46.30%	43.16%	46.40%	43.64%	47.61%	45.16%	46.37%	45.41%
para	3-0-20	3-1-20	4-0-10	4-1-10	4-0-15	4-1-15	4-0-20	4-1-20
Stoop	63.25%	59.83%	54.70%	55.56%	61.54%	51.28%	63.25%	54.70%
Running	51.52%	46.97%	48.48%	51.52%	56.06%	46.97%	56.06%	51.52%
Squat	21.99%	27.66%	35.46%	31.91%	21.28%	34.04%	23.40%	36.88%
Acc.	42.90%	43.21%	45.06%	44.44%	42.90%	42.90%	44.44%	46.30%
mAP	45.58%	44.82%	46.22%	46.33%	46.29%	44.10%	47.57%	47.70%

Evaluation Metrics. To evaluate the performance of proposed methods on human pose recognition, we calculate the accuracy of each category and entire testing images. Considering the number of testing images varies in different pose categories, mean average precision (mAP) is also included for evaluations.

Results. In the experiments, the iteration times T_{iter} and the number of top detection scores L are changed. At the same time, two selection modes (Max-Average/Average-Max selection) are also investigated. The experimental results are summarized in Table 1.

Figure 4 visualizes how following parameters will affect the recognition performance (mAP): (a) top detection scores number L; (b) iteration times T_{iter}; (c) selection mode Sel_{mode}. We can conclude that average-Max (AM) based visually similar k-poselets selection is superior to Max-Average (MA) based selection in term of mAP. The confusion matrix corresponding to the parameter selection that achieves the best recognition performance is shown in Fig. 5.

To highlight the superiority of proposed visually similar k-poselets, the experiments are repeated with original k-poselets (both 1-poselets and 2-poselets included). For fair comparison, the number of poselets collected for original k-poselets is constrained to 13 ($k = 1$)/12 ($k = 2$) while the average number

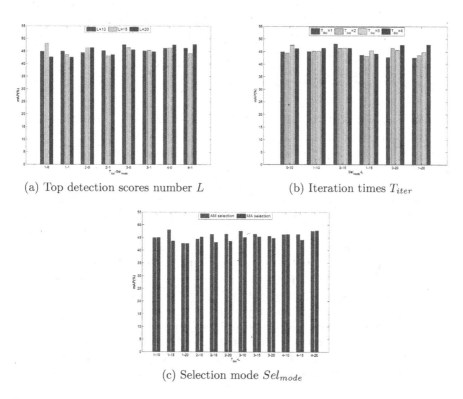

(a) Top detection scores number L (b) Iteration times T_{iter}

(c) Selection mode Sel_{mode}

Fig. 4. Performance comparisons concerning different parameter selections: (a) top detection scores number L; (b) iteration times T_{iter}; (c) selection mode Sel_{mode}.

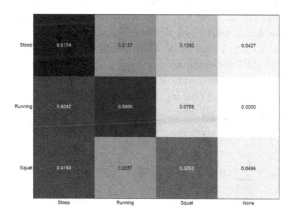

Fig. 5. Confusion matrix of the best recognition performance in Table 1.

Table 2. Comparisons of pose recognition performance achieved by proposed visually similar k-poselets and original k-poselets on our released database. Both 1-poselets/2-poselets with descending/equal weight are evaluated.

Methods	Acc. (%)	mAP (%)	$\bar{S}_{de>eq}/\bar{S}_{de\leq eq}$
Proposed 1-poselets + W_{O_de}	**56.79**	60.11	$0.0238(\frac{3.7660}{158})/0.0174(\frac{4.1432}{238})$
Proposed 1-poselets + W_{O_eq}	55.86	**60.74**	
Original 1-poselets + W_{O_de}	52.47	59.39	$0.0207(\frac{4.1069}{198})/0.0146(\frac{2.8987}{198})$
Original 1-poselets + W_{O_eq}	53.09	59.97	
Proposed 2-poselets + W_{O_de}	49.38	49.86	$0.0209(\frac{5.7150}{273})/0.0136(\frac{1.6756}{123})$
Proposed 2-poselets + W_{O_eq}	48.46	51.07	
Original 2-poselets + W_{O_de}	48.77	51.98	$0.0470(\frac{14.4224}{307})/0.0140(\frac{1.2435}{89})$
Original 2-poselets + W_{O_eq}	47.84	50.37	

of poselets of clustered 150 visually similar k-poselets is 12.27 ($k = 1$)/11.39 ($k = 2$). The weight W_{O_de} is previously set in descending order to favour higher ranked detection scores; equal weight $W_{O\_eq} \in \mathbb{R}^{1 \times L}$ with initial equal weight $w^l_{O\_eq} = 1/L\,(1 \leq l \leq L)$ is also investigated in comparison with descending weight. For each combination proposed/original 1-poselets/2-poselets + descending/equal weight, we iteratively alter other parameters (e.g., T_{iter}, L, Sel_{mode}) in the experiments, which results in 396 different parameter settings. The best recognition performance among 396 settings for each combination is summarized in Table 2.

As we can conclude, the recognition performance achieved by our proposed visually similar k-poselets is superior to original k-poselets, which further stress the importance of grouping visual similar/coherent instances in train strong detectors. Although more complicated models are expected to have better performance, 1-poselets based method outperforms 2-poselets based one in our experiments. It may result from the fact a 2-poselets considered to be detected requires corresponding 2 parts are detected with high confidence at the same time; however, due to great variations in appearance, a 2-poselets considered to be detected is more difficult compared with 1-poselets. At the same time, some of the generated 1-poselets are capable of capturing the visual property of given pose category, which leads to the high accuracy in pose recognition. To better analyze how descending/equal weight affects the recognition performance, we introduce 2 statistics $\bar{S}_{de>eq}$, $\bar{S}_{de\leq eq}$ that are defined as

$$
\begin{aligned}
\bar{S}_{de>eq} &= \sum_{i=1}^{396} (p^i_{de} - p^i_{eq})\Delta(p^i_{de} > p^i_{eq}) / \sum_{i=1}^{396} \Delta(p^i_{de} > p^i_{eq}), \\
\bar{S}_{de\leq eq} &= \sum_{i=1}^{396} (p^i_{eq} - p^i_{de})\Delta(p^i_{de} \leq p^i_{eq}) / \sum_{i=1}^{396} \Delta(p^i_{de} \leq p^i_{eq}),
\end{aligned}
\tag{14}
$$

where p_{de}^i, p_{eq}^i is the mAP of the i-th ($1 \leq i \leq 396$) experimental setting with descending/equal weight. $\bar{S}_{de>eq}$ represents the average mAP improvement of experimental settings that descending weight achieves better mAP than equal weight. By comparing these statistics, we can verify the effectiveness of adopted strategy of assigning more weight to higher ranked detection score.

5 Conclusions

We tackle the problem of human pose recognition in images and propose the recognition algorithm based on the pose dictionary of visually similar and discriminative k-poselets. These k-poselets are developed from original k-poselets, constrained by visual similarity and selected with discriminative criterion. To better evaluate the recognition performance and accelerate the research pace of related tasks, we release a new human pose database that contains 1080 annotated human images with varying poses, viewpoints and backgrounds. The experimental results on the released database demonstrate that the proposed visually similar k-poselets based pose recognition method is effective and promising.

Acknowledgement. This work was partly funded by NSFC (No. 61571297, No. 61527804, No. 61420106008), 111 Project (B07022), and China National Key Technology R&D Program (No. 2012BAH07B01).

References

1. Brulin, D., Benezeth, Y., Courtial, E.: Posture recognition based on fuzzy logic for home monitoring of the elderly. IEEE Trans. Inf. Technol. Biomed. **16**, 974–982 (2012)
2. Dalal, N., Triggs, B.: Histograms of oriented gradients for human detection. In: Proceedings of IEEE International Conference on Computer Vision and Pattern Recognition, vol. 1, pp. 886–893 (2005)
3. Endres, I., Shih, K.J., Jiaa, J., Hoiem, D.: Learning collections of part models for object recognition. In: Proceedings of the IEEE International Conference on Computer Vision and Pattern Recognition, pp. 939–946 (2013)
4. Fan, X., Zheng, K., Lin, Y., Wang, S.: Combining local appearance and holistic view: dual-source deep neural networks for human pose estimation. In: Proceedings of the IEEE International Conference on Computer Vision and Pattern Recognition, pp. 1347–1355 (2015)
5. Gkioxari, G., Hariharan, B., Girshick, R., Malik, J.: Using k-poselets for detecting people and localizing their keypoints. In: Proceedings of the IEEE International Conference on Computer Vision and Pattern Recognition, pp. 3582–3589 (2014)
6. Gouiaa, R., Meunier, J.: Human posture recognition by combining silhouette and infrared cast shadows. In: Proceedings of the IEEE International Conference on Image Processing Theory, Tools and Applications, pp. 49–54 (2015)
7. Jain, A., Gupta, A., Rodriguez, M., Davis, L.: Representing videos using mid-level discriminative patches. In: Proceedings of the IEEE International Conference on Computer Vision and Pattern Recognition, pp. 2571–2578 (2013)

8. Joachims, T., Finley, T., Yu, C.N.J.: Cutting-plane training of structural SVMs. Mach. Learn. **77**, 27–59 (2009)
9. Malisiewicz, T., Gupta, A., Efros, A.A.: Ensemble of exemplar-SVMs for object detection and beyond. In: Proceedings of the IEEE International Conference on Computer Vision, pp. 89–96 (2011)
10. Ni, S., Zhang, Q., Kamata, S., Zhang, C.: Learning discriminative and shareable patches for scene classification. In: Proceedings of the IEEE International Conference on Acoustic, Speech and Signal Processing, pp. 1317–1321 (2016)
11. Puwein, J., Ballan, L., Ziegler, R., Pollefeys, M.: Foreground consistent human pose estimation using branch and bound. In: Fleet, D., Pajdla, T., Schiele, B., Tuytelaars, T. (eds.) ECCV 2014. LNCS, vol. 8693, pp. 315–330. Springer, Heidelberg (2014). doi:10.1007/978-3-319-10602-1_21
12. Thurau, C., Hlaváč, V.: Pose primitive based human action recognition in videos or still images. In: Proceedings of the IEEE International Conference on Computer Vision and Pattern Recognition, pp. 1–8 (2008)
13. Toshev, A., Szegedy, C.: DeepPose: human pose estimation via deep neural networks. In: Proceedings of the IEEE International Conference on Computer Vision and Pattern Recognition, pp. 1653–1660 (2014)
14. Singh, S., Gupta, A., Efros, A.: Unsupervised discovery of mid-level discriminative patches. In: Fitzgibbon, A., Lazebnik, S., Perona, P., Sato, Y., Schmid, C. (eds.) Computer Vision – ECCV 2012. LNCS, vol. 7573, pp. 73–86. Springer, Heidelberg (2012). doi:10.1007/978-3-642-33709-3_6
15. Singh, V.K., Nevatia, R.: Action recognition in cluttered dynamic scenes using pose-specific part models. In: Proceedings of the IEEE International Conference on Computer Vision, pp. 113–120 (2011)
16. Zhao, R., Ouyang, W., Wang, X.: Learning mid-level filters for person reidentification. In: Proceedings of the IEEE International Conference on Computer Vision and Pattern Recognition, pp. 144–151 (2014)
17. Zhao, X., Liu, Y., Fu, Y.: Exploring discriminative pose sub-patterns for effective action classification. In: Proceedings of the ACM International Conference on Multimedia, pp. 273–282 (2013)

Public Security Video and Image Analysis Challenge: A Retrospective

Gengjian Xue[✉], Wenfei Wang, Jie Shao, Chen Liang, Jinjing Wu, Hui Yang, Xiaoteng Zhang, Lin Mei, and Chuanping Hu

The Third Research Institute of the Ministry of Public Security,
Shanghai 200031, China
xgjsword@163.com, wolfeiwang@139.com, jieshao.mail@gmail.com,
i_liang_chen@foxmail.com, xyzl.xt@163.com, dongjiyinxin@163.com,
zxt881108@126.com, gasswlw@126.com, cphu@vip.sina.com

Abstract. The Public Security Video and Image Analysis Challenge (PSVIAC) is a benchmark in object detection and instance search on public security surveillance videos. This challenge is first held in 2016, attracting participation from more than twenty institutions. This paper provides a review of this challenge, including tasks definition, datasets creation, ground truth annotation, and results comparison and analysis. We conclude the paper with some future improvements.

1 Introduction

Recently, many kinds of challenges for video and image analysis have been paid much attention [1–4], because they can provide large mount of data for specific tasks and build a platform for testing various algorithms fairly and publicly. For example, the PASCAL Visual Object Classes Challenge [1] has been an annual event sine 2006, which focuses on visual object classification, detection, segmentation, *etc.*; the ImageNet Large Scale Visual Recognition Challenge [2] has been run annually from 2010 to present, which can be considered doing objects recognition on large scale datasets; the TREC Video Retrieval Evaluation [3] aims to improve the content-based analysis and retrieval technologies. However, the datasets they used are often obtained in specific situations, whose original meaning is to model real world situations. As a result, researchers have got good performance on some tasks, but their algorithms may not work well in practice. Perhaps one of the main reasons is that the data is not obtained from real scenes.

To better deal with visual recognition problems in public security areas, the Public Security Video and Image Analysis Challenge (PSVIAC) was held for the first time. It was also as a special contest on the 2016 Symposium on Research and Application in Computer Vision. The data used in PSVIAC was obtained from real public security scenes, so this challenge could be considered as a benchmark in object detection and instance search for public security applications.

This paper is organized as follows: we start with a review of this challenge in Sect. 2, describing in brief tasks definition, datasets creation, annotation procedure, and evaluation measures. Section 3 provides an overview of the results. We

© Springer International Publishing AG 2017
C.-S. Chen et al. (Eds.): ACCV 2016 Workshops, Part III, LNCS 10118, pp. 441–452, 2017.
DOI: 10.1007/978-3-319-54526-4_32

then use these results for several additional analysis. Section 4 discusses some suggestions that may be useful for future challenges. Finally, we conclude this paper in Sect. 5.

2   Challenge Review

This challenge consists of two components: (1) an available dataset of images for training and test, and ground truth annotations for training; (2) a workshop for summarizing the challenge and discussing the results. This section describes in detail the tasks, datasets, annotations, and evaluation procedures.

2.1   Challenge Tasks

There are two principal challenges: (1) object detection – "does the image contain any instances of a particular object class and where are the instances of a particular object class"; (2) instance search – "given one example of the specific target, it is to find out more images that contain this target".

Object Detection. Three are three object classes for detection: **Non_vehicle**, **Vehicle**, and **Pedestrian**. For each of the three classes, participants are required to find each object of that class in a given test image (if any) and predict the bounding box of that object with associated real-valued confidence. In this task, participants may have to localize multiple object classes in the image, which makes the task more demanding. Any annotation provided in the PSVIAC training data could be used. Participants are not permitted to perform additional manual annotation of either training or test data.

Instance Search. Two classes of instances are used for this task: **Non_vehicle**, **Vehicle**. Given an instance target, participants are required to find out the images that most likely contain this instance and predict the bounding box of the instance. For each query instance, at most 100 candidate results are allowed to be submitted, arranging in descending order according to their possibilities. Thus, each result includes such information: the image name, the predicted bounding box coordinates, and its sorted number.

In this contest, the additional requirement to locate the instance in an image makes the task more challenging, since guessing the right answer is far more difficult to achieve. However, it is really needed in police practical applications.

2.2   Datasets Construction

For the purpose of challenge, the dataset is divided into two subsets: object detection dataset, and instance search dataset. In order to reduce the amount of calculation and ensure the data quality, we extract I-frames from collecting videos to construct the dataset.

Object Detection Dataset. In practical applications, **Non_vehicle**, **Vehicle**, and **Pedestrian** are the three common types. However, since real scenes are often complicated, some aspects should be considered for creating a valuable dataset.

1. *Weather Condition*: The dataset should contain images under various weather conditions, such as sunny, cloudy, and rainy.

2. *Scale Condition*: The dataset should contain images that are taken under a variety of distance, including long distance, middle distance, and close distance.

3. *Angle Condition*: The dataset should contain images that are taken from multiple views.

4. *Multiple Objects*: The dataset should contain images that multiple objects exist in an image.

5. *Occlusion Condition*: The dataset should contain images that contain occluded objects, including unrecognizable and recognizable objects.

Based on these requirements, we collect a large mount of candidate images and select high quality images from them to construct this dataset. Since too many or few objects in an image may be not useful for training and test, the high quality image should satisfy these conditions at the same time:

(a). The number of valid objects in an image is between 3 and 12.
(b). The number of invalid objects in an image is less than 10.
(c). The number of valid occluded objects in an image is less than 6.

The valid object is defined as the area of the object is bigger than 900 pixels and can be recognized by human eyes. The valid occluded object is defined as the ratio of the occluded area to its total area is smaller than 0.5, and the object can be recognized by human eyes.

Some examples of the object detection dataset are shown as follows. Figure 1 shows the examples in this dataset with various weather conditions. Figure 2 shows the examples in this dataset with various scale conditions. Figure 3 shows the examples in this dataset with various angle conditions. Figure 4 shows the examples in this dataset with multiple objects. Figure 5 shows the examples in this dataset with various occlusion conditions.

In this object detection dataset, there are total 39151 images. We first select 20000 images for training. The remaining 19151 images are for candidate test, then from which 10000 images are selected for formal object detection test.

Fig. 1. Examples of the object detection dataset with various weather conditions.

Fig. 2. Examples of the object detection dataset with various scale conditions.

Fig. 3. Examples of the object detection dataset with various angle conditions.

Fig. 4. Examples of the object detection dataset with multiple objects.

Fig. 5. Examples of the object detection dataset with various occlusion conditions.

Instance Search Dataset. For the instance search dataset creation, we have noticed some aspects in practical applications.

1. *Weather Condition*: The instance may appear in various weather conditions.

2. *Scale Condition*: The images containing the instance should be captured under a variety of distance, including long distance, middle distance, and close distance.

3. *Angle Condition*: The images containing the instance should be taken from multiple angles.

4. *Occlusion Condition*: The instance in the image may be occluded by other objects.

Based on these considerations, we collect a large mount of high quality images to construct the instance search dataset. Some example of this dataset are shown as follows. Figure 6 shows the examples in this dataset with various weather conditions. Figure 7 shows the examples in this dataset with various scale conditions.

Fig. 6. Examples of the instance search dataset with various weather conditions. The left image is the given instance. The middle and right images are the images containing this instance.

Fig. 7. Examples of the instance search dataset with various scale conditions. The left image is the given instance. The middle and right images are the images containing this instance.

Fig. 8. Examples of the instance search dataset with various angle conditions. The left image is the given instance. The middle and right images are the images containing this instance.

Fig. 9. Examples of the instance search dataset with various occlusion conditions. The left image is the given instance. The middle and right images are the images containing this instance.

Figure 8 shows the examples in this dataset with various angle conditions. Figure 9 shows the examples in this dataset with various occlusion conditions.

By statistics, the instance search dataset contains 47458 images, where the total number of images for **Vehicle** instance search is 24396, and the total number of images for **Non-vehicle** instance search is 23062. 100 candidate instances have been created, including 60 **Vehicle** instances and 40 **Non-vehicle** instances, then from which 10 **Vehicle** instances and 5 **Non-vehicle** instances are selected for formal instance search test.

2.3 Annotation Procedure

The annotation procedure consists of two steps:

(1) Sensitive information annotation. The sensitive information in an image includes: (a) The information from which we can determine the places that the image are taken, such as road names, place names. (b) The information from which we can determine the specific object, such as plate numbers, advertising messages. We annotate the bounding boxes that contain this information. Then, these areas are blurred by convolution.

(2) Valid objects and valid occluded objects annotation. For the object detection dataset, we annotate all the valid objects and valid occluded objects in each image. For the instance search dataset, we just annotate the object corresponding the same instance in an image. The ground truth area should be the smallest bounding box including the object. For occluded objects, we predict the occluded area, and label the ground truth area. The annotated deviation is within 4 pixels.

In the object detection dataset, statistics indicates that in the training dataset, the total number of **Vehicle, Non-vehicle**, and **Pedestrian** are 75653, 29725, and 9632 respectively. While in the formal test dataset, the total number of **Vehicle, Non-vehicle**, and **Pedestrian** are 42129, 11689, and 2767 respectively. In the instance search dataset, for each **Vehicle** instance query, the number of images containing this instance is about 20; and for each **Non-vehicle** instance query, the number of images containing this instance is about 15.

2.4 Evaluation Measures

Object Detection Evaluation. The criteria for objects detection is designed to penalize the algorithm for missing object instances, for duplicate detections of one instance, and for false positive detections.

Detections are assigned to groundtruth objects and judged to be true or false positives by measuring bounding box overlap. Let $IOU(B_p, B_{gt})$ be the overlap area between the predicted bounding box B_p and ground truth bounding box B_{gt}, it is computed as:

$$IOU(B_p, B_{gt}) = \frac{B_p \cap B_{gt}}{B_p \cup B_{gt}} \tag{1}$$

where $B_p \cap B_{gt}$ means the intersection area of the predicted and ground truth bounding boxes, and $B_p \cup B_{gt}$ denotes their union. A detection is considered as correct when its $IOU(B_p, B_{gt})$ value exceeds a given threshold T_{det}, where it is set to be 0.5 in this contest.

For each object class, we first compute its *precision-recall* curve from a method's rank output. The *precision* is defined as the fraction of correct detections out of the total detections returned by the algorithm, and the *recall* is defined as the fraction of the correct detections out of the total ground truth instances in the dataset. The interpolated average precision (denoted as AP_{det}) [5] is adopted as the average measure over one detection class, which summaries the shape of the precision-recall curve, and is defined as the mean precision at a set of uniformly-spaced *recall* value $[0, 0.1, 0.2, \ldots, 1]$:

$$AP_{det} = \frac{1}{11} \Sigma_{r \in 0,0.1,\ldots,1} P_{interp}(r) \tag{2}$$

The precision at each *recall* level r is interpolated by taking the maximum *precision* measured for a method for which the corresponding *recall* exceeds r:

$$P_{interp}(r) = \max_{r':r' \geq r} p(r') \tag{3}$$

where $p(r')$ is the measured *precision* at *recall* r'.

The overall performance on the object detection task is got by averaging the AP_{det} values on three object classes.

Instance Search Evaluation. For a given instance, at most 100 candidate results are allowed to be returned. One search result considered to be correct should satisfy these conditions: (1) The returned image name can be found in the ground truth image name list; (2) The overlap area between its predicted bounding box and the ground truth bounding box exceeds a given threshold T_{ins}, which is set to 0.1 in this contest, where the computation of the overlap area is according to Eq. (1).

The performance over one instance search is measured by computing the average precision in retrieval (denoted as AP_{ins}) [6,7], which is defined as follows:

$$AP_{ins} = \frac{1}{R} \Sigma_{j=1}^{n} I_j \times \frac{R_j}{j} \tag{4}$$

where R means the total number of ground truth images containing the specific instance, n stands for the total number of returned images by the algorithm (We set $n = 100$ in this contest), j is the index number, I_j is 1 when the j_{th} result is correct, otherwise I_j equals to 0, R_j means the total number of correct results in the first j results.

The overall performance on the instance search task is got by averaging the AP_{ins} values on all test instances.

Score and Ranking. In this challenge, the organization committee decides to give one final ranking according to both the object detection and instance search contests. Since different evaluation measures have been used for these two competitions, directly combining or adding the performance values is not feasible. To deal with this problem, we adopt the competition score instead of the performance value. The competition score of the individual team in each contest is computed according to its ranking order which is sorted based on its performance.

For each contest, all teams are first sorted according the performance in descending order. With a ranked number, the team's score in this contest is computed as:

$$DET_{score} = 25 * (2 - log(D_{th})) \tag{5}$$

$$INS_{score} = 25 * (2 - log(I_{th})) \tag{6}$$

where D_{th} and I_{th} are its ranked number in the object detection and instance search contests respectively, DET_{score} and INS_{score} denote the team's scores got in the object detection and instance search contests respectively.

The final score of this team in this challenge is computed by adding the above two scores:

$$TOTAL_{score} = INS_{score} + DET_{score} \tag{7}$$

where $TOTAL_{score}$ is the total score of this team in the challenge.

Finally, all teams are sorted according to their scores in descending order, and the winner of this challenge is the team that has the highest score.

We should mention that this is an empirical calculation method. According to this method, each team is encouraged to participate both two contests and submit their results. If it just submits one result, then the score of this team in the other contest is considered to be 0, which would have great impact on its final score.

2.5   Submission

Each team is allowed to submit one final result in each contest. Two contests are both just for automatic runs, not including interactive runs.

For data safety, we have provided virtual environments for operation. Each team had to sign a confidentiality agreement before using the data. The dataset was released in the virtual environment machine which was not permitted for downloading. But participants were allowed to upload their code and other sources to the virtual environment. Three operation systems have been provided for choosing: CentOS, Ubuntu, and Win 7. For each virtual environment, a GPU device has been provided as well. After inquiry and collection, 20 teams, 3 teams, and 5 teams have chosen Ubuntu, CentOS, and Win 7, respectively.

The ground truth training data and the submitted results are both saved as text file according to the required format. The test data was available for half a month days before final submission. During this period, we have encouraged teams to submit their periodical results for evaluation, which may help them to improve algorithms.

3   Results and Analysis

3.1   Results

To the end, only 13 teams have finished this challenge and submitted their results. Table 1 shows their performance on each object class, the overall performance, and rankings in the object detection contest. Table 2 shows their overall performance and rankings in the instance search contest. Table 3 shows their scores and final rankings in this challenge.

It can been that the TH-MIG team has got the best performance in the object detection contest, the HawkEye team and the ZJU teams have got the second and the third places, respectively. The overall performance values of these four teams were all higher than 0.75. In the instance search contest, the overall performance values were relatively low. However, the DongGua team has achieved outstanding performance. The ZJU and SkyWalker teams were the second and third respectively.

Overall, the DongGua and TH-MIG teams both got the highest score 84.95, and they tied for the first. The ZJU and HawkEye teams were the third and fourth. These four teams have been invited to attend the workshop and make a speech.

Table 1. The performance and ranking in the object detection contest.

Team name	Vehicle	Non_vehicle	Pedestrian	AP_{det}	Ranking
KAOYU	69.44	63.08	28.56	53.69	7
AHU_CVPR	86.17	75.63	53.22	71.67	5
DongGua	85.48	74.94	60.27	73.56	4
BaiPao	78.60	38.29	14.49	43.79	10
ZJU	86.56	82.17	60.57	76.43	3
Primary_CvVer	50.20	39.02	10.72	33.31	12
HawkEye	86.91	81.72	62.07	76.90	2
TeamAdelaide	84.37	71.62	32.54	62.85	6
TH-MIG	88.11	84.41	65.77	79.43	1
SkyWalker	68.46	48.92	28.45	48.61	8
Endless	0.29	0.01	0.00	0.10	13
FTD	60.45	48.54	22.87	43.95	9
HuanJing	64.11	44.66	12.06	40.27	11

3.2   Analysis

In this subsection, some analysis on the results will be given. Table 4 shows the average performance of each individual contest over all teams. We can seen that participants have got the best results in doing the **Vehicle** detections,

Table 2. The performance and ranking in the instance search contest.

Team name	AP_{ins}	Ranking
KAOYU	0.89	9
AHU_CVPR	0.00	10
DongGua	45.79	1
BaiPao	0.00	10
ZJU	6.07	2
Primary_CvVer	0.00	10
HawkEye	1.46	8
TeamAdelaide	4.00	5
TH-MIG	4.22	4
SkyWalker	5.20	3
Endless	0.00	10
FTD	2.01	6
HuanJing	1.69	7

Table 3. The final scores and rankings in this challenge.

Team name	DET_{score}	INS_{score}	$TOTAL_{score}$	Final ranking
KAOYU	28.87	26.14	55.02	9
AHU_CVPR	32.53	25.00	57.53	7
DongGua	34.95	50.00	84.95	1
BaiPao	25.00	25.00	50.00	11
ZJU	38.07	42.47	80.55	3
Primary_CvVer	23.02	25.00	48.02	12
HawkEye	42.47	27.42	69.90	4
TeamAdelaide	30.55	32.53	63.08	6
TH-MIG	50.00	34.95	84.95	1
SkyWalker	27.42	38.07	65.49	5
Endless	22.15	25.00	47.15	13
FTD	26.14	30.55	56.69	8
HuanJing	23.97	28.87	52.84	10

and better results have been obtained in the **Non_vehicle** detections. The **Pedestrian** detection seems to be the most difficult detection task. The reason may be as follows. First, the **Vehicle** detection is a common task. Much data could be used and researchers have studied it for years. Second, **Vehicle** is a rigid object. Although our scenes are complicated, its shape keeps relatively unchanged. Third, compared with other two classes, the area of **Vehicle** is bigger, which is easier to be detected. For the **Non-vehicle** detection, fewer publicly available data can be used. Furthermore, it is not absolutely a rigid object. Although many **Pedestrian** datasets have been released for training, many public algorithms may be not work well in our cases. The pedestrians in our dataset are often relatively small, with multiple views, and affected by various factors. Whereas most public datasets were obtained under ideal conditions.

For the instance search task, it is more demanding. It should mention that only two teams are actually well beyond the average value. One reason is that only one example per instance has been provided, so that the training process is challenging. Another reason is that the instance may appear in various forms, such as multi-scale variation, multi-view variation, and occlusion. The DongGua team has got the outstanding result by first using deep learning technologies [8] and other data to train many models off-line, and then fusing these models on our datasets. However, the performance may be further improved if small sample learning problems would be well resolved.

Table 4. The average performance of each individual contest

Name	Vehicle	Non-vehicle	Pedestrian	Instance search
Average performance	69.93	57.92	34.74	5.49

4   Discussions

This section discusses some topics for future improvement of the challenge.

The first topic is about the dataset augmentation. In this first challenge, we have totally released 86609 images for training and test. However, the mount of data seemed to be not big enough to support models training. Another problem is that the number of objects belonging to different classes is not uniform. For example, the number of pedestrian objects is much less than that of other two classes, which may affect the algorithm's performance. So it is necessary to effectively expand the dataset.

The second topic is about the evaluation measures. In this challenge, we have adopted an empirical approach to calculate scores and rankings. However, there may exist more reasonable methods. Besides, other objective evaluation measures could be adopted for evaluating algorithms from multiple aspects.

The third topic is that more in-depth analysis on the results and algorithms should be introduced, such as algorithms comparison, distribution analysis of results, and inter-class comparisons.

5   Conclusions

The PSVIAC has contributed to the development of video and image analysis technologies in public security. More than twenty institutes from home and abroad have participated this competition, and many effective methods have been proposed. We believe that this first challenge is a good start and it will be getting more and more better with continuous improvements.

Acknowledgement. Our research was sponsored by following projects: Program of Science and Technology Commission of Shanghai Municipality (No. 15530701300, 15XD1520200, 14DZ2252900); 2012 IoT Program of Ministry of Industry and Information Technology of China; Key Project of the Ministry of Public Security (No. 2014JSYJA007); Shanghai Science and Technology Innovation Action Plan (No. 16511101700).

References

1. Everingham, M., Eslami, S.M.A., Gool, L.V., Williams, C.K.I., et al.: The pascal, visual object classes challenge: a retrospective. Int. J. Comput. Vis. **111**, 98–136 (2015)
2. Russakovsky, O., Deng, J., Su, H., Krause, J., Satheesh, S., Ma, S., et al.: Imagenet large scale visual recognition challenge. Int. J. Comput. Vis. **115**, 211–252 (2015)
3. Over, P., Fiscus, J., Sanders, G., et al.: Trecvid 2014-an overview of the goals, tasks, data, evaluation mechanisms and metrics. In: Proceedings of TRECVID (2014)
4. Patino, L., Ferryman, J.: Pets: dataset and challenge. In: IEEE International Conference on Advanced Video and Signal Based Surveillance 2014, pp. 355–360 (2014)
5. Salton, G., Mcgill, M.J.: Introduction to Modern Information Retrieval. McGraw-Hill Inc., New York (1986)
6. Zhu, M.: Recall, precision, and average precision. Department of Statistics and Actuarial Science, University of Waterloo (2004)
7. Turpin, A., Scholer, F.: User performance versus precision measures for simple search tasks. In: Proceedings of the International ACM SIGIR Conference on Research and Development in Information Retrieval, pp. 11–18 (2006)
8. Bengio, Y., Courville, A., Vincent, P.: Representation learning: a review and new perspectives. IEEE Trans. Pattern Anal. Mach. Intell. **35**, 1798–1828 (2013)

Multiple-Branches Faster RCNN for Human Parts Detection and Pose Estimation

Kaiqiang Wei and Xu Zhao[✉]

Institute of Image Processing and Pattern Recognition,
Shanghai Jiao Tong University, Shanghai 200240, China
zhaoxu@sjtu.edu.cn

Abstract. In this work, we primarily address multiple people pose estimation challenge by exploring the performance of Faster RCNN on human parts detection. We develop a multiple-branches Faster RCNN model for our specific task of detecting persons and their parts. Our model can improve the performance of detecting human parts and the whole persons, meanwhile speeding up detection process with shared weights. A part-based method is proposed to estimate multiple people poses, bringing recent advances on object detection to this task. Experiments demonstrate that our model achieves better performance than the original Faster RCNN model on our task. Compared with other pose estimation approaches, our approach achieves fair or better results.

1 Introduction

Human pose estimation has long been a concentrated concerned research topic in computer vision. Fast and robust human pose estimation has extensive application prospects such as human-computer interaction, virtual reality and intelligent monitoring. Pose estimation is still a tough task despite the long history of efforts. Complex variation in limb orientation, large occlusions, truncation and distractions from clothes or other overlapping objects make human pose estimation a challenging problem.

Most of the work on human pose estimation focuses on single person, which makes the task less intractable because one does not need to search over people's positions. We try to deal with multiple people pose estimation, taking an image with multiple people as input and outputting people's poses in an end-to-end way. This is a much tougher task than single person pose estimation, because neighboring people frequently overlap with each other with strong interference.

In this work, we adopt Faster RCNN [17] and explore its performance on human parts detection. We develop a multiple-branches Faster RCNN model for simultaneously human parts and whole person detection. The multiple-branches model has two branches, one for person detection and the other for human parts detection. We take a three-stage training strategy to make the two branches sharing the basic convolutional layers. The multiple-branches model outperforms the original Faster RCNN on our task by a large margin, while maintaining

© Springer International Publishing AG 2017
C.-S. Chen et al. (Eds.): ACCV 2016 Workshops, Part III, LNCS 10118, pp. 453–462, 2017.
DOI: 10.1007/978-3-319-54526-4_33

the time efficiency of Faster RCNN as much as possible. In addition, a part-based method is proposed for multiple people pose estimation. We set a series of simple but practical rules to infer joints location from detected part boxes and connect the joints orderly to get human poses. We naturally take the predicted people location boxes as the basis of multiple people pose estimation, because person detection result is accurate and reliable. Experiments on public datasets demonstrate that our approach achieves fair or better results when compared with other pose estimation approaches.

2   Related Work

To deal with pose estimation problem, the Pictorial Strictures Model (PSM) [6] and many other PSM-based pose models [1] were developed in the early days. PSMs represent human parts with rectangle boxes and model parts connection relationship with tree structure. Yang [22] introduces part types based on PSM and uses several mixture part types to model human parts, which improves pose estimation performance. Part-based models such as poselets [3] and their variants [8] also play an import role in human pose estimation and action recognition. Random trees [18], graphical model [4] and convolutional neural networks [20] are also applied on pose estimation. A large amount of studies based on these learning-based methods had been presented and obtain competitive performance and good tractability. Most research focus on single person pose estimation only. There is not much literature addressing multiple people pose estimation, which is a more realistic problem.

Recently, tremendous progress has been achieved on object classification and detection due to convolutional neural network (CNN) [14]. The latest proposed Faster RCNN [17], one of the state-of-the-art object detection algorithms, shows amazing performance on PASCAL VOC object detection challenge. It dates back to RCNN [10], a region-based CNN detection algorithm, followed by SPP [12]. Fast RCNN [11] simplifies multi-level spatial pyramid pooling layers in SPP to be single-level RoI layer and adopts joint training with a multi-task loss. Faster RCNN introduces Region Proposal Network (RPN) to replace the selective search [21] to generate high-quality proposals and merges RPN with Fast RCNN to speed up detection process.

Many researchers apply convolutional neural networks on pose estimation and action recognition. DeepPose [20] trained a three-stage cascade of pose regressors to predict joints location in a holistic manner with high precision, each stage using the same 7-layer network and taking the output of previous stage as the input of next stage. DeepPose proves the feasibility and shows the power of convolutional neural networks on joints location task. Gkioxari et al. [9] train an RCNN-based detector with a multi-task loss to jointly optimize the task of key points location and action recognition. Recently, they [7] also proposed a part-based method by training part detectors for action and attributes classification, which shows that adding parts has essential contribution. Pfister et al. [15] introduce a novel network architecture to regress confidence heatmap of joint position to estimate human pose in videos, capitalizing the temporal information in videos.

3   Multiple-Branches Faster RCNN Model

3.1   Multiple-Branches Faster RCNN

In this work, we develop a multiple-branches Faster RCNN model for the task of detecting people's whole bodies and parts in the full image. The architecture of our model is shown in Fig. 1. Our model employs the 16-layer VGG network [19] as the basic model and has two branches above the shared convolutional layers, with each branch for one specific detection task. In the part branch, we detect human parts, including head, torso, forearm, upper arm, lower leg, upper leg as six different classes and train a Faster RCNN model to detect these parts. In the body branch, we train another Faster RCNN model to detect persons. The reason why we do not train one model to detect human body together with human parts is that they are not independent objects. Human body location box should cover all the human parts location boxes. This situation would affect the training of region proposal network, the proposal generating process, leading to inferior proposal quality and worse detection performance. The following experiments also prove our consideration, which will be described in detail next.

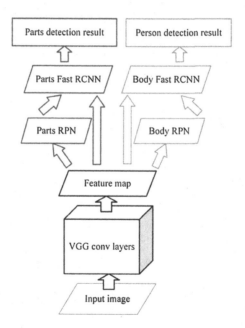

Fig. 1. Multiple-branches faster RCNN architecture.

Furthermore, we take similar ways as Faster RCNN to make our human parts detection model and human body detection model to share convolutional layers. In this way, the detection process can be accelerated. Our training strategy is elaborated as follows.

In the first stage, we train the initial human parts detection model. The human parts RPN model is trained to generate part proposals. Then these proposals are fed into Fast RCNN model to train part detection model. In this stage, both RPN and Fast RCNN are initialized with an ImageNet-pre-trained model.

In the second stage, human parts RPN and Fast RCNN are retrained in the same way as in the first stage, but initialized with the first stage Fast RCNN human parts detection network. We fix all the shared convolutional layers, only fine-tune the RPN and Fast RCNN's respective unique layers. Now the human parts detection model is ready, as the parts branch plotted in our network architecture (see Fig. 1).

In the third stage, we train human body detection model. The RPN is initialized with the first stage Fast RCNN human parts detection network. We only fine-tune the layers which are unique to RPN. Then we use the proposals generated by this stage's RPN to train our human body Fast RCNN detection model. The Fast RCNN in this stage is also initialized with the first stage's Fast RCNN human parts detection network, only updating the fully connected layers. Now we have the body detection branch.

Finally, the parts detection model and body detection model share the basic VGG convolutional layers. During testing process, we only need to do forward inference once to get human parts detection and person detection results, which would improve time efficiency. We call our model multiple-branches Faster RCNN model. Such modifications have advantages on typical tasks such as object and parts detection.

3.2 Multiple People Pose Estimation

For pose estimation, the task is to locate and connect body joints to form a kinematic tree structure. In order to transfer the task of pose estimation to object detection, we firstly set a series of rules to get parts location boxes to train our model. Then we try to reverse the process and set another series of inverse rules to get joints location from predicted parts location boxes. But the two processes are not completely reversible. Our rules are described in detail next.

From Joints to Boxes. To get parts location box annotations for training, we use two joints of a part to construct a tight box at first. If the ratio of the location rectangle box is less than 0.3, the shorter side is extended to be 0.3 times of the longer side. Then the box we get is still too tight to contain the whole part, so we resize the box to be 1.2 times larger, which leads to better detection performance by about 1% improvement of AP.

From Boxes to Joints. After the multiple-branches detection model is trained, we use it to detect human parts and get parts labels and location boxes. To restore joints location of parts, the box is resized to be 1/1.2 times smaller at first. Then, if the height-width ratio or width-height ratio is less than 0.5, we choose the middle point of the shorter side as the predicted joint. Otherwise, we take the image patch inside the box to judge the part orientation. We compute the image patch's gradient along its two diagonal to determine whether the part lies in the left-top to right-bottom diagonal direction or in the right-top to left-bottom diagonal direction. Then the two joints of the part are set to be the corresponding two vertexes of the image patch.

From Joints to Pose. The last step is to link the joints to get pose. We need to connect lower leg to upper leg to form leg, and connect forearm to upper arm to form arm. Since our model does not distinguish left leg from right leg, we match lower leg with its nearby upper leg according to their distance. Once a whole leg is obtained, we modify the knee joint to be the middle point of upper leg's low joint and lower leg's up joint, and connect the whole leg with hip, knee, and ankle joints. Same ways are taken to match forearm with upper arm. Then we get modified elbow joint and connect a whole arm with shoulder, elbow, and wrist joints.

Multiple People Pose Estimation. Since our model detects not only all persons' parts location boxes, but also all persons' whole body location boxes, we naturally take the predicted human body location box as one person's ground truth location, and regard the parts boxes that overlap with the body box as the person's parts. Then we can use the rules described above to get multiple people's poses. By the way, our model performs very well on detecting persons so it is reliable to use the predicted person boxes as the basis of multiple person pose estimation.

4   Experiments

To evaluate our model, we conduct experiments on three public datasets and report part detection and pose estimation performance respectively with standard metrics. We describe the details of our experiments in the next.

4.1   The MPII Human Pose Dataset

The first dataset we use to train our model is the MPII Human Pose Dataset [2], a state-of-the-art benchmark for evaluation of articulated human pose estimation. The dataset includes around 25K images containing over 40K people with annotated human part joints totally. We select images with enough annotated part joints to construct ground truth part boxes from the training set only, without data augmentation. Finally we pick out about 17K images in total, excluding some incomplete annotations. We split the data into two sets, four fifths for training and one fifth for testing. We quantify the performance of our model by computing the average precision (AP), a standard metric for object detection, on our test dataset.

Table 1. Parts and person detection performance on the MPII human pose dataset under different settings for train-time and test-time proposals.

AP(%)	Head	Torso	Upper arm	Forearm	Upper leg	Lower leg	Person
Train 2k, test 300	71.8	69.0	30.1	17.7	29.9	32.3	89.1
Train 6k, test 1k	75.1	71.3	32.5	18.7	31.7	34.4	89.0

We train our Faster RCNN model based on the 16-layer VGG network [19]. The parameters we fine-tuned are the number of train-time region proposals and test-time region proposals. The original setting outputs 2k RPN proposals for Fast RCNN training and selects 300 proposals for testing. We give more proposals during training and testing process because our model needs to output much more boxes of human parts in one single image comparing with the task of object detection on PASCAL dataset. During testing, we set the score thresh of person detection to be 0.75, while the threshes for parts remain as 0.6. We also set the nms threshes for arms and legs to be 0.4 because these limbs overlap with each other more easily, while keeping the nms threshes for head, torso and person as 0.3 (Table 2).

Table 2. Comparison of our multiple-branches Faster RCNN and original Faster RCNN on parts and person detection on MPII human pose dataset. Train-time proposals of both are set to be 2k and test-time proposals of both are both set to be 1k.

AP (%)	Head	Torso	Upper arm	Forearm	Upper leg	Lower leg	Person
Faster RCNN [17]	54.7	56.2	22.5	13.3	20.0	23.8	85.9
Ours	74.3	70.2	30.5	18.1	30.3	32.4	89.2

In order to prove that our Multiple-branches Faster RCNN outperforms the original Faster RCNN on our task of detecting parts and persons, we also train a original Faster RCNN model to treat head, torso, forearm, upper arm, lower leg, upper leg and person as seven different classes. We set the train-time proposals to be 2k and test-time proposals to be 1k, and report the AP metric in Table 1. As can be seen, we get significant improvement on parts and person detection. Figure 2 shows the head and torso detection comparison. The AP of our model is 74.3% for head and 70.2% for torso, while the AP of the original model is 54.7% for head and 56.2% for torso. The improvement of AP is mainly caused by higher recall of our multiple-branches Faster RCNN model, due to its better-trained Region Proposal Network.

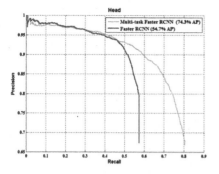 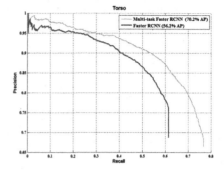

Fig. 2. AP comparison for head and torso detection between our multiple-branches faster-RCNN model and the original faster-RCNN model.

4.2   The PASCAL VOC Dataset

The second dataset we applied on to evaluate our trained model is PASCAL VOC09 person detection val dataset. This dataset has 1446 images with human joints annotated. So we are able to construct ground truth boxes with the annotated joints. The difference is that there are no head ground truth boxes in the PASCAL VOC09 dataset, but some facial points such as eyes, nose, ears and so on are provided. So we have to construct the head ground truth boxes with facial points. Torso box is also different. The MPII dataset provides pelvis annotation so we use pelvis joint and head's low joint to construct torso box, and the width box of ground truth box is about 0.3 times of its height. In PASCAL VOC09 dataset, we have to construct torso box using two joints of shoulders and two joints of hips, then we shrink the box's width to be 0.5 times of its original width.

Table 3 shows the AP comparison of our model and the original faster-RCNN model on this dataset. Our model gains better detection performance then the original faster-RCNN model on parts and person detection. Gkioxari et al. [7] also trained CNN models to detect human parts, including head, torso, legs on this dataset. We make comparison with their result in Table 4. Higher threshold of intersection-over-union (σ in Table 4) means stricter detection measure. Usually the threshold is set to be 0.5. As can be seen, our model outperforms Gkioxari's at high thresholds, which proves our model's good generalization ability.

Table 3. Comparison of our multiple-branches Faster-RCNN and original Faster-RCNN on parts and person detection on PASCAL VOC09 person detection dataset. Train-time proposals of both are set to be 2k and test-time proposals of both are set to be 1k.

AP (%)	Head	Torso	Upper arm	Forearm	Upper leg	Lower leg	Person
Faster RCNN [17]	32.6	13.0	14.6	7.7	8.6	11.9	54.2
Ours	39.7	16.5	17.8	9.6	14.5	17.9	56.0

Table 4. Comparison of parts detection performance on the PASCAL VOC09 person detection dataset.

AP(%)		$\sigma = 0.2$	$\sigma = 0.3$	$\sigma = 0.4$	$\sigma = 0.5$
Ours	Head	51.8	49.3	45.9	39.8
	Torso	47.0	41.9	31.2	17.5
	Upper leg	29.8	26.4	21.4	15.5
	Lower leg	32.9	29.6	24.8	17.5
Gkioxari's [7]	Head	55.2	51.8	45.2	31.6
	Torso	42.1	36.3	23.6	9.4
	Legs	34.9	27.9	20.0	10.4

4.3   The LSP Dataset

The third dataset we used is the Leeds Sport Dataset (LSP) [13] and its·extension. The LSP dataset contains about 2k pose images annotated with 14 joints, one half for training and the other half for testing. The LSP extension dataset contains 10k images, which can be used for training. We do data augmentation with image horizontal flip on the training set. The anchor size of the region proposal network is changed to [16, 32, 64] to fit the smaller image scale. We train our model and then get human pose according to the rules as described in the previous section. Percentage of Correct Parts (PCP) is reported on the dataset. PCP is a widely accepted metric to measure pose estimation performance. We report PCP for single person. We match left or right predicted part joints with left or right ground truth part joints according to their distance. Then the distance between two predicted joints of a part and two corresponding ground-truth joints is computed. A part is considered detected correctly if the distance is less than half of the part length.

We make comparison with other mainstream pose estimation approaches in Table 5. As can be seen, the average PCP of our approach is 63.0%, which is comparable to the other approaches. We achieve higher PCP on head (87.4%) and lower leg (74.2%) detection, which are the best among these approaches. Head and lower legs are large parts and overlap less easily with other parts. So our part-based approach is suitable. While upper arm and forearm are small

Table 5. Comparison of PCP performance at 0.5 of other approaches and ours on the LSP dataset.

Method	Head	Torso	Upper arm	Lower arm	Upper leg	Lower leg	Average
Andriluka [1]	74.9	80.9	46.5	26.4	67.1	60.7	55.7
Dontone [5]	79.2	81.6	45.1	24.7	66.5	61.0	55.5
Yang [22]	79.3	82.9	56.0	39.8	70.3	67.0	62.8
Pishchulin [16]	78.1	87.5	54.2	33.9	75.7	68.0	62.9
Ours	87.4	84.9	47.5	34.6	72.6	74.2	63.0

Fig. 3. Pose estimation results for single person and multiple persons.

and easily covered, other inference approaches [16,22] based on the relationship between parts achieves better results. In addition, the test time for one image is about 350 ms on K40 GPU. Figure 3 shows some pose estimation examples for single person and multiple persons. Images in the first row come from MPII dataset and images in the second row come from the LSP test set.

5 Conclusion

Our work explores the performance of Faster RCNN on human parts and whole body detection. The multiple-branches Fast RCNN model we developed shows comparative advantages on human parts and whole body detection. We speed up the detection process by making the parts detection model and person detection model to share weights. Our multiple-branches model can also be applied to other object and object part detection tasks with similar logic relationship of parts and wholes. We provide a practical part-based method to estimate multiple people poses. Compared with other pose estimation approaches, our method achieves comparative or better results.

Acknowledgement. We gratefully acknowledge that this work is supported by the fundings from National Natural Science Foundation of China (61273285, 61375019).

References

1. Andriluka, M., Roth, S., Schiele, B.: Pictorial structures revisited: people detection and articulated pose estimation. In: IEEE Conference on Computer Vision and Pattern Recognition, CVPR 2009, pp. 1014–1021. IEEE (2009)
2. Andriluka, M., Pishchulin, L., Gehler, P., Schiele, B.: 2D human pose estimation: new benchmark and state of the art analysis. In: 2014 IEEE Conference on Computer Vision and Pattern Recognition (CVPR), pp. 3686–3693. IEEE (2014)
3. Bourdev, L., Malik, J.: Poselets: body part detectors trained using 3D human pose annotations. In: 2009 IEEE 12th International Conference on Computer Vision, pp. 1365–1372. IEEE (2009)

4. Chen, X., Yuille, A.L.: Articulated pose estimation by a graphical model with image dependent pairwise relations. In: Advances in Neural Information Processing Systems, pp. 1736–1744 (2014)
5. Dantone, M., Gall, J., Leistner, C., Van Gool, L.: Human pose estimation using body parts dependent joint regressors. In: 2013 IEEE Conference on Computer Vision and Pattern Recognition (CVPR), pp. 3041–3048. IEEE (2013)
6. Felzenszwalb, P.F., Huttenlocher, D.P.: Pictorial structures for object recognition. Int. J. Comput. Vis. **61**, 55–79 (2005)
7. Gkioxari, G., Girshick, R., Malik, J.: Actions and attributes from wholes and parts. arXiv preprint arXiv:1412.2604 (2014)
8. Gkioxari, G., Hariharan, B., Girshick, R., Malik, J.: Using k-poselets for detecting people and localizing their keypoints. In: 2014 IEEE Conference on Computer Vision and Pattern Recognition (CVPR), pp. 3582–3589. IEEE (2014)
9. Gkioxari, G., Hariharan, B., Girshick, R., Malik, J.: R-CNNs for pose estimation and action detection. arXiv preprint arXiv:1406.5212 (2014)
10. Girshick, R., Donahue, J., Darrell, T., Malik, J.: Rich feature hierarchies for accurate object detection and semantic segmentation. In: 2014 IEEE Conference on Computer Vision and Pattern Recognition (CVPR), pp. 580–587. IEEE (2014)
11. Girshick, R.: Fast R-CNN. arXiv preprint arXiv:1504.08083 (2015)
12. He, K., Zhang, X., Ren, S., Sun, J.: Spatial pyramid pooling in deep convolutional networks for visual recognition. In: Fleet, D., Pajdla, T., Schiele, B., Tuytelaars, T. (eds.) ECCV 2014. LNCS, vol. 8691, pp. 346–361. Springer, Heidelberg (2014). doi:10.1007/978-3-319-10578-9_23
13. Johnson, S., Everingham, M.: Clustered pose and nonlinear appearance models for human pose estimation. In: BMVC, vol. 2, p. 5 (2010)
14. Krizhevsky, A., Sutskever, I., Hinton, G.E.: ImageNet classification with deep convolutional neural networks. In: Advances in neural information processing systems, pp. 1097–1105 (2012)
15. Pfister, T., Charles, J., Zisserman, A.: Flowing convNets for human pose estimation in videos. In: Proceedings of the IEEE International Conference on Computer Vision, pp. 1913–1921 (2015)
16. Pishchulin, L., Andriluka, M., Gehler, P., Schiele, B.: Poselet conditioned pictorial structures. In: Proceedings of the IEEE Conference on Computer Vision and Pattern Recognition, pp. 588–595 (2013)
17. Ren, S., He, K., Girshick, R., Sun, J.: Faster R-CNN: towards real-time object detection with region proposal networks. In: Advances in Neural Information Processing Systems, pp. 91–99 (2015)
18. Rogez, G., Rihan, J., Ramalingam, S., Orrite, C., Torr, P.H.: Randomized trees for human pose detection. In: IEEE Conference on Computer Vision and Pattern Recognition, CVPR 2008, pp. 1–8. IEEE (2008)
19. Simonyan, K., Zisserman, A.: Very deep convolutional networks for large-scale image recognition. CoRR abs/1409.1556 (2014)
20. Toshev, A., Szegedy, C.: DeepPose: human pose estimation via deep neural networks. In: 2014 IEEE Conference on Computer Vision and Pattern Recognition (CVPR), pp. 1653–1660. IEEE (2014)
21. Uijlings, J.R., Sande, K.E., Gevers, T., Smeulders, A.W.: Selective search for object recognition. Int. J. Comput. Vis. **104**, 154–171 (2013)
22. Yang, Y., Ramanan, D.: Articulated human detection with flexible mixtures of parts. IEEE Trans. Pattern Anal. Mach. Intell. **35**, 2878–2890 (2013)

SPID: Surveillance Pedestrian Image Dataset and Performance Evaluation for Pedestrian Detection

Dan Wang[1], Chongyang Zhang[1(✉)], Hao Cheng[1], Yanfeng Shang[2], and Lin Mei[2]

[1] Institute of Image Communication and Network Engineering,
Shanghai Jiao Tong University, Shanghai, China
sunny_zhang@sjtu.edu.cn
[2] The Third Research Institute of The Ministry of Public Security,
Shanghai, China

Abstract. Pedestrian detection is highly valued in intelligent surveillance systems. Most existing pedestrian datasets are autonomously collected from non-surveillance videos, which result in significant data differences between the self-collected data and practical surveillance data. The data differences include: resolution, illumination, view point, and occlusion. Due to the data differences, most existing pedestrian detection algorithms based on traditional datasets can hardly be adopted to surveillance applications directly. To fill the gap, one surveillance pedestrian image dataset (SPID), in which all the images were collected from the on-using surveillance systems, was constructed and used to evaluate the existing pedestrian detection (PD) methods. The dataset covers various surveillance scenes and pedestrian scales, view points, and illuminations. Four traditional PD algorithms using hand-crafted features and one deep-learning-model based deep PD methods are adopted to evaluate their performance on the SPID and some well-known existing pedestrian datasets, such as INRIA and Caltech. The experimental ROC curves show that: The performance of all these algorithms tested on SPID is worse than that on INRIA dataset and Caltech dataset, which also proves that the data differences between non-surveillance data and real surveillance data will induce the decreasing of PD performance. The main factors include scale, view point, illumination and occlusion. Thus the specific surveillance pedestrian dataset is very necessary. We believe that the release of SPID can stimulate innovative research on the challenging and important surveillance pedestrian detection problem. SPID is available online at: http://ivlab.sjtu.edu.cn/best/Data/List/Datasets.

1 Introduction

Pedestrians are the primary surveillance objects in security systems, and thus pedestrian detection is becoming the fundamental research area in intelligent surveillance systems. In the practical surveillance systems, pedestrian detection (PD) is still a challenging problem due to the visual appearance differences

© Springer International Publishing AG 2017
C.-S. Chen et al. (Eds.): ACCV 2016 Workshops, Part III, LNCS 10118, pp. 463–477, 2017.
DOI: 10.1007/978-3-319-54526-4_34

caused by the large-scale variations of surveillance scenes. Most PD algorithms require pedestrian datasets to train classifiers or learn discriminative features using machine learning. The training dataset is a strong dependency of PD algorithms. In the past few years, an increasing number of benchmarks have been proposed to push forward the performance of pedestrian detection, e.g., INRIA [1], ETH [2], Caltech [3] and KITTI datasets [4]. However, most existing pedestrian datasets are collected from non-surveillance videos. The comparisons of the self-collected data and practical surveillance data show that there exist significant data differences. The properties, like resolution, illumination, view point, and occlusion of pedestrians differ greatly. Despite an extensive set of ideas has been explored for pedestrian detection, most existing algorithms are trained on traditional datasets. Consequently the accuracy and robustness of the most existing PD algorithms may perform not so well in the real surveillance systems.

This paper introduces the surveillance pedestrian image dataset (SPID) that aims to fill the gap between the existing datasets and real surveillance data. We collected approximately 297 surveillance video clips (20 min per clip) from on-using surveillance cameras. The videos are collected from different areas and contain 8 typical surveillance scenes. The multiple scenes in SPID are shown in Fig. 2, including the highways, campuses, city roads and rural areas. Videos are recorded continuously 24 h per day, cover 6 various illumination conditions from morning to night. The differences of pedestrian distribution and appearance under various scenes are huge. We extracted the frames with at least one pedestrian, set up the annotation standard and labeled several properties for each pedestrian. After selection, about $110k$ ($k = 10^3$) pedestrian objects were collected and labeled. The pedestrian properties include: illumination, view, size, pose, attachment, occlusion and appearance. Our goal is to complement existing benchmarks by providing real-world surveillance data.

The main contributions of our work are three-fold: (a) The construction of a typical and diverse surveillance dataset, which requires significant efforts in collection and annotation. As far as we know, this dataset is the first released surveillance pedestrian dataset. SPID is available online at: http://ivlab.sjtu.edu. cn/best/Data/List/Datasets. We believe SPID can stimulate innovative research on the challenging and important surveillance PD problem; (b) Four traditional PD algorithms using hand-crafted features and one deep-learning-model based deep PD methods are evaluated using existing pedestrian dataset (such as INRIA [1] and Caltech [3]) and SPID in this work. Our experiments show that the PD performance on SPID is worse than that on traditional datasets. (c) From experiments, we also validate that the data differences between SPID and other pedestrian datasets influence the performances greatly. Main factors include scale, view point, illumination and occlusion.

The rest of the paper is organized as follows. Related works are reviewed in Sect. 2. Section 3 introduces the collection, annotation and properties of SPID. In Sect. 4 we choose seven PD methods to evaluate on INRIA dataset, Caltech dataset and SPID. We show the result ROC curves for each algorithm and analyze the performance differences, as well as some improvement methods. Finally, Sect. 5 summarizes our work and suggests future directions with SPID.

2 Related Works

Multiple public pedestrian datasets have been collected over the past decades, including commonly used INRIA, ETH [2], TUD-Brussels [5], Daimler [6], Caltech and KITTI datasets [4]. TUD-Brussels and ETH are medium-sized video datasets. Daimler is not frequently used by PD methods because it only contains grayscale images. Below are the datasets we consider in the paper. Table 1 shows the comparisons of several pedestrian detection datasets. The first column define the imaging setup method, and the next four columns indicate number of pedestrian and images in the training and test sets. Properties column summarizes additional characteristics of the datasets. The strengths and weaknesses of INRIA, Caltech and KITTI dataset are discussed in detail.

Table 1. Comparison of public pedestrian datasets

Dataset	imaging setup	Testing			Training			Properties								
		# pedestrians	# pos. images	# neg. images	# pedestrians	# pos. images	# neg. images	color images	occlusion labels	view labels	illumi. labels	pose labels	video seqs.	no select. bias	view point	publication
INRIA[1]	photo	1208	614	1218	566	288	453	√							hori.	2005
ETH[2]	mobile cam.	2388	499	-	12k	1804	-		√				√	√	-	2007
Daimler[6]	mobile cam.	15.6k	-	6.7k	56.6k	21.8k	-						√		-	2009
Caltech[3]	mobile cam.	192k	67k	61k	155k	65k	56k	√	√				√	√	hori.	2009
KITTI[4]	mobile cam.	-	7518	-	4445	7481	-		√				√	√	hori.	2012
SPID	fixed cam.	53.7k	15.4k	-	56.2k	14.5k	-	√	√	√	√	√	√	√	bird	2016

INRIA. INRIA dataset is the oldest pedestrian dataset with high quality annotations and high images resolution. Most pedestrians are captured horizontally in day time. The illumination is fine and contours of pedestrians are clear. The diverse scenes contain outdoor landscapes like city road, mountain, beach and indoor environment. However the number of images is not large. Figure 1(a) shows the height distribution of INRIA test set.

Caltech. Caltech dataset is one of the most popular PD datasets. The videos were captured by a vehicle driving through U.S. urban streets in sunny days. Camera is set on a moving car, therefore the view of pedestrians are horizontal and the sizes of pedestrians are small compared to INRIA. The test set contains 67k positive images and 192k pedestrians. Although Caltech is the largest pedestrian dataset, its images are extracted each frame per video, the sampling frequency is high to 30 fps. The difference between two sequential frames is small, therefore usually every 30th image in the Caltech dataset is used for training and testing [7]. The medium pedestrian height is [30, 80] pixels. Figure 1(b) shows the height distribution of Caltech test set.

KITTI. KITTI dataset contains videos captured by a moving car around city streets with good weather conditions, therefore the pedestrians are also captured at eyelevel. This object detection benchmark consists of 7481 training images and 7518 testing images, comprising a total of 80256 labeled objects. KITTI provides both flow and stereo data.

KITTI and Caltech are the predominant datasets for PD. The scales of these two benchmarks are both large and challenging. A large number of PD algorithms have been evaluated on INRIA and Caltech, meanwhile KITTI is being gradually adopted. Figure 1 shows the height distribution on INRIA, Caltech, and SPID test sets.

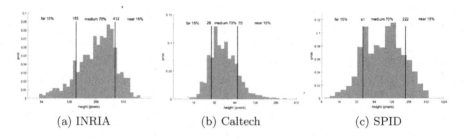

(a) INRIA   (b) Caltech   (c) SPID

Fig. 1. Pedestrian height distribution on INRIA, Caltech, and SPID test sets

The state-of-the-art pedestrian detectors developed in recent years mainly take two features: hand-crafted features and machine-learned deep features.

Hand-Crafted Features. In 2005 Dalal and Triggs introduced the HOG detector [1], which was a landmark for pedestrian detection. Later in 2009 Felzenswalb et al. put forward the classic deformation part based model (DPM) [8], a pedestrian is expressed as several parts with a deformable configuration. In 2009 Caltech dataset was introduced and a new evaluation method was proposed. Dollár et al. used FPPI (False Positive per Image) to compare the performance of the detectors. In the same year Dollár et al. proposed integral channel feature (ICF) method [9], in which Harr-like features are computed over multiple channels for each pedestrian. In 2014 the k-poselets [10] method was introduced as an improvement of DPM and poselets [11]. Aggregated channel features (ACF) [12] is an alternate approach to exploit approximate multi-scale features using ICF. Locally decorrelated channel features (LDCF) [7] is accomplished utilizing the ACF detector. LDCF uses the deeper trees and a denser sampling of the pedestrian data. In 2014, Informed Haar-like [13] features improve pedestrian detection. Pedestrian shapes are geared by three rectangles as models for different body parts. In 2015 [14] points out the link between ACF, ChnFtrs,

Informed Haar, and LDCF, and generates a series of filtered channel features. Checkerboards [9] is a naive set of filters that covers the same sizes, which get the best result on Caltech test set among all the algorithms up to 2015.

Deep Learning Method. The success of deep learning techniques in image classification has promoted researchers to try it on PD. Object detectors that out-perform others are generally based on variants of R-CNN model. A reduced set of detection proposals is created for an image, the proposals are evaluated by a convnet. Convolutional neural networks (CNN) perform best among the deep learning models. CNN optimizes the feature representation automatically in the detection task and regularizes the neural network. Chen et al. [15] used pre-trained deep CNN with ImageNet dataset to generate candidate windows, together with ACF detector to get final features. CifarNet [16] is a small network designed to solve the CIFAR-10 classification problem. AlexNet [17] is a network designed to solve the ILSVRC2012 classification problem. These two networks are both re-implemented in Caffe project [18]. GoogLeNet [19] was responsible for ILSVRC2014. Ren et al. designed Faster R-CNN [20], which used the Simonyan and Zisserman very deep model (VGG) [21]. Faster R-CNN achieved state-of-the-art object detection accuracy on both PASCAL VOC 2007 and 2012.

3   SPID: Surveillance Pedestrian Image Dataset

The images in Surveillance Pedestrian Image Dataset are extracted from the videos recorded by daily used surveillance cameras. All the pedestrians are well labeled in several different properties. Each image contains at least one pedestrian. The various collecting conditions cover various intensities, different time periods, multiple scenes and so on. SPID contains 29989 well-labeled images and about $110k$ pedestrian objects. We split the dataset into training and test sets roughly in half. The test set contains 14550 images and the training set contains 15439 images as well. For detailed statistics about the data, see the bottom row of Table 1. Although SPID is only second to Caltech dataset in terms of scale. Caltech contains all adjacent frames in the video while SPID dataset does not. Considering the effective size (containing not-so-similar frames), SPID may be even bigger than Caltech.

3.1   Dataset Collection

The images in the dataset are all collected from on-using surveillance systems. Some of them are collected from the monitoring cameras used in companies, university campus; some of them are got from public security cameras. The surveillance cameras are set along the streets in the cities and campus, or beside the squares and highways. Most cameras in cities are placed 10 m above the ground, the other cameras like those set at campus entrance are placed 1 m high. We gather 297 pieces of videos, totally about 61 h long. Basically each piece of the videos is about 20 min long, we extract the videos to images per second and pick up the images with at least one pedestrian to label. The resolution of the images are various from 352*288, 720*576, 1280*960 to 1920*1080. Our dataset contains both low resolution and HD images. Figure 1 shows multiple diverse examples of pedestrians in SPID.

Fig. 2. Examples of diverse pedestrians in SPID. The first four rows show the various view of pedestrians (right side, back side, left side and front side). The 5th row shows special pose pedestrians (stoop). Next row shows the pedestrians captured during night and illumination is bad. The 7th row shows pedestrians with various attachments. The last row shows the low resolution pedestrians.

3.2   Ground Truth Annotation

After referencing the annotations of the widely used datasets, like INRIA, Caltech, ETH and USC, the ground truth annotation file format for surveillance dataset is properly settled. Each annotation file, which is written in xml format, corresponds to an image. The xml file contains the total visible pedestrian number of a single image and the corresponding bounding box (BB) coordinate of each pedestrian. If a pedestrian is occluded, the BB only contains the visible part. If an image contains multiple pedestrian objects, we label them individually, i.e. this dataset only has person label and no group label.

Each BB describes a specific pedestrian object. For each pedestrian, we also provide the annotation file with several useful properties: the name of its source image, the specific size and level, scene, view, pose, intensity condition, occlusion level and attachment information. In addition, the top color, bottom color, top style, bottom style of pedestrian clothes are also labeled.

3.3   Dataset Statistics

Scene. The cameras are settled at different areas in cities and campuses. The scenes of the video are diverse, which contain roads, city highways, campus roads and school entrances (see Fig. 3). Therefore the scenes of our dataset represent the typical surveillance scenes in daily life.

Illumination. The videos we collected are recorded during continuous time periods, the basic illumination is high during the day and gradually becomes low during the night, as shown in Fig. 3. We divide the illumination into several different situations: normal, foggy, rainy, cloudy, dusk, night and other. Figure 4(a) shows specific statistics of illumination.

Scale. Due to the various height of the cameras, the distance between pedestrians and cameras covers a wide range. Figure 1 shows the pedestrians height distribution on INRIA, Caltech and SPID test sets. Pedestrian scales influence the detection performance greatly, as shown in [3].

View. Due to the variation of camera positions, pedestrians show various angles of view, such as front side, left side, back side and right side. The property of view is more useful for the granular pedestrian detection, due to that appearance features are different when a pedestrian stand with different views. Pedestrians

Fig. 3. Multiple scenes in SPID

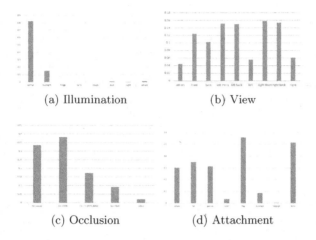

(a) Illumination   (b) View

(c) Occlusion   (d) Attachment

Fig. 4. Detail statistics of SPID data set.

with different view looks discrepant visually. For instance, when a pedestrian is standing at the front side, his face is a key info to detect the person (as detection in PASCAL). However, at the other three side, face is occluded. The division and label of this property may help to design some other features for this condition. Figure 4(b) shows the detail distribution of pedestrian views.

Pose. Pose is an important property for pedestrian detection, usually upright pedestrian is easy to be detected. In other pedestrian dataset like INRIA and Caltech, most pedestrian objects are also upright, barely contains other pedestrian poses. We consider the pose variation of surveillance pedestrians and fill the blank. Once we label the different poses, we could train specific features for the non-upright pedestrians. In our dataset we label different poses including upright, squat, bent, jump, lie down, swing, sit and others.

Occlusion. Occlusion influences the accuracy of pedestrian detection rapidly in [3]. Caltech dataset is the only one that labels this property. Unlike Caltech, we divide the occlusion levels by the ratio of occluded parts to the whole pedestrian body. Occluded ratio is grouped into 4 levels: Occ = none, Occ < 33%, Occ in [33%–66%], Occ > 66%, others. The levels are measured by human subjective vision, see Fig. 4(c) for detail.

Attachment. On one hand, basically other objects in surveillance videos do not contain attachment. Attachments may change the contour of pedestrian, which reduces the robustness of detection methods. On the other hand, we could use object (like bags, bicycles) detection to help with pedestrian detection. Common attachments of pedestrians contain hat, glasses, scarf, bag, dunnage, luggage and others, see Fig. 4(d).

Appearance. The color and style of clothes are important attributes for person identification. We choose some fundamental color and style to make a rough division. Colors include 11 types such as black, white, red, yellow and so on. The styles contain purity, horizontal stripes, vertical stripes, checks and pattern.

4 Evaluation of Pedestrian Detection Algorithms

We made an extensive evaluation of five pedestrian detection algorithms under various scenarios and different datasets to increase the scope of experiments. In Sect. 4.1 we introduce the selection and validation of algorithms. Next we give a brief description of the evaluation standard in Sect. 4.2 and report experiment performance curves in Sect. 4.3. In Sect. 4.4 we analyze and emphasize the influence of dataset discrepancy by showing the evaluation results under different conditions.

4.1 Algorithms Selection and Validation

The selection of the PD algorithms satisfies the rules that algorithms must be published and open source on the Internet. To compare the state-of-the-art PD methods, both hand-crafted detectors and deep learning methods are chosen. For hand-crafted features, we chose to evaluate the pretrained detectors with default parameters, which were obtained from online source codes. For Faster R-CNN, we used original network designed for multiple classification problems. The evaluated hand-crafted detectors are LDCF, ACF, DPM, k-poselets. ACFCaltech+ and LDCFCaltech indicates these two detectors pre-trained with Caltech training set. Similarly ACFInria and LDCFInria represents for detectors trained on INRIA training set. As for the deep learning methods, original Faster R-CNN, Faster R-CNN finetuned with Caltech training set and network finetuned with SPID training set are tested. Evaluation datasets include the test sets of INRIA, Caltech and SPID. The evaluation results are computed with the latest online released codes.

4.2 Evaluation Standard

Since Dollár et al. [22] proposed the evaluation methodology to use ROC curves compare the performance of different PD algorithms, we adopt this standard as well. The ROC curves show the relationship between miss rate and FPPI (False Positive Per Image). In INRIA dataset only the testing positive 288 images are considered. The test set (set06–set10) of Caltech is used, we extracted images with 30 frames interval and obtained totally 4024 images. The SPID test set contains 15439 images. For each image, the ground truth bounding boxes (BB_{gt}) and detected bounding boxes (BB_{dt}) are loaded. The aspect ratio of all the boxes is preprocessed to 0.41 and the overlap threshold of bouding boxes is set to 0.5. Specific accuracy calculation method is the same as that in [22].

4.3   Experiment Results

The experiments are grouped into two aspects: one is the comparison result of the pedestrian detection methods trained using existing datasets, the other is the comparison result of these methods retrained using SPID training data. The evaluation result ROC curves of five representative pedestrian algorithms on three proposed datasets under different settings are shown in Fig. 5.

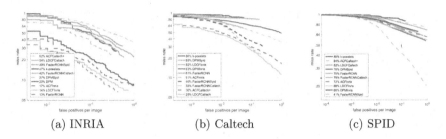

| (a) INRIA | (b) Caltech | (c) SPID |

Fig. 5. Evaluation results under reasonable condition on three test sets. (a) and (b) shows evaluation results on INRIA and Caltech. (c) performances upgrade on SPID.

Figure. 5a–c show performance for three test sets under reasonable setting. This setting serves as a filter, which selects pedestrians over 50 pixels tall to evaluate the performance. In Fig. 5 a Faster R-CNN and LDCFInria perform best with log-average miss rate of 13–14% On INRIA. ACFInria and DPM achieve the log-average miss rate about 20%. In Fig. 5b LDCFCaltech achieves the best result on Caltech-Test dataset and k-poselets is the worst one. Figure 5c plots performance on the entire SPID test set. Faster R-CNNSpid outperforms the other detectors remarkably. DPMInria and LDCFInria have close log-average miss rates about 68% on 0.1FPPI.

For all algorithms, performance is best on INRIA among all test sets. INRIA contains high-resolution pedestrians, with Faster R-CNN, LDCFInria and ACFInria achieving log-average miss rates of 13–17% (See Fig. 5a). Performance is also fairly high on Caltech (see Fig. 5b) with 25% log-average miss rate obtained by LDCFCaltech. This possibly due to that LDCFCaltech and ACFCaltech+ are both trained on Caltech training set. Faster R-CNNSpid and DPMInria perform better on SPID than Caltech-Test, which means the model of DPM and Faster R-CNN may be fit for SPID pedestrians when trained well.

The comparison result of the pedestrian detection methods using released codes (k-poselets, ACFCaltech+, LDCFCaltech, Faster R-CNN, ACFInria, LDCFInria and DPMInria), show the large performance decreasing when applying the existing methods on SPID directly; The other comparison result of these methods retrained (Faster R-CNNCaltech, DPMSpid, Faster R-CNNSpid), verifies that, the existing methods retrained using SPID or Caltech can't work very well on surveillance images. In other words, due to the remarkable discrepancy between the surveillance images and non-surveillance datasets, developing of specific pedestrian detection algorithms are needed for the surveillance applications.

4.4   What are the Factors Influencing the Performances?

We rank detector performance with multiple test sets to assess whether the discrepancy between data are influential to the results. Several data discrepancies are considered, including the pedestrian scale variation: near scale, medium scale and far scale; view point difference: horizontal and bird's eye view. We evaluate these two factors in detail respectively.

Scale. Figure 1 plots the height distribution for pedestrians on three test sets. However, the height distributions are somewhat dissimilar. Heights of SPID pedestrians have wider variations, basically cover the range of both INRIA and Caltech. About 70% pedestrians are 41–222 pixels tall. The wider pedestrian scale range makes SPID more challenging than INRIA and Caltech.

The results of Fig. 5 show that all the algorithms perform best on INRIA, which may due to the high resolution pedestrians in INRIA dataset. From Fig. 1, the discrepancy of pedestrian height among three datasets is large. We group pedestrians of SPID test set by their heights in pixels into three scales: near (283 or more pixels), medium (between 81–283 pixels), and far (81 pixels or less). This division to three scales is motivated by the distribution of heights in SPID, human performance and surveillance system requirements. To verify the influence of scale variation, we tested SPID pedestrians with three height constraints. Results for large, near and medium scale pedestrians, are shown respectively in Fig. 6. For near scale, Faster R-CNNSpid performs best; while other detectors, DPMSpid, LDCFInria, DPMInria and ACFInria achieve log-average miss rates about 40–50%. Under medium scale, performance degrages at least 15% for most algorithms. While Faster R-CNNSpid still achieves the best result. All algorithms degrade most on the far scale. ACFInria and LDCFInria are detectors trained on INRIA, the height of training pedestrians are tall, therefore for small pedestrians on SPID, the performances of these two detectors degrade rapidly. Faster R-CNNSpid achieves the best relative performance under this condition, but absolute performance is quite poor with log-average miss rate of 84%.

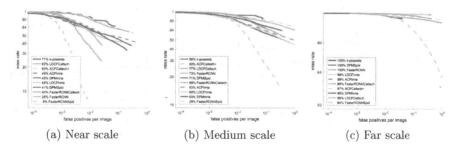

(a) Near scale   (b) Medium scale   (c) Far scale

Fig. 6. Evaluation results of various pedestrian scales on SPID. (a) shows the performance on unoccluded pedestrians over 283 pixels tall. (b) degrades for pedestrians about 81–283 pixels tall. (c) degrades the most for 32–80 pixel high pedestrians.

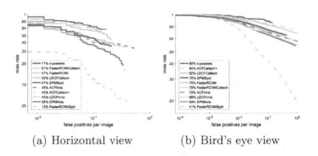

(a) Horizontal view   (b) Bird's eye view

Fig. 7. Evaluation results on SPID test set under camera angle variation.

View Point. During our experiment we notice that INRIA and Caltech pedestrians are all captured at eyelevel (i.e., the angle of a pedestrian head and camera is nearly horizontal). On the other hand, most cameras in SPID are set 10 m above the ground (i.e., the camera is like the bird's eye). Pedestrian body parts and ratios change greatly at different views. Up to 98.9% pedestrians in SPID are captured at bird's eye view, while none in INRIA and Caltech. Motivated by the great angle difference, we separate SPID to SPID-bird and SPID-horizontal. The evaluation results under view variation on SPID test set is shown in Fig. 7. Except for Faster R-CNN and k-poselets, other algorithms degrade severely, the log-average miss rate of DPM degrades 18%. LDCFCaltech, ACFInria, LDCFInria and ACFCaltech+ degrade most, these detectors are trained by pedestrians at eyelevel, thus perform not well on another view.

Summary. We evaluated 4 hand-crafted detectors and 1 deep learning methods on INRIA, Caltech-Test and SPID. The results show that all algorithms perform best on INRIA. ACFCaltech and LDCFCaltech, which are trained on Caltech training set, perform better on Caltech test set than SPID. While Faster R-CNNSpid performs best on SPID. This section shows the different influence of different properties. Figure 8 shows some typical detection results. FasterRC-NNSpid and DPMInria works better under bad conditions, while LDCFCaltech misses most pedestrians on low illumination images. However, on HD images with complex background, all the algorithms degrade remarkably.

Performance is far from perfect on SPID, even under the favorable conditions. Table 2 shows miss rates of all detectors under various conditions on SPID test set. The performance decrease is caused by the data differences of multiple datasets. Main factors include scale and view point, as well as illumination and occlusion. All have obvious influences of algorithm performances.

(a) 8% of all pedestrians are missed even at the near scale. For smaller pedestrians, performance degrades catastrophically. Table 2 shows that nearly half the detectors reach 100% log-average miss rate at far scales.

(b) The special bird's eye view, caused by surveillance cameras, also increases the detection difficulty, see Table 2.

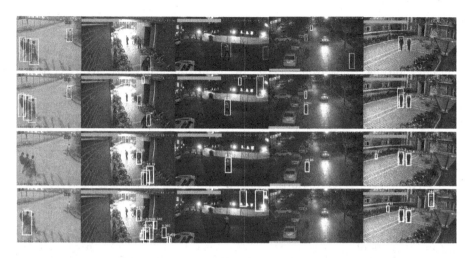

Fig. 8. Typical results of top four best detectors on SPID. The first row shows typical results of FasterRCNNSpid. And the second row shows results of DPMInria, the third row represents for LDCFInria, and the last row indicates the ACFInria performance.

Table 2. Miss rates of all detectors under various conditions on SPID test set.

Detector	Training dataset	Scales			View points	
		Near	Medium	Far	Horizontal	Bird's eye
FasterRCNNSpid	SPID	8%	29%	84%	13%	41%
FasterRCNN	ImageNet [23]	25%	73%	100%	57%	75%
FasterRCNNCaltech	Caltech [3]	40%	68%	50%	61%	75%
DPMSpid	SPID	41%	71%	100%	47%	76%
LDCFInria	INRIA [1]	45%	60%	99%	45%	69%
DPMInria	INRIA	45%	60%	96%	35%	69%
ACFInria	INRIA	49%	63%	99%	47%	72%
ACFCaltech+	Caltech	63%	80%	97%	45%	84%
LDCFCaltech	Caltech	63%	77%	96%	50%	82%
k-poselets	SPID	71%	88%	100%	71%	90%

(c) Low illumination interferes the detection result greatly, which increases the hardness for algorithms on surveillance data captured during night.

(d) Performance is abysmal under heavy occlusion, nearly all the pedestrians are missed even at high false positive rates.

The experiment results verify that, due to the dataset discrepancy, the algorithms trained on traditional datasets do not perform well on surveillance pedestrian detection problems, especially under far scale, bird's eye view, low illumination and heavy occlusion. However, the retrained algorithms on SPID training set still perform not good enough. There is considerable room for improvement in pedestrian detection. In addition, we validate the effectiveness and necessity of SPID.

5   Conclusion

This paper introduces the SPID: Surveillance Pedestrian Image Dataset, which contains multiple scenes from videos captured by on-using surveillance systems. The pedestrians in SPID are well labeled. Five latest pedestrian detection algorithms including hand-crafted detectors and deep learning methods are tested on INRIA, Caltech and SPID. Evaluation results show that the data differences, such as scale, view, illumination and occlusion between existing public datasets and SPID have large impact on the detection performances. Throwing new light on existing datasets, we hope that the proposed benchmarks will complement the gap. Some algorithms perform well under favorable conditions, however when it comes to complicated situations in surveillance images, performance degrades significantly. Our intention is to improve the algorithm performance on SPID.

Acknowledgement. This work was partly funded by NSFC (No. 61571297, No. 61527804), 111 Project (B07022), and China National Key Technology R&D Program (No. 2012BAH07B01). The authors also thank the following organizations for their surveillance data supports: SEIEE of Shanghai Jiao Tong University, The Third Research Institute of Ministry of Public Security, Tianjin Tiandy Digital Technology Co., Shanghai Jian Qiao University, and Qingpu Branch of Shanghai Public Security Bureau.

References

1. Dalal, N., Triggs, B.: Histograms of oriented gradients for human detection. In: IEEE Computer Society Conference on Computer Vision and Pattern Recognition, CVPR 2005, vol. 1, pp. 886–893. IEEE (2005)
2. Ess, A., Leibe, B., Gool, L.V.: Depth and appearance for mobile scene analysis. In: IEEE 11th International Conference on Computer Vision, ICCV 2007, pp. 1–8. IEEE (2007)
3. Dollár, P., Wojek, C., Schiele, B., Perona, P.: Pedestrian detection: 1 benchmark. In: IEEE Conference on Computer Vision and Pattern Recognition, CVPR 2009, pp. 304–311. IEEE (2009)
4. Geiger, A., Lenz, P., Urtasun, R.: Are we ready for autonomous driving? The kitti vision benchmark suite. In: Conference on Computer Vision and Pattern Recognition (CVPR) (2012)
5. Wojek, C., Walk, S., Schiele, B.: Multi-cue onboard pedestrian detection. In: IEEE Conference on Computer Vision and Pattern Recognition, CVPR 2009, pp. 794–801. IEEE (2009)
6. Enzweiler, M., Gavrila, D.M.: Monocular pedestrian detection: survey and experiments. IEEE Trans. Pattern Anal. Mach. Intell. **31**(12), 2179–2195 (2009)
7. Nam, W., Dollár, P., Han, J.H.: Local decorrelation for improved detection. arXiv preprint arXiv:1406.1134 (2014)
8. Felzenszwalb, P.F., Girshick, R.B., McAllester, D., Ramanan, D.: Object detection with discriminatively trained part-based models. IEEE Trans. Pattern Anal. Mach. Intell. **32**(9), 1627–1645 (2010)
9. Dollár, P., Tu, Z., Perona, P., Belongie, S.: Integral channel features (2009)

10. Gkioxari, G., Hariharan, B., Girshick, R., Malik, J.: Using k-poselets for detecting people and localizing their keypoints. In: Proceedings of the IEEE Conference on Computer Vision and Pattern Recognition, pp. 3582–3589 (2014)
11. Bourdev, L., Malik, J.: Poselets: body part detectors trained using 3D human pose annotations. In: 2009 IEEE 12th International Conference on Computer Vision, pp. 1365–1372. IEEE (2009)
12. Dollár, P., Appel, R., Belongie, S., Perona, P.: Fast feature pyramids for object detection. IEEE Trans. Pattern Anal. Mach. Intell. **36**(8), 1532–1545 (2014)
13. Zhang, S., Bauckhage, C., Cremers, A.B.: Informed Haar-like features improve pedestrian detection. In: IEEE Conference on Computer Vision and Pattern Recognition, pp. 947–954 (2014)
14. Zhang, S., Benenson, R., Schiele, B.: Filtered channel features for pedestrian detection. In: IEEE Conference on Computer Vision and Pattern Recognition, pp. 1751–1760 (2015)
15. Chen, X., Wei, P., Ke, W., Ye, Q., Jiao, J.: Pedestrian detection with deep convolutional neural network. In: Jawahar, C.V., Shan, S. (eds.) ACCV 2014. LNCS, vol. 9008, pp. 354–365. Springer, Heidelberg (2015). doi:10.1007/978-3-319-16628-5_26
16. Krizhevsky, A., Hinton, G.: Learning multiple layers of features from tiny images (2009)
17. Krizhevsky, A., Sutskever, I., Hinton, G.E.: ImageNet classification with deep convolutional neural networks. In: Advances in Neural Information Processing Systems, pp. 1097–1105 (2012)
18. Jia, Y., Shelhamer, E., Donahue, J., Karayev, S., Long, J., Girshick, R., Guadarrama, S., Darrell, T.: Caffe: convolutional architecture for fast feature embedding. In: Proceedings of the ACM International Conference on Multimedia, pp. 675–678. ACM (2014)
19. Szegedy, C., Liu, W., Jia, Y., Sermanet, P., Reed, S., Anguelov, D., Erhan, D., Vanhoucke, V., Rabinovich, A.: Going deeper with convolutions. In: Proceedings of the IEEE Conference on Computer Vision and Pattern Recognition, pp. 1–9 (2015)
20. Ren, S., He, K., Girshick, R., Sun, J.: Faster R-CNN: towards real-time object detection with region proposal networks. In: Advances in Neural Information Processing Systems, pp. 91–99 (2015)
21. Simonyan, K., Zisserman, A.: Very deep convolutional networks for large-scale image recognition. arXiv preprint arXiv:1409.1556 (2014)
22. Dollar, P., Wojek, C., Schiele, B., Perona, P.: Pedestrian detection: an evaluation of the state of the art. IEEE Trans. Pattern Anal. Mach. Intell. **34**(4), 743–761 (2012)
23. Deng, J., Dong, W., Socher, R., Li, L.J., Li, K., Li, F.F.: ImageNet: a large-scale hierarchical image database, pp. 248–255 (2009)

Actions Recognition in Crowd Based on Coarse-to-Fine Multi-object Tracking

Sixue Gong, Hu Han$^{(\boxtimes)}$, Shiguang Shan, and Xilin Chen

Key Lab of Intelligent Information Processing, Institute of Computing Technology,
Chinese Academy of Sciences (CAS), Beijing 100190, China
{sixue.gong,hu.han,shiguang.shan,xilin.chen}@vipl.ict.ac.cn

Abstract. Action recognition has wide applications from video surveillance, scene understanding to forensic investigation. While recent methods typically focus on a single action recognition from video clips, we investigate the problem of action recognition in crowd, which better replicates real video surveillance scenarios. We propose to perform actions recognition in crowd based on an efficient coarse-to-fine multi-object tracking algorithm. With Faster R-CNN as our human detector, we utilize a coarse-to-fine strategy for multi-object tracking in crowd, consisting of multi-object fast tracking and per-object fine tracking. The tracking results are used to extract the action cuboids, and spatial-temporal features are computed for action classification. We evaluate the proposed approach on a self-collected actions-in-crowd dataset, and two public domain databases (CMU and and MOT2015). The results show the effectiveness of the proposed approach for multi-action recognition in crowd.

1 Introduction

The recognition of both human and animal actions in videos plays a crucial role from automatic scene understanding to video surveillance. For example, in video surveillance, it is helpful to automatically discover suspects through action of interest detection. Also, a coach's workload can be alleviated if every players action statistics in a game can be automatically calculated. Considerable progress has been made in the past few years, particularly on action feature descriptors like Bag-of-Word (BoW) [1–3] and CNN features [4,5]. These feature representation methods significantly improve the accuracy of action classifiers. Nevertheless, these approaches mainly focus on a single action recognition from videos. However, in many applications with various arising actions, such as video surveillance and group sports analysis, it is required to recognize multiple actions of interest (see Fig. 1). Most of the existing action recognition methods are not designed for recognizing multiple actions in crowd scenarios.

Multi-action recognition in crowd is non-trivial because of the challenges from designing human detection and tracking algorithms, robust action representations, and classification models. To bridge the gap between the applicability of existing action recognition methods and the needs of emerging applications,

© Springer International Publishing AG 2017
C.-S. Chen et al. (Eds.): ACCV 2016 Workshops, Part III, LNCS 10118, pp. 478–490, 2017.
DOI: 10.1007/978-3-319-54526-4_35

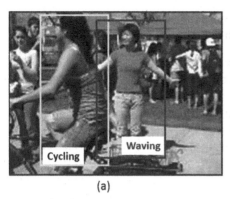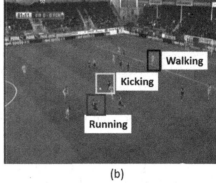

(a)   (b)

Fig. 1. In real applications, the scenarios often contain multiple actions in crowd: (a) actions including cycling and waving, and (b) actions including kicking, walking, and running. Multi-action recognition is necessary for exact scene understanding.

we propose an approach for multi-action recognition in crowd utilizing coarse-to-fine multi-object tracking. Given the individual subjects detected by a faster region-based convolutional neural networks (Faster R-CNN) [6], our tracking algorithm uses a linear quadratic estimation model, i.e., Kalman filter [7], for coarse but fast tracking of all the moving subjects, and a local sparse optical flow model for refined and stable tracking. Person-specific action tracks (cuboids) are then extracted based on the tracking results; each can be input to the traditional single action recognition algorithms for action recognition.

The main contributions of this work are as follows. (1) While existing action recognition approaches typically focus on a single action recognition per video clip, we explore a relatively underserved area of concurrent actions recognition per video clip; (2) We propose a coarse-to-fine multi-object tracking algorithm while balances the tracking speed and accuracy under crowd scenarios; (3) We also build an database to better replicate the action-in-crowd application scenarios.

2   Related Work

Action Recognition. There is a variety of work on action recognition from manually segmented video clips containing only one type of action [1,5,8–16]. The majority of these approaches aimed at utilizing trajectory [17–19] and spatial-temporal motion cues to get generative and discriminative features of actions. The early work on spatial-temporal feature extraction used smoothed and aggregated optical flow to model human motions [8]. Zhu et al. [13] used spatial-temporal interest points (STIP) as low-level motion features of the action segments, and used a multi-SVM classifier.

Jain et al. [10] trained a linear SVM classifier using the Divergence-Curl-Shear descriptor encoded by VLAD [20]. They employed the horizontal and vertical components of the flow field to compute the divergence, curl, and shear scalar values and consider all possible pairs of these kinematic features to capture more information through the joint distribution of the features. Ryoo et al. [11] categorized optical flows into multiple types based on their location and directions, and placed them into histogram bins as global features. These approaches mainly focused on designing an effective motion feature representation from a given video clip, but did not take into account the contextual information between the object and the background. However, such contextual information is of great importance in action recognition, particularly under scenarios with cluttered backgrounds (e.g., crowd). To address this issue, Gkioxari et al. [5] proposed a R-CNN method to perceive more visual information from a primary region of the subjects and a secondary region of the context areas.

The aforementioned approaches concentrate on action recognition with fixed background or with a single action inside [21]. To recognize actions in more complex scenes, Hu et al. [9] proposed a method that utilizes multi-instance learning (MIL) based SVM to handle the ambiguities in both spatial and temporal domains. They learned the SVM classifier from an action cuboid which is referred to as a bag containing more than one potential region and time slice. As for crowd scenes, Zhou et al. [12] proposed a statistical framework to detect abnormal behaviors of the crowd scene by using a multi-observation hidden Markov model (HMM) of pedestrian trajectories. Such unusual action detection methods concern more about discovering abnormal behavior, but not for recognition of multiple concurrent actions. Siva et al. [1] integrated both static appearance features by bag-of-word (BoW), and motion features by trajectory transition descriptor (TTD), and used SVM in a sliding-cuboid manner to detect particular actions from a video.

Visual Tracking. Significant progress of visual tracking has been made these years. Tracking-by-detection [22–24] is one of the popular frameworks in visual tracking, in which the tracker is used to follow the object from frame to frame, and the detector is used to localize all Such tracking methods may be time-consuming if the object detector is complicated. Approaches like correlation filtering was proposed for more efficient object tracking [25–28], in which the correlation can be computed using Fast Fourier Transform (FFT), making it a fast tracker. In recent years, there are some approaches exploiting the rich feature hierarchies in CNNs for robust tracking [25,29], which have shown state-of-the-art performance. As for multi-object tracking problem, Bae et al. [30] formulated a multi-object tracking problem based on the tracklet confidence, and used an incremental linear discriminant analysis for discriminating the appearances of objects to obtain reliable association between tracklets and detections. Xing et al. [31] proposed a two-stage framework to learn a tree-structured multi-view human detector to generate tracklets through particle filter in local stage.

3   Multi-action Recognition in Crowd

Given a video clip V in crowd with M subjects (each has an individual action), the objective of our approach is to obtain an action label l_i for each subject

$$L = \{l_i = C(F(N(T_i)))|T_i \in T(V)\}_{i=l}^{M}, \tag{1}$$

where $T(\cdot)$ is our coarse-to-fine multi-object tracker, which determines the candidate action tracks T_i with actions of interest, and T_i contains a series of bounding boxes $T_i = \{b_1, b_2, \cdots, b_{N_i}\}, i = 1, 2, \cdots, M$, where $b_j = [x_j, y_j, w_j, h_j]$ defines the bounding box's left-right location (x_j, y_j) and size (w_j, h_j); $N(\cdot)$ normalizes all the bounding boxes of each action track into the same size to form the 3D spatial-temporal cuboid; $F(\cdot)$ is to extract spatial-temporal features from each action cuboid $X_i = F(C_i), i = 1, 2, ..., M$; finally, the action recognition of one action cuboid is determined as

$$l_i = \arg\max_{\{1,2,...,M\}}(C(X_i)), \tag{2}$$

where $C(\cdot)$ calculates the confidences of all the action types that X_i belongs to.

Traditional action recognition approaches often extract spatial-temporal features from the entire frames, resulting in features that are helpful for one dominant action recognition, but not for recognizing multiple concurrent actions. The proposed approach of multi-action recognition in crowd scenario is built upon multi-object tracking to address the above issue. An overview of the proposed approach can be seen in Fig. 2. We provide the implementation details of the proposed approach in the next section.

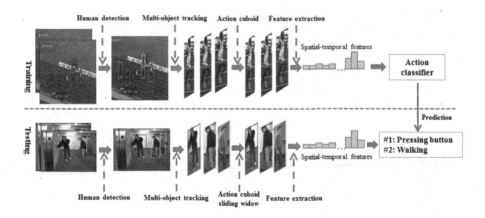

Fig. 2. Overview of the proposed approach for actions recognition in crowd.

4   Implementation Details

4.1   Coarse-to-Fine Multi-object Tracking

The proposed coarse-to-fine multi-object tracking consisting of fast subject trajectory estimation via a linear quadratic estimation model, i.e., Kalman filter,
and trajectory refinement using sparse optical flow. In particular, during our
fast subject trajectory estimation, we assume that the moving object is in uniform velocity within a frame interval due to its short time span. If an object O_i
appears at frame I_t, the state of the object O_i is denoted as $S_t^i = (p_t^i, s_t^i)$, where
p_t^i is the position, and s_t^i is the size. Then, we construct the system state model
of Kalman filter as

$$s(t) = \ddot{T}(t-1)s(t-1) + w(t), \tag{3}$$

where $s(t) = \begin{bmatrix} p_t^x \\ p_t^y \\ \Delta x_t \\ \Delta y_t \end{bmatrix}$, $s(t-1) = \begin{bmatrix} p_{t-1}^x \\ p_{t-1}^y \\ \Delta x_{t-1} \\ \Delta y_{t-1} \end{bmatrix}$; $\ddot{T}(t) = \begin{bmatrix} 1 & 0 & v_x & 0 \\ 0 & 1 & 0 & v_y \\ 0 & 0 & 1 & 0 \\ 0 & 0 & 0 & 1 \end{bmatrix}$ is the state

transition matrix; $w(t)$ is white Gaussian noise with zero mean.

Due to the unconstrained nature of crowd video surveillance scenario, objects
may disappear due to occlusion and reappear at a later time, and new subjects
may appear at any time. Thus, we need to detect new moving object every
frame. Given the newly detected moving objects at frame I_t and the estimated
positions of existing objects at frame I_t, we apply the Hungarian algorithm [32]
to calculate the optimal association matrix $A^* = \{a_i^*\}$ between subjects in frames

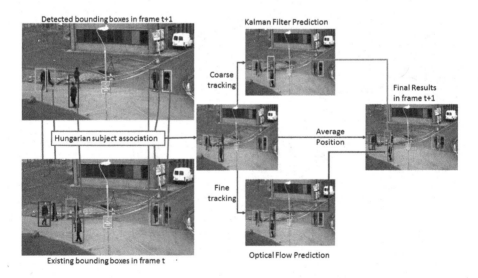

Fig. 3. A diagram of the proposed coarse-to-fine multi-object tracking algorithm.

I_{t-1} and I_t. In this process, each of the humans in I_t is either associated to one of the existing tracklets and used to update the system state model or grouped as a new tracking target (Fig. 3).

The fast tracking process is based on the assumption that the state posterior density of object motions follows a Gaussian distribution, which does not always hold in real tracking problems. As a result, abrupt appearance changes may occur in the candidate action tracks. To resolve this problem, sparse optical flow based on keypoints is employed to refine the fast tracking result. Based on the optimal assignment matrix A^*, if abrupt changes occur either in size or position of a subject, the sparse optical flow based tracking is applied to the subject. We compute the keypoints following the method in [33], and denote the computed keypoints at frame I_t as $\{b_{i_1}^t, b_{i_2}^t, ..., b_{i_M}^t\}$. Then, the average displacement field of all keypoints between frames I_t and I_{t-1} can be calculated as

$$D_i^t = \frac{1}{M} \sum_{j=1}^{M} |b_{ij}^t - b_{ij}^{t-1}| \tag{4}$$

Accordingly, the state of object O_i is updated as:

$$S_t^i = X_{t-1}^i + (D_t^i, 0) \tag{5}$$

Considering the success of Faster R-CNN [6] in a number of object detection tasks, we choose to use a Faster R-CNN [6] detector to detect the object of interest (persons) used by our tracking algorithm. In particular, we use a VGG-M model [34] with 5 convolutional (conv) layers in Faster R-CNN, and train it on PASCAL VOC 2007 dataset [35] for human detection. Such a detector takes about 70 ms to process one frame (resized to 1000×600 inside Faster R-CNN) on a Titan X GPU. Our coarse-to-fine tracking runs really fast, about 30 ms per frame (640×480) on an Intel i7 3.6 GHz CPU. Thus, the total detect and tracking time remains about 100 ms.

4.2 Action Recognition

Our coarse-to-fine tracker determines a number of candidate action tracks T_i from a video; each contains one subject if there is no tracking error. However, the number of actions in each track, and the action's spatial and temporal locations are still not known. Thus, the state-of-the-art action recognition methods designed for single action action recognition still cannot be applied directly. Another problem is that the sizes of the successive bounding boxes in T_i may differ from each other because of the subject appearance changes in scale and pose (also known as the heterogeneous issue in object recognition [36–38]). Thus, before extracting action informative features, we first resize all the bounding boxes in T_i into the same size of 20×35. We then do cuboid sliding inside the normalized T_i temporally, and each cuboid is used for action feature extraction. Following the settings in [1], we also restrict the length of acceptable cuboids between $L_{min} = 20$ and $L_{max} = 75$ frames. If N_i (the frame number of T_i) is larger than L_{max}, each sliding cuboid has 75 frames; if $L_{min} \leq N_i \leq L_{max}$, each sliding cuboid has N_i frames; otherwise, action track T_i is deserted directly and will not be used for action recognition.

Each sliding 3D spatial-temporal cuboid is supposed to have one dominant action of one subject, and thus single action recognition methods can be used for the final action recognition. Without loss of generality, we use two typical action feature extraction methods: Biologically Inspired Feature (BIF) [39] and bag-of-words (BoW) [1]. In its simplest form, the extraction of BIF consists of two layers of computational units, where simple S_1 units in the first layer are followed by complex C_1 units in the second layer. The S_1 units correspond to the classical simple cells in the primary visual cortex [40]. They are typically implemented with the convolution of a preprocessed image $\mathbf{\Gamma}$ with a family of Gabor filters [41],

$$\psi_{u,v}(\mathbf{z}) = \frac{\|\mathbf{k}_{u,v}\|^2}{\sigma^2} e^{-\frac{\|\mathbf{k}_{u,v}\|^2 \|\mathbf{z}\|^2}{2\sigma^2}} \left[e^{i\mathbf{k}_{u,v}\mathbf{z}} - e^{-\frac{\sigma^2}{2}} \right], \qquad (6)$$

where $\mathbf{z} = (x, y)$, σ is the relative width of the Gaussian envelope function w.r.t. the wavelength, and u and v are the orientation and scale parameters of Gabor kernels, respectively. The wave vector $\mathbf{k}_{u,v}$ is defined as,

$$\mathbf{k}_{u,v} = k_v e^{i\phi_u}, \qquad (7)$$

with $k_v = \frac{k_{max}}{f^v}$ defining the frequency, and $\phi_u = \frac{\pi u}{8}$ defining the orientation. k_{max} and f are constants specifying the maximum frequency and scaling factor between two neighboring kernels, respectively. The C_1 units correspond to cortical complex cells which are robust to shift and scale variations. They can be calculated by pooling over the preceding S_1 units with the same orientation but at two successive scales. To compute S_1 layer features, we build a family of Gabor filters similar to those in [41], but we use 8 orientations and 12 scales. We apply "MAX" pooling operator and "STD" normalization operator to extract C_1 features from the S_1 layer. The S_1 layer provides a multi-scale representation for face images, and the C_1 layer provides robustness against translation, rotation, and scaling changes of the subjects.

The calculation of our BoW feature is the same as the one used in [1], which was based on SIFT appearance descriptor. After the feature extractions, we used a SVM with an RBF kernel as the action classifier.

5   Experimental Results

5.1   Settings

During the training of the Faster R-CNN human detector, we set the learning rate base_lr as 0.01, and use a step learning rate policy with a weight of 0.1 every 40 K iterations (and a maximum iteration number of 100 K). The batch_size and iter_size were both set to 1. For divisions of the training and testing sets on different databases, we randomly use half clips of each scenario for training, and the remaining for testing.

5.2 Multi-action Recognition in Crowd

Our Dataset. Most crowd datasets such as KTH, YouTube, Hollywood2, and UIUC [42–45] are focusing one action recognition of the entire crowd, not for the purpose of multiple concurrent actions recognition. Therefore, we build a multi-action dataset based on public domain databases and videos from the Internet. As shown in Fig. 4, there are five typical scenarios in the collected dataset: Fig. 4(a) is walking of multiple subjects, Fig. 4(b) is running of multiple subjects, Fig. 4(c) is kicking and running, Fig. 4(d) is one sitting and one standing, Fig. 4(e) is one sitting and two standings. For each video clip, we manually provided the ground-truth bounding boxes and action labels of individual subjects.

We tested the performance of two baseline approaches using BIF and BoW, respectively, and their performance under the proposed framework (denoted as Proposed (BIF) and Proposed (BoW), respectively). We report the mean average precision (MAP) for each action category. When a video clip contains more than one actions or the same actions but of more than one subjects. The MAP is the average of these multiple actions or subjects.

As shown in Table 1, in the scenarios when there are more than one actions, e.g., scenarios (a, b, c, e), the proposed approach using BIF and BoW features significantly outperforms the baseline approach. This is understandable because these baseline methods does not explicitly use multi-object tracking, and are difficult in handling multi-actions. For scenario (e) with a single action, the baseline approaches making use of contextual information, still work better than our

(a) Walking of multiple sub- (b) Running of multiple sub- (c) Kicking and running
jects   jects

(d) Sitting and one standing   (e) Sitting and two standings

Fig. 4. Examples of the video clips from the multi-action in crowd dataset we collected.

Table 1. Comparisons (in MAPs) with the baseline action recognition methods on the multi-action in crowd dataset we collected under scenarios (a–e).

Scenarios	(a)	(b)	(c)	(d)	(e)
BIF	0.29	0.43	0.14	0.67	0.55
BoW	0.26	0.34	0.17	0.52	0.50
Proposed (BIF)	0.36	0.51	0.23	0.65	0.61
Proposed (BoW)	0.32	0.40	0.25	0.51	0.56

method. In cropped action cuboids separate the multiple actions into separate ones but each losses the contextual information which is also helpful for action recognition.

CMU Dataset. The CMU dataset [46] contains some crowd scenes, but most of the videos contain one a dominant action of one subject. So this database does NOT represent the scenarios that this work is focusing on. However, since there are not known multi-action in crowd databases, we still provide the results on this database as a reference. We use the state of the art results by QMUL method [1] as the baseline performance on CMU.

Not surprisingly, as shown in Table 2, the baseline methods work reasonably well under most scenarios except for the two-hand wave scenario. Such an observation actually provides a strong support of the usefulness of the proposed approach in recognizing multiple different actions or the same actions by multiple individuals (Fig. 5).

Fig. 5. An example of the multi-action recognition results by the proposed approach on the multi-action dataset we collected.

Table 2. Comparisons (in MAPs) with the baseline on the CMU dataset.

	Jump-jacks	Pick-up	Push button	1-hand wave	2-hand wave
QMUL [1]	0.36	0.68	0.94	0.45	0.54
Proposed (BIF)	0.36	0.51	0.23	0.65	0.61
Proposed (BoW)	0.32	0.40	0.25	0.51	0.56

5.3   Evaluation of the Tracking Module

Since the tracking module in the proposed approach plays an important role, we evaluate the tracking algorithm on the pubic database MOT2015 [47] using 9 video sequences (TUD-Stadtmitte, TUD-Campus, PETS09- S2L1, ETH-Bahnhof, ETH-Sunnyday, ADL-Rundle-6, ADL-Rundle-8, KITTI-17 and Venice-2). The experiment contains two parts: using the coarse tracking only, and using the entire tracking pipeline. We report the tracking performance in terms of precision, recall, MOTA, and MOTP. MOTA measures the multi-object tracking accuracy, and is a widely metric to evaluate multi-object tracking performance. MOTP measures the multi-object tracking precision, reflecting the average similarity between all true positives and their corresponding ground-truths. The most recent results on MOT2015 are available online (accessed in Sept. 2016)[1], and the state of the art performance is about 76.6% MOTP reported by NOMTwSDP. The results in Table 3 show that the proposed coarse-to-fine tracking works reasonable well (73.9% MOTP) to support the succeeding action cuboid extraction and recognition stages of our multi-action recognition approach.

Table 3. Evaluation of the tracking module in the proposed approach on MOT2015.

Measures	Recall	Precision	MOTA	MOTP
Coarse tracking	71.1	28.3	32.3	69.6
Coarse-to-fine	68.4	43.9	37.3	73.9

6   Conclusion

We investigate the multiple actions recognition problem in crowd scenarios, which is a problem lacking deep exploration. We use a divide and conquer strategy to handle this problem by proposing a coarse-to-fine multi-object tracking algorithm. Fast tracking using Kalman filter provides action trajectories for most objects with smoothing movement, and fine tracking with sparse optical flow refines the object trajectories with abrupt appearance changes. Such a coarse-to-fine tracking method allows us to obtain individual action tracks with balanced accuracy and speed. Then, traditional action recognition algorithms designed for single action recognition are applied on each of the action tracks in a sliding cuboid mode. Experimental results on a multi-action in crowd database we collected, and the public domain CMU ad MOT15 databases show the proposed approach is effective for actions recognition in crowd video surveillance scenarios.

Accurate spatial-temporal localization of actions still plays a very important role in action recognition. In our future work, we will investigate more robust object detection and tracking approaches for efficient action cuboid localization,

---

[1] https://motchallenge.net/results/2D_MOT_2015/.

and their applications from action recognition, person re-identification to human attribute prediction [48]. Additionally, we will extend the proposed approach towards handling single action recognition tasks in a flexible way.

Acknowledgement. This research was partially supported by 973 Program (grant No. 2015CB351802), and Natural Science Foundation of China (grant No. 61672496). The authors would like to thank Xiaoyan Li for her proofreading of this paper. H. Han gratefully acknowledges the support of NVIDIA Corporation with the donation of the Titan X GPU used for his research.

References

1. Siva, P., Xiang, T.: Action detection in crowd. In: BMVC, pp. 1–11 (2010)
2. Luo, Y., Cheong, L.F., Tran, A.: Actionness-assisted recognition of actions. In: ICCV, pp. 3244–3252 (2015)
3. Li, Y., Ye, J., Wang, T., Huang, S.: Augmenting bag-of-words: a robust contextual representation of spatiotemporal interest points for action recognition. Visual Comput. **31**, 1383–1394 (2015)
4. Simonyan, K., Zisserman, A.: Two-stream convolutional networks for action recognition in videos. In: NIPS, pp. 568–576 (2014)
5. Gkioxari, G., Girshick, R., Malik, J.: Contextual action recognition with R*CNN. In: ICCV, pp. 1080–1088 (2015)
6. Ren, S., He, K., Girshick, R., Sun, J.: Faster R-CNN: towards real-time object detection with region proposal networks. In: NIPS, pp. 91–99 (2015)
7. Fu, Z., Han, Y.: Centroid weighted Kalman filter for visual object tracking. Measurement **45**, 650–655 (2012)
8. Efros, A.A., Berg, A.C., G.M., Malik, J.: Recognizing action at a distance. In: ICCV, pp. 726–733 (2003)
9. Hu, Y., Cao, L., Lv, F., Yan, S., Gong, Y., Huang, T.S.: Action detection in complex scenes with spatial and temporal ambiguities. In: ICCV, pp. 128–135 (2009)
10. Jain, M., Jegou, H., Bouthemy, P.: Better exploiting motion for better action recognition. In: CVPR, pp. 2555–2562 (2013)
11. Ryoo, M.S., Matthies, L.: First-person activity recognition: what are they NG to me? In: CVPR, pp. 2730–2737 (2013)
12. Zhou, S., Shen, W., Zeng, D., Zhang, Z.: Unusual event detection in crowded scenes by trajectory analysis. In: ICASSP, pp. 1300–1304 (2015)
13. Zhu, Y., Nayak, N.M., Roy-Chowdhury, A.K.: Context-aware modeling and recognition of activities in video. In: CVPR, pp. 2491–2498 (2013)
14. Li, W., Wen, L., Choo Chuah, M., Lyu, S.: Category-blind human action recognition: a practical recognition system. In: ICCV, pp. 4444–4452 (2015)
15. Wu, J., Hu, D., Chen, F.: Action recognition by hidden temporal models. Visual Comput. **30**, 1395–1404 (2014)
16. Hoai, M., Zisserman, A.: Improving human action recognition using score distribution and ranking. In: Cremers, D., Reid, I., Saito, H., Yang, M.-H. (eds.) ACCV 2014. LNCS, vol. 9007, pp. 3–20. Springer, Heidelberg (2015). doi:10.1007/978-3-319-16814-2_1
17. Ni, B., Moulin, P., Yang, X., Yan, S.: Motion part regularization: improving action recognition via trajectory group selection. In: Proceedings of CVPR, pp. 3698–3706 (2015)

18. Wang, H., Kläser, A., Schmid, C., Liu, C.L.: Dense trajectories and motion boundary descriptors for action recognition. Int. J. Comput. Vision **103**, 60–79 (2013)
19. Wang, H., Schmid, C.: Action recognition with improved trajectories. In: ICCV, pp. 3551–3558 (2013)
20. Jegou, H., Perronnin, F., Douze, M., Sánchez, J., Perez, P., Schmid, C.: Aggregating local image descriptors into compact codes. IEEE Trans. Pattern Anal. Mach. Intel. **34**, 1704–1716 (2012)
21. Chen, W., Corso, J.J.: Action detection by implicit intentional motion clustering. In: ICCV, pp. 3298–3306 (2015)
22. Breitenstein, M.D., Reichlin, F., Leibe, B., Koller-Meier, E., Van Gool, L.: Robust tracking-by-detection using a detector confidence particle filter. In: ICCV, pp. 1515–1522 (2009)
23. Choi, W.: Near-online multi-target tracking with aggregated local flow descriptor. In: ICCV, pp. 3029–3037 (2015)
24. Chari, V., Lacoste-Julien, S., Laptev, I., Sivic, J.: On pairwise costs for network flow multi-object tracking. In: CVPR, pp. 5537–5545 (2015)
25. Ma, C., Huang, J.B., Yang, X., Yang, M.H.: Hierarchical convolutional features for visual tracking. In: ICCV, pp. 3074–3082 (2015)
26. Danelljan, M., Hager, G., Shahbaz Khan, F., Felsberg, M.: Learning spatially regularized correlation filters for visual tracking. In: ICCV, pp. 4310–4318 (2015)
27. Tang, M., Feng, J.: Multi-kernel correlation filter for visual tracking. In: ICCV, pp. 3038–3046 (2015)
28. Liu, T., Wang, G., Yang, Q.: Real-time part-based visual tracking via adaptive correlation filters. In: CVPR, pp. 4902–4912 (2015)
29. Wang, L., Ouyang, W., Wang, X., Lu, H.: Visual tracking with fully convolutional networks. In: ICCV, pp. 3119–3127 (2015)
30. Bae, S.H., Yoon, K.J.: Robust online multi-object tracking based on tracklet confidence and online discriminative appearance learning. In: CVPR, pp. 1218–1225 (2014)
31. Xing, J., Ai, H., Lao, S.: Multi-object tracking through occlusions by local tracklets filtering and global tracklets association with detection responses. In: CVPR, pp. 1200–1207 (2009)
32. Kuhn, H.W.: The Hungarian method for the assignment problem. Naval Res. Logistics Q. **2**, 83–97 (1955)
33. Horn, B.K., Schunck, B.G.: Determining optical flow. Artif. Intell. **17**, 185–203 (1981)
34. Chatfield, K., Karen Simonyan, A.: Return of the devil in the details: delving deep into convolutional nets. In: BMVC, pp. 2491–2498 (2014)
35. Everingham, M., Gool, L., Williams, C.K.I., Winn, J., Zisserman, A.: The Pascal visual object classes (VOC) challenge. Int. J. Comput. Vision **88**, 303–338 (2010)
36. Kang, D., Han, H., Jain, A.K., Lee, S.W.: Nighttime face recognition at large standoff: cross-distance and cross-spectral matching. Pattern Recogn. **47**, 3750–3766 (2014)
37. Klum, S.J., Han, H., Klare, B.F., Jain, A.K.: The FaceSketchID system: matching facial composites to mugshots. IEEE Trans. Inf. Forensics Secur. **9**, 2248–2263 (2014)
38. Han, H., Shan, S., Chen, X., Lao, S., Gao, W.: Separability oriented preprocessing for illumination-insensitive face recognition. In: Fitzgibbon, A., Lazebnik, S., Perona, P., Sato, Y., Schmid, C. (eds.) ECCV 2012. LNCS, vol. 7578, pp. 307–320. Springer, Heidelberg (2012). doi:10.1007/978-3-642-33786-4_23

39. Jhuang, H., Serre, T., Wolf, L., Poggio, T.: A biologically inspired system for action recognition. In: ICCV, pp. 1–8 (2007)
40. Hubel, D., Wiesel, T.: Receptive fields, binocular interaction and functional architecture in the cat's visual cortex. J. Physiol. **160**, 106–154 (1962)
41. Liu, C., Wechsler, H.: Gabor feature based classification using the enhanced fisher linear discriminant model for face recognition. IEEE Trans. Image Process. **11**, 467–476 (2002)
42. Schuldt, C., Laptev, I., Caputo, B.: Recognizing human actions: a local SVM approach. In: ICPR, pp. 32–36 (2004)
43. Liu, J., Luo, J., Shah, M.: Recognizing realistic actions from videos. In: CVPR, pp. 1996–2003 (2009)
44. Marszalek, M., Laptev, I., Schmid, C.: Actions in context. In: CVPR, pp. 2929–2936 (2009)
45. Tran, D., Sorokin, A.: Human activity recognition with metric learning. In: Forsyth, D., Torr, P., Zisserman, A. (eds.) ECCV 2008. LNCS, vol. 5302, pp. 548–561. Springer, Heidelberg (2008). doi:10.1007/978-3-540-88682-2_42
46. Ke, Y., Sukthankar, R., Hebert, M.: Event detection in crowded videos. In: ICCV, pp. 1–8 (2007)
47. Leal-Taixé, L., Milan, A., Reid, I., Roth, S., Schindler, K.: Motchallenge 2015: towards a benchmark for multi-target tracking. arXiv preprint arXiv:1504.01942 (2015)
48. Han, H., Otto, C., Liu, X., Jain, A.K.: Demographic estimation from face images: human vs. machine performance. IEEE Trans. Pattern Anal. Mach. Intell. **37**, 1148–1161 (2015)

Multi-view Multi-exposure Image Fusion Based on Random Walks Model

Xiao Xue and Yue Zhou[✉]

Institute of Image Processing and Pattern Recognition,
Shanghai Jiao Tong University, Shanghai 200240, China
zhouyue@sjtu.edu.cn

Abstract. By the constraints of the scenarios and cameras, we can hardly get a fully detailed image due to the preperties of exposure. Although some algorithms were proposed to deal with such problems these years, they still have some strict restrictions on the input images which must be captured from the same sight simultaneously. In this paper, we present a method which fuses multi-exposure images from different views. Some techniques in the field of stereo are introduced to deal with feature points matching, and a random walks framework is used to calculate the probabilities of one walking randomly from an unknown point to seed points. These probabilities reveal luminance changes of unknown pixels, and then we can enhance the intensities to make a fusion. Our experiments demonstrate that this method generates accurate results in most situations.

1 Introduction

We all know that one single picture can hardly express all the details due to the complexity of situation and luminance. Even though such pictures with all details exist, it can hardly be captured by one single professional camera. Such situation is not rare in our daily lives, just like one room with windows. When we capture pictures in such situation, we will find it hard to make either indoor or outdoor scenes clear simultaneously. Making the shutter faster or using a small aperture may result in dark indoor scene, while slowing down the shutter or trying a larger aperture will make outdoor scene too lightful so that one can hardly recognize the whole situation.

People began to seek for methods to deal with local image fusion these years. Goshtasby [1] first proposed image fusion based on blocks. When divided uniformly, Cheng and Basu [2] introduced a method which fused blocks by columns. This algorithm considers local information of image blocks, but it will generate halos by the edge. Vavilin and Jo [3] divided the fusion blocks unniformly, and divided step by step due to dynamic range. However, this method does not consider the background trend of the image, and results in ununiformly image. Jo and Vavilin [4] considered pixels under different background must present different intensity, but present similarly under the same background. Kakarala and

© Springer International Publishing AG 2017
C.-S. Chen et al. (Eds.): ACCV 2016 Workshops, Part III, LNCS 10118, pp. 491–499, 2017.
DOI: 10.1007/978-3-319-54526-4_36

Hebbalaguppe [5] introduced S-type function to increase the intensity of short-exposure blocks and replaced it in the long-exposure image. Although these methods generates great results in some situations, the informations between blocks have not been fully used.

Some pixel-based techniques have also generated good results. Mertens et al. [6] introduced multi-exposure fusion based on pyramid decomposition, which improved traditional image fusion techniques. Raman and Chaudhuri [7] set up generalized optimizing framework, converting fusion problems to optimizing models. Shen et al. [8] introduced generalized random walks model [9] into the field of image fusion creatively, which considered the weight of origin images as probabilities. These methods generate ralively fine results compared with ones formerly mentioned, however, they have strict restrictions about input images that they must have the same view. Troccoli et al. [10] used an exposure-invariant similarity statistic to establish correspondences before extracting camera radiometric response fuction, then converted all images to radiance space to get HDR texture recovery, but this method can hardly deal with saturated pixels and it will impose a restriction on the radiance. Sun et al. [11] firstly calculated the disparity map to generate camera response function, and eliminated edge artifacts. However, this algorithm can not deal with extreme situation. Our method considered such situation, and managed to fuse multi-exposure images from diffirent views.

2   The Algorithm

This sections presents our method for matching feature points of images from different views, then fusion problem can be converted to probabilities from one unknown point walking randomly to the seeds, as Fig. 1 shows. Details will be introducted below.

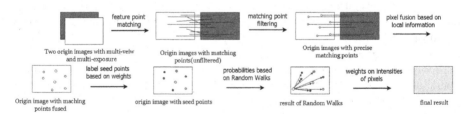

Fig. 1. Overview of our proposed algorithm.

2.1   Matching

As the most widely used algorithm in the field of feature matching, SIFT [12] manages to find extreme points in space scale with their positions, scales and rotational invariants. By searching for pixel positions in all scales, we can recognize the scale-invariant points using Gaussion function. A fitting model is introduced to ensure the position and scale, which iterates in each candidate position.

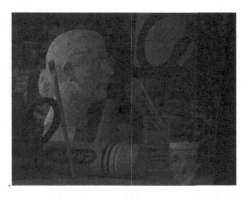

Fig. 2. Result of homograph. The distortion makes it impossible to fuse images directly.

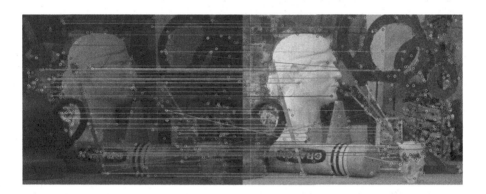

Fig. 3. Result of matching.

To these positions, we distribute one or more directions based on local gradient. All pixel oriented operations are fit to same directions, which comes to scale and rotation invariants. In our expriments, two origin images are captured by two cameras located in one baseline, which results in rotation. Considering rotation invariants, we choose SIFT to generate a matching result which contains mismatching. RANSAC [13] will be used to diminish some mismatching, and finally the matching can be fairly precise, as Fig. 3 shows.

2.2 Pixel Fusion Based on Background Trend

After homograph, we get the result of homegraph from one view to the other, as Fig. 2 shows. The distortion of the short-exposure input makes it impossible to do the image fusion by each pixel. What we have known is the information of matching points. Traditional methods fused pixels based on their intensities. The fused intensity map can be defined as

$$R_i = w_1 \times Y_{i1} + (1 - w_1) \times Y_{i2}, 0 < w_1 < 1 \qquad (1)$$

Y_{i1} is the intensity of pixel i in the long exposure image, while Y_{i2} is the same in the short exposure image. w_1 represents the contributions from Y_{i1} to R_i. Actually w_1 plays an important role in the fusion of whole image, considering local contrast of image. However, if the threshold value is set to a constant for the whole image, the intensities near the constant will display some choppy edge in the fused image, which result in unsatisfactory fusion. For single pixels, as we deal with matching points with Eq. 1, whatever w_1 is, the result will not show satisfactory details.

In fact, the adjacant pixels in the same background will have similar intensities in the fused image. Therefore, pixel intensities in the same background have higher similarity while lower in the different background, and we can use the variance to define the w_1. Our algorithm uses mean filter to construct the background trend M. To avoid the case of weights chop mentioned above, we define w_1 as a decreasing function of M, and

$$w_1 = \begin{cases} 1 & 0 < M < T_1, \\ (T_2 - M)/(T_2 - T_1) & T_1 < M < T_2, \\ 0 & T_2 < M < 1. \end{cases} \tag{2}$$

T_1 is defined as the average intensity of 5×5 neighborhood from the target pixel in the short-exposure image, while T_2 is the corresponding average intensity in the long-exposure image. The result from background trend shows smooth transition between different background. When we consider matching point, the weight of increasing intensity from short-exposure differs according to local information. Different enhancement provides different labels, and prepare for random walks.

2.3   Random Walks Model

When traversing a map from one point or a series of points, at a random point, the walker will move to the adjacent point with a probability of $1 - \alpha$, or jump to any point of the map with a probability of α, surely including adjacent points. We call α the teleport probability. After every walk, there will be a probability redistribution which reveals the probability of every point being visited. In the next step, this probability distrbution will be regarded as the input of next walk, and this process will repeat until reaching convergence, and final stable probability tells us the final probability one walking from one point to another.

We first define a graph G consisting of (V, E). $v \in V$ indicate the nodes and $e \in E \subseteq V \times V$ mean edges. e_{ij} indicates the edge, and the weight of the edges, w_{ij}, is calculated by the Euclidean distance of vertices i and j. The degree of a single vertice i is set as d_i, which can be defined as

$$d_i = \sum w_{ij} \tag{3}$$

where j indicates the adjacent vertice.

From Sect. 2.2, we get the respective intensity enhancement for each matching point, and these points are corresponding to vertices in the graph. We treat these points as seed points, and label them with their intensity enhancement ratio. Let $L = \{l_1, l_2, \cdots, l_k\}$ denote the set of labels, k the number of labels, which is far less than the number of matching points for some of them have the same enhancement ratio. As we define p_{ni} the probability of one walking from vertice i to label(seed) points n, for each unknown vertice, the enhancement ratio can be calculated by

$$p_i = \sum_{n=1}^{K} p_{ni} \times W_i \tag{4}$$

where W_i denotes the intensity enhancement of vertice i. Obviously, we can figure out $\sum_{n=1}^{K} p_{ni} = 1$, and if every vertice is given equal probability, all the pixels will have the same rate of intensity increasing, and some unclear details are still missing even some clear details will be gone. In that case, artifacts are introduced and we can hardly generate a satisfactory fusion result.

The main steps of the Random Walks will be introduced below:

Definition of Weights on Edges. In the graph mapped from image, we need to define a function to describe the features' change between pixels, which means the weights between vertices should be defined by image features like grayscale, color, texture et al. In our case, we define the weight based on pixel intensity. Guassian function is introduced to define the weight:

$$w_{ij} = \exp(-\beta(f_i - f_j)^2) \tag{5}$$

f denotes the feature value of the pixel, and β the free parameter, which is set to 80 in our expriments. When features in different pixels present similar, the weight between these two corresponding vertices are relatively large, which seems easy for walker to reach the destination.

Dirichlet Problems. We at first introduce the Laplacian matrix [14] as:

$$L_{ij} = \begin{cases} d_i & i = j, \\ -w_{ij} & i \ and \ j \ are \ attachable \\ 0 & otherwise. \end{cases} \tag{6}$$

where d_i has been define in Eq. 3. Based on electrical networks [15], the energy function can be defined as:

$$E = \frac{1}{2} \sum_{(v_i, v_j) \in E} w_{ij}(u(v_i) - u(v_j))^2 \tag{7}$$

u_i denotes electromotive force on vertice v_i, and all we want is to find a function u that minimizes E. Absolutely, the minimum value of E exists due to the matrix itself is positive semidefinite.

In our model, the labels of seed points are known while others' are not, so let V_L as the marked points and V_X as the unmarked points. Therefore, the Laplacian matrix can be rewriten as:

$$L = \begin{pmatrix} L_L & B \\ B^T & L_X \end{pmatrix} \tag{8}$$

Similarly, the energy function can be writen as:

$$E = \frac{1}{2} \begin{pmatrix} U_L^T & U_X^T \end{pmatrix} \begin{pmatrix} L_L & B \\ B^T & L_X \end{pmatrix} \begin{pmatrix} U_L \\ U_X \end{pmatrix} \tag{9}$$

Expanding Eq. 9, the equation turns:

$$E = \frac{1}{2}(U_L^T L_L U_L + 2U_X^T B^T U_L + U_X^T L_X U_X) \tag{10}$$

Differentiating E with respect to U_X, we can get:

$$L_X U_X = -B^T U_L \tag{11}$$

As we have assigned seed points x_L with label l_k, next we can solve Eq. 11 in iterations. The final result will show the probability of one walking randomly to seed points.

3   Experimental Result

We implement our image fusing algorithm in C++ language and evaluate its performance on sequences involving in art, laundry, book et al. All sequences are downloaded from http://vision.middlebury.edu/stereo/data/scenes2005/. This website provides many multi-exposure images captured from different views by Daniel Scharstein. Figure 4 shows our algorithm on the Art sequence.

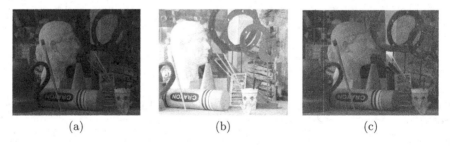

(a)   (b)   (c)

Fig. 4. Fusion result of art sequence. Two multi-exposure images with different views and their fusing result.

Figure 4(a) shows the art sequence with a short exposure, and Fig. 4(b) presents the same sequence with a long exposure. It can be seen that some details are ignored duo to under- or over-exposure, especially in the region of the face. Meanwhile, because of the input images from different views, some regions in the Fig. 4(a) are missing in the Fig. 4(b), such as the pixels in the teapot. There are surely not any matching point in the neighborhood, so the illumination compensation must rely on the seed points far from the teapot region. By introducing Random Walks, the intensity enhancement near the teapot can be estimated according to the distribution of the seed points. Figure 4(c) shows the final result, which presents details in input images and acts as a high dynamic range image.

We conduct another two different expriments on the sequence of books and computer, as Figs. 5 and 6 shows.

(a)   (b)   (c)

Fig. 5. Fusion result of book sequence. The details of the blue book in the left region acts better in the fusion result.

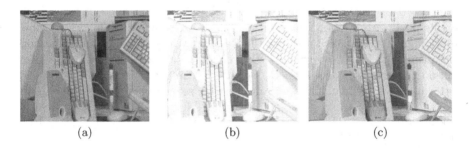

(a)   (b)   (c)

Fig. 6. Fusion result of computer sequence. Over-exposure on the yellow voltmeter is fixed and the final result presents smooth intensity trend.

Similarly as Fig. 4, it can be seen that there are great improvements achieved when our algorithm are used to enhance the intensity of unmarked pixels and estimated its true intensity. The blue book in Fig. 5(c) gives rich texture in the result while it occurs an occlusion in the input, and the yellow voltmeter in Fig. 6 is over-exposed whereas we manage to recover it using our methods.

As there are not an authoritative evaluation system on the multi-exposure images fusion form different views, we can only provide results with our sense of sight. We conducted expriments using all the sequences offered, and calculated

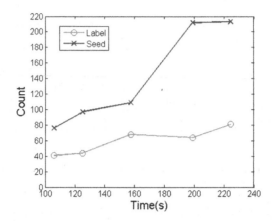

Fig. 7. Comparison of all the sequences about the numbers of seeds, labels, and time of random walks.

the numbers of seeds, labels, and time of Random Walks. Figure 7 reveals the relationship between these parameters, and the time may perform as a nonlinear function as the number of seeds and labels. Actually, we can also infer from the results that the large amounts of seeds (labels) and uniform distribution contributes to better fusion results.

4   Conclusion

In this paper, we present a method for fusing multi-exposure images captured from different views. We introduce the Random Walks model into our methods and estimate the intensity changes. Expriments show that our method performs well on the professional sequences. Further research can be done on exploring better methods for matching and random walk searching to get a better result.

Acknowledgement. The work is supported by National Program on Key Basic Research Project (973 Program).

References

1. Goshtasby, A.A.: Fusion of multi-exposure images. Image Vis. Comput. **23**, 611–618 (2005)
2. Cheng, I., Basu, A.: Contrast enhancement from multiple panoramic images. In: IEEE 11th International Conference on Computer Vision, ICCV 2007, pp. 1–7. IEEE (2007)
3. Vavilin, A., Jo, K.H.: Recursive HDR image generation from differently exposed images. In: Proceedings of Graphicon, pp. 23–27 (2008)
4. Jo, K.H., Vavilin, A.: HDR image generation based on intensity clustering and local feature analysis. Comput. Hum. Behav. **27**, 1507–1511 (2011)

5. Kakarala, R., Hebbalaguppe, R.: A method for fusing a pair of images in the JPEG domain. J. Real-Time Image Process. **9**, 347–357 (2014)
6. Mertens, T., Kautz, J., Van Reeth, F.: Exposure fusion. In: 15th Pacific Conference on Computer Graphics and Applications, PG 2007, pp. 382–390. IEEE (2007)
7. Raman, S., Chaudhuri, S.: A matte-less, variational approach to automatic scene compositing. In: IEEE 11th International Conference on Computer Vision, ICCV 2007, pp. 1–6. IEEE (2007)
8. Shen, R., Cheng, I., Shi, J., Basu, A.: Generalized random walks for fusion of multi-exposure images. IEEE Trans. Image Process. **20**, 3634–3646 (2011)
9. Grady, L.: Random walks for image segmentation. IEEE Trans. Pattern Anal. Mach. Intell. **28**, 1768–1783 (2006)
10. Troćcoli, A., Kang, S.B., Seitz, S.: Multi-view multi-exposure stereo. In: Third International Symposium on 3D Data Processing, Visualization, and Transmission, pp. 861–868. IEEE (2006)
11. Sun, N., Mansour, H., Ward, R.: HDR image construction from multi-exposed stereo LDR images. In: 2010 17th IEEE International Conference on Image Processing (ICIP), pp. 2973–2976. IEEE (2010)
12. Lowe, D.G.: Object recognition from local scale-invariant features. In: The Proceedings of the Seventh IEEE International Conference on Computer vision, vol. 2, pp. 1150–1157. IEEE (1999)
13. Fischler, M.A., Bolles, R.C.: Random sample consensus: a paradigm for model fitting with applications to image analysis and automated cartography. Commun. ACM **24**, 381–395 (1981)
14. Dodziuk, J.: Difference equations, isoperimetric inequality and transience of certain random walks. Trans. Am. Math. Soc. **284**, 787–794 (1984)
15. Doyle, P.G., Snell, J.L.: Random walks and electric networks. AMC **10**, 12 (1984)

Attributes and Action Recognition Based on Convolutional Neural Networks and Spatial Pyramid VLAD Encoding

Shiyang Yan[1(✉)], Jeremy S. Smith[2], and Bailing Zhang[1]

[1] Xi'an Jiaotong-Liverpool University, Suzhou, China
{shiyang.yan,bailing.zhang}@xjtlu.edu.cn
[2] University of Liverpool, Liverpool, UK
J.S.Smith@liverpool.ac.uk

Abstract. Determination of human attributes and recognition of actions in still images are two related and challenging tasks in computer vision, which often appear in fine-grained domains where the distinctions between the different categories are very small. Deep Convolutional Neural Network (CNN) models have demonstrated their remarkable representational learning capability through various examples. However, the successes are very limited for attributes and action recognition as the potential of CNNs to acquire both of the global and local information of an image remains largely unexplored. This paper proposes to tackle the problem with an encoding of a spatial pyramid Vector of Locally Aggregated Descriptors (VLAD) on top of CNN features. With region proposals generated by Edgeboxes, a compact and efficient representation of an image is thus produced for subsequent prediction of attributes and classification of actions. The proposed scheme is validated with competitive results on two benchmark datasets: 90.4% mean Average Precision (mAP) on the Berkeley Attributes of People dataset and 88.5% mAP on the Stanford 40 action dataset.

1 Introduction

Human attributes descriptions such as gender, clothing style, hair style and action categories such as using a computer, riding a horse or texting messages, are two popular yet challenging recognition problems in semantic computer vision. The tasks are particularly difficult in static images partly due to the lack of motion information. Large variances in illumination conditions, view point, human pose as well as occlusion add further obstacles to finding satisfactory solutions.

The description of human attributes and the classification of action categories all depends on local and global contextual information. On the one hand, the local regions that correspond to detailed, fine-grained appearance features may play critical roles in recognition; on the other hand, the global context of the surrounding objects and scenes is also instrumental to tackle the problem. As an example, human attributes like 'gender' not only depends on local features such

© Springer International Publishing AG 2017
C.-S. Chen et al. (Eds.): ACCV 2016 Workshops, Part III, LNCS 10118, pp. 500–514, 2017.
DOI: 10.1007/978-3-319-54526-4_37

as the face or hair style, but also relies on the global context, for example, the clothing style and body shape. As for action category classification, the pose, the objects a person interacts with, and the scene in which the action is performed, all contain useful information. This is better illustrated by the action types in sports. For example, for the action of 'playing basketball', the basketball and playground are both strong evidence for this action category.

A typical way for compactly representing image and incorporating global contextual information is to apply a patch feature encoding strategy such as the Bag-of-Visual-Words (BoVW) [1], Fisher Vectors (FV) [2], and Vector of Locally Aggregated Descriptors (VLAD) [3]. Among these, it is reported that the Fisher Vector outperforms many popular encoding methods previously published on benchmark image datasets. VLAD can be regarded as a simplified non-probabilistic version of FV and also shows comparable performance [3]. Recently, VLAD has continued to grow in popularity in computer vision, with many excellent demonstrations for problems including object detection, scene recognition and action recognition [4–7]. In this paper, we will further explore the potential of VLAD for the human attribute prediction and action classification in still images.

Conventional patch feature encoding strategies largely depend on local features such as Scale Invariant Feature Transform (SIFT) [8]. Such a mid-level feature description can be considered as 'hallow features', as no deep training is involved. Recently, convolutional neural networks (CNN) achieved breakthrough performance in many vision tasks [9–11]. Yet, it is noteworthy that CNNs are originally trained for the classification of objects [9], with the typical goal of correctly identifying a single, predominant object in an image. Hence, existing CNN models alone are limited in their capability in the description of attributes and the classification of actions despite their powerful feature representation capabilities. For action and attribute recognition, a rational strategy is to take advantages from both of the CNN and patch feature encoding strategies. Based on this intuition, we encode the CNN features for sub-regions of an image to generate a compact representation. Our approach is similar to [12] in which the Fisher Vector encoding scheme is applied on CNN features and each image is represented as a bag of windows. Likewise, our method can also be considered as bag of patches as the image patches are first extracted using region proposal algorithms such as Edgeboxes [13] and then represented with VLAD encoding.

As previously explained, VLAD encoding combines local and global features by generating image representations that better reflect the local information of the image patches. A region proposal algorithm would be instrumental to generate object-like local patches. Through VLAD encoding, the CNN features of many regions will be encoded into a higher level description closer to the images inherent signature. Such a compact signature preserves most of the information from CNNs while reducing the dimensionality. More importantly, the fine-grained properties of an image could be retained with the proposed VLAD encoding.

The only downside to VLAD encoding is its lack of preservation of spatial information [14]. To compensate for this, we firstly build spatial pyramids of the image which are matched to region level CNN features, and then perform VLAD encoding on the separate pyramid. The generated VLAD codes are concatenated into a final representation, which is subsequently forwarded to a classifier, e.g. a SVM, for final classification. We conducted extensive experiments including various comparisons, and achieved promising results on the Berkeley Attributes of People dataset [15] and the Stanford 40 action dataset [16]. To the best of our knowledge, we are the first to apply a spatial pyramid VLAD encoding scheme for attribute and action recognition in still images.

2   Related Works

2.1   Human Attributes and Action Recognition

Attributes as visual qualities of objects provide a bridge between lower level image representations and higher level semantic information. Accordingly, attribute learning has become important in many computer vision applications, for example, face verification [17], clothing description [18] and image retrieval [19]. At the same time, action recognition from still images has recently attracted more attention, because many action categories can be unambiguously defined in static images without motion information. The potential applications are obvious, for example, image annotation, image retrieval and human computer interaction.

The issue of the recognition of human attributes and actions has been researched for many years in computer vision and machine learning. A common practice is to apply the Bag-of-visual-words (BoVW) [20–22], which is advantageous in the global representation of an image. Vincent Delaitre et al. [23] handled the problem by applying a bag-of-features and part-based representation. Recently, as an extension of BoVW, Fisher Vector [2] and VLAD have been gaining ground in many vision problems, including video event classification [24] and action recognition [25].

For many vision problems, another influential train of thought is on part-based modelling, which has also witnessed some successes. Among them, the Deformable Part Model (DPM) [26] is a milestone in the development. Similarly, Poselets method [27]used a part-based template to interpolate a pose. Recently, Zhang et al. [28] proposed a Pose Aligned Networks Model (PANDA), which is a combination of deep learning model and Poselets, and demonstrated its capability in capturing human attributes.

For action recognition from still images, another common approach is centering on human-object interaction. Yao and Fei [29] detected objects in cluttered scenes and estimated articulated human body parts in human object interaction activities. Alessandro Prest et al. [30] introduced a weakly supervised approach for learning human actions modeled as interactions between humans and objects. In these previous studies, the global scene information has not been taken into account, which is one of the challenges this paper addresses.

2.2 Deep Learning Powered Approaches

Recently, deep learning methods, especially deep convolutional neural networks, have dramatically improved the state-of-the-art in visual object recognition [9], object detection [10,31], image segmentation [32] and many other vision tasks. For the task of action recognition, Oquab et al. [33] investigated transfer learning with a CNN model, showing that a CNN trained from a large dataset can be transferred to another visual recognition task when limited training data is available. Promising results were reported for action recognition [33], with the advantages of higher accuracy and a shorter training period. Gkioxari et al. [34] developed a part-based approach by leveraging trained deep body parts detectors for action and attribute classification. They showed state-of-the-art performance on the Berkeley Attributes of People dataset. Gkioxari et al. [11] proposed a joint training procedure for the automatic discovery of the top contributing contextual information for action recognition and attribute classification, which showed state-of-the-art results on several publicly available datasets. Recently, Ali Diba et al. proposed DeepCAMP [35], a scheme that utilizes CNN mid-level patterns for action recognition and attribute determination, which is also showing promising results.

Our work is different from these as we extract CNN features for postprocessing using spatial pyramid VLAD coding. VLAD coding [3,36], along with Fisher Vectors [37], is mostly applied in image classification or retrieval tasks [12,38]. Compared with BoVW and Fisher Vector, VLAD can be more balanced between memory usage and performance [3]. However, the main downside of VLAD, the lack of spatial information, has been less stressed. A well-known approach of encoding spatial information was proposed by Lazebnik et al. [39] by taking into account the spatial layout of keypoints in a pooling step, which divides the image into increasingly finer spatial sub-regions and creates histograms for each sub-region separately. In [40], the authors proposed to combine spatial pyramids and VLAD. Recently, Andrew Shin et al. [14] further examined the approach for image captioning. We applied the similar methods of [14,40] by extracting deep activation features from local patches at multiple scales, and coding them with VLAD. However, our approach extended much beyond the scene classification and object classification in [14,40], in which the significance of explicitly dealing with local objects and spatial information is less evident.

3 Methods

In this section, we describe the main components of our proposed pipeline, starting with region-based feature extraction after EdgeBoxes, and ending with Spatial Pyramid VLAD encoding for the attributes and actions to be classified. Pictorially, the overall workflow of our model can be described by Fig. 1.

3.1 Feature Extraction

Inspired by the recent successes of CNN-based object detection, which relies on category independent region proposals, we also start from a set of region proposals from images to pursue accuracy with an affordable computation cost [13].

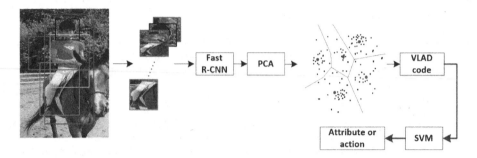

Fig. 1. Full pipeline of the proposed methods: each window is generated by region proposal algorithm and represented by fc6 features, Principle Component Analysis (PCA) is applied for dimension reduction, followed by k-means for centroid learning (the blue dots). Attributes and action can thus by classified with VLAD codes and a classifier. (Color figure online)

Among the recently published off-the-shelf technologies that follow this paradigm, we empirically choose EdgeBoxes [13] as the first component in our workflow to produce high-quality region proposals, mainly based on the computational efficiency and high-level performance in terms of localizing objects [41]. We consider each image as a bag, the object proposals from an image are considered as a bag of patches or windows. Thus a compact method for representing the information can be obtained through VLAD encoding. For the sake of efficiency, we simply extract 1000 patches per image.

While [14] directly exploited a CNN pre-trained on ImageNet as a generic extractor, we added an extra step of fine-tuning the pre-trained VGG16 model [42], which was conducted on top of the fast R-CNN [31] with candidate regions from EdgeBoxes. We annotated the target bounding boxes and corresponding labels, then performed training based on fast R-CNN. As verified by our experiments, this process yields a better category dependent feature representation capability.

We then use the CNN model as a generic feature extractor for 1000 image patches generated by EdgeBoxes. The top 1000 boxes have higher possibilities of containing objects. For the same reason in [14], we do not apply non-maximum suppression. The generated region proposals have arbitrary size, a common way to extract features using a CNN would be resizing them and forwarding them to a CNN model one by one. However, feature extraction of multiple regions in a CNN can be really time-consuming. Hence, we implemented our algorithm on top of the fast R-CNN in which the RoI projection and RoI pooling scheme [31] enable feature extraction of arbitrary size windows of one image in only one feed forward process, thus the computational cost can be much reduced.

3.2 Spatial Pyramid VLAD

While it is well-recognized that the VLAD encoding scheme performs well in preserving local features, it discards the spatial information. To tackle this problem,

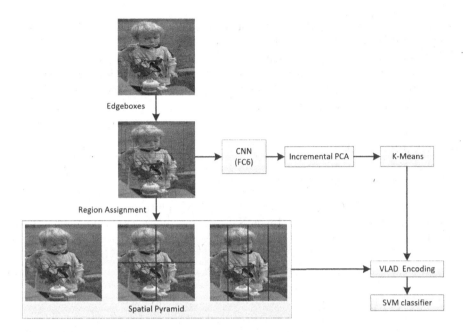

Fig. 2. VLAD encoding with a spatial pyramid: the image was divide with a 3 level spatial pyramid: 1×1, 2×2 and 4×1. Each pyramid is encoded separately with VLAD.

several recent papers [14,40] proposed spatial pyramid VLAD as a solution. In this paper, we followed the methodology with an efficient implementation for attribute prediction and action classification. More specifically, as shown in Fig. 2, a 3 level (1×1, 2×2, and 4×1) spatial pyramid is exploited. To allocate regions into each spatial grid, we assign each region according to the distribution of their centers. This simple yet effective approach can avoid overlapping and is discriminatively powerful.

For the 4096 dimensional features extracted from the CNN, one possibility is to perform VLAD encoding for each spatial pyramid separately. However, the 4096 dimensions would be too large to encode. As pointed out in [43], dimension reduction marginally affects the overall representation of VLAD. Hence, we perform dimensionality reduction with Principle Component Analysis (PCA) on the CNN features of each region. As the number of features is large, training a conventional PCA on all the features would be unrealistic. An alternative would be randomly selecting a certain number of features for training, and then performing PCA on all features. Here, we chose to implement incremental PCA [44] on all the features because of its high efficiency in terms of memory usage. We perform PCA on all the features with dimension of 256. Also, to study the influence of dimension on overall performance, we also reduced the features to 512 for comparison.

The last step of the encoding is similar to BoVW, i.e., code word learning implemented by unsupervised learning like k-means clustering. The number of clusters was set to 12, 16, 24, and 64 for testing. We also followed the practice of [14] by exploiting k-means++ [45] due to its improved speed and the accuracy from the randomized seeding technique. After obtaining the code words from k-means++, VLAD coding with L2-normalization can thus be implemented. The final dimensionality of a VLAD code is the number of clusters times the length of PCA reduced CNN features.

4   Experiments

4.1   Deep Learning Model

Experiments were conducted using the Caffe CNN platform [46]. A CNN model was first pre-trained using the ImageNet dataset which was subsequently fine-tuned using our specific datasets for the different tasks, as described in the previous section. We set the max training iteration at 40000, the other parameters were the same as for the original fast R-CNN.

4.2   VLAD Encoding

The incremental PCA and k-means++ are all implemented on top of the Scikit-learn Python machine learning package [47]. With the obtained dimensionality-reduced features and codewords, VLAD encoding was implemented in Matlab using the VLfeat toolbox [48]. For attribute classification, a SVM linear classifier was utilized using the LIBSVM toolbox [49], while for the action classification, we implemented a multi-layer perceptron based on the Matlab Neural Network Toolbox.

4.3   Attribute Recognition

We evaluated our method using the Berkeley Attributes of People Dataset [15]. The dataset includes 4013 images for training, and 4022 test images collected from the PASCAL and H3D datasets. This is a very challenging dataset as the people in the images often have large appearance variance and occlusion. Few reported methods worked well with this dataset [15,28].

As previous explained above, we first applied the spatial pyramid VLAD encoding, and then employed a SVM classifier for the final prediction. Specifically, the pre-trained VGG16 model [42] was used for fine-tuning. The training process was implemented in a fast R-CNN [31] framework. We then run Edgeboxes on each image, and extracted the features of the first fully connected layers (fc6) of each sub-region. VLAD encoding is then accomplished after PCA dimensionality reduction and codeword learning with k-means++.

More details about the experiments are explained in the following section. First of all, the first fully connected layers (fc6) CNN features of the ground truth region are extracted, and directly applied for attribute classification as a comparison baseline. As shown in Table 1, the mean average precision (mAP) is

Table 1. The average precision results on the Berkeley Attributes of People test Dataset and comparison with different approaches. The results show that with the combination of the original CNN fc6 features, there is a gain of 8.5%. Moreover, with the spatial pyramid, the mean average precision is improved by 3.6%.

Attribute	Male	Long hair	Glasses	Hat	Tshirt	Longsleeves	Shorts	Jeans	Long pants	Mean AP
fc6 features of ground truth region	90.1	80.8	77.6	80.6	57.4	84.2	64.9	71.1	96.5	78.1
PCA 256+16 clusters (No spatial pyramid)	88.9	76.4	74.7	68.2	68.5	88.5	73.3	71.8	94.2	78.3
PCA 256+16 clusters+fc6 features (No spatial pyramid)	92.5	87.4	85.2	90.4	68.3	89.7	85.5	83.9	98.0	86.8
PCA 256+16 clusters+fc6 features (With spatial pyramid)	**94.1**	**90.4**	**89.4**	**94.0**	**74.0**	**92.5**	**91.9**	**88.6**	**98.5**	**90.4**

78.1%, which means it is not very accurate in representing the attributes associated with the image based on the primitive CNN feature. Secondly, to evaluate the stand-alone performance of VLAD encoding, we encoded each images with CNN features of 256 dimensions and 16 learnt clusters into the VLAD code. The mean average precision is 78.3%. In this settings, we did not use the ground truth region for classification, and spatial pyramid are also not applied. Concatenating the VLAD code with CNN features yields an 8.5% rise in performance, which implies that the combination of local patches features and compact global representation all contribute to attribute recognition. Finally, to examine the influence of spatial pyramid coding, experiments were organized with results confirming that adding spatial pyramid coding does improve the overall performance, by 3.6% in the mean average precision.

To evaluate the influence of the number of k-means clusters, we performed VLAD encoding with 12, 16, 24, 64 centroids separately. It can be observed that 16 clusters works the best in our experiments as shown in Table 2. Also, we repeated the experiments with dimensionality-reduced CNN features of 512, which produced similar results. As noted in [43], dimension reduction plays a significant role in VLAD encoding, the lower dimensional CNN features may improve the performance. Therefore, we set the CNN features as 256 dimensionality in most of the experiments.

We also compared our results with existing methods. As illustrated in Table 2, our methods outperformed the published results listed in the table. In Fig. 3, we provide some examples of recognized attributes on the datasets.

4.4 Action Recognition

To evaluate the system performance on action recognition, we experimented on the Stanford 40 action dataset [16]. Which contains 9532 images corresponding to

Table 2. The average precision results of the Berkeley Attributes of People Dataset and comparison with previous methods. We provide results on 256 dimensionality of CNN features after PCA with 12, 16, 24 and 64 clusters of k-means. We also perform VLAD encoding on 512 dimensionality, there is not much differences in terms of performance. Performing VLAD encoding on 256 dimensionality CNN features with 16 clusters yields the best results.

Attribute	Male	Long hair	Glasses	Hat	Tshirt	Longsleeves	Shorts	Jeans	Long pants	Mean AP
Poselets [15]	82.4	72.5	55.6	60.1	51.2	74.2	45.5	54.7	90.3	65.0
PANDA [28]	91.7	82.7	70.0	74.2	49.8	86.0	79.1	81.0	96.4	79.0
R*CNN [11]	92.8	88.9	82.4	92.2	**74.8**	91.2	92.9	89.4	97.9	89.2
Gkioxari et al. [34]	92.9	90.1	77.7	93.6	72.6	**93.2**	**93.9**	**92.1**	**98.8**	89.5
Ours (PCA 256+12 clusters+fc6 features)	93.8	90.0	88.5	93.4	72.9	92.2	90.8	87.7	98.4	89.7
Ours (PCA 256+64 clusters+fc6 features)	93.8	**92.2**	89.1	93.8	73.1	92.1	91.4	87.8	98.4	90.0
Ours (PCA 256+24 clusters+fc6 features)	**94.1**	90.4	**89.5**	**94.0**	73.8	92.5	91.9	88.5	98.4	90.3
Ours (PCA 256+16 clusters+fc6 features)	**94.1**	90.4	89.4	**94.0**	74.0	92.5	91.9	88.6	98.5	**90.4**

(a)   (b)

Male:0.0064,	No	Male:0.8289,	Yes
Long-hair:0.7897,	Yes	Long-hair:0.0748,	No
Glasses:0.0643,	No	Glasses:not-certain,	Not-certain
Hat:0.0001,	No	Hat:0.9430,	Yes
T-shirt:0.6533,	No	T-shirt:0.0416,	No
Long-sleeves:0.0054,	No	Long-sleeves:0.2946,	No
Shorts:not-certain,	Not-certain	Shorts:0.0030,	No
Jeans:0.0276,	No	Jeans:0.0126,	No
Long-pants:0.0088	No	Long-pants:0.9929,	Yes

Fig. 3. Examples of attribute classification: the probabilities of certain attributes are provided, the blue text are the ground truth label. The red text shows an incorrect classification example. The threshold 0.5 was applied as a standard for classification. (Color figure online)

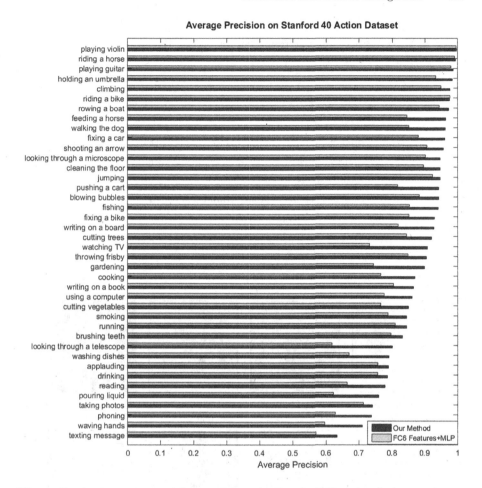

Fig. 4. Our best average precision results on the Stanford 40 action dataset and comparison with the baseline approach.

40 classes of actions. The dataset was split into a training set with 4000 images, and a testing set with 5532 instances. There are 180–300 images for each class. The images from each class have large variations in human pose, appearance, and background clutter.

We extract the CNN features (fc6) and forward them directly to a Multi-layer Perceptron (MLP) classifier as the baseline for comparison. As the parameters of 256 dimensionality of the CNN features and 16 clusters achieved the best results in the previous experiments for attribute recognition, we directly took the same parameters in action recognition. As shown in Fig. 4, our proposed spatial pyramid VLAD encoding scheme outperforms the primitive CNN features in all action classes except the 'riding a bike' class. In this action class, the performance results are similar. More importantly, VLAD performs better for

Table 3. Mean average precision results on the Stanford 40 action dataset and comparison with previous results.

Method	Mean AP
Object bank [50]	32.5
LLC [51]	35.2
EPM [52]	40.7
DeepCAMP [35]	52.6
Khan et al. [53]	75.4
Ours (fc6 features of ground truth region)	81.2
Ours (PCA 256+16 clusters)	85.9
Ours (PCA 256+16 clusters+fc6 features)	**88.5**

(a) Writing on a book: 0.9966   (b) Applauding: 0.9933   (c) Cleaning the floor: 1.0000   (d) Drinking: 0.9954

(e) Fixing a bike: 0.9969   (f) Riding a horse: 0.9979   (g) Brushing teeth: 0.6578   (h) Fishing: 0.9996

(i) Blowing bubbles: 0.8351   (j) Cutting trees: 0.9425   (k) Cooking: 0.9996   (l) Climbing: 0.9997

Fig. 5. Some examples of correct recognition in the Stanford 40 action dataset: the predicted label and corresponding confidence values are provided.

the more fine-grained action classes, for instance, 'writing on a board'. This is because VLAD encoding can retain the local information from small patches, and spatial pyramid coding can form a more compact representation of an image.

The comparison with previous reported methods is shown in Table 3, which demonstrates that our method outperforms all of the previously published work as listed in the table. It is noteworthy that Khan et al. [53] did not utilize a ground truth bounding box during action recognition, to impartially compare with their results, we also experimented without the ground truth region. With a setting of 256 dimensions from PCA and 16 word codes from clustering, our proposed method yields a 10.5% increase in the mean average precision compared to [53]. Some of the correctly recognized examples are shown in Fig. 5.

5 Conclusion

Human attribute and action recognition in static images are challenging tasks, with the main challenges being fine-grained recognition without motion information. How to efficiently exploit both the global features and local features is key to solving these problems. In this paper, we applied Vector of Locally Aggregated Descriptors on top of spatial pyramids to detect local information, not only from the ground truth region but also from nearby objects and scenes. Experiments confirmed that the combination of CNN features and VLAD codes is very effective in retaining both local and global information. As we encode CNN features on the first fully connected layer, the next step is to explore the possibility of directly encoding the original CNN features.

References

1. Fei-Fei, L., Perona, P.: A Bayesian hierarchical model for learning natural scene categories. In: 2005 IEEE Computer Society Conference on Computer Vision and Pattern Recognition (CVPR 2005), vol. 2, pp. 524–531 (2005)
2. Csurka, G., Perronnin, F.: Fisher vectors: beyond bag-of-visual-words image representations. In: Richard, P., Braz, J. (eds.) VISIGRAPP 2010. CCIS, vol. 229, pp. 28–42. Springer, Heidelberg (2011). doi:10.1007/978-3-642-25382-9_2
3. Jégou, H., Douze, M., Schmid, C., Pérez, P.: Aggregating local descriptors into a compact image representation. In: 2010 IEEE Conference on Computer Vision and Pattern Recognition (CVPR), pp. 3304–3311. IEEE (2010)
4. Sharma, G., Jurie, F., Schmid, C.: Discriminative spatial saliency for image classification. In: 2012 IEEE Conference on Computer Vision and Pattern Recognition (CVPR), pp. 3506–3513 (2012)
5. Delaitre, V., Laptev, I., Sivic, J.: Recognizing human actions in still images: a study of bag-of-features and part-based representations. In: Proceedings of the British Machine Vision Conference, pp. 97.1–97.11. BMVA Press (2010). doi:10.5244/C.24.97.
6. Wang, H., Kläser, A., Schmid, C., Liu, C.L.: Dense trajectories and motion boundary descriptors for action recognition. Int. J. Comput. Vis. **103**, 60–79 (2013)

7. Peng, X., Zou, C., Qiao, Y., Peng, Q.: Action recognition with stacked fisher vectors. In: Fleet, D., Pajdla, T., Schiele, B., Tuytelaars, T. (eds.) ECCV 2014. LNCS, vol. 8693, pp. 581–595. Springer, Heidelberg (2014). doi:10.1007/978-3-319-10602-1_38

8. Lowe, D.G.: Object recognition from local scale-invariant features. In: The Proceedings of the Seventh IEEE International Conference on Computer Vision, vol. 2, pp. 1150–1157. IEEE (1999)

9. Krizhevsky, A., Sutskever, I., Hinton, G.E.: Imagenet classification with deep convolutional neural networks. In: Neural Information Processing Systems (2012)

10. Girshick, R., Donahue, J., Darrell, T., Malik, J.: Rich feature hierarchies for accurate object detection and semantic segmentation. In: 2014 IEEE Conference on Computer Vision and Pattern Recognition, pp. 580–587 (2014)

11. Gkioxari, G., Girshick, R., Malik, J.: Contextual action recognition with R* CNN. In: Proceedings of the IEEE International Conference on Computer Vision, pp. 1080–1088 (2015)

12. Uricchio, T., Bertini, M., Seidenari, L., Bimbo, A.D.: Fisher encoded convolutional bag-of-windows for efficient image retrieval and social image tagging. In: 2015 IEEE International Conference on Computer Vision Workshop (ICCVW), pp. 1020–1026 (2015)

13. Zitnick, C.L., Dollár, P.: Edge boxes: locating object proposals from edges. In: Fleet, D., Pajdla, T., Schiele, B., Tuytelaars, T. (eds.) ECCV 2014. LNCS, vol. 8693, pp. 391–405. Springer, Heidelberg (2014). doi:10.1007/978-3-319-10602-1_26

14. Shin, A., Yamaguchi, M., Ohnishi, K., Harada, T.: Dense image representation with spatial pyramid VLAD coding of CNN for locally robust captioning. arXiv preprint arXiv:1603.09046 (2016)

15. Bourdev, L., Maji, S., Malik, J.: Describing people: a poselet-based approach to attribute classification. In: 2011 International Conference on Computer Vision, pp. 1543–1550 (2011)

16. Yao, B., Jiang, X., Khosla, A., Lin, A.L., Guibas, L., Fei-Fei, L.: Human action recognition by learning bases of action attributes and parts. In: 2011 IEEE International Conference on Computer Vision (ICCV), pp. 1331–1338. IEEE (2011)

17. Kumar, N., Berg, A.C., Belhumeur, P.N., Nayar, S.K.: Attribute and simile classifiers for face verification. In: IEEE International Conference on Computer Vision (ICCV) (2009)

18. Chen, H., Gallagher, A., Girod, B.: Describing clothing by semantic attributes. In: Fitzgibbon, A., Lazebnik, S., Perona, P., Sato, Y., Schmid, C. (eds.) ECCV 2012. LNCS, vol. 7574, pp. 609–623. Springer, Heidelberg (2012). doi:10.1007/978-3-642-33712-3_44

19. Cai, J., Zha, Z.J., Zhou, W., Tian, Q.: Attribute-assisted reranking for web image retrieval. In: Proceedings of the 20th ACM International Conference on Multimedia, pp. 873–876. ACM (2012)

20. Peng, X., Wang, L., Wang, X., Qiao, Y.: Bag of visual words and fusion methods for action recognition: comprehensive study and good practice. arXiv preprint arXiv:1405.4506 (2014)

21. Oneata, D., Verbeek, J., Schmid, C.: Action and event recognition with fisher vectors on a compact feature set. In: Proceedings of the IEEE International Conference on Computer Vision, pp. 1817–1824 (2013)

22. Ullah, M.M., Parizi, S.N., Laptev, I.: Improving bag-of-features action recognition with non-local cues. In: BMVC, vol. 10, pp. 95–1. Citeseer (2010)

23. Delaitre, V., Laptev, I., Sivic, J.: Recognizing human actions in still images: a study of bag-of-features and part-based representations (2010). http://www.di.ens.fr/willow/research/stillactions/
24. Sun, C., Nevatia, R.: Large-scale web video event classification by use of fisher vectors. In: 2013 IEEE Workshop on Applications of Computer Vision (WACV), pp. 15–22. IEEE (2013)
25. Jain, M., Jégou, H., Bouthemy, P.: Better exploiting motion for better action recognition. In: Proceedings of the IEEE Conference on Computer Vision and Pattern Recognition, pp. 2555–2562 (2013)
26. Felzenszwalb, P.F., Girshick, R.B., McAllester, D., Ramanan, D.: Object detection with discriminatively trained part-based models. IEEE Trans. Pattern Anal. Mach. Intell. **32**, 1627–1645 (2010)
27. Bourdev, L., Malik, J.: Poselets: body part detectors trained using 3D human pose annotations. In: 2009 IEEE 12th International Conference on Computer Vision, pp. 1365–1372. IEEE (2009)
28. Zhang, N., Paluri, M., Ranzato, M., Darrell, T., Bourdev, L.: PANDA: pose aligned networks for deep attribute modeling. In: Proceedings of the IEEE Conference on Computer Vision and Pattern Recognition, pp. 1637–1644 (2014)
29. Yao, B., Fei-Fei, L.: Recognizing human-object interactions in still images by modeling the mutual context of objects and human poses. IEEE Trans. Pattern Anal. Mach. Intell. **34**, 1691–1703 (2012)
30. Prest, A., Schmid, C., Ferrari, V.: Weakly supervised learning of interactions between humans and objects. IEEE Trans. Pattern Anal. Mach. Intell. **34**, 601–614 (2012)
31. Girshick, R.: Fast R-CNN. In: 2015 IEEE International Conference on Computer Vision (ICCV), pp. 1440–1448 (2015)
32. Zheng, S., Jayasumana, S., Romera-Paredes, B., Vineet, V., Su, Z., Du, D., Huang, C., Torr, P.H.S.: Conditional random fields as recurrent neural networks. In: 2015 IEEE International Conference on Computer Vision (ICCV), pp. 1529–1537 (2015)
33. Oquab, M., Bottou, L., Laptev, I., Sivic, J.: Learning and transferring mid-level image representations using convolutional neural networks. In: Proceedings of the IEEE Conference on Computer Vision and Pattern Recognition, pp. 1717–1724 (2014)
34. Gkioxari, G., Girshick, R., Malik, J.: Actions and attributes from wholes and parts. In: Proceedings of the IEEE International Conference on Computer Vision, pp. 2470–2478 (2015)
35. Diba, A., Pazandeh, A.M., Pirsiavash, H., Van Gool, L.: DeepCAMP: deep convolutional action & attribute mid-level patterns. In: 2016 IEEE Conference on Computer Vision and Pattern Recognition (CVPR), Las Vegas, NV, pp. 3557–3565 (2016). doi:10.1109/CVPR.2016.387
36. Arandjelovic, R., Zisserman, A.: All about VLAD. In: 2013 IEEE Conference on Computer Vision and Pattern Recognition (CVPR), pp. 1578–1585 (2013)
37. Sánchez, J., Perronnin, F., Mensink, T., Verbeek, J.: Image classification with the fisher vector: theory and practice. Int. J. Comput. Vis. **105**, 222–245 (2013)
38. Dixit, M., Chen, S., Gao, D., Rasiwasia, N., Vasconcelos, N.: Scene classification with semantic fisher vectors. In: 2015 IEEE Conference on Computer Vision and Pattern Recognition (CVPR), pp. 2974–2983 (2015)
39. Lazebnik, S., Schmid, C., Ponce, J.: Beyond bags of features: spatial pyramid matching for recognizing natural scene categories. In: 2006 IEEE Computer Society Conference on Computer Vision and Pattern Recognition, vol. 2, pp. 2169–2178. IEEE (2006)

40. Zhou, R., Yuan, Q., Gu, X., Zhang, D.: Spatial pyramid VLAD. In: 2014 IEEE Visual Communications and Image Processing Conference, pp. 342–345. IEEE (2014)
41. Hosang, J., Benenson, R., Schiele, B.: How good are detection proposals, really? In: 25th British Machine Vision Conference, pp. 1–12. BMVA Press (2014)
42. Simonyan, K., Zisserman, A.: Very deep convolutional networks for large-scale image recognition. CoRR abs/1409.1556 (2014)
43. Jégou, H., Douze, M., Schmid, C., Pérez, P.: Aggregating local descriptors into a compact image representation. In: IEEE Conference on Computer Vision & Pattern Recognition (2010)
44. Ross, D.A., Lim, J., Lin, R., Yang, M.: Incremental learning for robust visual tracking. Int. J. Comput. Vis. **77**, 125–141 (2008)
45. Arthur, D., Vassilvitskii, S.: k-means++: the advantages of careful seeding. In: Proceedings of the Eighteenth Annual ACM-SIAM Symposium on Discrete Algorithms, pp. 1027–1035. Society for Industrial and Applied Mathematics (2007)
46. Jia, Y., Shelhamer, E., Donahue, J., Karayev, S., Long, J., Girshick, R., Guadarrama, S., Darrell, T.: Caffe: convolutional architecture for fast feature embedding. In: Proceedings of the ACM International Conference on Multimedia, pp. 675–678. ACM (2014)
47. Pedregosa, F., Varoquaux, G., Gramfort, A., Michel, V., Thirion, B., Grisel, O., Blondel, M., Prettenhofer, P., Weiss, R., Dubourg, V., Vanderplas, J., Passos, A., Cournapeau, D., Brucher, M., Perrot, M., Duchesnay, E.: Scikit-learn: machine learning in python. J. Mach. Learn. Res. **12**, 2825–2830 (2011)
48. Vedaldi, A., Fulkerson, B.: VLFeat: an open and portable library of computer vision algorithms (2008)
49. Chang, C.C., Lin, C.J.: LIBSVM: a library for support vector machines. ACM Trans. Intell. Syst. Technol. **2**, 27:1–27:27 (2011). http://www.csie.ntu.edu.tw/cjlin/libsvm
50. Li, L.J., Su, H., Fei-Fei, L., Xing, E.P.: Object bank: a high-level image representation for scene classification & semantic feature sparsification. In: Advances in Neural Information Processing Systems, pp. 1378–1386 (2010)
51. Wang, J., Yang, J., Yu, K., Lv, F., Huang, T., Gong, Y.: Locality-constrained linear coding for image classification. In: 2010 IEEE Conference on Computer Vision and Pattern Recognition (CVPR), pp. 3360–3367. IEEE (2010)
52. Sharma, G., Jurie, F., Schmid, C.: Expanded parts model for human attribute and action recognition in still images. In: Proceedings of the IEEE Conference on Computer Vision and Pattern Recognition, pp. 652–659 (2013)
53. Khan, F.S., Xu, J., van de Weijer, J., Bagdanov, A.D., Anwer, R.M., Lopez, A.M.: Recognizing actions through action-specific person detection. IEEE Trans. Image Process. **24**, 4422–4432 (2015)

The Third Workshop on Computer Vision for Affective Computing (CV4AC)

Expression Recognition with Ri-HOG Cascade

Jinhui Chen[1(✉)], Zhaojie Luo[2], Tetsuya Takiguchi[2], and Yasuo Ariki[2]

[1] RIEB, Kobe University, Kobe 657-8501, Japan
ianchen@me.cs.scitec.kobe-u.ac.jp
[2] Graduate School of System Informatics, Kobe University, Kobe 657-8501, Japan

Abstract. This paper presents a novel classification framework derived from AdaBoost to classify facial expressions. The proposed framework adopts rotation-reversal invariant HOG as features. The Framework is implemented through configuring the Area under ROC curve (AUC) of the weak classifier with HOG, which is a discriminative classification framework. The proposed classification framework is evaluated with two very popular and representative public databases: MMI and AFEW. As a result, it outperforms the state-of-the-arts methods.

1 Introduction

Facial expression recognition (FER) is one of the most significant technologies for auto-analyzing human behavior. It can be widely applied to various application domains. Therefore, the need for this kind of technology in various different fields continues to propel related research forward every year.

In this paper, we propose a novel framework that adopts novel feature representation approach; namely, rotation-reversal invariant HOG (Ri-HOG) for learning boosting cascade. The proposed feature is reminiscent of Dalal and Triggs's HOG [1], but the proposed feature representation approach noticeably enhances the conventional HOG-type descriptors for the image local features on invariant representation. For robustness and speed, we carry out a detailed study of the effects of various implementation choices in descriptor performance. We subdivide the local patch into annular spatial bins, to achieve spatial binning invariance. Besides, inspired by Takacs *et al.*'s rotation-invariant image features [2], we apply polar gradient to attaining gradient binning invariance, which is derived from the theory of polar coordinate. By doing so, the proposed method can significantly enhance the features descriptors in regard to invariant representation ability and feature describing accuracy. Consequently, the proposed framework can robustly process out-of-plane head rotation cases.

The proposed learning model is derived from AdaBoost [3], but the proposed is implemented by configuring the area under the receiver-operating characteristic (ROC) curve (AUC) [4] to construct the weak classifier for expression classification. Adopting the AUC-based weak classifier, the false-positive-rate (FPR) of boosting training is adaptive to different stage, and it is usually much smaller than conventional approaches, which means its error rate is much smaller than the conventional approaches at each training iteration. Therefore, its convergence

© Springer International Publishing AG 2017
C.-S. Chen et al. (Eds.): ACCV 2016 Workshops, Part III, LNCS 10118, pp. 517–530, 2017.
DOI: 10.1007/978-3-319-54526-4_38

speed is much quicker than the conventional methods. Moreover, the accuracy of this classifier model is much better than conventional boosting classifiers. We experimentally evaluated the proposed method in two public expression databases *i.e.*, MMI [5,6], and AFEW [7], that together represent lab-controlled and real-world scenarios. The experimental results show that the proposed method can construct a robust FER system whose results outperform well-known state-of-the-art FER methods.

The main contribution of our study is the development of a novel framework, called Ri-HOG cascade, which can robustly process FER. In this paper, we are making the following original contributions: (1) we propose a robust local feature descriptor method called Ri-HOG, which is an appropriate similarity measure that can remain invariant for the rotated as well as reversed image representation; (2) We develop a novel cascade learning model that allows the FPR of boosting training is adaptive to different stage, in so doing, the convergence speed is quick and the accuracy of the classifiers is high.

The remainder of this paper is organized as follows: We describe the proposed framework in Sect. 2. In Sect. 3, we describe our experiments, and we draw our conclusions in Sect. 4.

2   Proposed Method

Our proposed framework has these components: Ri-HOG features for local patch description; logistic regression-based weak classifiers, which are combined with AUC as a single criterion for cascade convergence testing; and a cascade for boosting training.

2.1   Feature Description

Background and Problems: HOG is computed on a dense grid of uniformly-spaced cells and use overlapping local contrast normalization for improved accuracy. This feature is set based on *cells* and *blocks* representation system and it is widely used in classification applications, especially human detection. The describing ability of HOG features outperforms many existing features [8]. However, its robustness against image rotation is not satisfactory. Here one direct evidence is that the HOG feature is seldom applied to object tracking or image retrieval successfully. Giving a more scientific reason, see Fig. 1 for an example. Supposing Fig. 1(a) is an image with HOG block size, there are 4 cells in the block. Figure 1(b) is an image of Fig. 1(a) after making a quarter turn. HOG features are extracted from the two images individually. If the histogram of oriented gradients obtained from the regions 1, 2, 3, and 4 are severally denoted as x_1, x_2, x_3, x_4, then, the HOG features extracted from Fig. 1(a) and (b) are (x_1, x_2, x_3, x_4) and (x_3, x_1, x_4, x_2) respectively. This means that the rotation of image accompanies easily with the change of its HOG descriptors. Similarly, for reversal-image representation, HOG is also not invariant. Hence, we have to substantially enhance the robustness of HOG descriptors. Otherwise, applications of HOG features would be limited to some narrow ranges.

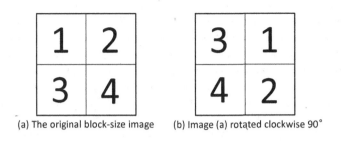

(a) The original block-size image   (b) Image (a) rotated clockwise 90°

Fig. 1. Analyzing the robustness of conventional HOG descriptors in regard to image rotation.

Our Approach: to robustly represent out-of-plane head rotation cases, we propose a novel feature descriptor on histograms of oriented gradients, *i.e.*, rotation-reversal invariant histograms of oriented gradients (Ri-HOG). We adopt annular spatial cells to replace rectangular cells (see Fig. 2(a)) and compute these cells on a dense polar gradient as feature descriptors. By doing so, the time complexity will not increase, but the invariant representation ability of the features will be extremely enhanced.

In this paper, we adopt polar gradient to represent the gradient for HOG descriptors, which is derived from Takacs *et al.*'s rotation-invariant image features [2]. But different from Takacs *et al.*'s approach, we only use the polar gradient to replace the Gaussian gradient function of conventional HOG. We subdivide the local patch into annular spatial cells (see Fig. 2(a)). How to calculate these descriptors is shown in Fig. 2. In Fig. 2(b), $\forall$ a point p in the circle c, the task is to compute the polar gradient magnitude of point p (x, y). Decompose the vector g into its local coordinate system as $(g^T r, g^T t)$, by projecting g into the r and t orientations as shown in Fig. 2(b). Since the component vectors of g in r and t orientations can be quickly obtained by $r = \frac{p-c}{\|p-c\|}$, $t = R_{\frac{\pi}{2}} r$, we can obtain the gradient g easily on the gradient filter. And, R_θ is the rotation matrix by angle θ.

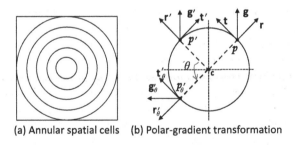

(a) Annular spatial cells   (b) Polar-gradient transformation

Fig. 2. Illustration of Ri-HOG descriptors.

Since Takacs *et al.* focus on image tracking applications, the speed is more important, they use Approximate Radial Gradient Transformation (ARGT) and

ROC curve to compute the feature descriptors [2]. However, in this way, it will decrease the distinctiveness of feature descriptors for recognition applications. In order to keep the distinctiveness of feature descriptors for recognition application, we do not follow Takacs *et al.*'s way to abandon gradient magnitudes, cells, and blocks representation system. Therefore, essentially, the feature (Ri-HOG) that we adopt here is an improved HOG feature, but the approach proposed by Takacs *et al.* is a very excellent and novel feature representation method for image tracking applications, which cannot be considered as a type of HOG feature. Ri-HOG persists and develops the discriminative representation of conventional HOG features. Meanwhile, it can also significantly enhances the descriptors with respect to rotation reversal invariant ability. Simply, we use the following four steps to extract the Ri-HOG descriptors:

1. Subdivide the local patch into annular spatial cells as shown in Fig. 2(a);
2. Calculate the polar gradient $(g^T r, g^T t)$ of each pixel in the cell;
3. Calculate the gradient magnitudes and the orientations of polar gradients using the Eq. 1:

$$M_{GRT}(x,y) = \sqrt{(g^T r)^2 + (g^T t)^2},$$
$$\theta(x,y) = arctan\frac{g^T t}{g^T r}; \tag{1}$$

4. Accumulating the gradient magnitude of polar gradient for each pixel over the annular spatial cells into 9 bins, which are separated according to the orientation of polar gradient. In this way, we can extract the feature descriptors from a dense annular spatial bin of these uniformly spaced cells.

Block Normalization: We tried all of 4 normalization approaches listed by Dalal *et al.* in [1]. In practice, $L_2 - Hys$, L_2 normalization followed by clipping is shown working best. The recognition template is 100×100 with 10 cells, and it allows the patch size ranging from 50×50 pixels to 100×100 pixels. We slide the patch over the recognition template with 5 pixels forward to ensure enough feature-level difference. We further allow different aspect ratio for each patch (the ratio of width and height). The descriptors are extracted according to the order from the inside to the outside of cells. Hence, concatenating descriptors in 10 cells together yield a 90-dimensional feature vector.

Now assume that the patch has been reversed and rotated by any given angle θ as shown in Fig. 2(b) (reversal: $p \to p'$; rotation: $p' \to p'_\theta$, the transformation orders can exchange). This yields a new local coordinate system and gradient: $p'_\theta = MR_\theta p$, $g'_\theta = MR_\theta g$, $r'_\theta = MR_\theta r$, $t'_\theta = MR_\theta t$, where M is the reversal matrix. As we known, the reversal matrix is a diagonal matrix with diagonal elements 1 or -1. Consequently, $M^T = M^{-1}$. The coordinates of the gradient in the local frame are invariant to reversal as well as rotation, which can be verified by

$$(g_\theta'^T r_\theta', g_\theta'^T t_\theta')$$
$$= ((MR_\theta g)^T MR_\theta r, (MR_\theta g)^T MR_\theta r)$$
$$= (g^T R_\theta^T M^T MR_\theta r, g^T R_\theta^T M^T MR_\theta t) \tag{2}$$
$$= (g^T r, g^T t).$$

Since the point $p(x, y)$ as well as the angle θ are any given ones, and all gradients are transformed via the same way; *i.e.*, they are one-to-one mapping. Thus, the set of gradients on any given point around the patch is invariant to reversal as well as rotation.

2.2   Training Weak Classifier

In this study, we build a weak classifier over each local patch described by the Ri-HOG descriptor, and select the optimum patches in each boosting iteration from the patch pool. Meanwhile, we construct the weak classifier for each local patch by logistic regression to fit our classifying framework, due to it being a probabilistic linear classifier.

On one hand, we build a weak classifier over each local patch, as described by the descriptor, and select optimum patches in each boosting iteration from the patch pool. On the other hand, we construct a weak classifier for each local patch by logistic regression to fit our classification framework, since it is a probabilistic linear classifier. Given a Ri-HOG feature $\mathbb{F}$ over a local patch, logistic regression defines the probability model:

$$P(q|\mathbb{F}, \mathbf{w}) = \frac{1}{1 + \exp(-q(\mathbf{w}^T \mathbb{F} + b))}, \tag{3}$$

when $q = 1$ means that the trained sample is a positive sample of the current class, $q = -1$ indicates negative samples, $\mathbf{w}$ is a weight vector for the model, and b is a bias term. We train classifiers on local patches from a large-scale dataset. Assuming, in each boosting iteration stage, that there are K possible local patches, which are represented by Ri-HOG feature $\mathbb{F}$, each stage is a boosting training procedure with logistic regression as weak classifiers. In this way, the parameters can be identified by minimizing the objective:

$$\sum_{k=1}^{K} \log(1 + \exp(-q_k(\mathbf{w}^T \mathbb{F}_k + b))) + \lambda \|\mathbf{w}\|_p, \tag{4}$$

where λ denotes a tunable parameter for the regularization term, and $\|\mathbf{w}\|_p$ is the L_p norm of the weight vector. Note that it is also applied to L_2-loss and L_1-loss linear support vector machines (SVMs) by the well-known open source code LIBLINEAR [9]. Therefore, this question can be solved using algorithms in [9]. In this study, the weak classifier is defined as:

$$h(\mathbb{F}) = 2P(q|\mathbb{F}, \mathbf{w}) - 1. \tag{5}$$

We trained the boosting cascade on local patches from a large-scale dataset. In practice, AdaBoost is not skilled at processing the vector-descriptor feature directly. Inspired by Li et al.'s SURF cascade [10], we found that the AUC score [11] can solve the problem. Therefore, by innovating the AUC score, we can avoid the difficult convergence risk.

Given the weak classifiers h_n for cascade iteration n, the strong classifier is defined as $H_N(\mathbb{F}) = \frac{1}{N}\sum_{n=1}^{N} h_n(\mathbb{F})$. Assuming there are a total of N boosting iteration rounds, in the round n, we will build K weak classifiers $[h_n(\mathbb{F}_k)]_{k=1}^{K}$ for each local patch in parallel from the boosting sample subset. Meanwhile, we also test each model $h_n(\mathbb{F}_k)$ in combination with previous $n-1$ boosting rounds. In other words, we test $H_{n-1}(\mathbb{F}) + h_n(\mathbb{F}_k)$ for $H_n(\mathbb{F})$ on the all training samples, and each test model will produce a highest AUC score [4,11] $J(H_{n-1}(\mathbb{F}) + h_n(\mathbb{F}_k))$. i.e.,

$$S_n = \max_{k=1,\cdots K} J(H_{n-1}(\mathbb{F}) + h_n(\mathbb{F}_k)). \tag{6}$$

This procedure is repeated until the AUC scores converge, or the designated number of iterations N is reached.

The whole procedure involves a forward selection and inclusion of a weak classifier over possible local patch temples that can be adjusted using different temple configurations, according to the processing images. To enhance both the speed of learning convergence and robustness, our algorithm further introduces a backward removal approach. For more details on including backward removal or even a floating searching capability into the boosting framework, please refer to [12]. In this study, we implement backward removal on Algorithm 1 step 4, to extend the procedure with the capability to backward remove redundant weak classifiers. In so doing, it is not only able to reduce the number of weak classifiers in each stage, but also able to improve the generalization capability of the strong classifiers. The details of how to implement these learning approaches are indicated in Algorithm 1.

Boosting Cascade Training: To the best of our knowledge, almost all existing cascade detection frameworks are trained based on two conflicting criteria, i.e. the false-positive-rate (FPR) f_j and the hit-rate (or recognition rate) r_j for the detection-error tradeoff. The overall FPR of a $T-$stage cascade is $F = \prod_{j=1}^{T} f_j$, while the overall hit-rate is $R = \prod_{j=1}^{T} r_j$. Usually, the maximum suggested setting of f_j is 0.5 [13]. Therefore, for the system to reach an overall FPR$= 10^{-6}$, it requires at least 20 stages $(0.5^{20} \approx 10^{-6})$ by the given global setting. Note that some stages may reach this goal without convergence. Hence, it is better that the FRP be adaptive among different stages so that we could easily reach the overall· training goal. Some automatic scheme methods [14–17] tune the intermediate thresholds of each stage. These approaches may alleviate painful manual tuning efforts, but do not address the convergence speed. Therefore, we do not consider these appropriate for implementing our cascade-type ensemble of weak classifiers.

Algorithm 1. Learning Boosting Classifiers.

Require:

1. Given: the number of label categories M and the overall sample set $\mathbf{S} = \{(x_1, y_1), \cdots, (x_\tau, y_\tau)\}$, where τ is the number of the samples;

2. Initialize the weight parameter w_0 for positive (labeled as "+") samples and negative (labeled as "-") samples:

   a. $w_0^+ = 1/(M \times \tau_+)$ for those $q = 1$;

   b. $w_0^- = 1/(M \times \tau_-)$ for those $q = 1$;

3.

for $(n = 0; n < N; n = n + 1)$ **do**

   a. Sampled $30 \times p$ (in this paper, $p = 3$) positive samples and $30 \times p$ negative samples from training set;

   b. Parallel replace each Ri-HOG template to train a series of logistic regression models $[h_n(\mathbb{F}_k)]_{k=1}^K$;

   c. In order to obtain the AUC score, calculate $H_{n-1}(\mathbb{F}) + h_n(\mathbb{F}_k)$ on the best model of previous stage: S_{n-1} and each $h_n(\mathbb{F}_k)$;

   d. Choose the best model S_n which contains the best weak classifier $h_n(\mathbb{F}_k)$, according to the Eq. 6;

   f. Update weight

$$w_{n+1} = \frac{w_n \exp(-q_n h_n(\mathbb{F}_k))}{Z_n},$$

where Z_j is a normalization factor, on which it can make the weight follow to $M \sum w^+ = 1$ and $M \sum w^- = 1$;

   g. If AUC value S_n is converged, break the loop;

end for

4. In order to ensure the overall AUC score to be the highest one, test all learned models during the current iteration process:

for $(k = 0; k < K; k = k + 1)$ **do**

  if $H_{n-1}(\mathbb{F}) + h_n(\mathbb{F}_k) > S_n$ **then**

   a. $S_n = H_{n-1}(\mathbb{F}) + h_j(\mathbb{F}_k)$;

   b. Empty those unnecessary data to free the memory;

  end if

end for

5. Output final strong model H_N for this stage.

Inspired by [4,10], here we introduce AUC as a single criterion for cascade convergence testing, which realizes an adaptive FPR among different stages. Hence, combined with logistic regression-based weak classifiers to adopt Ri-HOG features, this approach can yield a fast convergence speed and a cascade model with much shorter stages.

To avoid overfitting, we restricted the number of samples used during training, as in [14]. In practice, we sampled an active subset from the whole training set according to the boosting weight. It is generally good practice to use about $30 \times p$ samples of each class, where p is a multiple coefficient (Algorithm 1 step 3.a).

Within one stage, no threshold for intermediate weak classifiers is required. We need only determine each decision threshold θ_i. In our case, using the ROC curve, the FPR of each emotional category is easily determined when given

---

Algorithm 2. Training Multithreaded Boosting Cascade

Require:
  1. Over all FPR: F_N for i-th category data;
  2. Minimum hit-rate per stage $d_i^{(min)}$;
  3. Current class samples: $\mathbf{X}_i^+$;
  4. Non-current class samples: $\mathbf{X}_i^-$;
  5. The number of expression labels: M;
Initialize: $j = 0, F_i^{(j)} = 1, D_i^{(j)} = 1$;
for $(i = 0; i < M; i = i + 1)$ **do**
  while $(F_i^{(j)} > F_i^{(n)})$ **do**
    1. j=j+1;
    2. Train a stage classifier $H_i^{(j)}(\mathbb{F})$ by samples of $\mathbf{X}^+$ and $\mathbf{X}^-$ via Algorithm 1;
    3. Evaluate the model $H_i^{(j)}(\mathbb{F})$ on the whole training set to obtain ROC curve;
    4. Determine the threshold $\theta_i^{(j)}$ by searching on the ROC curve to find the point $(d_i^{(j)}, f_i^{(j)})$ such that $d_i^j = d_i^{(min)}$, but when existing the mimimum one $d_i^{(j)}$ that follows to the condition: $d_i^{(j)} < d_i^{(min)}$, set $d_i^{(min)} = d_i^{(j)}$ to update the minimal hit-rate;
    5. Update: $F_i^{(j)} = F_i^{(j-1)} \times f_i^{(j)}$,
             $D_i^{(j)} = D_i^{(j-1)} \times d_i^{(j)}$;
    6. Empty the set $\mathbf{X}_i^-$;
    7. **while** $(F_i^{(j)} > F_i^{(j-1)}$ and size $|\mathbf{X}_i^+| \neq |\mathbf{X}_i^-|$) **do**
    Adopt current cascade detector to scan non-target images with sliding window and put false-positive samples into $\mathbf{X}_i^-$;
    end while
  end while
end for
  8. Output the boosting cascade detector $\{H_i^{(j)} > \theta_i^{(j)}\}$ and overall training accuracy F and D.

---

the minimal hit-rate $d_i^{(min)}$. We decrease $d_i^{(j)}$ from 1 on the ROC curve, until reaching the transit point $d_i^j = d_i^{(min)}$. The corresponding threshold at that point is the desired θ_i, i.e., the FPR is adaptive to different stage, and it is usually much smaller than 0.5.

After one stage of classifiers learning is converged via Algorithm 2, we continue to train another one with false-positive samples coming from the scanning of non-target images with the partially trained cascade. We repeat this procedure until the overall FPR reaches the stated goal. In ding so, the FPR is usually much smaller than 0.5 and it is adaptive for different stages. Therefore, this approach can result in a model size that is much smaller, and has the recognition speed and accuracy that is dramatically increased.

3   Experiments

In this section, we provide details of the dataset and evaluation results for the proposed method. We implemented all training and recognition programs in

C++ on Win 10 OS, processed with a PC with a Core i7-6700K 4.0 GHz CPU and 32 GB RAM.

3.1 Databases and Protocols

We evaluated the proposed method on two reference databases, *i.e.* MMI, and AFEW, which include the lab-controlled database and the database in the wild.

MMI DB. The MMI DB is a public database that includes more than 30 subjects, in which the female-male ratio is roughly 11:15. The subjects' ages range from 19 to 62, and they are of European, Asian or South American descent. This database is considered to be more challenging than CK+ [18], because there are many side-view images and some posers have worn accessories such as glasses. To evaluate the out-of-plane head rotation cases clearly, we adopt MMI database representing the lab-controlled to test proposed approaches in this paper. In the experiments, we used all 205 effective image sequences of the six expressions in the MMI dataset. In the recognition stage, the images of MMI were made into videos according to the person-independent.

AFEW DB. For the AFEW DB, which is a much more challenging database, evaluation experiments also have been done [7]. All of the AFEW sets were collected from movies to depict so-call wild scenarios. In experiments, the videos in training set are decomposed into images for training. We trained the training set and the results are reported for its validation set, in the same way as for the latest FER work [19].

We used all training samples in AFEW training set and collected training samples from according to the person-independent 10-fold cross-validation rule. In order to reduce the process time of training, the samples from two datasets were trained together. All of training samples were normalized to 100×100-pixel facial patches. In order to enhance the generalization performance of boosting learning, we dealt with the training samples by some transformations (mirror reflection, rotate the images *etc.*), finally, the original samples were increased by a factor of 64. The testing sample sequences were not done on any normalization. In the training stages, the training data of current processing expression were adopted as positive sample data; the other expressions' data were used for negative data.

3.2 Training Speed Evaluation Results

We replaced 40 types of the local patches on the 100×100 detection template as described in Subsect. 2.1. The proposed method used 377 min to converge at the $16th$ iteration stage. The cascade detector contained $2,394$ classifiers of all categories, and only need to evaluate 1.5 HOG per window. After training, we observed that the top-3 picked local patches for FER laid in the regions of two eyes and mouth. This situation is similar to Haar-based classifiers [20], see the examples in Fig. 3.

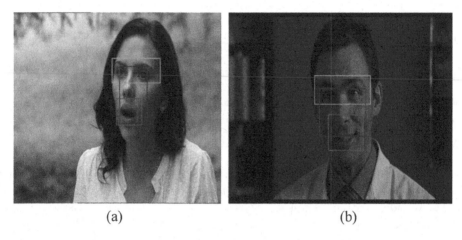

(a)   (b)

Fig. 3. Top-3 local patches picked by training procedure in the green-red-blue order on AFEW database. (Color figure online)

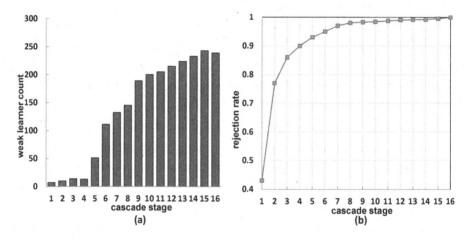

Fig. 4. (a) The number of weak classifiers at each cascade stage; (b) the accumulated rejection rate over all stages.

More details for cascade of FER are illustrated in Fig. 4(a) and (b), which include the number of weak learners in each stage and the average accumulated rejection rate over the whole cascade stages. It shows that the first 8 stages have rejected 98% of the non-current class samples.

In order to evaluate the convergence speed of the AUC model, we determined the FPR at each boosting stage. The results show that, in the AUC model, the FPR f_j at each cascade stage is adaptive among the different stages, ranging from 0.04101 to 0.22337, is much smaller than the conventional model FPR of 0.5. In almost all existing cascade frameworks FPR $\prod_{j=1}^{T} f_j$ (T denotes the total cascade stages) reaches the goal (It is usually set as 10^{-6}). This means that conventional models require more iterations and that the AUC model cascade can converge much faster. These relate directly to training efficiency and recognition speed. Therefore, these experimental results confirm that the AUC cascade model is much more efficient than conventional cascade models. However, since the proposed framework makes the classifiers parallel recognize the multiclass expressions, the peak of memory cost is nearly six times more than the conventional one.

3.3   Recognition Results Comparison

In this paper, all the labels of the expression categories were named the same as they are in the original databases. Since the proposed is a binary classification framework, we test the expression class one by one. The facial region is detected by V-J framework [3] implemented by Open CV, and expression in face is recognized on proposed approaches. Here we show the recognition results on MMI, and AFEW.

Adopting Ri-HOG features, we evaluated almost of existing classifiers proposed for cascade learning and top ones of them are reported in Table 2. The results show that the proposed classifier is more suitable for processing FER. The reason why we have to adopt Ri-HOG as features is also shown in Table 2; *i.e*, it dominates others on the accuracy. Meanwhile, its recognition speed can meet the real-time recognition. However, adopting SIFT as features, the real-time recognition is an impossible task (speed: only 18 frames per second), although the performance of the proposed framework with SIFT is also quite excellent.

The comparison moths were selected to represent the state-of-the-art level of this field, which includes proposing for the improvement of local spatiotemporal descriptors: such as LBP-TOP [22], HOE [21], HOG 3D [23], which are very popular for FER, while 3DCNN-DAP [26] and STM [19] are the latest ones; also including those methods that focus on enhancing the robustness of their classifying frameworks or making the frameworks can be encoded robustly, like, ITBN [24], 3D LUT [20] and LSH-CORF [25] *etc.* For fair comparison with them, we used the same databases, which were evaluated via the standardized items what they had done.

Table 1 compares our method with these state-of-the-art methods. Furthermore, almost of these methods were conducted using their released codes and the parameters had been tuned to better-adapt for our experiments. However, about some methods, because we cannot obtain their source codes until now (*e.g.* STM [19] and 3DCNN-DAP [26], *etc.*), thus, we have to cite the reported results from the related works. The precisions of our framework (Ri-HOG cascade) were 72.4% on MMI database, and 56.8% on AFEW whose baseline is 30.9% (*the results are cited from the work [7], yet we donot test the Neutral class in this paper). The state-of-the-art levels were improved 7% and 25.1% respectively by the proposed framework on MMI and AFEW. In addition, the recognition speed of the proposed framework reached 55 frames per second.

Table 1. Recognition results on MMI and AFEW.

Method	Accuracy on MMI (%)							Accuracy on AFEW(%)						
	An	Di	Fe	Ha	Sa	Su	Ave.	An	Di	Fe	Ha	Sa	Su	Ave.
HOE [21]	46.4	58.3	33.2	62.6	60.8	65.1	55.5	11.2	16.5	9.0	33.5	15.3	28.3	19.0
LBP-TOP [22]	58.1	56.3	53.6	78.6	46.9	50.0	57.2	11.7	**19.6**	17.9	42.3	**23.8**	33.6	24.8
HOG 3D [23]	61.3	53.1	39.3	78.6	43.8	55.0	55.2	–	–	–	–	–	–	26.9
ITBN [24]	46.9	54.8	57.1	71.4	65.6	62.5	59.7	91.1	94.0	83.3	89.8	76.0	91.3	86.3
LSH [25]	59.6	**71.4**	62.3	68.9	**70.3**	75.1	61.8	23.1	12.8	**38.6**	9.7	21.1	10.9	19.4
3D LUT [20]	43.3	55.3	56.8	71.4	28.2	77.5	47.2	45.7	0	0	**62.0**	13.2	48.6	28.2
3DCNN-DAP [26]	64.5	62.5	50.0	**85.7**	53.1	57.5	62.2	–	–	–	–	–	–	–
STM [19] –	–	–	–	–	–	–	65.4	–	–	–	–	–	–	31.7
Baseline [7]	–	–	–	–	–	–	–	50.0	25.0	15.2	57.1	16.4	21.7	33.2*
Ours	**70.2**	60.4	**76.5**	81.2	62.1	**84.2**	**72.4**	**56.2**	36.3	**48.5**	74.6	36.0	89.1	**56.8**

Table 2. Average precision using different classifiers and features.

Database	Precision of classifiers (%)				Precision of feature (%)				
	BinBoost [27]	JC [28]	SC [15]	Proposed	SIFT	SURF	Haar	HOG	Ri-HOG
MMI	62.6	55.9	50.2	**72.4**	65.4	46.0	42.2	58.8	**72.4**
AFEW	43.9	40.6	26.8	**56.8**	41.5	35.8	17.3	32.4	**56.8**

4   Conclusion

In this paper, we have proposed a novel cascade framework called rotation-reversal invariant HOG cascade for robust FER. The proposed framework adopts Ri-HOG for robustly process out-of-plane head rotation cases. Meanwhile, in the cascade learning, the proposed method use AUC as a single criterion for cascade convergence testing to enhance the classifiers learning. We used two representative public databases in FER research field, to experimentally confirm the validity of the proposed method. These issues are important to those with related research interests.

About the future work, we will attempt to study the question about how does the feature representation error impact on recognition frameworks.

References

1. Dalal, N., Triggs, B.: Histograms of oriented gradients for human detection. In: Proceedings of IEEE Conference on Computer Vision and Pattern Recognition (CVPR), vol. 1, pp. 886–893 (2005)
2. Takacs, G., Chandrasekhar, V., Tsai, S., Chen, D., Grzeszczuk, R., Girod, B.: Fast computation of rotation-invariant image features by an approximate radial gradient transform. IEEE Trans. Image Proc. (TIP) **22**, 2970–2982 (2013)
3. Viola, P., Jones, M.: Robust real-time face detection. Int. J. Comput. Vis. (IJCV) **57**, 137–154 (2004)

4. Ferri, C., Flach, P.A., Hernández-Orallo, J.: Learning decision trees using the area under the ROC curve. In: Proceedings of International Conference Machine Learning (ICML), pp. 139–146 (2002)
5. Valstar, M.F., Pantic, M.: Induced disgust, happiness and surprise: an addition to the MMI facial expression database. In: Proceedings of International Conference on Language Resources and Evaluation, Workshop on Emotion, pp. 65–70 (2010)
6. Pantic, M., Valstar, M.F., Rademaker, R., Maat, L.: Web-based database for facial expression analysis. In: Proceedings of IEEE International Conference on Multimedia and Expo (ICME), pp. 317–321 (2005)
7. Dhall, A., Goecke, R., Lucey, S., Gedeon, T.: Collecting large, richly annotated facial-expression databases from movies. IEEE MultiMedia **19**, 34–41 (2012)
8. Thurau, C., Hlavac, V.: Pose primitive based human action recognition in videos or still images. In: Proceedings of IEEE Conference on Computer Vision and Pattern Recognition (CVPR), pp. 1–8 (2008)
9. Fan, R.E., Chang, K.W., Hsieh, C.J., Wang, X.R., Lin, C.J.: LIBLINEAR: a library for large linear classification. J. Mach. Learn. Res. **9**, 1871–1874 (2008)
10. Li, J., Wang, T., Zhang, Y.: Face detection using SURF cascade. In: Proceedings of IEEE International Conference on Computer Vision (ICCV) Workshops, pp. 2183–2190 (2011)
11. Long, P., Servedio, R.: Boosting the area under the ROC Curve. In: Proceedings of Advances in Neural Information Processing Systems (NIPS), pp. 945–952 (2007)
12. Li, S.Z., Zhang, Z., Shum, H.Y., Zhang, H.: FloatBoost learning for classification. In: Proceedings of Advances in Neural Information Processing Systems (NIPS), pp. 993–1000 (2002)
13. Li, J., Zhang, Y.: Learning SURF cascade for fast and accurate object detection. In: Proceedings of IEEE Conference on Computer Vision and Pattern Recognition (CVPR), pp. 3468–3475 (2013)
14. Xiao, R., Zhu, H., Sun, H., Tang, X.: Dynamic cascades for face detection. In: Proceedings of IEEE International Conference on Computer Vision (ICCV), pp. 1–8 (2007)
15. Bourdev, L., Brandt, J.: Robust object detection via soft cascade. In: Proceedings of IEEE Conference on Computer Vision and Pattern Recognition (CVPR), vol. 2, pp. 236–243 (2005)
16. Sochman, J., Matas, J.: WaldBoost - learning for time constrained sequential detection. In: Proceedings of Conference on Computer Vision and Pattern Recognition (CVPR), vol. 2, pp. 150–156 (2005)
17. Brubaker, S., Wu, J., Sun, J., Mullin, M., Rehg, J.: On the design of cascades of boosted ensembles for face detection. Int. J. Comput. Vis. (IJCV) **77**, 65–86 (2008)
18. Lucey, P., Cohn, J.F., Kanade, T., Saragih, J., Ambadar, Z., Matthews, I.: The extended Cohn-Kanade dataset (CK+): a complete dataset for action unit and emotion-specified expression. In: Proceedings of IEEE Conference on Computer Vision and Pattern Recognition (CVPR) Workshops, pp. 94–101 (2010)
19. Liu, M., Shan, S., Wang, R., Chen, X.: Learning expressionlets on spatio-temporal manifold for dynamic facial expression recognition. In: Proceedings of IEEE Conference on Computer Vision and Pattern Recognition (CVPR), pp. 1749–1756 (2014)
20. Chen, J., Ariki, Y., Takiguchi, T.: Robust facial expressions recognition using 3d average face and ameliorated adaboost. In: Proceedings of ACM Multimedia Conference (MM), pp. 661–664 (2013)
21. Wang, L., Qiao, Y., Tang, X.: Motionlets: mid-level 3d parts for human motion recognition. In: Proceedings of IEEE Conference on Computer Vision and Pattern Recognition (CVPR), pp. 2674–2681 (2013)

22. Zhao, G., Pietikainen, M.: Dynamic texture recognition using local binary patterns with an application to facial expressions. IEEE Trans. Pattern Anal. Mach. Intell. (TPAMI) **29**, 915–928 (2007)

23. Klaeser, A., Marszalek, M., Schmid, C.: A spatio-temporal descriptor based on 3d-gradients. In: Proceedings of British Machine Vision Conference (BMVC), pp. 99.1–99.10 (2008)

24. Scovanner, P., Ali, S., Shah, M.: A 3-dimensional sift descriptor and its application to action recognition. In: Proceedings of ACM Multimedia Conference (MM), pp. 357–360 (2007)

25. Rudovic, O., Pavlovic, V., Pantic, M.: Multi-output Laplacian dynamic ordinal regression for facial expression recognition and intensity estimation. In: Proceedings of IEEE Conference on Computer Vision and Pattern Recognition (CVPR), pp. 2634–2641 (2012)

26. Liu, M., Li, S., Shan, S., Wang, R., Chen, X.: Deeply learning deformable facial action parts model for dynamic expression analysis. In: Cremers, D., Reid, I., Saito, H., Yang, M.-H. (eds.) ACCV 2014. LNCS, vol. 9006, pp. 143–157. Springer, Heidelberg (2015). doi:10.1007/978-3-319-16817-3_10

27. Trzcinski, T., Christoudias, M., Lepetit, V.: Learning image descriptors with boosting. IEEE Trans. Pattern Anal. Mach. Intell. (TPAMI) **37**, 597–610 (2015)

28. Chen, D., Ren, S., Wei, Y., Cao, X., Sun, J.: Joint cascade face detection and alignment. In: Fleet, D., Pajdla, T., Schiele, B., Tuytelaars, T. (eds.) ECCV 2014. LNCS, vol. 8694, pp. 109–122. Springer, Heidelberg (2014). doi:10.1007/978-3-319-10599-4_8

The LFW-Gender Dataset

Ahsan Jalal[(⊠)] and Usman Tariq

Department of Electrical Engineering, College of Engineering,
American University of Sharjah, Sharjah, UAE
b00067887@aus.edu

Abstract. Gender identification is a precursor for context specific emotion recognition. Small but significant differences have been noted across different gender groups in terms of emotion expressiveness. Apart from facial expressions and security applications, gender recognition is becoming increasingly relevant after the rise of applications involving social media platforms. Labelled Faces in the Wild (LFW) dataset is designed for studying the problem of face recognition under unconstrained environment. However, it is used to study other facial attributes as well, including gender. In this paper, we propose a standardized subset of LFW database (LFW-gender) that can be used as a benchmark for gender recognition algorithms. We also provide a baseline for performance on the dataset for gender recognition with various algorithms and some results may suggest that this is a harder subset to classify.

1 Introduction

Gender Recognition from faces has wide ranging applications, including but not limited to; human computer interaction, intelligent user interfaces, surveillance, demographic studies, e-shopping choices and so on [1]. Gender recognition is also important for context specific emotion recognition. Different stimuli may elicit different intensities of emotions, if not different emotions, from different genders. Although small, but significant differences in emotion expressiveness have been reported across the male and female population in several studies conducted in USA and some European countries [2]. For instance, women may show greater emotion expression overall [3], particularly for the positive emotions [4]. However, they may inhibit negative emotions, e.g. sadness [5]. And men, on the other hand, may show a greater degree of aggressiveness and anger in certain contexts [6]. Hence, gender identification can be thought of as a precursor for context specific emotion recognition.

In this paper, we introduce a standardized subset of the Labelled Faces in the Wild (LFW) database for gender recognition purposes. We term this subset as the LFW-gender database. The database has equal number of male and female images with clearly defined four randomized sets of training, validation and testing subsets. The labels of all the images are manually verified to avoid any errors in labelling. The faces are detected, aligned and cropped. The standardized subset will help researchers to compare algorithms with each other. This will

© Springer International Publishing AG 2017
C.-S. Chen et al. (Eds.): ACCV 2016 Workshops, Part III, LNCS 10118, pp. 531–540, 2017.
DOI: 10.1007/978-3-319-54526-4_39

particularly help in making fair comparisons between algorithms when difference amongst performance is otherwise less than a couple of percentage points, as in such situations the subset selection can introduce a bias in performance as well. We also present some baseline results towards the end of paper. The results are inferior to some similar algorithms tested on the LFW database in the past, signalling thereby that this may be a harder subset to classify. Hence, it gives the researchers a greater room for improvement. The proposed LFW-gender database can be downloaded from [7].

In the following, we briefly review some of the literature in gender recognition, along with the databases used by the researchers for the said purpose. This is then followed by the details of the LFW-gender database and baseline results.

2   Literature Review

Gender recognition, like other classification tasks, can be divided into several steps [8]: face detection, preprocessing such as illumination and contrast normalization, feature extraction and then classification using a binary classifier. Like other computer vision tasks, gender recognition algorithms require datasets for the algorithms to train on. There are several face image datasets that have been used for gender recognition in the literature. For instance, consider some of the publicly available datasets such AR [9], XM2VTS [10], FERET [11], BioID [12], PIE [13], FRGC [14], MORPH-2 [15], LFW [16], CAS-PEAL-R1 [17], Multi-PIE [18] and Adience [19] datasets. All the afore-mentioned datasets, with the exception of the LFW [16] and Adience [19] databases, were captured in constrained environments. However, these present a variety of illumination, pose, age, expression and gender variations.

The AR dataset [9] consists of over 4,000 images from 126 individuals with various expression and illumination variations. The XM2VTS dataset [10] contains 5,900 images from 295 persons with pose and lighting variations. The very popular FERET database [11] contains 14,126 images from 1,199 unique individuals with pose, lighting and expression changes. FRGC database contains 50,000 images from 688 persons with various lighting, expression and background variations. Morph-2 dataset contains 55,285 images from 13,660 individuals in different age groups. CAS-PEAL-R1 dataset [17] is a publicly available subset of a very large Chinese face database. The subset contains 30,900 images from 1,040 individuals.

Two other very popular face image databases are the PIE [13] and Multi-PIE [18] databases. The PIE dataset contains 41,368 images from 68 subjects while Multi-PIE consists of more than 750,000 images from 305 individuals. The images are captured in studio lighting but with a lot of variation across illumination, expression and pose.

Apart from the datasets captured in constrained environments, the two unconstrained face image datasets are Adience [19] and LFW [16]. The Adience dataset is designed for age and gender classification and has around 26,000 images from 2,284 subjects. The LFW dataset, on the other hand, contains

13,233 images from 5,749 persons and contains variations typically encountered in real life. The number of unique individuals in the LFW database is much higher than that in the Adience. It is designed to study the problem of face recognition under unconstrained conditions. However, it is also used for gender recognition in several studies such as [20–25]. Hence, it would be befitting to provide a standardized protocol for comparing gender recognition algorithms.

Gender recognition has been done with a variety of algorithms. The first application was introduced in 1990 by Golomb et al. [26] using Neural Networks. Since then, a variety of features and classifiers have been reported in the literature. For instance, [25–31], use raw pixels; [32] uses Independent Component Analysis (ICA) features; [33–35] use Principal Component Analysis (PCA) features; [20,36,37] use Local Binary Pattern (LBP) features; [38,39] use SIFT features; [40] uses Histogram-of-Gradients (HOG) features; [37] employs Gabor features and so on. Apart from features, some popular classifiers for gender recognition include Neural Networks [26]; Support Vector Machines (SVM) [28,37,40,41]; Adaboost [20,30,36]; Linear Discriminant Analysis [32,34]; and more recently Deep Convolutional Neural Networks (CNNs) [25,31]. For a detailed survey of methodologies, please refer to [42].

In the following we describe the proposed LFW-gender dataset, which is then followed by description of the baseline performance.

3 LFW-Gender Dataset

The LFW dataset is originally designed to study the problem of unconstrained face recognition. However, as noted earlier, it is also used in a number of works on gender recognition. In some cases it is also used for cross-database evaluation for gender recognition, such as [25,31]. It contains 13,233 face images, of mostly celebrities, collected from the web. However, the dataset is male dominated, with much more images from male class (10256), compared to the female class (2977). There are 5749 unique individuals; out of them, 4263 are males and 1486 are females. Now, when a dataset is imbalanced in terms of number of classes, it may lead to problems during training, particularly if the classifier's objective function is not used carefully. For instance, if the classifier is minimizing the classification error rate, the classifier may begin to learn to classify the dominant class more accurately, unless the weights for the less dominant class are not adjusted carefully. Another possibility can be to sub-sample the dominant class. Apart from this, the original images in the LFW dataset, also have a lot of background clutter, hence the classifiers may learn some things from the background as well. However, similar to the LFWcrop [43,44], we can just crop the face and work with it.

In the newly proposed subset, this is what we do. We use the aligned LFW images with the deep funneling algorithm [45]. We then detect faces using the Viola-Jones face detector and crop the faces to 200 × 200 dimensions. To label for gender; we follow a two-step process; automatic labeling and manual verification. We first use a Python API that queries a web-server with the first name of

Fig. 1. Each horizontal pair has same first name with opposite gender

the person in a filename and filters the results into male faces and female faces. However, there may be unisex names, as shown in Fig. 1. Hence, we manually verify the image labels. We then randomly sub-sample the male class with one image per individual and then divide the images into four sets of training, validation and testing folds. The resulting LFW-gender dataset has the following characteristics.

- The faces are aligned using the deep funneling algorithm [45].
- The faces are detected and cropped using the Viola-Jones face detector.
- There are 5810 images in total; 2905 each from the male and female classes.
- There are 4365 individuals in total; 2905 unique males and 1460 unique females.
- The train, validation and test partitions are disjoint, in terms of persons as well.

The LFW-gender database (aligned and cropped face images; train, test and validation partitions and labels) can be downloaded from the link given in [7]. Some example images are shown in Fig. 2. And the breakdown of the four test, train and validation partitions is given in Table 1.

Fig. 2. Sample images from LFW-gender dataset. First two rows are from male cate-. gory and other two rows from the female category.

Table 1. Distribution of the randomized partitions of LFW-gender

Fold #	Train set	Validation set	Test set	Total
1	1886	1556	2368	5810
2	2296	1284	2230	5810
3	1932	1550	2328	5810
4	2108	1302	2400	5810

4 Establishing the Baseline

To establish baseline on the proposed LFW-gender dataset, we select few algo-rithms, some of which have been previously used in the literature for gender recognition as outlined in Sect. 2. We experiment with various feature-classifier combinations. For features we use raw pixel values, PCA (eigen) projections, LDA projections, and Random projections. For classifiers we use k-Nearest Neighbors (k-NN), Support Vector Machines (SVM) with linear and rbf kernels and Deep CNNs. In the following, we briefly describe some of them.

Before any further processing, we convert all the images into grayscale and resize the images to 50×50 pixels. And for each image, we subtract out its mean pixel value and divide each pixel value by the standard deviation of all the pixel values in the same image. The motivation here is to reduce the effect of lightness and contrast changes across different images. The raw pixel features are then simply obtained by the concatenation of image pixels in a vectorial form, hence each has 2500×1 dimensions.

For PCA features, we do the eigen analysis of the covariance matrix of the train+validation sets. We experiment with various number of PCA features and test on the validation set to find the optimum number of principal directions. We found this number to be 50.

For LDA features, we project the train+validation and the test set using the learnt optimum direction. Since there are just two classes; this will result in single dimensional points after projection.

We also experiment with Random Projections which has not been used for gender recognition before (to our best knowledge). The motivation comes from [46] who apply it for dimensionality reduction on images and text data. The basic idea for random projections is the Johnson-Lindenstrauss lemma that states that if points in a vector space are projected onto a random subspace of suitably high dimension then the distances amongst points are approximately preserved. The key idea is quite simple. We build a matrix W whose values are sampled from a uniform random distribution. In our case, the dimensionality of W should be $2500 \times d$, where d is the dimensionality to which we want to reduce our data vectors to. We then normalize the column of W to have unit norm. Then if x is the input raw vector, the feature vector after projection onto columns of W is given by $y = W'x$. We repeat the randomization process 200 times and select the one that gives the best performance on the validation set of each fold. We then combine training and validation set and test on the testing set.

We test using k-NN with 10 neighbours, SVM with linear kernel, SVM with rbf kernel and Deep CNN. The parameters for SVM are validated on the validation set. The Deep CNN in our experiments had 6-layers; consisting of convolution, sub-sampling and fully connected layers. The number of layers was restricted to avoid overfitting. We did not do any data augmentation. The input to the Deep CNN were the 50×50 images. We used TensorFlow library [47] for our Deep CNN.

Results and Discussion

The baseline results are outlined in Table 2. Overall Deep CNN comes out to be the best performer. However, SVM with rbf kernel is not lagging too far behind. kNN gave the worst overall performance. If we leave the Deep CNN aside, the features can be ordered in the following sequence in decreasing recognition rate; PCA features > Raw features > LDA features > Random projections.

It is interesting to note that the paper by Shan [20], reports the mean overall recognition rate for five-folds for raw pixels and SVM classification to be 91.27%. However, they did the five fold cross validation on the original LFW dataset.

Table 2. Performance evaluation - average across the four partitions

	kNN	SVM (linear)	SVM (rbf)	Deep CNN
Raw features	72.35%	78.28%	85.81%	87.95%
LDA features	69.44%	76.01%	83.82%	-
PCA features	76.51%	83.88%	86.11%	-
Random projections	69.83%	74.08%	75.09%	-

We could only achieve a classification rate of 78.28% with linear SVM and 85.81% with non-linear SVM. This gives a strong indication that, the chosen subset is harder compared to the complete LFW dataset. We can also notice from the results of [20] that the classification results on male images are significantly higher than those on the female images. This may primarily be because of the dataset imbalance amongst the two classes. However, such a problem may not arise in the proposed dataset, as the two classes are well balanced.

The results for Deep CNN are also lower than those reported in [25,31], however, it may worthwhile to note that their Deep CNNs were trained on other datasets, hence this is not a fair comparison.

5 Concluding Remarks

In short, we have presented a balanced and well-structured subset of the LFW database for gender recognition. The proposed LFW-gender database has shown inferior performance compared to some similar algorithms applied for gender recognition on the complete LFW dataset in the past, signalling thereby that this may be a harder subset to classify which gives researchers a greater room for improvement. LFW-gender with all the partitions, aligned and cropped face images can be downloaded from [7]. We sincerely hope that the proposed database will farther research and help researchers in gender recognition and related fields, including context specific emotion recognition.

Acknowledgements. This work was supported in part by a faculty research grant, FRG15-R-42. We thank the efforts of Mohamed, Riaz, Siyam and Zeid who helped in manual verification of gender labels.

References

1. Baluja, S., Rowley, H.A.: Boosting sex identification performance. Int. J. Comput. Vis. **71**, 111–119 (2007)
2. Chaplin, T.M.: Gender and emotion expression: a developmental contextual perspective. Emot. Rev. **7**, 14–21 (2015)
3. Brody, L., Hall, J.: Gender and emotion in context. In: Lewis, M., Haviland-Jones, J.M., Barrett, L.F. (eds.) Handbook of Emotions, pp. 395–408. Guilford Press, New York (2010)

4. LaFrance, M., Hecht, M.A., Paluck, E.L.: The contingent smile: a meta-analysis of sex differences in smiling. Psychol. Bull. **129**, 305 (2003)
5. Allen, J.G., Haccoun, D.M.: Sex differences in emotionality: a multidimensional approach. Hum. Relat. **29**, 711–722 (1976)
6. Archer, J.: Sex differences in aggression in real-world settings: a meta-analytic review. Rev. Gen. Psychol. **8**, 291 (2004)
7. https://sites.google.com/site/usmantariq/
8. Ng, C.B., Tay, Y.H., Goi, B.-M.: Recognizing human gender in computer vision: a survey. In: Anthony, P., Ishizuka, M., Lukose, D. (eds.) PRICAI 2012. LNCS (LNAI), vol. 7458, pp. 335–346. Springer, Heidelberg (2012). doi:10.1007/978-3-642-32695-0_31
9. Martínez, A., Benavente, R.: The AR face database (1998)
10. Messer, K., Matas, J., Kittler, J., Luettin, J., Maitre, G.: Xm2vtsdb: the extended m2vts database. In: Second International Conference on Audio and Video-based Biometric Person Authentication, vol. 964, pp. 965–966. Citeseer (1999)
11. Phillips, P.J., Moon, H., Rizvi, S.A., Rauss, P.J.: The feret evaluation methodology for face-recognition algorithms. IEEE Trans. Pattern Anal. Mach. Intell. **22**, 1090–1104 (2000)
12. Jesorsky, O., Kirchberg, K.J., Frischholz, R.W.: Robust face detection using the hausdorff distance. In: Bigun, J., Smeraldi, F. (eds.) AVBPA 2001. LNCS, vol. 2091, pp. 90–95. Springer, Heidelberg (2001). doi:10.1007/3-540-45344-X_14
13. Sim, T., Baker, S., Bsat, M.: The CMU pose, illumination, and expression (PIE) database. In: 2002 Fifth IEEE International Conference on Automatic Face and Gesture Recognition, Proceedings, pp. 46–51. IEEE (2002)
14. Phillips, P.J., Flynn, P.J., Scruggs, T., Bowyer, K.W., Chang, J., Hoffman, K., Marques, J., Min, J., Worek, W.: Overview of the face recognition grand challenge. In: 2005 IEEE Computer Society Conference on Computer Vision and Pattern Recognition (CVPR 2005), vol. 1, pp. 947–954. IEEE (2005)
15. Ricanek, K., Tesafaye, T.: MORPH: a longitudinal image database of normal adult age-progression. In: 7th International Conference on Automatic Face and Gesture Recognition (FGR06), pp. 341–345. IEEE (2006)
16. Huang, G.B., Ramesh, M., Berg, T., Learned-Miller, E.: Labeled faces in the wild: a database for studying face recognition in unconstrained environments. Technical report 07-49, University of Massachusetts, Amherst (2007)
17. Gao, W., Cao, B., Shan, S., Chen, X., Zhou, D., Zhang, X., Zhao, D.: The CAS-PEAL large-scale Chinese face database and baseline evaluations. IEEE Trans. Syst., Man, Cybern.-Part A: Syst. Hum. **38**, 149–161 (2008)
18. Gross, R., Matthews, I., Cohn, J., Kanade, T., Baker, S.: Multi-PIE. Image Vis. Comput. **28**, 807–813 (2010)
19. Eidinger, E., Enbar, R., Hassner, T.: Age and gender estimation of unfiltered faces. IEEE Trans. Inf. Forensics Secur. **9**, 2170–2179 (2014)
20. Shan, C.: Learning local binary patterns for gender classification on real-world face images. Pattern Recognit. Lett. **33**, 431–437 (2012)
21. Shih, H.C.: Robust gender classification using a precise patch histogram. Pattern Recognit. **46**, 519–528 (2013)
22. Tapia, J.E., Perez, C.A.: Gender classification based on fusion of different spatial scale features selected by mutual information from histogram of lbp, intensity, and shape. IEEE Trans. Inf. Forensics Secur. **8**, 488–499 (2013)
23. Bekios-Calfa, J., Buenaposada, J.M., Baumela, L.: Robust gender recognition by exploiting facial attributes dependencies. Pattern Recognit. Lett. **36**, 228–234 (2014)

24. Jia, S., Cristianini, N.: Learning to classify gender from four million images. Pattern Recognit. Lett. **58**, 35–41 (2015)
25. Antipov, G., Berrani, S.A., Dugelay, J.L.: Minimalistic CNN-based ensemble model for gender prediction from face images. Pattern Recognit. Lett. **70**, 59–65 (2016)
26. Golomb, B.A., Lawrence, D.T., Sejnowski, T.J.: SEXNET: a neural network identifies sex from human faces. In: NIPS, vol. 1, p. 2 (1990)
27. Gutta, S., Wechsler, H., Phillips, P.J.: Gender and ethnic classification of face images. In: 1998 Third IEEE International Conference on Automatic Face and Gesture Recognition, Proceedings, pp. 194–199. IEEE (1998)
28. Moghaddam, B., Yang, M.H.: Learning gender with support faces. IEEE Trans. Pattern Anal. Mach. Intell. **24**, 707–711 (2002)
29. Kim, H.C., Kim, D., Ghahramani, Z., Bang, S.Y.: Appearance-based gender classification with gaussian processes. Pattern Recognit. Lett. **27**, 618–626 (2006)
30. Baluja, S., Rowley, H.A.: Boosting sex identification performance. Int. J. Comput. Vis. **71**, 111–119 (2007)
31. Levi, G., Hassner, T.: Age and gender classification using convolutional neural networks. In: Proceedings of the IEEE Conference on Computer Vision and Pattern Recognition Workshops, pp. 34–42 (2015)
32. Jain, A., Huang, J.: Integrating independent components and linear discriminant analysis for gender classification. In: 2004 Sixth IEEE International Conference on Automatic Face and Gesture Recognition, Proceedings, pp. 159–163. IEEE (2004)
33. Khan, A., Majid, A., Mirza, A.M.: Combination and optimization of classifiers in gender classification using genetic programming. Int. J. Knowl.-Based Intell. Eng. Syst. **9**, 1–11 (2005)
34. Bekios-Calfa, J., Buenaposada, J.M., Baumela, L.: Revisiting linear discriminant techniques in gender recognition. IEEE Trans. Pattern Anal. Mach. Intell. **33**, 858–864 (2011)
35. Tariq, U., Hu, Y., Huang, T.S.: Pattern recognition, machine intelligence and biometrics. In: Wang, S.P. (ed.) Gender and Race Identification by Man and Machine, 1st edn. Springer, Berlin, Heidelberg (2011). doi:10.1007/978-3-642-22407-2
36. Sun, N., Zheng, W., Sun, C., Zou, C., Zhao, L.: Gender classification based on boosting local binary pattern. In: Wang, J., Yi, Z., Zurada, J.M., Lu, B.-L., Yin, H. (eds.) ISNN 2006. LNCS, vol. 3972, pp. 194–201. Springer, Heidelberg (2006). doi:10.1007/11760023_29
37. Xia, B., Sun, H., Lu, B.L.: Multi-view gender classification based on local Gabor binary mapping pattern and support vector machines. In: 2008 IEEE International Joint Conference on Neural Networks (IEEE World Congress on Computational Intelligence), pp. 3388–3395. IEEE (2008)
38. Toews, M., Arbel, T.: Detection, localization, and sex classification of faces from arbitrary viewpoints and under occlusion. IEEE Trans. Pattern Anal. Mach. Intell. **31**, 1567–1581 (2009)
39. Wang, J.G., Li, J., Yau, W.Y., Sung, E.: Boosting dense sift descriptors and shape contexts of face images for gender recognition. In: 2010 IEEE Computer Society Conference on Computer Vision and Pattern Recognition-Workshops, pp. 96–102. IEEE (2010)
40. Guo, G., Dyer, C.R., Fu, Y., Huang, T.S.: Is gender recognition affected by age? In: 2009 IEEE 12th International Conference on Computer Vision Workshops (ICCV Workshops), pp. 2032–2039. IEEE (2009)
41. Chen, C., Ross, A.: Evaluation of gender classification methods on thermal and near-infrared face images. In: 2011 International Joint Conference on Biometrics (IJCB), pp. 1–8. IEEE (2011)

42. Reid, D., Samangooei, S., Chen, C., Nixon, M., Ross, A.: Soft biometrics for surveillance: an overview. In: Machine Learning: Theory and Applications, pp. 327–352. Elsevier (2013)
43. Sanderson, C., Lovell, B.C.: Multi-region probabilistic histograms for robust and scalable identity inference. In: Tistarelli, M., Nixon, M.S. (eds.) ICB 2009. LNCS, vol. 5558, pp. 199–208. Springer, Heidelberg (2009). doi:10.1007/978-3-642-01793-3_21
44. http://conradsanderson.id.au/lfwcrop/
45. Huang, G., Mattar, M., Lee, H., Learned-Miller, E.G.: Learning to align from scratch. In: Advances in Neural Information Processing Systems, pp. 764–772 (2012)
46. Bingham, E., Mannila, H.: Bingham, E., Mannila, H.: Random projection in dimensionality reduction: applications to image and text data. In: Proceedings of the Seventh ACM SIGKDD International Conference on Knowledge Discovery and Data Mining, pp. 245–250. ACM (2001)
47. https://www.tensorflow.org/

Thermal Imaging Based Elderly Fall Detection

Somasundaram Vadivelu[1], Sudakshin Ganesan[1], O.V. Ramana Murthy[1(✉)],
and Abhinav Dhall[2]

[1] Department of Electrical and Electronics Engineering,
Amrita School of Engineering, Amrita Vishwa Vidyapeetham,
Amrita University, Coimbatore, India
ovr_murthy@cb.amrita.edu
[2] Department of Computer Science, University of Waterloo, Waterloo, Canada

Abstract. Elderly fall detection is very special case of human action recognition from videos and has very practical application in old age home and nursing centers. Fall detection in its simplest form is a binary classification of fall event or other daily routine activities. Hence, the current trend of sophisticated techniques being developed for human action recognition, particularly with scenarios of large number of classes may not be required in elderly fall detection. However, other design considerations such as simplicity (ready to be deployed), privacy issues (not revealing the identity) are to focused and are the major contributions of this paper. The Spatio-Temporal Interest Points (STIP) and Fisher vector framework for human action recognition is established as baseline in this work. A novel optical flow based technique is proposed that yields better performance than the baseline. Further, a very economical thermal imaging based input modality is proposed. Along with the thermal images not revealing the identity of the persons, thermal images also aid human detection from backgrounds – a useful solution in computing the optical flow of human movements. The proposed solution is also validated on the **KUL Simulated Fall** dataset showing its generalization capability.

1 Introduction

Statistically, about one third of the population over the age of 65, tend to fall at least once a year[1], and the risk of falls increases with further ageing. Often, such falls can lead to injury and death in extreme cases among the elderly. It has been identified that 87% of the fractures in the elderly are primarily due to the falls. Fall incidents lead to hospital or nursing home admissions, making it further expensive in charges. Sadly, 40% of such admitted do not return to independent living and worse, one-fourth die within a year. Even if there are no injuries sustained from a fall, a large percentage of fallers (47%) cannot get up without assistance. The gap period of time spent immobile between a fall and assistance received and crucial. Longer the gap the higher is the risk of health

---

S. Vadivelu and S. Ganesan—Equal contribution

[1] http://www.alohahomecarefl.net/orthopedic-care/fall-reduction-and-home-safety/?.

© Springer International Publishing AG 2017
C.-S. Chen et al. (Eds.): ACCV 2016 Workshops, Part III, LNCS 10118, pp. 541–553, 2017.
DOI: 10.1007/978-3-319-54526-4_40

affects consequence such as dehydration, hypothermia, pneumonia and pressure sores. Apart from physical injuries, fall can also lead to psychological issues such as fear of future falls, feeling lonely and depression [1]. Getting immediate help after a fall largely reduces many such health risks and also enhances the confidence to return to independent living.

Timely detection of falls by automation can be used to trigger alarm to the caring personnel. Further, the past fall incidents can be analyzed to avoid similar situations leading to falls in the future. Often, the fall incidents also aid in identifying health related issues that happened to be unnoticed by the subject until then. In this way the quality of the elderly can be enhanced.

There are three major approaches for automatic fall detection as follows

1. Wearables. These are body-worn containing sensors such as MEMS accelerometer [2]. The data being collected directly from the boy contact, is rich and yield very high accurate performance. However, these devices need to be always worn all the time. Usually, old people tend to forget things easily and further the wearable requires regularly constant recharge; posing a challenge to the user.
2. Ambient analyzers. These can be pressure sensors installed on the floors [3]. These are non-invasive and also conceal the person's identity. However, it is costly – requiring installation on all the floor. Further, false alarm rate is high due to any nonhuman object falls.
3. Vision/camera based approach [4]. They are non-invasive and economical. However, their solutions can be computationally complex and further the privacy of the individual is not ensured.

In human action recognition (HAR), only one event – fall or one non-fall event – is assumed to constitute the entire segment. However, in the current task, non-fall video segment contains more than one non-fall event such as walking, sitting down, arranging bed and so on; hence challenging than HAR. In this paper the two limitations of the vision based approaches are addressed as follows

– Thermal imaging based input modality to conceal the individual's identity.
– Optical flow based feature extraction.

Thermal imaging captures human data even during the night or absence of very less ambient light; times when probability of fall is high. The cost of Thermal cameras is also decreasing and we have selected the most economical – Flir ONE thermal imaging camera – as to make our solution practically viable. We have also created a Thermal imaging Fall event dataset which will be released to the research community for further progress in fall detection. The layout of the rest of the paper is as follows. Section 2 contains a literature survey on vision based fall detection techniques and the datasets existing in the literature. Section 3 contains the overall framework and our proposed optical based solution. The datasets used to conduct our experiments is discussed in Sect. 4. Section 5 contains the results obtained by applying our proposed techniques and comparisons with baseline framework of human action recognition. Finally we summarize in Sect. 6.

2 Related Literature

This section contains a brief survey on the vision based techniques and the datasets proposed in the literature to conduct simulation studies.

2.1 Vision-Based Techniques

A good survey on fall detection systems was recently given in [5]. Being non-invasive and economical set-up, camera based solutions are attracting vast majority of researchers. Debard *et al.* [4] used background subtraction and a particle filter to get best estimate for the background. The person is then located in the frame by fitting the biggest ellipse in the foreground section of the frame. The particle filter was based on three different measuring functions computed using – an ellipse that fitted the foreground binary image best, histogram correlation of the ellipse divided into four different parts of the body; and an upper body detector. Any portions inside the fitted ellipse were updated very slowly while the rest of the image (as background) was updated very fast. When the person location in the video frame is thus finalized, a fall feature vector is comprised consisting of values (fitted ellipse's aspect ratio, change in aspect ratio, fall angle, centre speed and head speed) calculated over different time slots before, during and after the fall incident. The feature vector is subsequently used for learning a Support Vector Machine (SVM) model to discriminate fall from non-fall scenario. Auvinet *et al.* [6] computed a set of features extracted from human body silhouette tracking. The features are based on height and width of human body bounding box, the users trajectory with her/his orientation, Projection Histograms and moments of order 0, 1 and 2. They investigated several combinations of usual transformations of the features (Fourier Transform, Wavelet transform, first and second derivatives). Charfi *et al.* [7] detected fall events by analyzing the volume distribution along the vertical axis. An alarm is triggered when the major part of this distribution is abnormally near the floor during a predefined period of time, which implies that a person has fallen on the floor.

2.2 Datasets

Due to ethical and privacy related reasons, the videos of real fall incidents are not shared to research public at large. Hence, majority of the algorithms are evaluated on simulated falls performed in artificial surroundings [4]. However, Kangas *et al.* [8] reported significant differences between falls in artificial surroundings and real fall incidents in terms factors such as illumination and occlusions. Auvinet *et al.* [6] shared dataset containing video segments ranging from 30 s to 4 min using eight calibrated cameras. The video segments contain 22 fall events and 24 other events such as sitting, crouching and lying on a sofa. Charfi *et al.* [7], used a single camera setup to record 249 video segments. These video segments ranged from 10 to 45 s. The datasets contained 192 fall video segments and 57 non-fall/normal activities such as sitting down, walking, standing up and housekeeping. Baldewijns *et al.* [9] created and shared publicly **KUL Simulated Fall** dataset consisting of 55 fall events and 14 normal daily activities. This dataset was created by re-enacting real-life fall incidents that were observed during previous studies [10].

A limitation observed with all these techniques is the generalization capacity of their proposed techniques is not validated i.e. not tested on multiple datasets. To avoid situations of overfitting to one dataset, we investigate state-of-the-art human action recognition framework and propose novel optical flow technique on two datasets with completely different settings – RGB and Thermal videos.

3   Overall Framework and Background

The overall layout of the proposed framework is shown in Fig. 1.

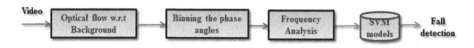

Fig. 1. Overall framework: starting with an input video segment, optical flow vectors are computed w.r.t the first frame (taken as background image). Angles of the flow vectors are computed and categorized into 10 bins. The distribution of each bin over time is analysed in frequency domain to generate a temporal feature vector. Finally, a SVM classifier is used to predict the action label in the given video segment – fall or non-fall.

The action recognition framework of Fisher vector computation from spatio-temporal interest points is taken as baseline to compare performance of our proposed technique. Firstly, interest points, (Harris 3D corners), are detected. Local descriptors are computed around these detected interest points. Gaussian Mixture Modelling is applied to each of these descriptors to yield GMM centres. Fisher vectors are generated from these GMMs to learn a classifier (for each action class). This baseline benchmark is described below.

3.1   Harris 3D Corner Interest Points

Laptev and Lindeberg [11] proposed the Harris 3D corners as an extension of traditional (2D) Harris corner points for spatio-temporal analysis and action recognition. These interest points are local maxima of a function of space-time gradients. They compute a spatio-temporal second-moment matrix at each video point in different spatio-temporal scales. The interest points are obtained as local maxima of a function of this second-moment matrix. We use the original implementation[2] with standard parameter settings. The interest points are extracted at multiple scales based on a regular sampling of spatial and temporal scale

---

[2] http://www.di.ens.fr/~laptev/download.html/#stip.

values. They are defined in 5 dimensions (x, y, t, σ, τ), where x, y and t are spatial and temporal axes, respectively; while σ and τ are the spatial and temporal scales, respectively. We then compute the local descriptors – histograms of oriented gradients (HOG) and histograms of optical flow (HOF) with default parameter setting. While the former captures the local motion and appearance, the latter captures the temporal changes.

3.2 Feature Encoding

We construct Fisher vectors (FV) [12] for each HOG and HOF descriptors separately. Initially, 250,000 descriptors are randomly from the training set Gaussian Mixture Modeling (GMM) is applied. We define the parameters obtained from the GMM fitting as $\theta = (\pi_j, m_j, \sum_j; j = 1, 2, ..., k)$ where π_j, m_j and $\sum_j$ are the prior probability, mean and covariance of each distribution. The GMM ascribes each descriptor x_i to a mode j in the mixture with a strength given by its corresponding posterior probability

$$q_{ij} = \frac{(x_i - m_j)^T \sum_j^{-1} (x_i - m_j)}{\sum_{t=1}^{k} (x_i - m_t)^T \sum_t^{-1} (x_i - m_t)}. \tag{1}$$

The mean (u_{jk}) and deviation vectors (v_{jk}) in each mode k are computed as follows

$$u_{jk} = \frac{1}{N\sqrt{\pi_k}} \sum_{i=1}^{N} q_{ik} \frac{x_{ji} - m_{ik}}{\sigma_i} \tag{2}$$

$$v_{jk} = \frac{1}{N\sqrt{2\pi_k}} \sum_{i=1}^{N} q_{ik} [(\frac{x_{ji} - m_{ik}}{\sigma_i})^2 - 1] \tag{3}$$

where $j = 1, 2, ..., D$ spans the local descriptor vector dimensions. The resultant vectors (u_{jk}) and (v_{jk}) for each of the k modes in the Gaussian mixtures are concatenated to yield w, and then normalised by the 'power-law normalisation' [13] defined as $w_j = |w_j|^\alpha \times sign(w_j)$ with $\alpha = 0.5$. Finally, the vector is L_2-normalised as $w = \frac{w}{||w||}$ to yield the FV vector. We use $k = 256$ for each descriptor type in all our experiments.

For classification, we concatenate all Fisher Vectors (of different descriptors) and use linear SVM LIBLINEAR [14] to select the class with the highest score.

3.3 Frequency Analysis of Optical Flow Angle Bins

In this section, we propose a novel approach to model the temporal information of the optical flow angular bins. We take motivation from the recent work on temporal modelling in human action recognition in [15]. We compute optical flow of each frame w.r.t to the background and hypothesize that angles of optical flow vector will hold discriminating information for a fall action from non-fall action. Hence, we compute the angles of the optical flow vectors and assign them into

10 bins; $0 - 360^0$ is divided into 10 bins each of size 36°. Thus, 10 histograms are obtained for each frame optical flow vector. The distribution of each histogram intensity over time dimension contains the action information in the given video segment. We then apply frequency analysis to derive the feature vector from that time distribution curve.

We perform frequency analysis in Fourier transform domain and as outlined in Algorithm 1. Initially, the temporal occurrence of each angular bin (of the optical flow vector) over entire duration of video is collected as a distribution-time curve. Fourier transform [16,17] is applied to detect which frequencies occur most. High frequencies are indicative of large angular changes in the optical flow vector indicating actions such as sudden fall of people. Thus, we collect amplitudes of the constituent frequencies and use it to learn a classifier.

Algorithm 1. Computing temporal feature vectors from 10 Angular bins of optical flow

---

    Input: Video
    Output: Temporal feature vectors for the video
1 $N \leftarrow$number of frames in the video
2 $I_{Bgrnd} \leftarrow$First frame
3 **for** $j \leftarrow 2$ *to* N **do**
4       $Opt =$Optical flow vector w.r.t I_{Bgrnd}
5       $Ang =$Angles of the Opt
6       $Bin[j][0...1] =$ Divide Ang into 10 bins

7 **for** $i \leftarrow 1$ *to* *10* **do**
       /* Apply N-point Fast Fourier Transform   */
8       $Y = \dfrac{1}{N}||fft(Bin[][i])||^2$
       /* Omit the d.c component; Select the first $\dfrac{N-1}{2}$ values   */
       /* Rearrange amplitude values in descending manner   */
9       $[a,b] = descendsort(Y)$
       /* Select amplitude components only as the Temporal feature vector   */
10     $TFA = a$

11 **return** *Temporal Feature Vectors TFA*

---

4   Datasets

The proposed technique is applied on two datasets – benchmark **KUL Simulated Fall** [9] dataset and in-house created **Thermal Simulated Fall** dataset. The details are given below

KUL Simulated Fall dataset: This dataset contains a total of 55 simulated fall scenarios and 17 video segments containing non-fall activities. Real-life fall incidents recorded and described as in [4] were studied and then re-enacted by 10 different actors. A room was set-up similar to a nursing-home room environment.

Table 1. Overview of different fall scenarios in **KUL Simulated Fall dataset**

Using walking aid	Scenarios
Walker	13
Wheelchair	4
No walking aid	38
Fall speed	
Slow falls	22
Fast falls	33
Moving objects during the fall	
Walker	8
Wheelchair	4
Blanket	3
Chair	10
None	30
Starting pose	
Standing	26
Sitting	10
Bending over	5
Squatting	6
Transitions (sit stand)	8
Ending positions	
Lying on the floor	49
Getting back-up after fall	6

Five IP-cameras (12 fps, resolution: 640 × 480) were installed in this room to record fall and normal daily activities. The average length of fall event is around 2 min 45 s, with a minimum length of 50 s and a maximum length of 4 min 58 s. Fall scenarios differed from each other in terms of using walking aid, other moving objects during the fall, fall speed, starting pose and ending pose. Different types of fall scenarios are summarized in Table 1.

The video segments containing non-fall activities had an average length of 20 min 39 s per scenario of which the shortest segment was 11 min 38 s and the longest was 35 min 30 s duration. Each segment consisted of a set of normal activities. Non-fall segments varied from each other in terms of using different walking aids, changing the order in which the activities were performed, changing the pace of the person performing the actions, not including all activities in each scenarios and finally not performing all activities in the same positions (e.g. sleeping on top of the blankets, under the blankets or in the chair). The normal activities included were as follows [9]:

- Walking (with and without walking aid);
- Transitions from sitting to standing and vice-versa;

- Sitting;
- Eating and drinking;
- Getting into and out of bed;
- Sleeping;
- Changing clothes;
- Removing and putting on shoes;
- Reading;
- Transfers from wheelchair to chair and vice-versa;
- Making the bed;
- Coughing and sneezing violently;
- Picking something up from the floor.

Thermal Simulated Fall dataset. This dataset is based on **KUL Simulated Fall** dataset. It contains 9 video segments with non-fall scenarios and 35 segments with fall scenarios. These videos are having spatial resolution of 640×480 and captured in a room setting, single view using FLIR ONE thermal camera mounted to a Android phone[3]. FLIR ONE is very economical and explored towards practical deployment. Thermal camera has the advantages of protecting the privacy/identity of the individual, capturing even during no-light conditions. Sample snapshots of Real-incident, simulated (RGB) [9] and contributed Thermal dataset are show in Fig. 2. It can be easily observed that the human detection is easier in thermal imaging. This dataset is made publicly available[4].

5   Results and Discussions

As per the experimental settings set forth in [4,9], 10-fold cross-validation is used in each camera, each technique to compute the performance metrics – the precision-recall curves. To draw these performance curves, we need two parameters: (1) Precision and (2) Recall. In pattern recognition and information retrieval with binary classification, precision (also called positive predictive value) is the fraction of retrieved instances that are relevant, while recall (also known as sensitivity) is the fraction of relevant instances that are retrieved. Both precision and recall are therefore based on an understanding and measure of relevance (Table 2).

The precision curve is obtained by varying the margin value (the distance between the nearest point and the hyperplane in SVM). As the margin value is varied from 0 to 1 in very small increments of 0.01, the values of TP,TN,FP and FN are computed for each margin value. From these values, Precision and Recall are computed for each margin value as follows

---

[3] http://www.flir.com/flirone/android/.

[4] https://drive.google.com/open?id=0ByBHFkIRDnx6S2M2WllKaVg5eGc.

(a)   (d)

(b)   (e)

(c)   (f)

Fig. 2. Beginning of fall (a) Real-life (b) Enacted RGB, (c) Enacted, thermal; Middle of fall (d) Real-life (e) Enacted, RGB (f) Enacted thermal. Source of (a), (b), (d), (e) [9]

Table 2. AUC computation

Predicted actual	Positive	Negative
Positive	True Positive (TP)	False Negative (FN)
Negative	False Positive (FP)	True Negative (TN)

$$precision = \frac{TP}{TP + FP} \tag{4}$$

$$recall = \frac{TP}{TP + FN} \tag{5}$$

The results obtained are shown in Fig. 4 and Table 3.

Table 3. Performance (AUC) of the proposed approach

Approach	Cam1	Cam2	Cam3	Cam4	Cam5	Thermal
Proposed approach	0.71	0.69	0.67	0.65	0.67	0.64
STIP + FV	0.71	0.62	0.65	0.64	0.61	0.60
Baldewijins *et al.* [9]	0.56	0.35	0.40	0.56	0.38	-

5.1 On the Performance of Different Techniques

Baldewijins *et al.* [9] performed the least in all the 5 camera views. Their technique is based on ellipse fitting to human detected in each frame. During occlusions or partial view, human detection fails completely and hence their least performance. The state-of-the-art framework used for human action recognition –STIP+FV – performed better than their technique by 0.08–0.27% (absolute). This technique is based on space-time interest points and not dependent on human localization. Thus they are free from occlusions and partial views; hence their better performance than Baldewijins *et al.* [9]. Our proposed technique is based on optical flow w.r.t initial frame as background image. This technique is based on global motion changes and hence contains more information that the STIP+FV framework, which is dependent on local information (STIP). Hence better than STIP+FV performance by 0–0.07% (absolute). Consistent better performance is observed on **Thermal Simulated Fall** dataset too, supporting the generalization capability of the proposed technique.

Further there is no need of any codebook construction unlike STIP+FV. When any new daily activity has to be considered, STIP+FV framework needs to rebuild codebook; our technique doesn't need such.

5.2 On the Choice of 'N' While Computing FFT

The number of points for computing N-FFT is taken as 1500 because the minimum length of all the videos in the dataset was around 50 s; with frame rate

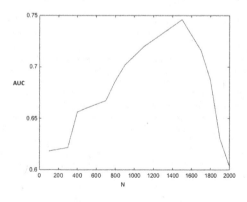

Fig. 3. AUC for different values of 'N' while computing N-point FFT

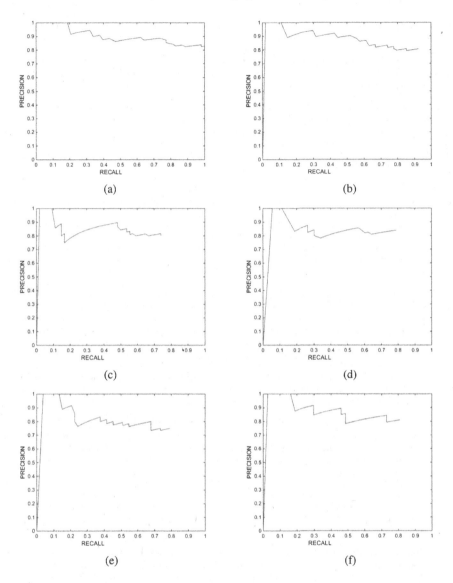

Fig. 4. AUC for (a) Cam 1, STIP (b) Cam 1, proposed technique, (c) Cam 2, proposed, (d) Cam 3, proposed, (e) Cam 4, proposed and (f) Thermal camera, proposed technique.

as 30 fps; totaling to 1500 frames. To investigate further, we also repeated the experiment with our proposed technique on Cam 1 for different values of N. The AUC obtained in each case is plotted as shown in Fig. 3. It can be observed that best AUC is obtained when 'N' is chosen properly according to the video segment length and frame rate of the dataset. In other words, 'N' is dependent on the length of the video segment/step for which we want to compute the decision continuously in real-time implementation.

6 Conclusions

In this paper, we investigated the elderly fall detection using vision based techniques. Skipping the codebook and tracking based techniques, we proposed a simple optical flow based technique. In this technique, angles of the optical flow vectors are computed and categorized into 10 angular bins. The distribution of each bin over temporal dimension is analyzed in frequency domain and a temporal feature vector is constructed for (SVM) classifier training. To conceal the identity of the individual, the proposed solution is validated on a created thermal imaging dataset. The generalization capacity of the proposed technique is thus tested on two datasets. In future, we would like to investigate optical flow w.r.t previous frames and include more daily/routine activities to build enhanced fall detection model.

References

1. Fleming, J., Braynel, C.: Inability to get up after falling, subsequent time on floor, and summoning help: prospective cohort study in people over 90. Br. Med. J. (BMJ) **337**, 1279–1282 (2008)
2. Bagal, F., Becker, C., Cappello, A., Chiari, L., Aminian, K., Hausdorff, J.M., Zijlstra, W., Klenk, J.: Evaluation of accelerometer-based fall detection algorithms on real-world falls. PLoS ONE **7**, 1–9 (2012)
3. Rimminen, H., Lindstrm, J., Linnavuo, M., Sepponen, R.: Detection of falls among the elderly by a floor sensor using the electric near field. IEEE Trans. Inf. Technol. Biomed. **14**, 1475–1476 (2010)
4. Debard, G., et al.: Camera-based fall detection on real world data. In: Dellaert, F., Frahm, J.-M., Pollefeys, M., Leal-Taixé, L., Rosenhahn, B. (eds.) Outdoor and Large-Scale Real-World Scene Analysis. LNCS, vol. 7474, pp. 356–375. Springer, Heidelberg (2012). doi:10.1007/978-3-642-34091-8_16
5. Mubashir, M., Shao, L., Seed, L.: A survey on fall detection: principles and approaches. Neurocomputing **100**, 144–152 (2013)
6. Auvinet, E., Multon, F., Saint-Arnaud, A., Rousseau, J., Meunier, J.: Fall detection with multiple cameras: an occlusion-resistant method based on 3-D silhouette vertical distribution. IEEE Trans. Inf Technol. Biomed. **15**, 290–300 (2011)
7. Charfi, I., Miteran, J., Dubois, J., Atri, M., Tourki, R.: Definition and performance evaluation of a robust SVM based fall detection solution. In: 2012 Eighth International Conference on Signal Image Technology and Internet Based Systems (SITIS), pp. 218–224 (2012)
8. Kangas, M., Vikman, I., Nyberg, L., Korpelainen, R., Lindblom, J., Jms, T.: Comparison of real-life accidental falls in older people with experimental falls in middle-aged test subjects. Gait & Posture **35**, 500–505 (2012)
9. Baldewijns, G., Debard, G., Mertes, G., Vanrumste, B., Croonenborghs, T.: Bridging the gap between real-life data and simulated data by providing a highly realistic fall dataset for evaluating camera-based fall detection algorithms. Healthc. Technol. Lett. **3**(5), 6–11 (2016)
10. Vlaeyen, E., Deschodt, M., Debard, G., Dejaeger, E., Boonen, S., Goedemé, T., Vanrumste, B., Milisen, K.: Fall incidents unraveled: a series of 26 video-based real-life fall events in three frail older persons. BMC Geriatr. **13**, 103 (2013)

11. Laptev, I., Lindeberg, T.: Space-time interest points. In: International Conference on Computer Vision (ICCV), pp. 432–439 (2003)
12. Perronnin, F., Dance, C.: Fisher kernels on visual vocabularies for image categorization. IEEE (2007)
13. Perronnin, F., Sánchez, J., Mensink, T.: Improving the fisher kernel for large-scale image classification. In: Daniilidis, K., Maragos, P., Paragios, N. (eds.) ECCV 2010. LNCS, vol. 6314, pp. 143–156. Springer, Heidelberg (2010). doi:10.1007/978-3-642-15561-1_11
14. Fan, R.E., Chang, K.W., Hsieh, C.J., Wang, X.R., Lin, C.J.: LIBLINEAR: a library for large linear classification. J. Mach. Learn. Res. **9**, 1871–1874 (2008)
15. Murthy, O.V.R., Goecke, R.: The influence of temporal information on human action recognition with large number of classes. In: International Conference on Digital Image Computing: Techniques and Applications (DICTA) (2014)
16. Fourier, J.B.J.: Thorie Analytique de la Chaleur (1822)
17. Cooley, J.W., Tukey, J.W.: An algorithm for the machine calculation of complex fourier series. Math. Comput. **19**, 297–301 (1965)

Workshop on Interpretation and Visualization of Deep Neural Nets

Multi-Scale Hierarchy Deep Feature Aggregation for Compact Image Representations

Zhenbing Zhao, Guozhi Xu$^{(\boxtimes)}$, and Yincheng Qi

School of Electrical and Electronic Engineering,
North China Electric Power University, 619 Yonghua North Street,
Baoding 071003, Hebei, China
xuguozhi0124@foxmail.com

Abstract. Deep Convolutional Neural Networks have set remarkable milestones in the field of computer vision, especially in image classification tasks. However, training a deep network is heavily depending on massive labeled data and expensive computation resource. A number of studies have shown that utilizing a pre-trained model for deep feature extraction can achieve excellent performance. While most of these methods only consider the features from fully connected layers, we delve deep into the intermediate convolution layers. We propose the Selected Multi-Scale Convolution feature (SMSC) for compact deep representations. A convolutional feature map selection and deep descriptor aggregation method are proposed, and a fusion method of the multi-layer features for compact representation is introduced. The experimental results on the known MIT-Indoor dataset have demonstrated the effectiveness and efficiency of the proposed method.

1 Introduction

Since the emergence of the Deep Convolutional Neural Networks (DCNNs) [1,2], the limits of many computer vision tasks have been pushed forward by a large step based on deep learning methods. With the ability to effectively capture rich semantic information based on image global representations, state-of-the-art performances can be achieved by training the deep convolutional neural networks with large scale image datasets [3]. However, in real world applications, the labeled data is of limited number and scarce. In order to generate a discriminative representation for specific tasks, many researchers have proposed hand-crafted local features combined with Bag-of-Feature (BoF) model [4] for global image representations. Despite the success of the BoF and its improved methods as Vector of the Locally Aggregated Descriptors (VLAD) [5] and Fisher Vector (FV) [6], local features are still shallow, and lack high-level semantic information [7]. Meanwhile, the DCNN features are richer in mid-level representation, but vulnerable to geometric variations [8–10].

Inspired by recent advances in utilizing intermediate activations extracted from pre-trained DCNN model [11,12], instead of tuning a complex end-to-end network with millions of parameters on the small image dataset, we take the

© Springer International Publishing AG 2017
C.-S. Chen et al. (Eds.): ACCV 2016 Workshops, Part III, LNCS 10118, pp. 557–571, 2017.
DOI: 10.1007/978-3-319-54526-4_41

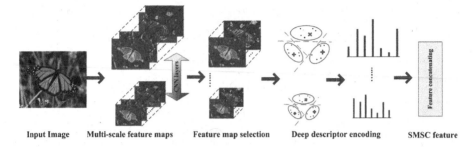

Input Image   Multi-scale feature maps   Feature map selection   Deep descriptor encoding   SMSC feature

Fig. 1. The flow chart of the proposed method: (1) extract convolutional feature maps, (2) compute importance score of each feature map and select top-Q, (3) extract deep local descriptors from selected feature maps, (4) aggregate these descriptors and fuse the feature vectors.

advantages of the pre-trained model based on large-scale image dataset [3]. We delve deep into the convolution layers, and dig the multi-scale hierarchies for compact image representation aggregation. Our primary goal is to extract all the useful information from low-level to high-level hierarchies, and combine them to generate a more powerful and robust representations.

We denote features from multi-scale convolution layers as Selected Multi-Scale Convolution feature, and SMSC for short. The strategy is presented in Fig. 1. We want to convey a new view to look upon these deep feature hierarchies and propose a novel method to extract deep features for image representation.

The contributions of this paper are:

(1) We propose a sparse deep representation by spatial aggregating multi-scale hierarchies from DCNN model for rich semantic representation.
(2) We take advantage of the generalization ability of DCNN, and extract convolution activations from different layers with spatial weighting that achieves the trade-off between low-level details and high-level rich semantic DCNN activations.
(3) A feature map selection strategy, which utilizes the meaningful activations on convolution layers is introduced.

2   Related Works

In the literature, there are two major methods to apply DCNN in specific tasks. One is an end-to-end optimizing strategy that directly trains or fine-tunes a model on large-scale datasets. Another is to utilize a pre-trained model as a global feature extractor [11,12] or a local feature extractor [7]. Since Krizhevsky et al. [1] achieve impressive performance in large-scale image classification using DCNN, many kinds of network architectures have been proposed, such as GoogleNet [13], VGG [14], ResNet [15]. Because there are millions of parameters to be tuned, the end-to-end training needs enormous training dataset and expensive computing hardware like GPU. A whole optimizing process usually takes days even weeks and lots of trick are needed during training. However, in real world applications, labeled image data are scarce and expensive which cannot meet the required amount for CNN training.

Recently, the second method has drawn much attention for image representation. Deep feature activations extracted from a pre-trained CNN model have been successfully utilized as a general feature extractor for image representations. CNN activations are extracted for classifier training and have been successfully applied in various image recognition tasks. To get a generic representation, after a series of convolutional filtering and pooling, usually the neural activations from first or second fully connected layers are extracted from a pre-trained CNN model.

However, the deep activations from the fully connected layers are sensitive to spatial transformations such as global translation, rotation, and scaling. In order to enhance the invariance of deep feature, some encoding methods are applied. Gong et al. [9] propose to pool multi-scale deep features to improve the rotation invariance of deep representations. Yoo et al. [10] propose to apply Fisher Vector to encode multi-scale CNN descriptors. These strategies require dense CNN feature extraction on multi-scale images, and the applicability of very deep architectures will be fundamentally limited due to memory constraints. Mohedano et al. [16] propose a bag of words based CNN feature encoding method to generate high-dimensional sparse image representations. Very recently, Salvador et al. [17] propose to extract convolutional features from both global and local scale. They take advantage of the object proposals learned by a Region Proposal Network (RPN). In [18] CNN features are extracted from a set of image windows for better representations. However, only the activations from the first fully connected layer fc7 were utilized.

As studied in [19,20], the ensemble of features from different layers could boost the performance. A DCNN network contains multiple levels of image abstraction, and it can be seen as rich semantic feature hierarchies. Hariharan et al. [19] take the advantage of features from different layers with various scales for semantic segmentation. We note that the aforementioned CNN feature extractors only exploit the features of one single layer and ignore the selectivity and sparse characteristic of CNN activations [21]. In this paper, we extract CNN activations from multiple layers based on pre-trained model, and in-layer feature map selection. Pooling strategy and a layer wise spatial weighting are proposed.

3   Proposed Method

Our main idea is to represent an image into a compact and spatial invariant representation. Instead of extracting features from the fully connected layers, we focus on the intermediate convolution layers.

Compared with the activations from the fully connected layers, the convolutional features are embedded with more spatial information. In this section, we start by introducing the DCNN model applied in our work and the convolution features among each layers, and then describe the in-layer feature map selection method and layer-wise spatial weighting and pooling strategy. To encode the extracted CNN features for classification, we adopt VLAD to aggregate the CNN descriptors into a compact representation.

3.1 Convolutional Neural Network Activations

Our approach is applicable to various convolutional neural network architectures. The main objective of the DCNN is to extract high level semantic representation of the input images. Recently, a number of DCNN pre-trained models have been released and most of them are publicly available. We experiment with the well known VGG-16 network [14] based on the Caffe framework [22], which has previously achieved the state-of-the-art accuracy on ILSVRC classification and localization tasks.

The network consists of 13 convolutional layers and 3 fully connected layers, and the convolutional layer is of 3×3 small convolutional filters and 5 max-pooling layers are embedded between convolutional layers. The detailed parameters of the network architecture are presented in Table 1.

On each convolutional layer l, a convolution operation of its M_{l-1} input maps from previous layer l-1, with a filter of size $k_l \times k_l$ is conducted. The resulting output is the summations of the responses with a non-linear function:

$$F_j^l = \sigma \left(\sum_{i=1}^{M^{l-1}} F_i^{l-1} * W_{ij}^l + b_j^l \right) \tag{1}$$

Table 1. Details of the feature maps from VGG-16

Layer	Output size		
	Width	Height	Depth
conv1_1	224	224	64
conv1_2	224	224	64
conv2_1	112	112	128
conv2_2	112	112	128
conv3_1	56	56	256
conv3_2	56	56	256
conv3_3	56	56	256
conv4_1	28	28	512
conv4_2	28	28	512
conv4_3	28	28	512
conv5_1	14	14	512
conv5_2	14	14	512
conv5_3	14	14	512

where l indicates the layer, F is the feature maps and W^l is a filter of size $k_l \times k_l$. b indicates the bias, and $\theta(\cdot)$ is the ReLU (Rectified Linear Unit) function:

$$\theta(x_i) = \begin{cases} x_i, & if \ x_i > 0 \\ 0, & if \ x_i \leq 0 \end{cases} \tag{2}$$

The feature map indicates the convolution results of one channel, which can reveal the pattern of distribution of the neural activity. Suppose I is an image of size H $\times$ W as the input of a DCNN, the extracted features from a convolution layer are formulated as an order-3 tensor T with size of $h \times w \times d$ [23], which includes a set of 2-D feature maps $\mathbf{S}= (S_1, S_2,...,S_n)(n{=}1,...,d)$. S_n of size h $\times$ w is the nth feature map of the corresponding channel as illustrated in Table 1. Different from [7,23], we derive deep representations from the distribution of neural activations. These feature maps from different layers can be seen as the high-level non-linear projections of the input image. By applying the pre-trained VGG-16 model, we extract the feature maps from low-level to high-level convolution layers.

In Fig. 2, we present five images from large-scale ImageNet dataset [3]. We visualize the feature maps from lower level (conv1_1) to higher level (conv5_1) of the convolutional layers, and we will release the deep feature map visualization toolbox in the future.

From the visualization of these feature maps, we can see that the lower convolutional layers tend to capture detailed spatial information, and the features in the higher layers are more abstract. Same as the observations from [24], a lower layer provides more detailed local features and higher layer captures more abstract and high-level semantic features. They are strong at distinguishing objects of different classes.

3.2   In-Layer Feature Map Selection

Based on the observations from the previous results, we find that not all the feature maps contain useful information. Same as [8], a feature map is usually sparse and indicates some semantic regions. The patterns of the activation distribution on a feature map remain quite amount of redundancy.

To delve deep for the useful representations, we carefully exam the patterns among the feature maps from lower to higher convolutional layers. Some patterns of the neurons located in a feature map are sensitive to the main instance as illustrated in the first row of Fig. 3(b). While some neurons are interested in the background noise as illustrated in the second row of Fig. 3(b), and some are not activated, as shown in the last row of Fig. 3(b). The reason for this phenomenon might due to the pre-training on large-scale image dataset which covers a variety of objects, thus the neurons are interested in all kinds of objects. There are a lot of feature maps that are not related to the target object, and it is necessary to select useful feature maps. We propose a feature map importance ranking strategy based on the activation patterns of neurons, and adopt it for feature map selection. The first step is to quantify the importance of feature maps.

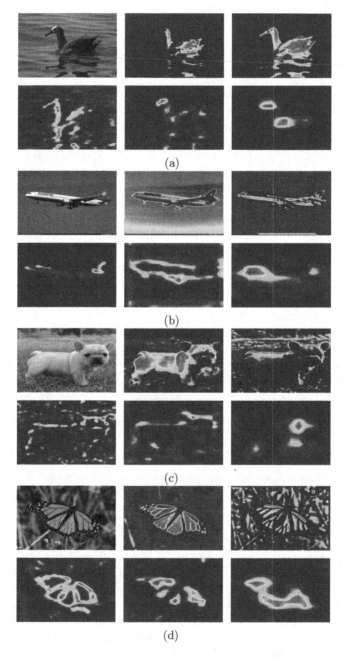

(a)

(b)

(c)

(d)

Fig. 2. Visualizations of randomly sampled feature maps from intermediate convolution layers of conv1_1, conv2_1, conv3_1, conv4_1 and conv5_1 of images from large scale ImageNet datasets, (a) bird, (b) plane, (c) french bulldog, and (d) butterfly.

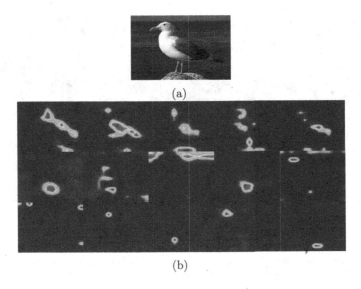

(a)

(b)

Fig. 3. Visualization results of selected feature maps from conv5_1 layer.

We use the classic image entropy [25] as the quantification method. Image entropy is a statistic pattern of features, that represents the average information of an image. The entropy indicates the distribution of gray-scale image values, and the entropy of a feature map S_i can be computed as:

$$H\left(S_i\right) = -\sum_{n=0}^{255} p_n \log_2\left(p_n\right) \tag{3}$$

where p_n means the probability of the gray-scale value n emerging within the image. All the elements within a feature map are first normalized to integer ranging 0–255. By computing the entropy of feature map, we can acquire the quantification of it. Moreover, standard deviation is also applied in the composition of feature importance, as it indicates the degree of dispersion within a feature map. The standard deviation of a feature map S_i can be acquired as:

$$\sigma\left(S_i\right) = \sqrt{\frac{1}{N}\sum_{j=1}^{N}\left(x_j - \mu\right)^2} \tag{4}$$

where x_j means the value of the j-th element, N is the total number of elements and μ is the mean. By combining the two quantification methods, the importance of convolutional feature map S_i can be presented as:

$$M\left(S_i\right) = H\left(S_i\right) + \lambda\sqrt{\sigma\left(S_i\right)} \tag{5}$$

where λ is empirically set to be 0.01, and we compute the importance score of each extracted feature map. Based on the computed importance score, we sort

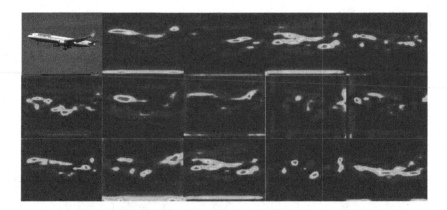

Fig. 4. Top ranking feature maps with high importance scores from conv5_1 layer.

all the feature maps from the same layers. Part of the sorting results of conv5_1 layer are illustrated in Fig. 4.

We select the top-Q feature maps from M convolutional feature maps. The selected feature maps contain most of the useful information while the depth is half of the unselected feature tensor. We then stack all the selected feature maps as the newly generated tensor, and apply the new tensor for the following feature pooling.

3.3   Deep Convolutional Local Descriptor Aggregation

After extracting deep descriptors from a convolutional layer, one can directly perform traditional max-pooling or sum-pooling to obtain the image-level representation. There are several methods for integrating features into compact high-level feature representation. For example, Bag-of-Features method [4] used an orderless collection of local features to represent an image. Benefiting from the power of local invariant features, such methods have achieved excellent performance. However, they still suffer from the drawback of the limited semantic description of local descriptors. Jégou et al. [5] introduced the VLAD, and this feature aggregation method was designed to be low dimension yet high discriminative.

Intuitively, we apply VLAD for the selected feature pooling. We first perform the deep convolutional feature map selection on the convolutional layer l, and the codebook $C^l = (c_1, c_2, ..., c_k)$ is generated by k-means clustering on the selected deep descriptor sets. When the clustering is finished, the centers are assigned as the visual words. The codebook is a $k \times D$ matrix, composed of k visual words with dimension D. Given an input image I, firstly, the selected feature map can be seen as a set of deep descriptors $X = (x_1, x_2, ..., x_n)$. Then, each descriptor x_n is associated with its nearest visual word $c_n = NN(\text{xn})$, and NN indicates the nearest neighbor search. The nearest $c(x_n)$ can be indexed by Eq. 6, where $d|\cdot|$ denotes the distance between two features.

$$c(x_n) = \arg\min_{ci} d\,|c_i, x_n| \qquad (6)$$

VLAD encodes feature x_n by considering the residuals:

$$v_i^l = \sum_{s.t. \ NN(x)=c_i} x_n - c_i \tag{7}$$

Then the residuals are stacked together to obtain the final vector:

$$\omega^l(I) = \left[v_1^l, ..., v_k^l \right] \tag{8}$$

3.4 Layer-Wise Spatial Weighting

A deep convolutional neural network can be seen as a hierarchical feature extractor, and each convolutional feature map is the non-linear projection of the input image. From our point of view, the whole deep structure is a multi-scale feature pyramid. The features in higher level convolutional layers preserve rich semantic information and lower level layers preserve more local and instance-level characteristics of objects. Adding spatial information through a spatial pyramid [26] has been shown to significantly improve image recognition performance when dense SIFT features are used, and we want to take all the advantages from global representation and local deep descriptors.

After the layer-wise deep descriptors aggregation, all the deep activations $X = (x_1, x_2, ..., x_n)$ within each convolutional layer l can be represented by a feature vector $\omega^l(I)$. Based on the aforementioned observations, we propose to pool the multi-scale deep feature hierarchies of input image I as:

$$\pi(I) = [\alpha^1 \cdot \omega^1(I), \alpha^2 \cdot \omega^2(I), ..., \alpha^l \cdot \omega^l(I)]$$
$$s.t. \ \sum_i^L \alpha^i = 1 \tag{9}$$

where the weight α^i depends on the contribution of the i-th convolutional layer. Note that this weighting strategy takes place in multi-scale space, and the final pooled representation $\pi(I)$ is the simple pooling of the spatial weighted feature vector among lower to higher layers. In this paper, we apply a set of convolutional layers (conv5_1, conv5_2 and conv5_3) for compact image representation generation. And we consider all the layers to contribute equally to the final representation. The pipeline of the proposed method is summarized in Algorithm 1.

4 Experiments

We evaluate our proposed method using publicly available classification dataset. First we perform an in-depth evaluation of the performance of the proposed feature aggregation method, and comparison with other related methods. Then, we conduct exploration experiments to determine some crucial factors. In the following, we evaluate the feature maps selection and different layers successively.

Algorithm 1. *SMSC* feature generation

Input: Pretrained model, image I
Output: *SMSC* feature vector $\pi(I)$
Procedure:
1. Extract deep feature maps from layer l $S=[S_1, ..., S_i, ...S_d]$
2. Compute importance of each feature maps $M(S_i)=H(S_i)+\lambda \sqrt{\sigma(S_i)}$
3. Select Top-ranked Q feature maps
4. Extract deep descriptors from the feature map tensor $X = (x_1, x_2, ..., x_n)$
5. k-means clustering for codebook $C = (c_1, c_2, ..., c_k)$
6. Aggregating deep descriptors
    for i=1 to n **do**
      $t = $ index argmin $d|c_j, x_i|$, $j\in\{1, 2, ...,k\}$
      $v_t = v_t + (x_i - c_t)$
    end for
  $\omega=[v_1, v_2, ..., v_k]$
7. Layer-wise spatial pooling of multi-scale vepctors
  $\pi(I) = [\alpha^1 \cdot \omega^1(I), \alpha^2 \cdot \omega^2(I),..., \alpha^l \cdot \omega^l(I)]$
Return: $\pi(I)$

4.1 Dataset

MIT Indoor [27], also known as MIT-67, is used for scene classification task. The dataset contains 15,620 images with 67 indoor scene classes in total.

Sample images are shown in Fig. 5. The classification accuracy is the evaluation criteria for the performance. It is a challenging task as it contains complex background and various transformations. The standard training and test split for the Indoor dataset consists of 80 training and 20 test images per class.

4.2 Experiment Details

Feature extraction and encoding. The DCNN model VGG-16 [14] consists 13 convolutional layers and it is pre-trained on ImageNet [3]. In this experiment we simply extract the conv5_1, conv5_2 and conv5_3 layers and select the feature maps for descriptor extraction. VLAD is a popular encoding strategy. To evaluate the performances, we extracted the selected deep descriptors from convolutional layers and apply VLAD encoding to aggregate these selected descriptors.

VLAD Centers and Normalization. In VLAD encoding, the number of centers k determines the final feature dimension. We perform feature map selection and deep descriptor extraction from convolutional layers in MIT-67 dataset to explore the best k. The distinctiveness of the generated features improves when k increases until 256. As a result, we fix the number of VLAD centers to 256 in our experiments to obtain good performances. We first apply component-wise l_2 normalization on each feature vector v_k to divide its norm $\|v_k\|_2$, and then use global l_2 normalization on the VLAD descriptor $\omega(I)$ by dividing its norm $\|\omega(I)\|_2$.

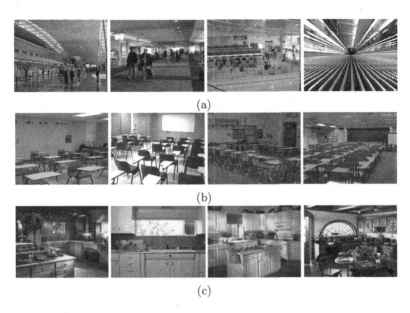

(a)

(b)

(c)

Fig. 5. Sample images from MIT-67 [27]. This classification task is challenging due to the complex environment and variant transformations.

Linear Classifier. In image classification, the generated feature dimension is usually very high. So we apply one-versus-all multi-class linear Support Vector Machine (SVM) as the classifier. The LIBLINEAR [28] implementation is adopted in our experiments. As to the parameter C, we set it to 0.1. Our experiments use the following open source libraries: VLFeat [29], Caffe [22] and LIB-LINEAR [28].

4.3 Classification Results

We evaluate the classification results on the MIT-67 dataset. During the training and testing procedure, we follow the standard image dataset splits provided by [27]. The classification results are shown in Table 2. Note that CNN × t means that t times of CNN forward calculation is performed. From the results above, we can see that our proposed method can get promising performance which is compatible to the state-of-the-art methods [8]. The DCNN based methods have outperformed the traditional methods which is based on hand-crafted features. Due to the limited number of labeled training data, end-to-end training method [32] is 4.9% inferior to our method. Directly extracting the activations from the fully connected layer for SVM training is not the best method (58.40%). From our point of view, the activations from the fully-connected layers are sensitive to spatial transformations and the images in MIT-67 share large amount of global transformations. Discovering the treasure from the convolutional layers [8] can be a useful strategy for better feature extraction.

Table 2. Classification results on MIT-67.

Method	Accuracy	Details
SPM [26]	34.40%	Baseline
FV+Bag of parts [30]	63.18%	Dim.221,550
DPM [31]	37.60%	-
VLAD multi-scale [9]	66.12%	Dim.4096
VLAD level 2 [9]	65.52%	Dim.4096
MOP-CNN [9]	68.88%	Dim.12288 CNN×53
Fine-tuning [32]	66.00%	AlexNet
CNN-FC-SVM [11]	58.40%	Dim.4096
CL+CNN-Jitter [8]	71.50%	CNN×15
SMSC5-1	65.52%	VGG-16
SMSC5-2	69.05%	VGG-16
SMSC5-3	68.33%	VGG-16
SMSC [5-1,5-2,5-3]	**70.9%**	VGG-16

Our method is 0.6% slightly inferior to the CL+CNN-Jitter [8] (71.50%). However, we extract all the useful features through only one forward calculation, and the methods in [9] and [8] take ×53 and ×15 forward calculations respectively. Compared with the time cost our method greatly outperforms the multiple forward strategy, which is efficient and simple for deployment. Our main differences to the method in [8] is that we take the moderate sparsity nature of deep features into consideration, and apply feature map selection before the feature encoding (pooling).

From our observation, not all the activations with a feature map contribute to the final representation of an image. Much irrelevant activations will have a negligible effect to the performance. Our proposed method could be a better solution for the trade-off between computation cost and recognition performance.

4.4   Different Number of Selected Feature Maps

We first perform layer-wise importance ranking of the feature maps from convolutional layers, and select top-Q feature maps for the convolution descriptor extraction. The selection of different Q can slightly affects the classification results. The ranking results of sample images are shown in Fig. 6 and the classification results are illustrated in Fig. 7.

From the results in Fig. 6(b) we find that the top ranked feature maps contain rich information while the neural activations of low importance feature maps are most zero, as shown in Fig. 6(c). And best number of Q is 256, which is the half number of the total convolutional feature maps. And the results confirm the observations that the deep neural activations are moderate sparse.

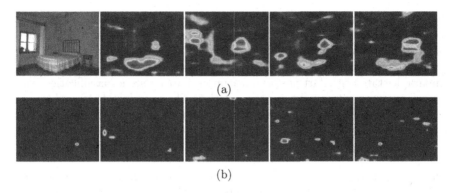

(a)

(b)

Fig. 6. Feature map ranking results, (a) input image and top ranked feature maps, (b) feature maps with lower importance score.

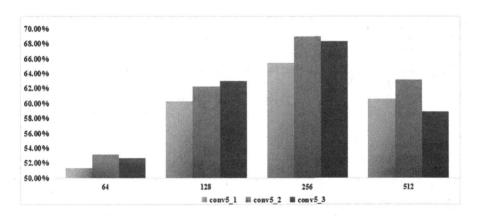

Fig. 7. Results of different selected feature maps number Q

The classification results also reveal the difference between convolutional layers from different depth. Deep descriptors extracted from conv5_2 layer are slightly better than that from other two layers (conv5_1 and conv5_3). The higher layers contain rich semantic information while the lower layers contain local detailed information. The intermediate layers tend to achieve the trade-off between high-level semantic and low-level detailed information. After single layer representation is encoded, our spatial weighting method will fuse these representations into a more compact feature vector, the performance can be improved by 1.85% compared with the best single layer performance (69.05%).

5   Conclusion

In this paper, we delve deep into the convolution layers, and dig the multi-scale hierarchies for compact image feature aggregation. Our primary goal is to extract all the useful information from low-level to high-level hierarchies, and combine them to generate a more powerful and robust representations. We propose a Selected Multi-Scale Convolution feature generation method, and a convolutional feature map selection and encoding strategy is introduced. Our method achieves comparable state-of-the-art performance on MIT-67 for scene classification based on single network with one forward calculation. The experiments have shown the effectiveness of the proposed method. Future work will focus on integrating the feature map selection and deep descriptor aggregation strategy into an end-to-end training procedure, and speeding up the processing time for real-time use.

Acknowledgement. This work was supported in part by the National Natural Science Foundation of China under grant number 61401154, by the Natural Science Foundation of Hebei Province under grant number F2016502101, and by the Fundamental Research Funds for the Central Universities under grant number 2015ZD20.

References

1. Krizhevsky, A., Sutskever, I., Hinton, G.E.: Imagenet classification with deep convolutional neural networks. In: Advances in Neural Information Processing Systems, pp. 1097–1105 (2012)
2. Chatfield, K., Simonyan, K., Vedaldi, A., Zisserman, A.: Return of the devil in the details: delving deep into convolutional nets. arXiv preprint arXiv:1405.3531 (2014)
3. Deng, J., Dong, W., Socher, R., Li, L.-J., Li, K., Fei-Fei, L.: ImageNet: a large-scale hierarchical image database. In: CVPR (2009)
4. Sivic, J., Zisserman, A.: Video google: a text retrieval approach to object matching in videos. In: ICCV (2003)
5. Jegou, H., Douze, M., Schmid, C., Perez, P.: Aggregating local descriptors into a compact image representation. In: CVPR (2010)
6. Perronnin, F., Sánchez, J., Mensink, T.: Improving the fisher kernel for large-scale image classification. In: Daniilidis, K., Maragos, P., Paragios, N. (eds.) ECCV 2010. LNCS, vol. 6314, pp. 143–156. Springer, Heidelberg (2010)
7. Gao, B. Bin Wei, X.S.: Deep spatial pyramid: the devil is once again in the details. arXiv preprint arXiv:1504.05277 (2015)
8. Liu, L., Shen, C., van den Hengel, A.: The treasure beneath convolutional layers: cross-convolutional-layer pooling for image classification. In: CVPR (2015)
9. Gong, Y., Wang, L., Guo, R., Lazebnik, S.: Multi-scale orderless pooling of deep convolutional activation features. In: Fleet, D., Pajdla, T., Schiele, B., Tuytelaars, T. (eds.) ECCV 2014. LNCS, vol. 8695, pp. 392–407. Springer, Heidelberg (2014)
10. Yoo, D., Park, S., Lee, J.Y., Kweon, I.: Multi-scale pyramid pooling for deep convolutional representation. In: CVPR Workshops (2015)
11. Razavian, A., Azizpour, H., Sullivan, J., Carlsson, S.: CNN features off-the-shelf: an astounding baseline for recognition. In: CVPR Workshops (2014)
12. Xiao, T., Xu, Y., Yang, K., Zhang, J., Peng, Y., Zhang, Z.: The application of two-level attention models in deep convolutional neural network for fine-grained image classification. In: CVPR (2015)
13. Szegedy, C., Liu, W., Jia, Y., Sermanet, P., Reed, S., Anguelov, D., Rabinovich, A.: Going deeper with convolutions. In: CVPR (2015)

14. Simonyan, K., Zisserman, A.: Very deep convolutional networks for large-scale image recognition. In: ICLR (2015)
15. He, K., Zhang, X., Ren, S., Sun, J.: Deep residual learning for image recognition. arXiv preprint arXiv:1512.03385 (2015)
16. Mohedano, E., Salvador, A., McGuinness, K., Marques, F., O'Connor, N.E., Giró-i-Nieto, X.: Bags of local convolutional features for scalable instance search. arXiv preprint arXiv:1604.04653 (2016)
17. Salvador, A., Giró-i-Nieto, X., Marqués, F., Satoh, S.I.: Faster R-CNN features for instance search. In: CVPR Workshops (2016)
18. Uricchio, T., Bertini, M., Seidenari, L., Bimbo, A.: Fisher encoded convolutional bag-of-windows for efficient image retrieval and social image tagging. In: CVPR Workshops (2016)
19. Hariharan, B., Arbelez, P., Girshick, R., Malik, J.: Hypercolumns for object segmentation and fine-grained localization. In: CVPR (2015)
20. Kulkarni, P., Zepeda, J., Jurie, F., Perez, P., Chevallier, L.: Hybrid multi-layer deep CNN/aggregator feature for image classification. In: IEEE International Conference on Acoustics, Speech and Signal Processing (ICASSP) (2015)
21. Sun, Y., Wang, X., Tang, X.: Deeply learned face representations are sparse, selective, and robust. In: CVPR (2015)
22. Jia, Y., Shelhamer, E., Donahue, J., Karayev, S., Long, J., Girshick, R., Darrell, T.: Caffe: convolutional architecture for fast feature embedding. In: Proceedings of the ACM International Conference on Multimedia (2014)
23. Wei, X.S., Luo, J.H., Wu, J.: Selective convolutional descriptor aggregation for fine-grained image retrieval. arXiv preprint arXiv:1604.04994 (2016)
24. Wang, L., Ouyang, W., Wang, X., Lu, H.: Visual tracking with fully convolutional networks. In: ICCV (2015)
25. Gonzalez, R.C., Woods, R.E., Eddins, S.L.: Digital Image Processing Using MATLAB. McGraw Hill Education, New York City (2010)
26. Lazebnik, S., Schmid, C., Ponce, J.: Beyond bags of features: spatial pyramid matching for recognizing natural scene categories. In: CVPR (2006)
27. Quattoni, A., Torralba, A.: Recognizing indoor scenes. In: CVPR (2009)
28. Fan, R.E., Chang, K.W., Hsieh, C.J., Wang, X.R., Lin, C.J.: LIBLINEAR: a library for large linear classification. J. Mach. Learn. Res. 9, 1871–1874 (2008)
29. Vedaldi, A., Fulkerson, B.: VLFeat: an open and portable library of computer vision algorithms. In: Proceedings of the 18th ACM International Conference on Multimedia (2010)
30. Juneja, M., Vedaldi, A., Jawahar, C.V., Zisserman, A.: Blocks that shout: distinctive parts for scene classification. In: CVPR (2013)
31. Pandey, M., Lazebnik, S.: Scene recognition and weakly supervised object localization with deformable part-based models. In: ICCV (2011)
32. Azizpour, H., Razavian, A., Sullivan, J., Maki, A., Carlsson, S.: From generic to specific deep representations for visual recognition. In: CVPR Workshops (2015)

Glance and Glimpse Network: A Stochastic Attention Model Driven by Class Saliency

Mingming Li, Shuzhi Sam Ge$^{(\boxtimes)}$, and Tong Heng Lee

Department of Electrical and Computer Engineering, The Social Robotics Lab,
Interactive and Digital Media Institute (IDMI), National University of Singapore,
Singapore 117576, Singapore
samge@nus.edu.sg

Abstract. We present a hybrid model named Glance and Glimpse Network (GGNet) for visual classification, which includes an attention-based recurrent neural network (Glimpse Network) and a convolutional neural network (Glance Network). The Glimpse Network is trained to deploy a sequence of glimpses at different image patches and then output classification results. On the other hand, the Glance Network is designed to take the downsampled input image and generates an image-specific class saliency map to provide hints for training the Glimpse Network. We show that training the Glimpse network with such cues can be interpreted under both frameworks of probabilistic inference and reinforcement learning, therefore establishing high-level connections between these two separate fields. We evaluate the performance of our model on Cluttered Translated MNIST benchmark datasets and show that the GGNet can achieve the state-of-the-art results compared to other recently proposed attention models.

1 Introduction

Convolutional neural network (CNN) has recently achieved substantial success in tasks of visual recognition and classification [1–3]. However, one of the common features of these successful models is the usage of increasingly deep architecture [2,3], which inevitably brings expensive computation cost that grows linearly with the image resolution as CNN has to process every region of an image at a uniform scale. This drawback has motivated recent researches in attention-based network models that are trained to selectively process different parts of an image, which shows several appealing advantages over traditional CNNs in terms of both efficiency and interpretability [4–10]. The mechanism of attention allows the network to focus its effort locally on a sequence of image patches and therefore the number of network parameters and computational operations can be controlled independently to the image resolution. On the other hand, the focus of the network also indicates where and what the network is looking at, increasing the interpretability of the model as it shows what specific information the model has taken into account before determining its final output [8].

The mechanism of attention models is analog to that of humans when they are perceiving an image cluttered with rich visual cues, e.g. a complex scene. That is,

© Springer International Publishing AG 2017
C.-S. Chen et al. (Eds.): ACCV 2016 Workshops, Part III, LNCS 10118, pp. 572–587, 2017.
DOI: 10.1007/978-3-319-54526-4_42

the visual information is first collected through a sequence of glimpses focusing on different attractive parts of the image and then aggregated to produce the final output. Formally, a glimpse is crop of the image and therefore it is of limited horizon, which means only a fraction of the whole image is attended. Note that apart from observing an image carefully via glimpses, humans can also take a quick glance to have an overall understanding of the image's visual context, which can provide guidance to task-relevant spatial attention during perception [11]. In the community of computer vision, these cues are usually modeled as visual saliency [12,13]. However, saliency is mainly drawn in a bottom-up manner and according to low-level features such as local contrast. Thus, it cannot provide much insight to locate the potentially informative regions, which are strongly task-specific. This motivates us to make use of the class saliency [14] obtained in a top-down manner to model the glance process and provide hints to learn an effective glimpse strategy for our attention model. To this end, we propose a hybrid neural network model called Glance and Glimpse Network (GGNet), which combines (1) a lightweight CNN (Glance network) providing cues on the possibly informative regions and (2) a deep stochastic recurrent network (Glimpse network) trained to actively deploy a sequence of glimpsess and then classify the image. Note that though a CNN is employed, the drawbacks of traditional CNN can be avoided as the glance network can be kept small by firstly down-sampling the high-resolution input into a coarse one, which is actually similar to human's quick glance behavior.

Training the Glance network is as straightforward as training a typical supervised CNN and class saliency map can be generated efficiently based on the image gradients obtained from a single forward and backward propagation [14]. However, optimizing the Glimpse network is difficult as it requires computing gradients with high-dimensional latent variables and intractable posteriors. We will show in the following sections that, by making use of the cues from the class saliency map, an end-to-end training algorithms for the Glimpse network can be derived under either the framework of probabilistic inference or reinforcement learning. Particularly, the Glance network serves as a teacher to the Glimpse network by providing hints on whether a certain glimpse sequence has covered the potentially informative regions that need to be attended. By viewing our training algorithm from two independent learning frameworks, we further address the relationship between probabilistic inference using importance sampling [15] and policy optimization via REINFORCE [16], which are two mainstream training frameworks for the recent attention models [4–6,10].

We evaluate the performance of our GGNet on the benchmark dataset Cluttered Translated MNIST (CT-MNIST) [4]. The results suggest that our model outperforms the original Recurrent Attention Model [4] with a wide margin and achieves the state-of-the-art performances compared to other recently proposed attention models [7,10].

2   Related Work

Neural networks with attention mechanism can be classified into two categories: the soft and hard attention models. The soft attention models [7–9] adopts a differentiable attention mechanism by applying learnable filters (e.g. Gaussian filters in Differentiable Recurrent Attention Model (DRAM) [7]) onto the whole image to extract the corresponding attended region. The advantage of using soft attention model is that learning the attention mechanism can be easily integrated into the standard back propagation framework due to its differentiability. However, it also requires processing the whole image in each time, which inevitably decreases the computational efficiency during learning and inference as the traditional CNN does. Conversely, the hard attention models learn to select a sequence of glimpses in a discrete and stochastic manner [4–6,10]. This hard mechanism has higher efficiency during inference as only the glimpsed region is processed in each time. Though intuitively simple, the training of hard attention model is indeed nontrivial as the error gradient cannot flow through the discrete glimpse selections and thus the standard back propagation paradigm cannot work as a plug-in method. Instead, the objective function is first marginalized over the high-dimensional space of glimpse selections. Then, its stochastic gradient estimation is obtained by different sampling methods [4–6,10].

Our proposed GGNet model adopts a hard attention mechanism for the Glimpse network, which is a recurrent network that learns to deploy a glimpse at a certain part of the image at each recurrence. Particularly, our Glimpse network is closely related to the Recurrent Attention Models (RAM) proposed in [4] but uses a novel training algorithm to learn the attention mechanism from the cues provided by the Glance network. The training of RAM [4] is formulated as a glimpse policy optimization problem in the context of reinforcement learning, where a positive reward will be given to the model if it makes a correct prediction and therefore the objective is to maximize the expected reward. The gradient of the glimpse policy is approximated via REINFORCE rule, which is a Monte Carlo sampling-based method. In [10], the authors propose the Decoupled Controller-recognizer Model (DCRM), whose attention and classification policies are learned by the two decoupled modules called "controller" and "recognizer", respectively. In [5], another recurrent attention model of structure similar to RAM is proposed and it is trained via variational inference, which tries to maximize the variational lower bound of the original objective function. The authors of [5] show that the gradient estimation derived by variational inference is equivalent to that in RAM [4] if the highly variant log likelihood in the gradient approximation is replaced with a discrete reward indicator. In fact, such relationship between variational inference and reinforcement learning is investigated with a more general formulation [17], which shows that the concepts and ideas in these two separate fields can be regarded as counterparts to each other. However, as analyzed in [6,18], the Monte Calro gradient estimation in [4,5] can be ineffective when the gradient samples are drawn from a location prior that may not be well-conditioned, especially in the early stage of learning. As a remedy, [6] proposes a new training algorithm to directly optimize the performance of their recurrent attention model based on the reweighted wake-sleep (RWS) approach, which uses a separate inference network to learn the proposal distribution for constructing an importance sampling-based gradient estimator. As a result, the attention model tends to learn from the sampled glimpses that carry more relevant information.

3   The Glance and Glimpse Network

Our proposed Glance and Glimpse network is trained end-to-end but it has a switching structure between training and inference. During inference, an input image $I \in \mathbb{R}^{m \times n \times c}$ is fed only to the Glimpse network for classification, while in training, the input and its label are fed to both subnetworks so that the Glance network can act as a teacher to the Glimpse network with the knowledge of the true label. The network structure is illustrated in Fig. 1.

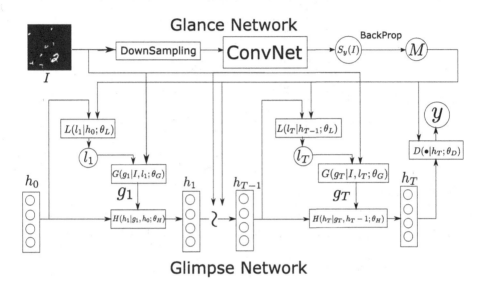

Fig. 1. Network model: the proposed network model is a hybrid of a CNN (Glance network) and a RNN (Glimpse Network). During training, an input image I will be fed to these two network in parallel. The Glance network takes a down-sampled coarse input $I' \in \mathbb{R}^{m' \times n' \times c}$ and outputs an image-specific class saliency map $M \in \mathbb{R}^{m \times n}$ after a forward and backward process. This map is used to generate signals to train the Glimpse network so that it learns to sequentially deploy its glimpse sensor g at different locations $l_1, \cdots, l_T$. Finally, the Glimpse network will make a prediction of the input label y through a prediction network $D(\bullet|h_T; \theta_D)$ after the last recurrence of the network.

In the following, we will elaborate the detailed architecture and purposes of the components in Fig. 1.

Glance Network: The Glance network is optimized by minimizing the Cross Entropy Loss with input image downsampled by a factor $d = \frac{m}{m'}$ using average pooling. The structure of the Glance network and the process of generating an image-specific class saliency map are shown in Fig. 2.

For an input image I_0 with class label y, we consider its class scores $S_y(I_0)$ as the value return by the 2-layer fully connected (FC) network in Fig. 2. According

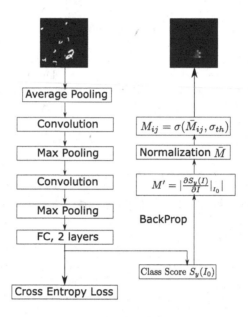

Fig. 2. Glance network

to [14], an image-specific class saliency map can be computed by taking the absolute value of the derivative of $S_y(I_0)$ with respect to the input image I_0, i.e.,

$$M' = |\frac{\partial S_y(I_0)}{\partial I}|_{I=I_0}| \in \mathbb{R}^{m \times n \times c} \qquad (1)$$

As only a single class saliency value is expected for each pixel (i, j), only the maximum value across the c channels will be taken and they are further normalized with its maximum element $\max_{i,j,c} M_{ijc}$ as

$$\bar{M}_{ij} = \frac{\max_c M'_{ijc}}{\max_{i,j,c} M'_{ijc}} \qquad (2)$$

Then, a threshold function $\sigma(x, \sigma_{th})$ is applied element-wisely to $\bar{M}$ and the class saliency map M is obtained with $M_{ij} = \sigma(\bar{M}_{ij}, \sigma_{th})$ so as to filter out the less salient pixels.

$$\sigma(x, \sigma_{th}) = \begin{cases} x, \text{if } x > \sigma_{th} \\ 0, \text{if } x < \sigma_{th} \end{cases} \qquad (3)$$

Glimpse Network: The Glimpse network extends the recurrent structure of RAM [4]. It is a RNN $H(h_t|g_t, h_{t-1}; \theta_H)$ that, at each time step t, takes the output g_t of Glimpse sensor $G(g_t|I, l_t; \theta_G)$ and updates its internal state vector h_t. At the next time step $t+1$, h_t is output to a locator network $L(l_{t+1}|h_t; \theta_L)$, which is stochastic unit that outputs the Cartesian coordinates of a location

$l_{t+1} \in \mathbb{R}^2$ according to a parameterized probability distribution, i.e., $l_{t+1} \sim p(\bullet | L(h_t; \theta_L))$. At the last time step T, the internal state h_T is output to a prediction network $\log p(y|I) = D(\bullet | h_T; \theta_D)$ to predict the label of the input I. Specifically, the modules mentioned above are described as follows:

Glimpse Sensor: Given an image I and a location l_t, the glimpse sensor will extract multi-resolution foveation-like representations of a set of square sub-patches $\chi = \{x_1, x_2, \cdots, x_n\}$ centered at l_t, whose widths are determined as $w_i = w \times \delta^{i-1}$ and δ is the scale to the size of different subpatches. These sub-patches χ and the glimpse location l_t are encoded into two feature vectors h_g and h_l with parameters θ_{G1} and θ_{G2}, respectively. Finally, they are combined as the glimpse feature vector g_t with parameters θ_{G3} and therefore g_t is formulated as $g_t = G(g_t | I, l_t; \theta_G)$, with $\theta_G = [\theta_{G1}^T, \theta_{G2}^T, \theta_{G3}^T]^T$.

Locator Network: The locator network $L(l_{t+1} | h_t; \theta_L)$ is a stochastic network that outputs the glimpse location l_t according to a learnable Gaussian distribution parameterized as $l_t \sim \mathcal{N}(\mu, \Sigma)$, where $\mu = \bar{L}(\mu | h_{t-1}; \theta_L)$ is the mean learned from $\bar{L}(\mu | h_{t-1}; \theta_L)$, a fully-connected layer with parameter θ_L and input h_{t-1}. Σ is the fixed covariance matrix.

Internal State: The internal state h_t of the network is updated through a recurrent network $H(h_t | g_t, h_{t-1}; \theta_H)$ with one linear layer of parameters θ_H. By the nature of RNN, it aggregates information collected from each glimpse.

Prediction Network: The prediction network $D(\bullet | h_T; \theta_D)$ is a simple fully connected feed-forward network with 2 linear layers and a LogSoftMax layer to predict the label of an input I.

4   Training

As described in the last section, our Glance network is simply trained to minimize the cross entropy loss given the true label y of a training sample. On the other hand, the Glimpse network has to learn a policy of allocating its glimpses at the most informative locations and make a prediction on the label y based on the collected information. In the following, we first derive the learning rule of Glimpse network by solving a probabilistic inference problem. This result will then be interpreted from the angle of policy optimization with reinforcement learning, which extends the results in [5,17] by further establishing links between reinforcement learning and direct probabilistic inference.

4.1   Learning as a Probabilistic Inference Problem

In this subsection, the training algorithm of the Glimpse network is derived in the context of a supervised classification problem with only the class labels y available, as it is best tailored to the nature of category recognition. Specifically, we aim to maximize the log probability $J = \log(y|I; \theta)$ of the correct class label by marginalizing it over all possible glimpse sequences:

$$J = \log(y|I; \theta) = \log \sum_{1} p(1|I; \theta) p(y|1, I; \theta) \tag{4}$$

where $1 = [l_1, \cdots, l_T]^T$ is the glimpse location vector and $\theta = [\theta_H^T, \theta_G^T, \theta_L^T, \theta_D^T]^T$ is the vector of model parameters. Then, we differentiate the objective function in Eq. (4) with respect to the model parameters and obtain

$$\nabla_\theta J = \frac{1}{p(y|I; \theta)} \nabla_\theta \sum_{1} p(1|I; \theta) p(y|1, I; \theta) \tag{5}$$

$$= \sum_{1} \frac{p(1|I; \theta) p(y|1, I; \theta)}{p(y|I; \theta)} [\nabla_\theta \log p(y|1, I; \theta) + \nabla_\theta \log p(1|I; \theta)] \tag{6}$$

$$= \sum_{1} p(1|y, I; \theta) [\nabla_\theta \log p(y|1, I; \theta) + \nabla_\theta \log p(1|I; \theta)] \tag{7}$$

where $p(1|y, I; \theta)$ is the posterior. Note that in Eq. (7) both the summation and the posterior are intractable to evaluate as the number of possible glimpse sequences is exponential. Thus, a practical way to estimate the gradients $\nabla_\theta J$ is to use importance sampling [15]. We choose the conditional prior $p(1|I; \theta)$ as the distribution proposal for importance sampling. With this, a sampling-based estimation of $\nabla_\theta J$ is obtained as

$$\nabla_\theta J \approx \frac{1}{S} \sum_{i=1}^{S} R^i [\nabla_\theta \log p(y|1^i, I; \theta) + \nabla_\theta \log p(1^i|I; \theta)] \tag{8}$$

where $1^i \sim p(1|I; \theta)$ and

$$R^i = \frac{p(1^i|y, I; \theta)}{p(1^i|I; \theta)} = \frac{p(y|1^i, I; \theta)}{p(y|I; \theta)} \tag{9}$$

is the importance weights of the i-th samples drawn from the distribution $p(1|I; \theta)$. One still need to get an estimation of Eq. (9) as it either contains $p(1^i|y, I; \theta)$ or $p(y|I; \theta)$, which are all intractable. While in [18] the term $p(y|I; \theta)$ is simply approximated as the average $\frac{1}{S} \sum_{i=1}^{S} p(y|1^i, I; \theta)$ to obtain an estimation of R^i, in this paper, we propose to directly estimate the ratio $\frac{p(y|1^i, I; \theta)}{p(y|I; \theta)}$ in a more task-specific manner by using the class saliency map M. To do this, we first split the two gradients into the following:

$$\nabla_\theta J \approx \frac{1}{S} \sum_{i=1}^{S} [R_y^i \nabla_\theta \log p(y|1^i, I; \theta) + R_l^i \nabla_\theta \log p(1^i|I; \theta)] \tag{10}$$

Then, R_y^i and R_l^i are determined using the class saliency maps obtained from the Glance network as follow:

$$R_y^i = \frac{p(y|1^i, I; \theta)}{p(y|I; \theta)} \approx \frac{\sum_{(i', j')}^{(i', j') \in \mathbb{D}(1^i, w)} M_{i'j'}}{\sum_{i'}^{m} \sum_{j'}^{n} M_{i'j'}} \tag{11}$$

$$R_l^i \approx \begin{cases} R_y^i, & \text{if } y = \arg\max_{y'} p(y'|\mathbf{l}^i, I; \theta) \\ 0, & \text{otherwise} \end{cases} \qquad (12)$$

where $\mathbb{D}(\mathbf{l}^i, w)$ denotes a region representing what a specific glimpse sequence located at $l_1^i, \cdots, l_T^i$ has focused on. Particularly, let $\mathbb{D}(l_t^i, w)$ denote the truncated horizon of a glimpse g_t centered at the location l_t^i, which is defined as

$$\mathbb{D}(l_t^i, w) = \{z || z - l_t|_\infty \le \frac{1}{2}w\} \qquad (13)$$

with $z = (i', j')$ representing the coordinates of an arbitrary element of M. Equation (13) corresponds to a square region where the glimpse sensor is deployed with full resolution. Then, the truncated horizons of different glimpses are merged as $\mathbb{D}(\mathbf{l}^i, w)$, representing the truncated glimpsed region:

$$\mathbb{D}(\mathbf{l}^i, w) = \bigcup_{t=1}^{T} \mathbb{D}(l_t^i, w) \qquad (14)$$

Intuitively, the class saliency map M can be regarded as a quantification of how much useful information a certain region (pixel) is likely to provide when classifying an input I as a specific category y, where pixels of higher saliency values represent more informative ones. Thus, (11) calculates a ratio between (1) the amount of informative resources collected after deploying the glimpse sequence with glimpse location $\mathbf{l}^i$ and (2) the total of informative resources available in the input I. As only the potentially informative regions can help predict the true label y, it is natural to see that the more a sequence of glimpses have seen, the closer $p(y|\mathbf{l}^i, I, \theta)$ is to likelihood $p(y|I, \theta)$, which is an ideal prediction the network makes after accessing all informative resources.

Note that in (12), $R_l^i = 0$ when $y \ne \arg\max_{y'} p(y'|\mathbf{l}^i, I; \theta)$, which means that the network makes a wrong prediction based on the glimpses located at $\mathbf{l}^i$. In this situation, the gradient update $\frac{p(y|\mathbf{l}^i, I;\theta)}{p(y|I;\theta)} \nabla_\theta \log p(\mathbf{l}^i|I; \theta)$ tends to learn the invalid glimpse policy and can introduce high variance to the gradient estimation (8), especially when the specific sampled glimpse locations $\mathbf{l}^i$ are off the object location. Thus, in (12), the sampled gradient $\nabla_\theta \log p(\mathbf{l}^i|I; \theta)$ will be discarded, which ensures that the locator network does not learn from the potentially invalid glimpse samples. Upon a correct prediction, the gradients will be scaled with a factor $R_l^i = R_y^i$ such that the locator network $L(l_t|h_t; \theta_L)$ is encouraged to deploy the glimpse sensor at areas that (1) have more informative resources indicated by M and (2) help predict the correct label.

On the other hand, all sampled gradients $\nabla_\theta p(y|\mathbf{l}^i, I; \theta)$ are simply multiplied with the importance weight R_y^i defined in (11) as it is generally well behaved [19]. The importance weights R_y^i can suppress the gradient when the sampled glimpse sequence does not provide much useful information and therefore encourage the network to focus its efforts on learning to predict the label from glimpses that are believed to be more informative.

Finally, as analyzed in [4,5,16,20], the convergence of the gradient ascent algorithm can be significantly speed up by introducing the baseline (or control variates in [6]) to reduce the high variance of the stochastic gradient estimations $R_l^i \nabla_\theta \log p(\mathbf{l}^i | I; \theta)$ in (10). Particularly, the baseline variable is chosen as the expectation of R_l, i.e., $b = \mathbb{E}[R_l]$, which is learnt by minimizing the square error $(R_l - b)^2$ using a simple feed-forward neural network with 1 linear layer, and the variance-reduced gradient estimation is obtained as:

$$\nabla_\theta J \approx \frac{1}{S} \sum_{i=1}^{S} [R_y^i \nabla_\theta \log p(y | \mathbf{l}^i, I; \theta) + (R_l^i - b) \nabla_\theta \log p(\mathbf{l}^i | I; \theta)] \quad (15)$$

Note that variance reduction is not needed for the first term in (15) as it well approximates the true gradient $\nabla_\theta \log p(y | \mathbf{l}, I; \theta)$. In addition, subtracting the baseline from R_y may also change the direction of the resulted gradient.

4.2 Interpretation with Reinforcement Learning

As investigated in [5,17], optimizing an attention model by variational inference with a reparameterized lower bound of the original objective function (4) can be mapped as a policy optimization problem using reinforcement learning when the learning process involves inference through latent variables (i.e., l) and stochastic gradient algorithms. On the other hand, the learning algorithm for our Glimpse network in last section is derived by directly maximizing the log likelihood $\log p(y | I; \theta)$. In this case, optimizing the stochastic Glimpse network can be regarded as a direct probabilistic inference problem without reparameterization of the objective function, which is a formulation that is also adopted in [6,18]. With these observations and motivations, we propose to explore the connections between direct probabilistic inference and reinforcement learning to further extend the results in previous attempts [5,17]. Particularly, in this subsection, the derived learning algorithm (15) for our Glimpse network will be reexamined under the framework of reinforcement learning.

When viewed in the context of reinforcement learning, the Glimpse network is considered as an agent subject to the partially observable Markov decision process (POMDP). The Glimpse network is supposed to actively execute a sequence of actions, which include choosing a sequence of glimpse locations $l_1, \cdots, l_T$ and predicting the label y based on the information aggregated from these glimpses. A pair of stochastic action policies $\pi_l(l_t | s_{0:t-1}; \theta)$ and $\pi_y(\hat{y} | s_{0:T-1}, l_T; \theta)$ are to be learnt by the agent, where s_t is the history of past actions up to time t, $\pi_l(l_t | s_{0:t-1}; \theta)$ corresponds to the policy of choosing glimpse location and $\pi_y(\hat{y} | s_{0:T-1}, l_T; \theta)$ is an action policy that predicts the label of the input as $\hat{y}$ based a sequence of T glimpses. In our Glimpse network model, s_t for $t = 0, \cdots, T-1$ is compactly summarized as $s_t = h_t$ and $s_T = \hat{y}$, which is the label predicted by the agent. After performing a certain action, a reward will be generated and given to the agent by the unobserved environment and the total reward $R(\hat{y}, \mathbf{l})$ is formulated as follows:

$$R(\hat{y}, \mathbf{l}) = r_y(\hat{y}) + \sum_{t=1}^{T} r_i(l_t) = r_y(\hat{y}) + r_l(\mathbf{l}) \quad (16)$$

where $r_y(\hat{y})$ is the final reward obtained after the agent makes a prediction according to the policy $\pi_y(\hat{y}|s_{0:T-1}, l_T; \theta)$, r_i corresponds to the instantaneous reward gained after deploying the glimpse sensor at a location l_t and r_l is the sum of all instantaneous reward. Given the true label y and the class saliency map M of the input, the two rewards signals are designed as:

$$r_l = R_y = \frac{\sum_{(i,j)}^{(i,j) \in \mathbb{D}(l,w)} M_{ij}}{\sum_i^m \sum_j^n M_{ij}} \tag{17}$$

$$r_y = \begin{cases} R_y, & \text{if } \hat{y} = y \\ -R_y, & \text{else} \end{cases} \tag{18}$$

where r_l is actually the approximation of the importance weight R_y we defined in (11). The learning objective is to maximize the expectation of the total reward $R(\hat{y}, \mathbf{l})$, i.e.,

$$J_1 = \mathbb{E}_{p(s_{0:T};\theta)}[R(\hat{y}, \mathbf{l})] = \mathbb{E}_{p(s_{0:T-1};\theta)}[r_l] + \mathbb{E}_{p(s_{0:T};\theta)}[r_y] \tag{19}$$

Intuitively, in order to maximize the total reward, the agent must (1) cover the most informative areas with T glimpses so as to get as much total instantaneous reward r_l as possible; (2) predict the correct label (i.e., $\hat{y} = y$) to get positive final reward. Otherwise, $r_y = -r_l$ and thus $R(\hat{y}, \mathbf{l}) = 0$, which means the agent will get zero reward at the end if it predicts a wrong label.

Taking the gradient with respect to the model parameters θ and considering the fact that $\nabla_\theta R(\mathbf{l}, y) = 0$, we have

$$\nabla_\theta J_1 = \mathbb{E}_{p(s_{0:T-1};\theta)}[r_l \sum_{t=1}^T \nabla_\theta \log \pi_l(l_t|s_{0:t-1}; \theta)]$$
$$+ \mathbb{E}_{p(s_{0:T};\theta)}[r_y \nabla_\theta \log \pi_y(\hat{y}|s_{0:T-1}, l_T; \theta) + r_y \sum_{t=1}^T \nabla_\theta \log \pi_l(l_t|s_{0:t-1}; \theta)] \tag{20}$$

An approximation to the policy gradients can be obtained using to the REIN-FORCE rule [16]:

$$\nabla_\theta J_1 \approx \frac{1}{S} \sum_{i=1}^S [(r_l^i + r_y^i) \sum_{t=1}^T \nabla_\theta \log \pi_l(l_t^i|s_{0:t-1}^i; \theta)$$
$$+ \mathbb{E}_{\pi_y}[r_y^i \nabla_\theta \log \pi_y(\hat{y}|s_{0:T-1}^i, l_T^i; \theta)]] \tag{21}$$

Let $\bar{y}$ denote an arbitrary wrong prediction, i.e., $\bar{y} \in \mathbb{Y}/y$, where $\mathbb{Y}$ is the set of all possible label predictions by $\pi_y(\hat{y}|s_{0:T-1}, l_t; \theta)$. According to (18), the expectation term in (21) can be approximated as

$$\mathbb{E}_{\pi_y}[r_y^i \nabla_\theta \log \pi_y(\hat{y}|s_{0:T-1}^i, l_T^i; \theta)] = R_y^i \pi_y(y|s_{0:T-1}^i, l_T^i; \theta) \nabla_\theta \pi_y(y|s_{0:T-1}^i, l_T^i; \theta)$$

$$- \sum_{\bar{y}} R_y^i \pi_y(\bar{y}|s_{0:T-1}^i, l_T^i; \theta) \nabla_\theta \pi_y(\bar{y}|s_{0:T-1}^i, l_T^i; \theta)$$

$$\approx R_y^i \nabla_\theta \pi_y(y|s_{0:T-1}^i, l_T^i; \theta) \tag{22}$$

Let $\bar{r}_l = r_l + r_y$ and introduce the reward baseline $b_1 = \mathbb{E}[\bar{r}_l]$ [4] to reduce the variance of the term $\bar{r}_l^i \nabla_\theta \log \pi_l(l_t^i | s_{0:t-1}^i; \theta)$. According to (22), the gradient approximation in (21) can be rewritten as:

$$\nabla_\theta J_1 \approx \frac{1}{S} \sum_{i=1}^{S} [(\bar{r}_l^i - b_1) \sum_{t=1}^{T} \nabla_\theta \log \pi_l(l_t^i | s_{0:t-1}^i; \theta) + R_y^i \nabla_\theta \log \pi_y(y | s_{0:T-1}^i, l_T^i; \theta)]$$

$$(23)$$

where

$$\bar{r}_l = \begin{cases} 2R_y, & \text{if } \hat{y} = y \\ 0, & \text{else} \end{cases} \tag{24}$$

Comparing (15) and (23), we can see that these two approximated gradients are essentially equivalent as $\nabla_\theta \log p(y | l^i, I, \theta) = \nabla_\theta \log \pi_y(y | s_{0:T-1}^i, l_T^i; \theta)$ and $\nabla_\theta \log p(l^i | I, \theta) = \sum_{t=1}^{T} \nabla_\theta \log \pi_l(l_t^i | s_{0:t-1}^i; \theta)$, which are the corresponding gradients computed by backpropagating the error/reward signals through the Glimpse network. The coefficient '2' in (24) can be regarded as a scaling factor of the locator network's learning rate.

The above derivation demonstrates that the training algorithms for our proposed Glimpse network can also be interpreted using the ideas and techniques from reinforcement learning. Particularly, Table 1 shows how the concepts and variables in learning rule (15) can be mapped to those in (23):

Table 1. High-level connections between the learning rules (15) and (23) derived by direct probabilistic inference and reinforcement learning, respectively

Direct probabilistic inference	Reinforcement learning
Maximizing $J = \log p(y | I; \theta))$	Maximizing $J_1 = \mathbb{E}[R(\hat{y}, 1)]$
Conditional prior $p(l | I; \theta)$	Action policy $\pi_l(l_t | s_{0:t-1}; \theta)$
Likelihood $p(y | l, I; \theta)$	Action policy $\pi_y(y | s_{0:T-1}, l_T; \theta)$
Importance weight R_y	Reward r_l
Importance weight R_l	Reward $\bar{r}_l = r_y + r_l$
Baseline $b = \mathbb{E}[R_l]$	Baseline $b_1 = \mathbb{E}[\bar{r}_l]$

According to Table 1, there is similarity between the roles of the importance weights R_y, R_l and the reward signals r_y, r_l in their corresponding stochastic gradient estimators (15) and (23), which are actually derived from different objective functions. Our approximations of the importance weights (11) and (12) with the class saliency map as independent priors are actually inspired by reinforcement learning and its usage of reward signals that are drawn by an external environment, which is the Glance network in our proposed model. Thus, this motivates us to investigate and summarize the relationship between the concepts in the two discussed learning frameworks, hopefully providing more insights for how to leverage ideas from both sides to develop more advanced probabilistic models and learning algorithms in the future.

5 Experiments

In this section, performance of the proposed GGNet is evaluated on the benchmark CT-MNIST datasets proposed in [4]. Each sample in this dataset is generated by placing the original MNIST 28×28 digit at a random location of a larger canvas cluttered with several other 8×8 digit fragments. We adopt the same experiment setup as those in [4]. Specifically, we evaluate our GGNet on 60×60 and 100×100 CT-MNIST, with 4 and 8 irrelevant digit fragments, respectively. The design choices for the Glance and Glimpse networks are described as follows:

Glance Network: Down-sample an input image with a factor of 2/4 for 60×60 and 100×100 CT-MNIST, respectively, followed by Convolution (32 5×5 kernels, 1×1 stride)-Max Pooling (3×3 size, 3×3 stride)-Convolution (64 5×5 kernels, 1×1 stride)-Max Pooling (2×2 size, 2×2 stride)-Linear (200 output neurons)-Linear (10 output neurons). The threshold value $\sigma_{th} = 0.4$.

Glimpse Network: The dimensions of the feature vectors $h_g \in \mathbb{R}^{512}, h_l \in \mathbb{R}^{128}, g \in \mathbb{R}^{256}$ and the internal state $h \in \mathbb{R}^{256}$. The prediction network D has two fully connected layers of output size 200 and 10, respectively, followed by a LogSoftMax layer. With this network configuration, our Glimpse network has roughly the same size of RAM. The variances on both dimensions of the locator network's output are chosen as 0.05.

Activation functions of all layers in our model are rectifier non-linearities. Stochastic gradient descent with minibatch of 16 samples and momentum 0.9 is used to train both the Glance and Glimpse network. The learning rate is designed to decay linearly along the numbers of training epochs and Dropout [21] with a rate of 0.2 is applied only to the prediction network. We compare our GGNet with, the Glance network itself with down-sampled (DS) inputs, RAM [4] and two state-of-the-art attention models on the two CT-MNIST datasets, namely the DCRM [10] and DRAM [7]. The glimpse size for RAM, DRAM and GGNet are all 12×12, i.e., $w = 12$. Note that the recurrent activation function of DCRM and DRAM is either GNU or LSTM unit. Our Glimpse network adopts the simplest ReLU without using any gating structure and hence has less parameters and higher computational efficiency. Though we also believe that these gated unites may further boost our model's performance, we keep the structural complexity and training setup of the Glimpse network as close to RAM as possible. This control can well demonstrate that our results come from innovations in learning algorithm. The classification results on both datasets are shown in Table 2.

As shown in Table 2, our GGNet achieves superior performances than RAM with significantly less glimpses. In addition, GGNet outperforms DCRM on 60×60 CT-MNIST and its performance on 100×100 CT-MNIST is similar to that of

Table 2. Classification performance on the CT-MNIST datasets

CT-MNIST 60×60		CT-MNIST 100×100	
Model	Error	Model	Error
Glance network, $2 \times$ DS	9.08%	Glance network, $4 \times$ DS	26.96%
RAM, 4 Glimpses, 3 scales	4.96%	RAM, 4 Glimpses, 4 scales	9.41%
RAM, 8 Glimpses, 3 scales	4.08%	RAM, 8 Glimpses, 4 scales	8.11%
DCRM	3.42%	DCRM	-
DRAM	-	DRAM, 4 Glimpses	4.18%
DRAM	-	DRAM, 8 Glimpses	**3.36%**
GGNet, 4 Glimpses, 3 scales	3.65%	GGNet, 4 Glimpses, 4 scales	4.64%
GGNet, 6 Glimpses, 3 scales	**3.33%**	GGNet, 6 Glimpses, 4 scales	3.80%
GGNet, 8 Glimpses, 3 scales	3.55%	GGNet, 8 Glimpses, 4 scales	3.85%

DRAM, which, as analyzed in Introduction, is actually a much more complicated soft attention model with two interactive deep RNNs and requires processing the full image in each iteration.

As our proposed GGNet is closely related to RAM, a more detailed comparison of their performances can better demonstrate how GGNet can learn a more effective attention policy. Particularly, for 100×100 CT-MNIST dataset, with the same glimpse configuration, our GGNet outperforms the best result of RAM with 8 glimpses by a wide margin with only half of the number of glimpses. We can observe that the performance of RAM drops significantly when tested on 100×100 CT-MNIST, which has a larger canvas and more distractive digit fragments. Conversely, GGNet can effectively retain its classification capacities between the two tested CT-MNIST datasets. This suggests that our GGNet is able to retain its robustness against increasing numbers of distracters by centering its attention around the true digit without excessively exploring the irrelevant areas. In other words, GGNet tends to learn a more cost-effective attention policy than RAM, as the learning process of the Glimpse network is guided by the Glance network such that it tries to focus on the potentially informative areas. Finally, although the Glance network itself does not provide satisfactory classification results in Table 2, the Glimpse network is still able to learn an efficient and effective attention policy from the class saliency map it provides.

Figure 3 provides some illustrative examples of our trained GGNet classifying the 10 input images with 4 glimpses. The square boxes of different sizes represent the horizons of our multi-resolution glimpse sensor. Figure 3(e) shows that the Glance network can provide noisy hints on the location of the true digit. However, the Glimpse network is able to distill the valid information from the class saliency map and learn to deploy its glimpses around the true digit in Figs. 3(a) to (d).

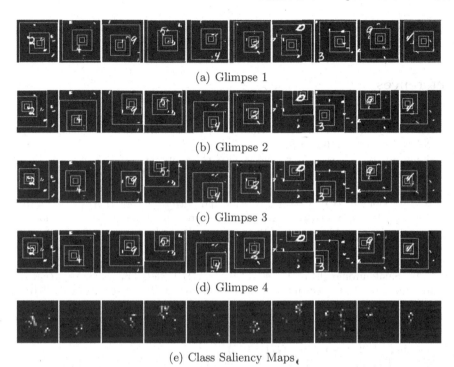

(a) Glimpse 1

(b) Glimpse 2

(c) Glimpse 3

(d) Glimpse 4

(e) Class Saliency Maps

Fig. 3. Exemplar glimpse sequences with corresponding visualized class saliency map (regions with higher pixel values denote more informative areas)

6   Conclusion

We have presented the Glance and Glimpse Network (GGNet), a hybrid network model that learns to perceive an image in both coarse and fine scales, which is in analog to human's glance and glimpse behaviors. A Glance network has been designed to efficiently generate a top-down class saliency map as indicators of the potentially informative regions in an image. The recurrent Glimpse network has been proposed to learn an effective and efficient hard attention policy using the hints from the Glance network. The learning rule of Glimpse network has been first derived under the context of probabilistic inference. It has then been interpreted and reexamined under the framework of reinforcement learning, therefore addressing high-level connections between these two independent fields when learning probabilistic models with stochastic gradient estimator. The performance of GGNet has been evaluated on the CT-MNIST datasets and state-of-the-art results have been obtained by learning a more cost-effective attention policy that drives the Glimpse network to deploy its capacities only on regions that tends to be informative.

Acknowledgement. This work is supported by the A*STAR Industrial Robotics Program of Singapore, under grant number R-261-506-007-305.

References

1. Krizhevsky, A., Sutskever, I., Hinton, G.E.: ImageNet classification with deep convolutional neural networks. In: Advances in Neural Information Processing Systems (NIPS), pp. 1097–1105 (2012)
2. Simonyan, K., Zisserman, A.: Very deep convolutional networks for large-scale image recognition. arXiv preprint arXiv:1409.1556 (2014)
3. He, K., Zhang, X., Ren, S., Sun, J.: Deep residual learning for image recognition. In: IEEE Conference on Computer Vision and Pattern Recognition (CVPR) (2016)
4. Mnih, V., Heess, N., Graves, A., et al.: Recurrent models of visual attention. In: Advances in Neural Information Processing Systems (NIPS), pp. 2204–2212 (2014)
5. Ba, J., Mnih, V., Kavukcuoglu, K.: Multiple object recognition with visual attention. In: International Conference on Learning Representations (ICLR) (2015)
6. Ba, J., Salakhutdinov, R.R., Grosse, R.B., Frey, B.J.: Learning wake-sleep recurrent attention models. In: Advances in Neural Information Processing Systems (NIPS), pp. 2575–2583 (2015)
7. Gregor, K., Danihelka, I., Graves, A., Rezende, D., Wierstra, D.: Draw: a recurrent neural network for image generation. In: International Conference on Machine Learning (ICML), pp. 1462–1471 (2015)
8. Xu, K., Ba, J., Kiros, R., Cho, K., Courville, A., Salakhudinov, R., Zemel, R., Bengio, Y.: Show, attend and tell: neural image caption generation with visual attention. In: International Conference on Machine Learning (ICML), pp. 2048–2057 (2015)
9. Jaderberg, M., Simonyan, K., Zisserman, A., et al.: Spatial transformer networks. In: Advances in Neural Information Processing Systems (NIPS), pp. 2008–2016 (2015)
10. Moczulski, M., Xu, K., Courville, A., Cho, K.: A controller recognizer framework: how necessary is recognition for control? In: International Conference for Learning Representations (ICLR) Workshops (2016)
11. Chun, M.M., Jiang, Y.: Contextual cueing: implicit learning and memory of visual context guides spatial attention. Cogn. Psychol. **36**, 28–71 (1998)
12. Judd, T., Ehinger, K., Durand, F., Torralba, A.: Learning to predict where humans look. In: International Conference on Computer Vision (ICCV), pp. 2106–2113 (2009)
13. He, H., Ge, S.S., Zhang, Z.: A saliency-driven robotic head with bio-inspired saccadic behaviors for social robotics. Auton. Robot. **36**, 225–240 (2014)
14. Simonyan, K., Vedaldi, A., Zisserman, A.: Deep inside convolutional networks: visualising image classification models and saliency maps. In: International Conference on Learning Representations (ICLR) Workshops (2014)
15. Melchers, R.: Importance sampling in structural systems. Struct. Saf. **6**, 3–10 (1989)
16. Williams, R.J.: Simple statistical gradient-following algorithms for connectionist reinforcement learning. Mach. Learn. **8**, 229–256 (1992)
17. Weber, T., Heess, N., Eslami, A., Schulman, J., Wingate, D., Silver, D.: Reinforced variational inference. In: Advances in Neural Information Processing Systems (NIPS) Workshops (2015)

18. Tang, Y., Salakhutdinov, R.R.: Learning stochastic feedforward neural networks. In: Advances in Neural Information Processing Systems (NIPS), pp. 530–538 (2013)
19. Mnih, A., Rezende, D.J.: Variational inference for monte carlo objectives. In: International Conference on Machine Learning (ICML) (2016)
20. Mnih, A., Gregor, K.: Neural variational inference and learning in belief networks. In: International Conference on Machine Learning (ICML) (2014)
21. Srivastava, N., Hinton, G., Krizhevsky, A., Sutskever, I., Salakhutdinov, R.: Dropout: a simple way to prevent neural networks from overfitting. J. Mach. Learn. Res. **15**, 1929–1958 (2014)

Fine-Tuning Deep Neural Networks in Continuous Learning Scenarios

Christoph Käding[1,2](✉), Erik Rodner[1,2], Alexander Freytag[1,2], and Joachim Denzler[1,2]

[1] Computer Vision Group, Friedrich Schiller University Jena, Jena, Germany
christoph.kaeding@uni-jena.de
[2] Michael Stifel Center Jena, Jena, Germany

Abstract. The revival of deep neural networks and the availability of ImageNet laid the foundation for recent success in highly complex recognition tasks. However, ImageNet does not cover *all* visual concepts of *all* possible application scenarios. Hence, application experts still record new data constantly and expect the data to be used upon its availability. In this paper, we follow this observation and apply the classical concept of fine-tuning deep neural networks to scenarios where data from known or completely new classes is continuously added. Besides a straightforward realization of continuous fine-tuning, we empirically analyze how computational burdens of training can be further reduced. Finally, we visualize how the network's attention maps evolve over time which allows for visually investigating what the network learned during continuous fine-tuning.

1 Introduction

"How would you train a deep neural network when new data from potentially new categories is continuously added to the training set?"

Machine learning and vision have significantly benefited from benchmarking on fixed datasets, since they allowed for comparison between algorithms and developed models [1–5]. However, our world is an environment which undergoes ongoing change. Instead, both the semantic space of object categories as well as the visual appearance of known categories are not fixed. To handle this, we humans are able to continuously learn and adapt our knowledge. Both aspects, fixed models and changing environments, are in contrast with each other. Therefore, incremental learning is an important field of research aiming at developing visual recognition systems that are able to deal with new data from known or

This research was supported by grant DE 735/10-1 of the German Research Foundation (DFG).

Electronic supplementary material The online version of this chapter (doi:10.1007/978-3-319-54526-4_43) contains supplementary material, which is available to authorized users.

© Springer International Publishing AG 2017
C.-S. Chen et al. (Eds.): ACCV 2016 Workshops, Part III, LNCS 10118, pp. 588–605, 2017.
DOI: 10.1007/978-3-319-54526-4_43

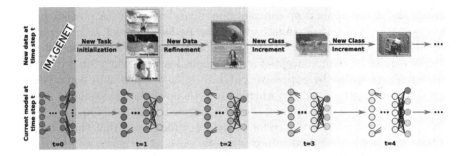

Fig. 1. At the beginning of the learning process, we have a pre-trained network as initialization for parameters. The first step is the classical fine-tuning where the final layer is replaced to fit the novel task (shown as colored dots). Upon the availability of more data, the relative importance of initial weight estimates is further reduced (indicated by reduced blue color). When even novel categories are discovered, also the network architecture needs to be adapted, *e.g.*, by adding new output variables. (Color figure online)

even completely new classes by performing learning in a continuous fashion. Furthermore, it is an essential element for active learning [6] and active discovery [7] approaches, which strictly require continuously changing models. Incremental learning aspects have been studied for a great variety of different models *e.g.*, [8–10], but not so far for deep neural networks (DNN).

In addition to large and fixed datasets, the resurrection of DNNs lead to the latest innovation pitch in computer vision research. Besides training deep models from scratch, additional benefits have become apparent when looking at the common use of fine-tuning to small datasets. Originally, fine-tuning has been referred to the process of pre-training neural networks with a generative objective followed by an additional training phase with a discriminative objective on the same dataset [11]. More recently, fine-tuning refers to re-using parameter values estimated on potentially large datasets as initialization in applications with limited access to labeled data. This approach has paved the way to significant performance gains in many applications, *e.g.*, [12–16].

From the perspective of continuous learning, the latter fine-tuning scenario can be seen as the extreme case of continuous learning which is restricted to only two time steps: pre-training and update.

More general forms of continuous learning for deep convolutional neural networks (CNN) have hardly been studied before. The question remains how continuous learning with a series of update steps can be performed robustly and efficiently. To study this question, we continuously fine-tune convolutional neural networks and empirically evaluate the effect of individual hyperparameters on the robustness of learning. A visualization of the process of continuous learning is given in Fig. 1.

The concept of continuous fine-tuning is general and applies to any form of a deep neural network. In this paper, we especially focus on image understanding scenarios and hence apply deep convolutional neural networks. A possible application scenario is active learning, where the updated model is immediately required for the sub-sequent selection [6]. Similarly, automated visual monitoring scenarios require efficient concepts for continuous learning when labeled data is incrementally provided by expensive but rather slow experts. In consequence, we specifically focus on scenarios where only little novel data is available in each update step but an updated model should be immediately available.

We provide empirical evidence for guidelines which show that the computation time for parameter updates in continuous learning scenarios can be significantly reduced. Although our empirical findings are intuitive, we believe that sharing our obtained insights is beneficial for a broader audience in several application areas. Furthermore, we investigate how CNNs evolve over time by visualizing how the attention of the nets shifts towards discriminative parts of newly added categories. This visualization allows for controlling and analyzing the continuous learning process.

2   Related Work

Since the necessity for dealing with continuous data streams is vast, update techniques for a variety of classification techniques have been proposed. While a complete overview is far beyond the scope of this paper, we briefly present examples for some well-established classification techniques.

Continuous Learning for Various Model Types. The frequently used approach of *online learning* can be viewed as a special case of continuous learning. Here, the training dataset is not presented at once (as in batch learning) but with only a single example in every time step. Support vector machines (SVMs) have been the presumably most frequently used model in the last two decades. For SVMs, Cauwenberghs and Poggio presented online learning with fixed categories [17]. Based on these results, Tax and Laskov [9] introduced online learning of Support Vector Data Description (SVDD) models for novelty detection scenarios. Later on, incremental learning for the frequently used SVM solver SMO (sequential minimal optimization) with varying numbers of categories has been presented by Yeh and Darrell [18]. Gaussian Process (GP) models, which are closely related to SVMs, have been used for continuous learning of object categories by Freytag et al. [8]. Mensink et al. propose update techniques for nearest class mean (NCM) classifiers in [19] followed by Ristin et al., who present continuous extensions of a hybrid model consisting of NCM and RDF [20]. Sillito and Fisher derive update rules for GMM models in [10] which laid the foundation for the work of Hospedales et al. for the task of continuous class discovery [21]. Despite all reported benefits, these techniques have been rarely used recently due to their rather low model complexity compared to the one provided by deep neural networks. Let us therefore briefly investigate continuous learning approaches for deep nets.

Continuous Learning of Deep Neural Networks. A decade ago, Wilson and Martinez compared batch learning and online learning of neural networks in [22].

According to their argumentation, online learning should be always preferred over batch learning, especially for large datasets. While the currently preferred mini-batch learning [23] is somewhat in-between, the authors of [22] found no practical advantage of learning with mini-batches over online learning. Similar arguments have been put forward by LeCun et al. [24]. However, online learning assumes the number of categories within the data stream to be known in advance. Furthermore, multiple cyclic passes (epochs) through the dataset are required for robust estimation of parameters.

For the more general problem of continuous learning, also novel categories have to be included into the existing network appropriately. One example is the work in [25] for training of deep networks with incrementally added categories. The authors approach the necessity of increased network capacity by duplicating the existing network, assigning available classes equally to one of both networks, fine-tune each net individually, and adding a third network on top to predict which net to use. Thus, this strategy can be seen as training a decision tree with a deep network in every node. Besides this relatively complex, memory-intensive, and computationally demanding approach, little is known so far how to continuously train deep networks efficiently.

Fine-Tuning of Deep Neural Networks. As mentioned before, fine-tuning of pre-trained networks to new tasks can be viewed as a special case of continuous learning with only two time steps: one initial learning step and one update step. A variety of publications underlines the benefits which arise from pre-training deep networks on large datasets. As an example, Agarwal et al. stated that *"pre-training significantly improves performance"* for the task of object recognition [12]. Similarly, Girshick et al. draw the conclusion that *"We conjecture that the 'supervised pre-training/domain-specific fine-tuning' paradigm will be highly effective for a variety of data-scarce vision problems."* [16]. Further benefits have been reported for image retrieval [13], semantic segmentation [14], fine-grained recognition [15], or object localization [26]. However, fine-tuning is only used with a fixed dataset of a new task. In this paper, we empirically investigate the ongoing process of "continuous fine-tuning" with increasing data and number of categories for improving the model of a single task.

Further Related Topics. Among the previously shown works, there is a growing set of approaches which aim at replacing fixed models with systems that are able to adapt themselves continuously. A closely related research area is domain adaptation (*e.g.,* [27–29]). Different domains can either occur from different recording techniques (*e.g.,* a change of camera technology) or from a change of data distributions between two data collections. In contrast to our application scenario, domain adaptation techniques aim at estimating the differences between domain distributions to optimally leverage information between data collections. Instead, we take incoming data as-is and allow the network to adapt itself smoothly over time in case of occurring domain shifts.

Similarly related is the area of transfer learning [29,30], which is also known as learning to learn [31]. In transfer learning scenarios, new categories are incrementally added to previously known data. The underlying assumption is that transferring model parameters from known but semantically related categories is beneficial to represent the novel category. Instead of explicitly finding support tasks or categories to transfer parameters, we rely on the learned representational power of a single network shared by all categories.

3   Deep Neural Network Learning in a Nutshell

Before we investigate continuous learning of deep neural networks, we briefly review the learning of these models and define hyperparameters used in our experiments.

Batch Learning of Deep Networks. Let $f(x; \theta)$ be the output of a neural network with parameters θ for a given image $x \in \Omega$. Learning a network from a given labeled training set $\mathcal{D} = (X, y) = (x_i, y_i)_{i=1}^{N} \subset (\Omega \times \mathcal{Y})$ boils down to minimizing a desired learning objective. Results of computational learning theory tell us that the objective should be comprised by the loss of the training set and a regularization term ω:

$$\bar{\mathcal{L}}(\theta; \mathcal{D}) = \frac{1}{N} \sum_{i=1}^{N} \mathcal{L}(f(x_i; \theta), y_i) + \omega(\theta). \tag{1}$$

In contrast to the overall training objective $\bar{\mathcal{L}}$, the loss term $\mathcal{L}$ operates on individual examples and compares the obtained with the desired model outputs. Common choices for $\mathcal{L}$ are the quadratic loss for regression tasks or the softmax loss for multi-class classification scenarios. The term ω is usually an elastic-net regularization [32] that combines L_2 and L_1-regularization of the parameters θ.

As in many other application domains, vision tasks require model functions of high complexity. Therefore, $f(x_i; \theta)$ is commonly a composition of functions, usually referred to as layers, with their individual parameters:

$$f(x; \theta) = f_L (\dots (f_2 (f_1 (x; \theta_1); \theta_2) \dots); \theta_L). \tag{2}$$

The underlying idea of "deep" learning is to train these layered models $f(\cdot; \theta)$ directly from input data by optimizing all involved parameters $\theta = (\theta_1, \dots, \theta_L)$ jointly with respect to the single loss function specified in Eq. (1). Unfortunately, this learning objective is non-convex and highly non-linear, except for trivial architectures. Thus, closed-form solutions for minimization, such as the ones existing for simpler models like least-squares regression [33, Sect. 3.1.4] can by no means be expected. Instead, optimization techniques such as gradient descent need to be applied which iteratively refine initial parameter guesses and ideally converge to suitable (local) optima. For layered models in Eq. (2), applying the chain rule allows for calculating partial derivatives for parameters of hidden layers which lead to the backpropagation algorithm [34].

Learning with Mini-Batches. Since gradient descent requires the entire "batch" of training examples at once, it is rarely feasible for large-scale datasets.

In consequence, today's standard optimization techniques are stochastic gradient descent (SGD) [35] and mini-batch gradient descent [36]. Both techniques approximate the true gradient $\nabla_{\boldsymbol{\theta}}\bar{\mathcal{L}}(\boldsymbol{\theta};\mathcal{D})$ of the objective function using a randomly drawn subset $\mathcal{S}^k \subseteq \mathcal{D}$ in every iteration k. SGD originally referred to the special case of a single example in each iteration (*i.e.*, $|\mathcal{S}| = 1$), whereas mini-batch gradient descent uses larger sets ($1 < |\mathcal{S}| \ll N$). Since both techniques rely on randomized approximations of the underlying gradient, both are commonly called stochastic gradient descent. In the remainder of this paper, we follow this naming convention and refer with SGD to every gradient descent which is not using the entire training set for gradient calculations.

In every iteration, SGD computes an approximated gradient $\tilde{\nabla}_{\boldsymbol{\theta}}$ based on the currently drawn subset:

$$\tilde{\nabla}_{\boldsymbol{\theta}}\bar{\mathcal{L}}(\boldsymbol{\theta};\mathcal{S}^k) = \frac{1}{|\mathcal{S}^k|}\sum_{i \in \mathcal{S}^k}\nabla_{\boldsymbol{\theta}}\mathcal{L}(f(\boldsymbol{x}_i;\boldsymbol{\theta}),y_i) + \nabla_{\boldsymbol{\theta}}\omega\,(\boldsymbol{\theta}). \tag{3}$$

Based on that, parameter estimates are updated using a gradient descent step of length $\gamma > 0$ which is often referred to as learning rate:

$$\boldsymbol{\theta}^{k+1} = \boldsymbol{\theta}^k - \gamma \cdot \tilde{\nabla}_{\boldsymbol{\theta}}\bar{\mathcal{L}}(\boldsymbol{\theta}^k;\mathcal{S}^k). \tag{4}$$

There are plenty of further optimization modifications including momentum [37], weight decay [34], dropout [38], as well as annealing schemes for the learning rate [39]. Since we are not empirically analyzing the training deep neural networks per se or the influence of such modifications on continuous learning, we refer to [36,37,40] for an evaluation of the different strategies.

4   Continuous Learning of Deep Neural Networks

In the following, we first define continuous learning scenarios and then propose different strategies how to cope with them with straightforward fine-tuning when using deep neural networks.

Continuous Learning Scenarios. In contrast to standard batch learning applications, we are interested in continuous learning of deep neural networks in this paper. Thus, we are given a *series* of learning datasets $\mathcal{D}^t \subset (\Omega \times \mathcal{Y}^t)$. The goal is to learn the network $f(\cdot;\hat{\boldsymbol{\theta}}^{(t)})$ for each time step $t \geq 0$ with dataset $\mathcal{D}^t$ continuously over time. Depending on the overlap between the sets $\mathcal{D}^t$ for different t, we can differentiate between three scenarios:

(F) *Classical fine-tuning for new tasks*: Only two disjoint datasets are given and the task changed, *i.e.*, $\mathcal{D}^0 \cap \mathcal{D}^1 = \emptyset$ and $\mathcal{Y}^0 \cap \mathcal{Y}^1 = \emptyset$. This is for example used in [12–16] when a convolutional neural network pre-trained on ImageNet is fine-tuned in one step by either re-learning the last layer only or by performing optimization steps for all layers. For simplicity in notation, we ignore that fractions of $\mathcal{D}^1$ can already be included in $\mathcal{D}^0$. This can happen in special cases when the one-step fine-tuning is performed for new datasets covering a part of the original one.

(C1) *Continuous learning of known classes*: In this case, we get additional training examples for the classes we already know from the initial learning set
$\mathcal{D}_0$, *i.e.*, $\forall t$: $\mathcal{Y}^0 = \mathcal{Y}^t$ and $\mathcal{D}^{t-1} \subset \mathcal{D}^t$. This case describes a continuously
growing training dataset and resembles the classical online learning setup.

(C2) *Continuous learning of known and new classes*: In addition to the previous
scenario, we might also get examples of completely new classes, *i.e.*, $\forall t$:
$\mathcal{Y}^{t-1} \subseteq \mathcal{Y}^t$ with both sets being not necessarily equal and $\mathcal{D}^{t-1} \subset \mathcal{D}^t$.

In the following, we refer to $\mathcal{U}_t = \mathcal{D}^t \backslash \mathcal{D}^{t-1}$ as the *update set* of time step t.

Continuous Learning with Incremental Fine-tuning. As stated before,
the learning objective in Eq. (1) is in general non-convex and highly non-linear.
In consequence, there are no closed-form update rules available like in the case
of Gaussian process regression [41]. However, we can make use of the technique
of warm-start optimization [42,43], where we use the parameters $\boldsymbol{\theta}^{t-1}$ of the
previous time step as initialization for the optimization of the current parameters
$\boldsymbol{\theta}^t$. This strategy is also applied for standard use of fine-tuning and assumes that
the network's parameters $\boldsymbol{\theta}^t$ vary smoothly with the extensions of the training
dataset. Furthermore, it expresses the expectation that the optimization is more
robust against bad local minima when started from a parameter vector which is
likely close to an appropriate solution. Since case (F) is already well studied, we
focus on (C1) and (C2) in the remainder of this paper.

Whereas (C1) can be directly tackled with fine-tuning without any modification of the network, scenario (C2) requires extending the last layer of a deep
neural network for new categories. In particular, we need to add an additional
output node in the last layer along with parameters in the corresponding previous fully-connected layer. We initialize those new parameters randomly using
the normalization technique by [44].

In both settings, (C1) and (C2), a number of questions remains in practice.
For example, it is unclear whether the solver needs to be run until convergence
in each update step. It might be even beneficial to perform an early stopping of
the optimization to increase generalization performance [45]. Furthermore, there
is no clear guidance for how many layers we can adapt robustly when small sets
of novel data arrives consecutively. Finally, the influence of label noise on the
success of model updates is important to know for real-world applications. In
Sect. 5, we empirically investigate these questions. In addition, we visualize in
Sect. 6 the network's region of attention which changes most strongly when new
data is continuously added.

5   Empirical Evaluation of Continuous Learning of DNNs

We conducted experiments in continuous learning scenarios with known and
unknown classes. For different update strategies, we evaluated the classification
accuracy of the continuously learned models. Our main findings can be summarized as follows:

1. The number of required SGD iterations can be significantly reduced in comparison to standard batch learning and set to small constants during training without a notable loss in test accuracy (Sect. 5.3).
2. Continuous fine-tuning by neglecting already known data leads to overfitting towards the update samples (Sect. 5.4).
3. Continuous fine-tuning is robust with respect to small amounts of label noise (Sect. 5.5).

We also studied the influence of the SGD batch size $|\mathcal{S}|$ and the learning rate γ on the quality of continuous fine-tuning but found no surprising behavior. The results can be found in the supplementary material. In the following, we briefly explain the experimental setup before inspecting each of the previous three aspects in detail. Furthermore, we investigate the principal applicability of continuous learning for CNNs in Sect. 5.2 before we show more efficient realizations in the following sections.

5.1   Experimental Setup, Datasets, and Implementation Details

Network Architectures and Learning. Since our main interest is in image categorization and understanding, we restrict our experiments to CNNs. Thereby, the desired invariance with respect to translation is explicitly encoded into the network layout. In all our experiments, we use the BVLC version of the classical AlexNet architecture [23] with network weights which have been pretrained using the ImageNet ILSVRC-2010 challenge dataset [46]. Experiments were conducted with MatConvNet [47].

It is well known and extensively studied that a careful choice of the learning rate is among the most crucial aspects for learning [48]. Hence, a intensive evaluation of different learning rates is beyond the scope of this paper. Instead, we use results of preliminary experiments and fix the learning rate to 0.001 in all scenarios. This choice is consistent with reported observations [22] and allows for comparable results in all experiments. Furthermore, we use small update sets $\mathcal{U}_t$ of size $|\mathcal{U}_t| = 25$ which contain only samples from a single category. We further follow the default parameter settings of MatConvNet [47] and apply a weight decay of 0.0005 and a momentum of 0.9.

Datasets for Evaluations. For the evaluation of our proposed learning strategies, we use two state-of-the-art datasets. First, we use MS-COCO-full-v0.9 of Lin et al. [4] which provides a challenging setup similar to real-world applications. Since we are interested in classification tasks, we consider all ground truth bounding boxes with at least 256 pixel height and width. Furthermore, we only use categories that consist of 500 to 1,000 examples. Thereby, we obtain 15 categories which provide enough training examples for initial training and several update sets. The total number of images of this subset is approximately 11,000 for training and test set each.

As second evaluation dataset, we chose the Stanford40Actions dataset introduced in [49]. The dataset contains still images of different activities and provides a split into 4,000 images for training and 5,532 hold-out images for testing. Hence, adapting pre-trained models to the new task is clearly required.

In comparison with large-scale datasets like ImageNet, both datasets contain only small numbers of examples. Thereby, they are well suited for evaluating continuous learning with little data and allow for insights in a rather uncommon application scenario for deep neural networks.

Experimental Setup. To evaluate continuous learning, we randomly pick 10 of the given classes to initially fine-tune the CNN. From the remaining data, we chose 5 more classes randomly which serve as novel data to be added during the process of continuous learning. For every category, we randomly select 100 examples. Those samples are either used as initial known data and or as update sets depending on if the category is selected to be added during the experiments or to be known initially. Classification accuracy after every update step is measured on the corresponding test data of the dataset. To reduce randomization effects during evaluation, we repeat the process of random sample selection three times. Furthermore, the entire setup is conducted for three different class splits, which yields 9 different setups to average results over.

5.2   Comparison of Continuous and One-Step Fine-Tuning

In a first experiment, we analyze whether continuous fine-tuning of deep neural networks is possible without a loss in accuracy. As baseline serves the classical one-step fine-tuning paradigm, where we perform only a single update step when all data $\mathcal{D}_T$ is available. This is similar to case (F) described in Sect. 4. We compare this baseline against continuous fine-tuning, for which we conduct an fine-tuning cycle as soon as an update set $\mathcal{U}_t$ becomes available which corresponds to case (C1) and (C2). Due to the design of our experiments, we evaluate both cases in a mixed manner because not every update set contains new classes. An additional evaluation of case (C1) can be found in the supplementary material.

Arguably, if one-step fine-tuning and continuous fine-tuning have been finally conducted with exactly the same amount of data, it sounds intuitively plausible that estimated parameters lead to the same classification accuracy. However, the path (θ^t) of parameters the continuous learning follows are end in a completely different initialization for θ^T in the last step compared to one-step fine-tuning. For the importance of initialization, see also the initialization-as-regularization discussion in [11]. We follow the previously described experimental setup and perform 10 epochs with and a SGD batch size of $|\mathcal{S}| = 64$ per update to fine-tune all layers. Although we also performed analyses with adapting parameters of an arbitrary number of layers, experimental results lead to similar findings. Results are shown in Fig. 2.

It can clearly be seen that both learning approaches obtain similar accuracy after all training data has been processed. Hence, we conclude that continuous fine-tuning of deep neural networks is possible without significant loss in accuracy. However, conducting an entire fine-tuning cycle whenever a new update set is available is computationally expensive. In the next experiment, we analyze how to reduce this computational burden.

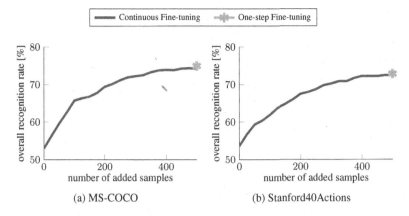

Fig. 2. Comparison of classical one-step fine-tuning (*i.e.,* waiting for all data to be available) and continuous fine-tuning (*i.e.,* fine-tuning as soon as new datasets are available).

5.3   Speeding up Continuous Fine-Tuning

The simplest solution for reducing the computation time of each fine-tuning step is to reduce the total number of SGD iterations from entire epochs to few steps done at each update. However, this comes at the risk of interrupting gradient descent too early, *i.e.,* far off the local optimum. In the following, we empirically investigate how many iterations we need in each step without reducing the classification accuracy of learned models. Therefore, we keep the previous experimental setup and only vary the number of SDG iterations T_{sgd} conducted in each update step but keep the SGD mini-batch size of $|\mathcal{S}| = 64$.

It is common practice to express the number of SGD iterations relative to the total number of known training examples, *i.e.,* in epochs e. The resulting number of SGD iterations, *i.e.,* gradient descent updates, is then given by $T_{\text{sgd}} = \left\lceil \frac{n_t}{|\mathcal{S}|} \right\rceil \cdot e$. In the first part of this analysis, we investigate how many epochs we require performing continuous fine-tuning in a stable fashion for all layers. Hence, we evaluated how many epochs we need to successfully learn deep neural networks continuously. Results for different epochs in each step are shown in Fig. 3.

As can be seen, the resulting accuracy remains surprisingly unchanged even with extremely few iterations in each update step. In fact, even a half epoch can be sufficient and we only observe a clear drop in accuracy for the extreme case of 0.1 epochs. We attribute this observation to the effect that 0.1 epochs can easily miss selecting a representative subset for SGD optimization.

Based on the previous results, the question arises whether it is indeed necessary to see every training example in each update. To answer this question, we go beyond the previous analysis and further explore the possible range of fast fine-tuning. Whereas the number of SGD steps grew previously due to the increasing number of known examples after each step, we now fix the number of SGD update steps T_{sgd} to a constant number. This leads to a further significant

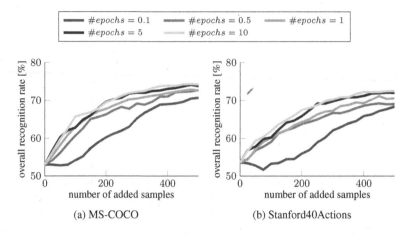

(a) MS-COCO   (b) Stanford40Actions

Fig. 3. Performance comparison for different numbers of epochs and therefore an increasing number of SGD iterations used during fine-tuning.

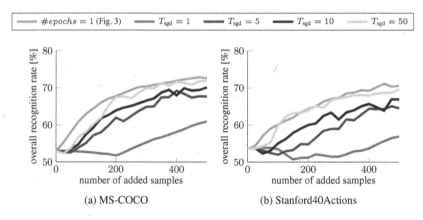

(a) MS-COCO   (b) Stanford40Actions

Fig. 4. Performance comparison for continuous fine-tuning with a fixed number T_{sgd} of SDG iterations compared to an increasing number of SGD iterations depending on the number of training examples (fixed number of epochs).

speed-up of DNN learning. Each update step is comprised of a single mini-batch of size 25 which equals the update set size $|\mathcal{U}_t|$. The results can be seen in Fig. 4.

Similarly to our previous analysis with a varying number of epochs (*i.e.,* with an increasing number T_{sgd} of SGD steps), we observe again that we can drastically reduce the required computations without a strong loss in accuracy. As previously, we attribute the significant accuracy drop for $T_{\mathrm{sgd}} = 1$ to the fact that no representative subset could be randomly sampled. In the next analysis, we investigate whether alternative sampling techniques can improve the accuracy of continuous fine-tuning.

5.4   Continuous Learning with Varying Update Influence

A majority of techniques that incrementally update classifier parameters involves a hyperparameter λ that weights the influence of the new data of the current update set. In the following, we empirically answer the natural question whether different weights for old and new data also lead to an improved continuous fine-tuning accuracy for DNNs. Therefore, we sample examples during SGD iterations non-uniformly and dependent on an example-specific probability p_i:

$$p_i = \frac{1-\lambda}{|\mathcal{U}_t|} \quad \text{if} \quad x_i \in \mathcal{U}_t \quad \text{and} \quad \frac{\lambda}{|\mathcal{D}^{t-1}|} \quad \text{otherwise,} \tag{5}$$

with $0 \leq \lambda \leq 1$. For the extreme case of $\lambda = 0$, training examples of previous time steps are completely ignored during sampling and their information is only indirectly used through parameter initializations. This could be seen as a naive realization of continuous learning since it is similar to online updates of common classification models. Furthermore, $\lambda = 1$ is equivalent to the previous continuous fine-tuning analyses but postpones the usage of examples from the current update set to the next update step. For values of λ within that range, we can control the relative importance of update set data by their probability of being sampled. To investigate the influence of λ, we performed $T_{\text{sgd}} = 10$ iterations with a fixed SGD size of $|\mathcal{S}| = 25$ which corresponds to the number of added samples $|\mathcal{U}_t|$. The results for different values of λ can be seen in Fig. 5 where parameters of all layers are adapted. Note that we also investigated to learn only upper layers, which lead to similar results (can be found in the supplementary material).

We conclude that a CNN can only be fine-tuned robustly in an continuous fashion if a fraction of new and old data is considered during SGD iterations. The balance of the data has to be chosen carefully because already known data prevents the net from overfitting to new data and too few new data is not

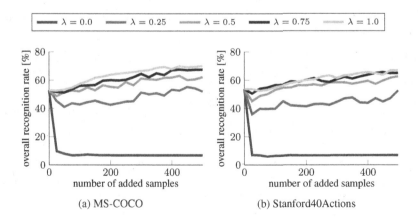

(a) MS-COCO   (b) Stanford40Actions

Fig. 5. Incremental fine-tuning of all layers with different choices of λ.

sufficient to prevent this behavior. In our case, new samples are considered as known, one step after they are added. This explains the similar results $\lambda = 0.75$ and $\lambda = 1.0$. For the extreme case $\lambda = 0$, iterative fine-tuning completely fails which can be attributed to complete overfitting to the new update set $\mathcal{U}_t$.

5.5   Continuous Learning with Label Noise

In contrast to the previous analyses, object labels are hardly perfect in real-world applications, especially with non-experts as annotators in the loop. We were thus interested in how robust the process of continuous learning is against wrongly provided label information. To investigate this question, we perform 10 epochs in each update step, and optimize parameters of all layers. We work with a mini-batch size of $|\mathcal{S}| = 64$. To analyze the influence of label noise, a specific fraction of the newly added data is replaced with examples from remaining categories while keeping the label as in the previous experiments. The results for different amounts of the induced label noise are shown in Fig. 6.

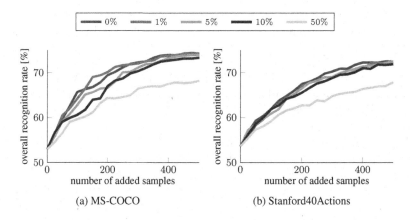

Fig. 6. Impact of label noise in update sets on continuous learning performance.

As expected, the resulting accuracy is constantly degraded with an increasing level of wrongly labeled examples. However, 10% noise results in only 2% loss in accuracy which is likely acceptable for real-world applications.

6   What Was New? Visualizing Network Changes

Especially when new categories are added to the training set, we can expect that filters in convolutional layers become specifically tuned to characteristic patterns of the new category. To visually investigate where and how our network changed over time, we first pick the filter that underwent the most drastic change and then visualize it's corresponding attention region in the image.

We expect that the chosen filter is likely to pay attention to a pattern which is discriminative for the category that has been added during the learning step. An intuitive possibility to measure the "change" of a filter is to calculate differences of filter weights before and after the update. However, this strategy might easily lead to wrong interpretations, since large weight differences in one layer could be partially compensated by changes in previous layers. Hence, large changes of filter entries do not need to correspond to changes in the resulting learned representation. Due to this reason, we follow an alternative selection strategy: we determine the filter with the largest overall change of its outputs rather than in its parameters. The output change is measured by L_2-norm of differences of outputs obtained from single filters before and after an update with new samples. We only consider the conv5 layer of AlexNet, since it is known to be related to semantic part information [50,51].

Fig. 7. Visualization of network changes during continuously learning categories from the Stanford40Actions dataset. We visualized attention regions of the single filter in conv5 which changed most strongly during a single learning step. Regions are shown before and after the respective category became known (*magenta box* and *cyan box*). *Left*: the attention shifts towards the action-related objects. *Middle*: no visible changes in attention. *Right*: the attention region shifted to contextual related areas. (Color figure online)

For visualizing the corresponding attention region of the filter in the image, we could use directly the conv5 activations. However, the resolution of the conv5 channel is rather low. For a spatially precise localization analysis of the filter's attention, we use the gradient technique of [51,52] to calculate attention maps. We then compute bounding boxes from these maps, by considering the maximum response location as center and twice the standard deviation in x and y direction as width and height. This allows us to visualize the main attention region of the chosen conv5 filter.

Visualization Results. We apply our visualization technique to the previously investigated scenarios of continuous learning and visualize the attention maps after a single category has been added. We use update sets of sizes $|\mathcal{U}_t| = 100$ and learn parameters of all layers. The output change of conv5 filters is evaluated on all examples of the novel category. We show the result of our visualization in Fig. 7 for different categories.

As can be seen on the left hand side of the figure, the conv5 channel with the highest output change often shifts focus to parts of the newly added class, although there was no localization information provided during training. Furthermore, it is interesting to see that the representation within the convolutional neural network adapts already for a small update set size. More visualizations can be found in the supplementary material.

7   Conclusions

In this paper, we studied continuous learning of deep neural networks with incoming data streams. While previous work focused on one-step fine-tuning pre-trained networks to novel tasks, we empirically studied more general scenarios of frequent and small update steps. Based on our analyses, we conclude that continuous learning can be directly achieved by continuous fine-tuning. We further investigated how continuous fine-tuning can be speeded up by reducing the number of SGD iterations. In fact, not even a single epoch is required during each update step to train models robustly. Furthermore, we also analyzed the influence of label noise as it is typically present in real-world applications. While heavily noisy data severely degrades the accuracy of learned models, we found that even with 10% incorrectly labeled data continuous learning can still be successful. Finally, we visualized the change of the network's attention during continuous learning. Thereby, we were able to visualize how well the model shifts its focus to semantically plausible regions of recently learned categories.

References

1. Griffin, G., Holub, A., Perona, P.: Caltech-256 object category dataset. Technical report 7694, California Institute of Technology (2007)
2. Everingham, M., Van Gool, L., Williams, C.K.I., Winn, J., Zisserman, A.: The PASCAL visual object classes challenge 2008 (VOC2008) results (2008). http://www.pascal-network.org/challenges/VOC/voc2008/workshop/index.html

3. Deng, J., Dong, W., Socher, R., Li, L.J., Li, K., Fei-Fei, L.: ImageNet: a large-scale hierarchical image database. In: CVPR, pp. 248–255 (2009)
4. Lin, T.-Y., Maire, M., Belongie, S., Hays, J., Perona, P., Ramanan, D., Dollár, P., Zitnick, C.L.: Microsoft COCO: common objects in context. In: Fleet, D., Pajdla, T., Schiele, B., Tuytelaars, T. (eds.) ECCV 2014. LNCS, vol. 8693, pp. 740–755. Springer, Heidelberg (2014). doi:10.1007/978-3-319-10602-1_48
5. Cordts, M., Omran, M., Ramos, S., Scharwächter, T., Enzweiler, M., Benenson, R., Franke, U., Roth, S., Schiele, B.: The cityscapes dataset. In: CVPR Workshop on the Future of Datasets in Vision (CVPR-WS) (2015)
6. Freytag, A., Rodner, E., Denzler, J.: Selecting influential examples: active learning with expected model output changes. In: Fleet, D., Pajdla, T., Schiele, B., Tuytelaars, T. (eds.) ECCV 2014. LNCS, vol. 8692, pp. 562–577. Springer, Heidelberg (2014). doi:10.1007/978-3-319-10593-2_37
7. Käding, C., Freytag, A., Rodner, E., Bodesheim, P., Denzler, J.: Active learning and discovery of object categories in the presence of unnameable instances. In: CVPR, pp. 4343–4352 (2015)
8. Freytag, A., Rodner, E., Bodesheim, P., Denzler, J.: Rapid uncertainty computation with Gaussian processes and histogram intersection kernels. In: Lee, K.M., Matsushita, Y., Rehg, J.M., Hu, Z. (eds.) ACCV 2012. LNCS, vol. 7725, pp. 511–524. Springer, Heidelberg (2013). doi:10.1007/978-3-642-37444-9_40
9. Tax, D., Laskov, P.: Online SVM learning: from classification to data description and back. In: Workshop on Neural Networks for Signal Processing (NNSP), pp. 499–508 (2003)
10. Sillito, R.R., Fisher, R.B.: Incremental one-class learning with bounded computational complexity. In: Sá, J.M., Alexandre, L.A., Duch, W., Mandic, D. (eds.) ICANN 2007. LNCS, vol. 4668, pp. 58–67. Springer, Heidelberg (2007). doi:10.1007/978-3-540-74690-4_7
11. Erhan, D., Bengio, Y., Courville, A., Manzagol, P.A., Vincent, P., Bengio, S.: Why does unsupervised pre-training help deep learning? JMLR 11, 625–660 (2010)
12. Agrawal, P., Girshick, R., Malik, J.: Analyzing the performance of multilayer neural networks for object recognition. In: Fleet, D., Pajdla, T., Schiele, B., Tuytelaars, T. (eds.) ECCV 2014. LNCS, vol. 8695, pp. 329–344. Springer, Heidelberg (2014). doi:10.1007/978-3-319-10584-0_22
13. Babenko, A., Slesarev, A., Chigorin, A., Lempitsky, V.: Neural codes for image retrieval. In: Fleet, D., Pajdla, T., Schiele, B., Tuytelaars, T. (eds.) ECCV 2014. LNCS, vol. 8689, pp. 584–599. Springer, Heidelberg (2014). doi:10.1007/978-3-319-10590-1_38
14. Hariharan, B., Arbeláez, P., Girshick, R., Malik, J.: Simultaneous detection and segmentation. In: Fleet, D., Pajdla, T., Schiele, B., Tuytelaars, T. (eds.) ECCV 2014. LNCS, vol. 8695, pp. 297–312. Springer, Heidelberg (2014). doi:10.1007/978-3-319-10584-0_20
15. Branson, S., Van Horn, G., Belongie, S., Perona, P.: Improved bird species categorization using pose normalized deep convolutional nets. In: BMVC (2014)
16. Girshick, R., Donahue, J., Darrell, T., Malik, J.: Rich feature hierarchies for accurate object detection and semantic segmentation. In: CVPR, pp. 580–587 (2014)
17. Cauwenberghs, G., Poggio, T.: Incremental and decremental support vector machine learning. In: NIPS, pp. 409–415 (2001)
18. Yeh, T., Darrell, T.: Dynamic visual category learning. In: CVPR, pp. 1–8 (2008)
19. Mensink, T., Verbeek, J., Perronnin, F., Csurka, G.: Distance-based image classification: generalizing to new classes at near-zero cost. TPAMI 35, 2624–2637 (2013)

20. Ristin, M., Guillaumin, M., Gall, J., Gool, L.V.: Incremental learning of NCM forests for large-scale image classification. In: CVPR, pp. 3654–3661 (2014)
21. Hospedales, T.M., Gong, S., Xiang, T.: Finding rare classes: active learning with generative and discriminative models. TKDE **25**, 374–386 (2013)
22. Wilson, D.R., Martinez, T.R.: The general inefficiency of batch training for gradient descent learning. Neural Netw. **16**, 1429–1451 (2003)
23. Krizhevsky, A., Sutskever, I., Hinton, G.E.: ImageNet classification with deep convolutional neural networks. In: NIPS, pp. 1097–1105 (2012)
24. LeCun, Y., Bottou, L., Orr, G.B., Müller, K.-R.: Efficient BackProp. In: Orr, G.B., Müller, K.-R. (eds.) Neural Networks: Tricks of the Trade. LNCS, vol. 1524, pp. 9–50. Springer, Heidelberg (1998). doi:10.1007/3-540-49430-8_2
25. Xiao, T., Zhang, J., Yang, K., Peng, Y., Zhang, Z.: Error-driven incremental learning in deep convolutional neural network for large-scale image classification. In: International Conference on Multimedia, pp. 177–186 (2014)
26. Oquab, M., Bottou, L., Laptev, I., Sivic, J.: Learning and transferring mid-level image representations using convolutional neural networks. In: CVPR (2014)
27. Hoffman, J., Darrell, T., Saenko, K.: Continuous manifold based adaptation for evolving visual domains. In: CVPR, pp. 867–874 (2014)
28. Tzeng, E., Hoffman, J., Darrell, T., Saenko, K.: Simultaneous deep transfer across domains and tasks. In: ICCV, pp. 4068–4076 (2015)
29. Pan, S.J., Yang, Q.: A survey on transfer learning. TKDE **22**, 1345–1359 (2010)
30. Jie, L., Tommasi, T., Caputo, B.: Multiclass transfer learning from unconstrained priors. In: ICCV, pp. 1863–1870 (2011)
31. Thrun, S.: Lifelong learning: a case study. Technical report, DTIC Document (1995)
32. Zou, H., Hastie, T.: Regularization and variable selection via the elastic net. J. R. Stat. Soc.: Ser. B (Stat. Methodol.) **67**, 301–320 (2005)
33. Bishop, C.M.: Pattern Recognition and Machine Learning. Information Science and Statistics. Springer, New York (2006)
34. Rumelhart, D.E., Hinton, G.E., Williams, R.J.: Learning representations by back-propagating errors. Nature **323**, 533–536 (1986)
35. Bottou, L.: Stochastic gradient descent tricks. In: Montavon, G., Orr, G.B., Müller, K.-R. (eds.) Neural Networks: Tricks of the Trade. LNCS, vol. 7700, pp. 421–436. Springer, Heidelberg (2012). doi:10.1007/978-3-642-35289-8_25
36. Ngiam, J., Coates, A., Lahiri, A., Prochnow, B., Le, Q.V., Ng, A.Y.: On optimization methods for deep learning. In: ICML, pp. 265–272 (2011)
37. Sutskever, I., Martens, J., Dahl, G., Hinton, G.: On the importance of initialization and momentum in deep learning. In: ICML, pp. 1139–1147 (2013)
38. Srivastava, N., Hinton, G., Krizhevsky, A., Sutskever, I., Salakhutdinov, R.: Dropout: a simple way to prevent neural networks from overfitting. JMLR **15**, 1929–1958 (2014)
39. Zeiler, M.D.: ADADELTA: an adaptive learning rate method. arXiv preprint arXiv:1212.5701 (2012)
40. Larochelle, H., Bengio, Y., Louradour, J., Lamblin, P.: Exploring strategies for training deep neural networks. JMLR **10**, 1–40 (2009)
41. Lütz, A., Rodner, E., Denzler, J.: I want to know more: efficient multi-class incremental learning using gaussian processes. Pattern Recogn. Image Anal. Adv. Math. Theory Appl. (PRIA) **23**, 402–407 (2013)
42. Tsai, C.H., Lin, C.Y., Lin, C.J.: Incremental and decremental training for linear classification. In: SIGKDD, pp. 343–352 (2014)
43. Chu, B.Y., Ho, C.H., Tsai, C.H., Lin, C.Y., Lin, C.J.: Warm start for parameter selection of linear classifiers. In: SIGKDD, pp. 149–158 (2015)

44. Glorot, X., Bengio, Y.: Understanding the difficulty of training deep feedforward neural networks. In: AISTATS, pp. 249–256 (2010)
45. Perronnin, F., Akata, Z., Harchaoui, Z., Schmid, C.: Towards good practice in large-scale learning for image classification. In: CVPR (2012)
46. Russakovsky, O., Deng, J., Su, H., Krause, J., Satheesh, S., Ma, S., Huang, Z., Karpathy, A., Khosla, A., Bernstein, M., Berg, A.C., Fei-Fei, L.: ImageNet large scale visual recognition challenge. IJCV **115**, 211–252 (2015)
47. Vedaldi, A., Lenc, K.: MatConvNet - convolutional neural networks for MATLAB. In: International Conference on Multimedia (2015)
48. Orr, G.B., Müller, K.R.: Neural Networks: Tricks of the Trade. Springer, Heidelberg (2003)
49. Yao, B., Jiang, X., Khosla, A., Lin, A.L., Guibas, L., Fei-Fei, L.: Human action recognition by learning bases of action attributes and parts. In: ICCV, pp. 1331–1338 (2011)
50. Bolei, Z., Khosla, A., Lapedriza, A., Oliva, A., Torralba, A.: Object detectors emerge in deep scene CNNs. In: ICLR (2015)
51. Simon, M., Rodner, E., Denzler, J.: Part detector discovery in deep convolutional neural networks. In: Cremers, D., Reid, I., Saito, H., Yang, M.-H. (eds.) ACCV 2014. LNCS, vol. 9004, pp. 162–177. Springer, Heidelberg (2015). doi:10.1007/978-3-319-16808-1_12
52. Simonyan, K., Vedaldi, A., Zisserman, A.: Deep inside convolutional networks: visualising image classification models and saliency maps. In: ICLR-WS (2014)

Dense Residual Pyramid Networks for Salient Object Detection

Ziqin Wang[1(✉)], Peilin Jiang[2], and Fei Wang[1]

[1] The Institute of Artificial Intelligence and Robotics,
Xi'an Jiaotong University, Xi'an, China
wangziqin@stu.xjtu.edu.cn, wfx@mail.xjtu.edu.cn
[2] The School of Software Engineering, Xi'an Jiaotong University,
Xi'an 710049, China
pljiang@mail.xjtu.edu.cn

Abstract. We introduce a coarse-to-fine method for salient object detection. In fully convolutional networks (FCN), pooling operation generates downsampled feature maps, while full size estimation is required for salient objet detection. Our *Dense Residual Pyramid Networks* (*DRPN*) attends to generating high-resolution and high-quality results. However, in order to provide enough local information, we extract extra local features from pre-trained networks. Finally, the proposed dense residual blocks learn to merge all the information and generate full size saliency maps.

In our work, the thought of reconstructing Gaussian pyramids is first introduced into the frameworks of convolutional neural networks. We employ dense residual learning to learn residual maps. We hope these feature maps can be used to refine the upsampled feature maps, as Laplacian images can be used to reconstruct images in Gaussian pyramids.

Experiments show that our *DRPN* has huge improvement over previous state-of-the-art methods on all the datasets. Especially, our *DRPN* outperforms previous state-of-the-art over 11.6% on ECSSD.

1 Introduction

Saliency detection aims at identifying salient objects in images and segmenting their object boundaries. The attention mechanism of human visual system helps a lot in finding the most important things. In many computer vision tasks, saliency detection can serve as a pre-processing step.

In previous works [1–5], low-level features, that include color, intensity, contrast and so on, are important to find salient objects. Though handcrafted features base method tend to perform well in standard scenarios, they are not robust for challenging cases. Although machine learning has been used in many saliency detection models [6–11], most of them are based on the hand-crafted features, which are not robust for all the cases.

To obtain more robust features for salient object detection, learned features have recently been employed. Recently, deep learning based methods [12–17] have

© Springer International Publishing AG 2017
C.-S. Chen et al. (Eds.): ACCV 2016 Workshops, Part III, LNCS 10118, pp. 606–621, 2017.
DOI: 10.1007/978-3-319-54526-4_44

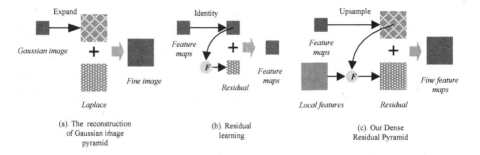

(a). The reconstruction of Gaussian image pyramid

(b). Residual learning

(c). Our Dense Residual Pyramid

Fig. 1. Compare the reconstruction of Gaussian image pyramid, residual learning and our dense residual pyramid networks.

achieved new state-of-the-art results [6,9] in salient object detection, However, CNNs are used to predict the saliency score of a region instead of a pixel. As a result, saliency maps are blurry without fine details, especially the boundary of salient objects.

We aim to design convolutional neural networks for salient object detection which predicts a dense map. Pooling layers in deep networks subsample feature maps, thus the feature maps could be very small. Though deconvolutional layers can be used to refine the small feature maps, most of the information is lost. In order to make use of more local information for refinement, we propose Dense Residual Networks for salient object detection.

Our work is inspired by the reconstruction of Gaussian image pyramid and the residual networks. Denote the images in a Gauss-pyramid and the corresponding Laplacian-pyramid respectively as G_i and L_i ($i = 0, 1...n$), then we have:

$$G_i = F_{expand}(G_{i-1}) + L_i \qquad (1)$$

where F_{expand} is a *Expand* function which doubles the height and width of an image, and G_i is the reconstructed image from G_{i-1} with the residual information from L_i.

In deep neural networks, we can treat G_{i-1} as low-resolution feature maps, and treat G_i as the refined feature maps. But unfortunately, we can't get a residual L_i directly. As a solution, we propose *dense residual blocks* to learn the residual L_i, and we can get the refined feature maps.

The residual networks were introduced by He *et al.* [17], and a simplify of a residual block is shown in the center of Fig. 1. Denote the input features as G'_{i-1} and denote the desired underlying mapping as G'_i, a residual block learns

$$F(G'_{i-1}) := G'_i - G'_{i-1} \qquad (2)$$

where $F(G'_{i-1})$ is residual features, and F is a residual function, which is implemented by convolutional neural networks in [17]. For better compared with Eqs. 1 and 2 can be written as

$$G'_i = G'_{i-1} + F(G'_{i-1}) \qquad (3)$$

We can find the similarity between these two equations.

In this paper, we proposed a novel method based on deep learning and the thought of traditional image processing methods. We build an image pyramid analogous network which employs residual learning to learn the residual maps. In Sect. 5.3, we evaluate the performance of our method, and our results outperform over previous state-of-the-art methods. We also visualize our networks in Sect. 6, results show that the residual pyramid networks successfully recover the feature maps.

2   Related Works

Salient object detection methods can be categorized into bottom-up and top-down methods. Many previous methods are based on low-level visual cues like color, contrast, orientation and texture. For top-down methods, high-level knowledge is used.

Recently, deep learning methods have achieved new state-of-the–art on many computer vision tasks, such as image classification [17], object detection [15], scene labeling [13,18] and semantic segmentation [19,20]. Especially, deep residual [17] networks achieves 3.57% error on the ImageNet test set, and this result won the first place on ILSVRC 2015 classification task.

Since the advantage of deep networks, many deep learning based methods for salient object detection are proposed and achieve state-of-the-art results. But most of them predict a region-wise saliency score. MC [6] and MDF [9] are based on superpixel segmentation [21], thus the final saliency map is blurry.

Similar to many other computer vision tasks, salient object detection can be treated as a dense estimation task. In [19], the authors employ a FCN for semantic segmentation, however, the results are blurry. [18] introduce recurrent convolutional network for scene labeling, the results are refined by recurrently inputting the previous outputs and the source images, but the final results only have a few improvements over the first results. [18] proposed convolutional-deconvolutional networks for semantic segmentation, as most of local information is lost, the boundaries of objects are not clear.

Nevertheless, we propose a novel pixel-wise coarse-to-fine approach for saliency detection. With additional local information, dense residual pyramid networks are able to generate fine feature maps.

3   Dense Residual Networks

3.1   Over View

In our model, we employ pre-trained deep networks (we use VGG16 [22] model in our work) for extracting deep features. We first cut all the fully connected layers in the pre-trained networks, and fine-turn fully convolutional networks (FCN) which can generate small saliency maps as intermediate results. Denote the size of input images as $3 \times W \times H$, the size of intermediate features before the intermediate output is $256 \times \frac{W}{8} \times \frac{H}{8}$. As shown in Fig. 2, the width and

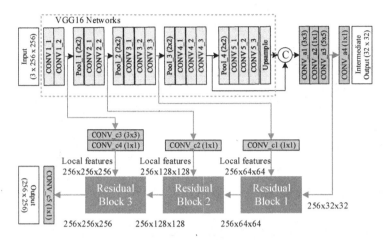

Fig. 2. The illustration of our dense residual pyramid networks. We used a pre-trained VGG16 networks in our framework as shown in the gray dotted box, but the last pooling layer of VGGnet is replaced by an upsample layer. The node "C" stands for a layer which concatenates the inputs in depth.

height of feature maps are doubled after passing though each residual block, and finally, the output saliency maps have the same sizes with the input images. For refinement, we propose residual blocks to combine local features. In many approaches [20], unpooling and deconvolution are used to generate larger feature maps, but the information lost by pooling layers can't be completely recovered. However, in our approach, the local features before pooling layers are extracted and merged into our residual block.

3.2 Residual Block for Refinement

As shown in Fig. 3, as residual networks achieved the state-of-the-art results on image classification, we employ residual architecture in our deep networks. Denoting the input and output respectively as x_{i-1} and x_i, and formula of the original residual block is

$$x_i = F(x_{i-1}) + x_{i-1} \tag{4}$$

where $F(x)$ is a residual function. In our dense residual block, we add deep local features extracted from pre-trained networks as extra inputs. Denote local features as u, the output of dense residual block is

$$x_i = R_i + Up(x_{i-1}) \tag{5}$$

$$R_i = F(x_{i-1}, u_i) \tag{6}$$

where Up is a upsample layer. x_{i-1} are upsampled by scale 2, and x_{i-1} and u_i are concatenated in depth. Then we use four convolutional layers to learn the residual R_i.

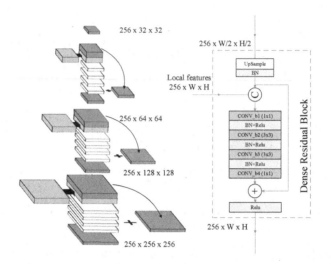

Fig. 3. Left: The illustration of our dense residual pyramid refinement. The nonlinearities and batch normalization layers are not shown. Right: The proposed dense residual block.

For refinement, we aim to generate large and high quality feature maps after each step. In our dense residual block, the width and height of output x_i are twice over the size of x_{i-1}, and additional local information u_i are used. Unlike the unpooling layer which only recovers limited information, we directly extract the local features before the pooling layers in the pre-trained networks. In this way, we hypothesize that all the lost information can be recovered and used to generate high quality feature maps.

4   The Complete Algorithm

4.1   Fine-Tuning for Deep Features

As described in Sect. 3.1, the pre-trained VGG16 networks are used in our method. We only use the convolutional layers and remove all the fully connected layers, so those FC layers are not shown in Fig. 2. Besides, the last pooling layer is replaced with an upsample layer, which generates larger feature maps instead of the smaller. We use a scale rate of 2, so that the width and height of the output feature maps are $\frac{W}{8}$ and $\frac{H}{8}$, which is the same with the width and height of feature maps before Pool_4 layer.

Our method contains two steps, fine-tuning the deep networks to generate features and refining by dense residual blocks. As shown in the top of Fig. 2, for fine-tuning, we extract and concatenate features from two layers of the pre-trained networks, and only the additional four convolutional layers are trained. In our work, the input size is $3 \times 256 \times 256$, so that both the widths and heights of the feature maps output from these four layers are 32.

We use a binary crossentropy function as our loss function, which can be written as Eq. 7

$$L = t_0 \times log(p_0) - (1 - t_0) \times log(1 - p_0) \tag{7}$$

where p_0 is the intermediate prediction of the networks, and t_0 is the target map which is resize to 32×32. As we use a sigmoid function as the activation function of the intermediate output layer, p_0 can be written as

$$p_0 = (1 + EXP(f_{FCN}(x, \Phi, \Theta)))^{-1} \tag{8}$$

where Θ is the parameters of the pre-trained networks, and Φ is the parameters of the added four convolutional layers (from $conv_{a1}$ to $conv_{a4}$) for fine-tuning. x means a input image.

In saliency detection, the resolution of a saliency map is important, as a low-resolution saliency map cannot convey all the information. Even though inputting a huge image lead to a high resolution saliency map as the fully convolutional networks do not limit the input size, the receptive field of a FCN is limited.

4.2 Dense Residual Refinement

Pooling operation loses local information, and we aim to recover these information for generating high-resolution and high-quality saliency maps.

We extract deep features from $conv_{1\_2}$, $conv_{2\_2}$ and $conv_{3\_3}$ in pre-trained networks, and before those features are input into dense residual blocks, they respectively pass through a batch normalization layer and one (for features from $conv_{2\_2}$ and $conv_{3\_3}$) or two (for features from $conv_{1\_2}$) convolutional layers.

As described in Sect. 3.2, local features are added into dense residual blocks. And after each dense residual block, the widths and heights are doubled. Finally, we can get a full size saliency map.

Similar to Eq. 7, we also use a binary crossentropy loss function in training, the difference is the prediction and target maps. The loss function can be written as

$$L = t \times log(p) - (1 - t) \times log(1 - p) \tag{9}$$

where p is the intermediate prediction of the networks, and t is the target map with the size of 256×256. And a sigmoid function is used as the activation function of the final output layer. Thus, p can be written as

$$p = (1 + EXP(f_{DRPN}(x, \Psi)))^{-1} \tag{10}$$

$$\Psi = \{\psi, \Phi', \Theta\} \tag{11}$$

where Ψ is the parameters of the whole $DenseResidualNetworks$, including the parameters of the pre-trained networks(Θ), the parameters of the first three convolutional layers for fine-tuning (Φ', and the intermediate output layer $conv_{a4}$ is not used here), and the parameters of the layers for refinement (ψ, parameters of layers $conv_{bi}$ and $conv_{ci}$).

5 Experiments

5.1 Dataset

We evaluate the performance of our method on eight datasets, including MSRA10K [23], HKU-IS [9], ASD [24], iCoseg [25], ECSSD [26], DUT-OMRON [27], ImgSal [28] and Pascal-S [29]. MSRA10K is a large dataset including 10000 images. Most of the images has a single salient object, and we use many training instances from this dataset. We randomly divided this dataset into three parts, 8000 for training, 1000 for validation and 1000 for testing. HKU-IS was recently introduced by Li and Yu [9]. Images in his dataset contains many multi-object, big and small salient instances. iCoseg was firstly introduced for image segmentation, and recently it was used for evaluating saliency detection. ASD is popularly used for evaluating the performance of salient object detection approaches, but there are some overlaps between the instance of ASD and our training set of MSRA10K. We remove the overlapped images in the ASD dataset, thus about 950 images are used for evaluation. ECSSD [26] is also widely used for evaluating the performance of algorithm, which is more challenging than MSRA10K.

In our experiments, we use the training part from MSRA10K and the training set of the HKU-IS. We use the split provided by [9], thus in HKU-IS dataset, there are 2,500 images for training, 500 for validation and 1447 for testing. MSRA10K is a large dataset but most of the salient objects in this dataset are single big objects, while HKU-IS contain many images with multi salient objects. Thus we combine the training set of these two datasets to train our networks.

5.2 Implement Details

The details of our Dense Residual Networks are shown in Table 1, and all the three residual blocks have same configurations in our Implement. The details of the proposed dense residual block is shown in Table 2. For training, we resize and pad all the images to the size of 256×256 keeping the aspect ratio.

Table 1. The details of the proposed dense residual networks. The type of a layer, size and number of the kernels are listed. Where DRB means dense residual block. Batch normalization is used in $conv_{c1}$, $conv_{c2}$ and $conv_{c4}$

Layers	upsample	concat	$conv_{a1}$	$conv_{a2}$	$conv_{a3}$	$conv_{a4}$	$conv_{c1}$
Type	upsample	concat	conv/relu	conv/relu	conv/relu	conv/relu	conv/relu/bn
Size	/	/	3	1	5	1	1
Num	/	/	1024	512	256	1	256
Layers	Block1	$conv_{c2}$	block2	$conv_{c3}$	$conv_{c4}$	block3	$conv_{c5}$
Type	DRB	conv/relu/bn	DRB	conv/relu	conv/relu/bn	DRB	conv/sigmoid
Size	/	1	/	3	1	/	1
Num	/	256	/	128	256	/	1

Table 2. The details of the proposed dense residual blocks, where *sum* is a elementwise summation layer.

Layers	*upsample*	*concat*	$conv_{b1}$	$conv_{b2}$	$conv_{b3}$	$conv_{b4}$	*sum*
Type	upsample/bn	concat	conv/bn/relu	conv/bn/relu	conv/bn/relu	conv/linear	sum/relu
Size	/	/	1	3	3	1	/
Num	/	/	256	256	256	256	/

Given training images and their target saliency map, we first train the neural network on the upper of Fig. 2 alone to obtain its initial weights ($conv_{a1}$, $conv_{a2}$ and $conv_{a3}$). Deep features are extracted using the VGG16 network [22] pre-trained over the ImageNet dataset. During training, we use a fixed learning rate 0.001 to fine-turn the $conv_{a1-a4}$ layers, and we use a learning rate 1×10^{-5} to train the layers in the VGG16 networks. After this initialization, we fix the parameters that we trained in the first step, and train the refinement streams ($conv_{c1-c5}$ and layers in the three residual blocks) of our dense residual networks. We also use a learning rate 0.001 to train this stream.

5.3   Comparison with the State of the Art

We compare the performance of our method with several state of the art methods. Including MDF [9], MC [6], DRFI [8], RRWR [30], RC [23], HS [26] and FT [24]. We use eight datasets described in Sect. 5.1 to evaluate the performances, including HKU-IS, MSRA10K, iCoseg, ASD, ECSSD, DUT-OMRON, ImgSal and Pascal-S. Due to that MC used MSRA10K for training, we don't show the results of MC on MSRA10K.

We randomly select some saliency maps of these methods and show them in Fig. 4. The proposed method *DRPN* can better detect salient objects, and our method is more similar to the ground truth.

The same as [9], we draw the PR-curve and calculate F-Measure to evaluate the performances of salient object detection methods. The results are shown in Figs. 5 and 6. And following [24,29], we use $\beta^2 = 0.3$ to calculate the F-Measure from PR curve.

$$F_\beta = \frac{(1 + \beta^2) Precision \times Recall}{\beta^2 \times Precision + Recall} \qquad (12)$$

As shown in Figs. 5 and 6, our *DRPN* has a huge improvement over previous state-of-the-art methods. On ECSSD, our method outperforms MDF over 11.6%, and better than MC about 9%. And on HKU-IS, we still have an advantage of 6.2% over MDF. And on average, our *DRPN* outperforms these state-of-the-art methods over 5%.

And according to [6,9], with the same input size, MC cost 1600 ms to process an image on a TITAN GPU, and MDF cost 8000 ms on a TITAN Black GPU. While our *DRPN* cost 150 ms on a TITAN X GPU. It's obvious our method is more efficient than these two deep learning based methods.

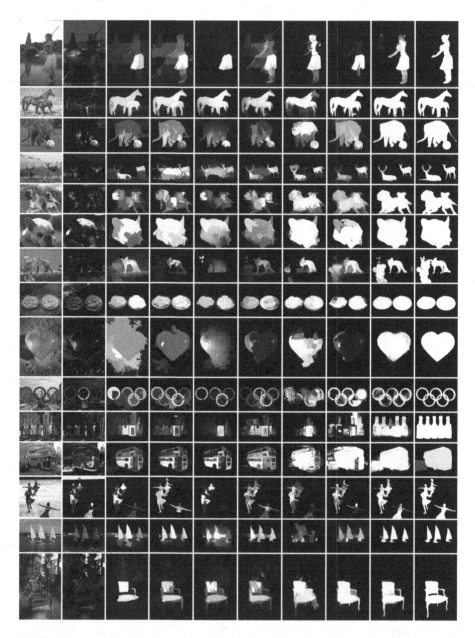

Fig. 4. Compare the saliency maps generate by eight algorithms. From left to right: source images, FT [24], HS [26], RC [23], RRWR [30], DRFI [8], MC [6], MDF [9], our *DRPN* and the ground truth.

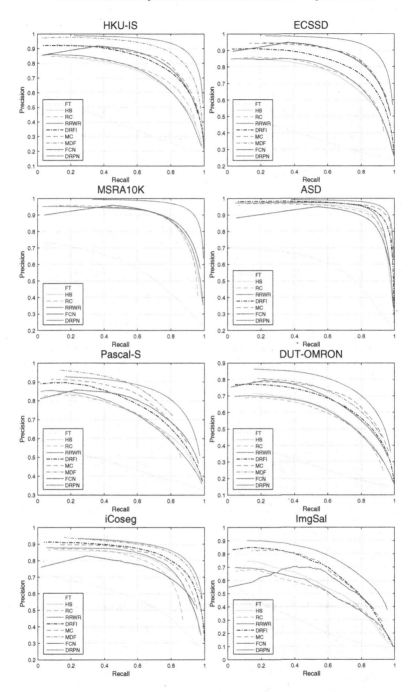

Fig. 5. Compare the performance of our method and other seven algorithm using PR-curves on eight datasets. From top left to bottom right, including HKU-IS [9], ECSSD [26], MSRA10K [23], ASD [24], Pascal-S [29], DUT-OMRON [27], iCoSeg [25] and ImgSal [28].

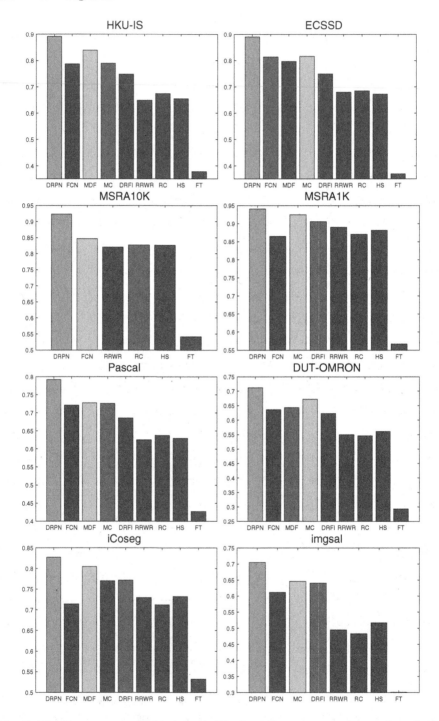

Fig. 6. The F-measure on eight dataset. The best F-measure results are in red, the second are in green, and the third are in light blue. (Color figure online)

6   Visualization and Discussion

In this section, we visualize our networks. With the input image as shown on the top of Fig. 7, we extract and visualize the feature maps from layer $conv_{b3}$

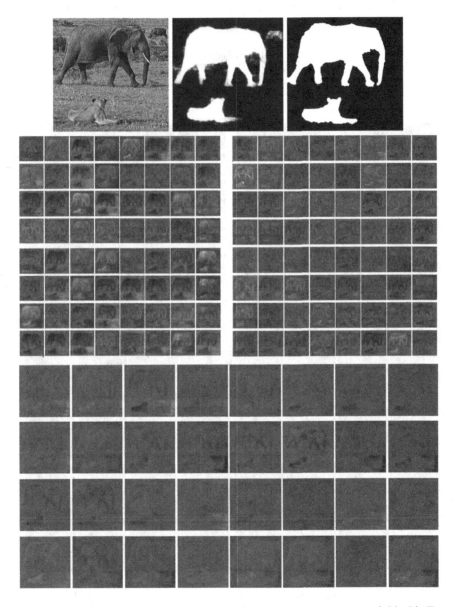

Fig. 7. Visualizing the residual feature maps (output from $conv_{b4}$ in each block). From top left to right bottom: the input image, the final output of our networks, the ground truth, the 64 residual maps in block1, the 64 residual maps in block2, the 32 residual maps in block3.

and the output feature maps from each block. The former are residual feature maps which are shown in Fig. 7, and the latter are refined feature maps shown in Fig. 8.

As shown in Fig. 7, the dense residual maps tend to recover the feature maps in multi-scale space. The first group of residual maps learn both region level and coarse edge features; the second group of residual maps mainly learn coarse edge residuals, while the third residual block learns fine edge information.

The outputs of three dense residual blocks are shown in Fig. 8. Even though the last results are 4 times larger then the outputs of the first residual block, they are not at all blurry.

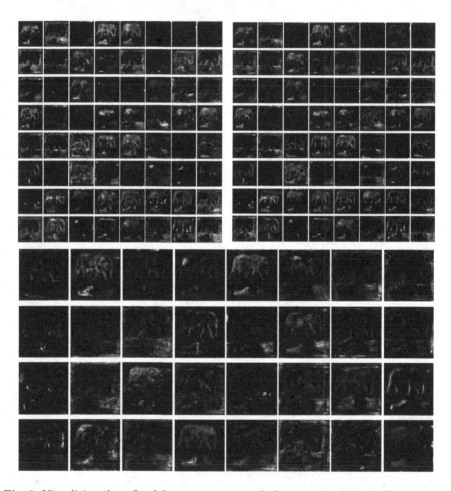

Fig. 8. Visualizing the refined feature maps in each dense residual block. From top left bottom right: 64 fine feature maps output from block1, 64 fine feature maps output from block2, 32 fine feature maps output from block3. The size of the feature maps increase after getting through a upsample layer, but with the learned residual information, the feature maps won't be blurry.

7 Conclusion

We propose a novel method for salient object detection. As pooling layers generates down sampled feature maps, we introduce dense residual pyramid networks to refine the feature maps. The feature maps are doubled and refined after getting through each residual block. Via visualization, we can find that the residual feature maps in different *denseresidualblock* concentrate on different scale, and the most importantly, the final feature maps is as clear as the previous small feature maps, even though they are magnified many times. And we evaluate the performance and the speed of our method, results shows our *DRPN* has huge improvement over previous state-of-the-art methods.

Acknowledgement. This work was supported by the National Natural Science Foundation of China under Grant Nos: 61273366 and the program of introducing talents of discipline to university under grant no: B13043 and the National Key Technology R&D Program: 2015BAH31F01.

References

1. Han, S.H., Jung, G.D., Lee, S.Y., Hong, Y.P., Lee, S.H.: Automatic salient object segmentation using saliency map and color segmentation. J. Cent. South Univ. **20**, 2407–2413 (2013)
2. Sun, J., Lu, H., Li, S.: Saliency detection based on integration of boundary and soft-segmentation. In: 2012 19th IEEE International Conference on Image Processing (2012)
3. Klein, D.A., Frintrop, S.: Center-surround divergence of feature statistics for salient object detection. In: 2011 International Conference on Computer Vision (2011)
4. Tong, N., Lu, H., Zhang, Y., Ruan, X.: Salient object detection via global and local cues. Pattern Recogn. **48**, 3258–3267 (2015)
5. He, S., Lau, R.W.H.: Saliency detection with flash and no-flash image pairs. In: Fleet, D., Pajdla, T., Schiele, B., Tuytelaars, T. (eds.) ECCV 2014. LNCS, vol. 8691, pp. 110–124. Springer, Heidelberg (2014). doi:10.1007/978-3-319-10578-9_8
6. Zhao, R., Ouyang, W., Li, H., Wang, X.: Saliency detection by multi-context deep learning. In: 2015 IEEE Conference on Computer Vision and Pattern Recognition (CVPR) (2015)
7. Liu, T., Yuan, Z., Sun, J., Wang, J., Zheng, N., Tang, X., Shum, H.Y.: Learning to detect a salient object. IEEE Trans. Pattern Anal. Mach. Intell. **33**, 353–367 (2011)
8. Huaizu, J., Jingdong, W., Zejian, Y., Yang, W., Nanning, Z., Shipeng, L.: Salient object detection: a discriminative regional feature integration approach. In: 2013 IEEE Conference on Computer Vision and Pattern Recognition (2013)
9. Li, G., Yu, Y.: Visual saliency based on multiscale deep features. In: 2015 IEEE Conference on Computer Vision and Pattern Recognition (CVPR) (2015)
10. Rahtu, E., Kannala, J., Salo, M., Heikkilä, J.: Segmenting salient objects from images and videos. In: Daniilidis, K., Maragos, P., Paragios, N. (eds.) ECCV 2010. LNCS, vol. 6315, pp. 366–379. Springer, Heidelberg (2010). doi:10.1007/978-3-642-15555-0_27

11. Khuwuthyakorn, P., Robles-Kelly, A., Zhou, J.: Object of interest detection by saliency learning. In: Daniilidis, K., Maragos, P., Paragios, N. (eds.) ECCV 2010. LNCS, vol. 6312, pp. 636–649. Springer, Heidelberg (2010). doi:10.1007/978-3-642-15552-9_46
12. LeCun, Y., Boser, B., Denker, J.S., Henderson, D., Howard, R.E., Hubbard, W., Jackel, L.D.: Backpropagation applied to handwritten zip code recognition. Neural Comput. 1, 541–551 (1989)
13. Farabet, C., Couprie, C., Najman, L., LeCun, Y.: Learning hierarchical features for scene labeling. IEEE Trans. Pattern Anal. Mach. Intell. 35, 1915–1929 (2013)
14. Krizhevsky, A., Sutskever, I., Hinton, G.E.: ImageNet classification with deep convolutional neural networks. In: Pereira, F., Burges, C.J.C., Bottou, L., Weinberger, K.Q. (eds.) Advances in Neural Information Processing Systems, vol. 25, pp. 1097–1105. Curran Associates, Inc., New York (2012)
15. Girshick, R., Donahue, J., Darrell, T., Malik, J.: Rich feature hierarchies for accurate object detection and semantic segmentation. In: 2014 IEEE Conference on Computer Vision and Pattern Recognition (2014)
16. He, S., Lau, R.W.H., Liu, W., Huang, Z., Yang, Q.: SuperCNN: a superpixelwise convolutional neural network for salient object detection. Int. J. Comput. Vis. 115, 330–344 (2015)
17. He, K., Zhang, X., Ren, S., Sun, J.: Deep residual learning for image recognition (2015)
18. Pinheiro, P., Collobert, R.: Recurrent convolutional neural networks for scene labeling. In: International Conference on Machine Learning, pp. 82–90 (2014)
19. Long, J., Shelhamer, E., Darrell, T.: Fully convolutional networks for semantic segmentation. In: 2015 IEEE Conference on Computer Vision and Pattern Recognition (CVPR) (2015)
20. Noh, H., Hong, S., Han, B.: Learning deconvolution network for semantic segmentation. In: 2015 IEEE International Conference on Computer Vision (ICCV), pp. 1520–1528 (2015)
21. Achanta, R., Shaji, A., Smith, K., Lucchi, A., Fua, P., Susstrunk, S.: SLIC superpixels compared to state-of-the-art superpixel methods. IEEE Trans. Pattern Anal. Mach. Intell. 34, 2274–2282 (2012)
22. Simonyan, K., Zisserman, A.: Very deep convolutional networks for large-scale image recognition. CoRR abs/1409.1556 (2014)
23. Cheng, M.M., Mitra, N.J., Huang, X., Torr, P.H.S., Hu, S.M.: Global contrast based salient region detection. IEEE Trans. Pattern Anal. Mach. Intell. 37, 569–582 (2015)
24. Achanta, R., Hemami, S., Estrada, F., Susstrunk, S.: Frequency-tuned salient region detection. In: 2009 IEEE Conference on Computer Vision and Pattern Recognition (2009)
25. Batra, D., Kowdle, A., Parikh, D., Luo, J., Chen, T.: iCoseg: interactive cosegmentation with intelligent scribble guidance. In: 2010 IEEE Computer Society Conference on Computer Vision and Pattern Recognition (2010)
26. Yan, Q., Xu, L., Shi, J., Jia, J.: Hierarchical saliency detection. In: 2013 IEEE Conference on Computer Vision and Pattern Recognition (2013)
27. Yang, C., Zhang, L., Lu, H., Ruan, X., Yang, M.H.: Saliency detection via graph-based manifold ranking. In: 2013 IEEE Conference on Computer Vision and Pattern Recognition (2013)
28. Li, J., Levine, M.D., An, X., Xu, X., He, H.: Visual saliency based on scale-space analysis in the frequency domain. IEEE Trans. Pattern Anal. Mach. Intell. 35, 996–1010 (2013)

29. Li, Y., Hou, X., Koch, C., Rehg, J.M., Yuille, A.L.: The secrets of salient object segmentation. In: 2014 IEEE Conference on Computer Vision and Pattern Recognition (2014)

30. Li, C., Yuan, Y., Cai, W., Xia, Y., Feng, D.D.: Robust saliency detection via regularized random walks ranking. In: 2015 IEEE Conference on Computer Vision and Pattern Recognition (CVPR) (2015)

Quantitative Analysis of a Bioplausible Model of Misperception of Slope in the Café Wall Illusion

Nasim Nematzadeh$^{(\boxtimes)}$, David M.W. Powers, and Trent Lewis

CSEM, Flinders University, Adelaide, Australia
{nasim.nematzadeh,david.powers,trent.lewis}@flinders.edu.au

Abstract. This paper presents a model explaining tilt illusion effect in the Café Wall pattern. In this geometric illusion, we perceive horizontal edges as tilted. We explain this as the result of *innate retinal/gangliar* visual processing of the pattern. Our bioplausible model is based on a simple early layer using Difference of Gaussian over simple ON-center and OFF-center receptive fields, with a quantification module replacing later layers of a Deep Neural Network. The experimental results show that this bioplausible filtering technique can explain the tilt illusion of the Café Wall pattern. Our statistical analysis of tilt provides a quantitative measurement and an empirically testable prediction for the degree of tilt. This shows that the Difference of Gaussian reveals cues for perception and clues about the illusions we perceive.

1 Introduction

Geometric or tilt illusions are optical illusions that are perceived differently from their actual geometry, e.g. parallel lines are perceived as not being parallel or straight lines as bulges or bows. The Café Wall illusion is studied here with its illusory perception of convergent and divergent mortar lines rather than parallel lines.

The geometrical interpretation of visual input is the end result of a process involving both low level retinal and cortical processing and high level cortical processing including feedback from previous experience and interactions (binding) with other modalities, and is thus a natural framework for bioplausible understanding of deep neural processing in vision. We present the earliest stages of a deep neural network (DNN) using a Difference of Gaussian (DoG) model, and replace the later layers with a standard (Hough) edge detection pipeline to discover and quantify the parameters of the perceived edges. Our assumption here is that DNNs (involving autoassociation/autocorrelation/autoencoding, e.g. ART, RBM, LSTM) encode features that are edge-, curve- or stroke-like [1, 10, 13, 18, 34, 35].

Visual processing starts in the retina as light from the image is focused on the photoreceptors in the retina and generates synaptic signals that feed into the middle layer, where *'lateral inhibition'* (LI) [30] enhances the synaptic signal

© Springer International Publishing AG 2017
C.-S. Chen et al. (Eds.): ACCV 2016 Workshops, Part III, LNCS 10118, pp. 622–637, 2017.
DOI: 10.1007/978-3-319-54526-4_45

as activated cells inhibit the activations of nearby cells. This retinal convolution is specified as a pulse response or *point spread function* (PSF) that has an edge enhancement effect [12] and acts as a band pass filter. In the final layer we find retinal ganglion cells (RGCs) whose axons carry the visual signal to the cortex. The contrast sensitivity of RGCs in layer two is the underlying mechanism which is implemented in our model with circular center and surround receptive fields (RFs) [6,25,26,32].

Physiological and psychophysical findings support multiscale representation and processing in the visual cortex and the retina [7,8,13]. Earlier, it was assumed that orientation detection takes place solely in the cortex, but recent retinal research has significantly extended our understanding of RGCs. Field and Chichilnisky [7] published a comprehensive study about retinal circuitry and coding, mentioning at least 17 distinct RGC types in the retina, each with a specific encoding role. The intra-retinal circuitry, with the variations of RFs type and their size change in relation to distance from the fovea (eccentricity), are all indicating the underlying multiscale encoding inside the retina. Some retinal cells have the orientation selectivity property similar to the cortical cells [7,8], all support the raw to full primal sketch of Marr's idea in vision [22].

Several authors have offered explanations of the Café Wall illusion, such as 'border locking' [9] and 'phenomenal model' [16] at a high level, and 'brightness assimilation and contrast' [15] and 'band-pass spatial frequency' [24] at a low level. While they appear to explain the illusion at different levels of processing, many have common features including lateral inhibition and the *suppression effect* [12,29,30]. A hybrid retino-cortical explanation for Café Wall illusion has been proposed by Westheimer [36] recently, which considers light spread, compressive nonlinearity and center-surround transformation (retinal) and sharp straight borders, pointed corners and angle shifts (cortical processing).

Our contribution is to demonstrate computationally that early DoG processing does lead to tilted features, which we analyze quantitatively to predict specific angles at specific resolutions (or visual angles) that are capable of empirical validation with human subjects.

2   Café Wall Illusion and Edge Detection

The more general class of tile illusion patterns have tilt, bulge, or wave inducing effects and their explanation requires techniques to highlight these cues. Applying simple edge detection and edge sharpening reveals the outlines, for instance in the Café Wall the tile's borders and mortar, but no tilt is detected to explain the illusion.

A satisfactory explanation of an optical illusion requires both computational feasibility and biological plausibility and a computational model thus can be a neurological theory of how the world is really represented in the brain. The tilt illusion in Café Wall pattern is characterized by the appearance of slanted line segments along the mortar lines [5,24], making the tiles seem wedge-shaped [9]. At a global level we see alternating converging and diverging mortar lines.

Because the effect of tilt illusion in Café Wall is highly directional, it raises the question of whether lateral inhibition and point spread function of RFs can explain the tilt effect in the pattern or not. We will show that a bioplausible model [25,26], with a circular center and surround organization for retinal GCs, a Classical RF (CRF) model [6,32], is able to predict the tilt illusion in Café Wall pattern. There is no need to utilize complex models of non-CRFs [2,28] for the retinal/cortical orientation selective cells implementation for explaining this illusion.

The significance of our contribution is to show that at lower levels of visual processing, the underlying geometrical cues of illusions start to appear. Many tile illusions are commonly considered to be high level illusions [9,16,36] and our low level explanation [25,26] thus has significant implications for human and computer vision and may facilitate development of more general model of illusion.

3   The Bioplausible Model

There is physiological evidence [7,8] that retinal cells have a diverse range of receptive fields of different size and type as a function of eccentricity (centrality) [20]. This suggests a multiscale retinal encoding of the visual scene. The scientific understanding of retinal receptive field processing is that the multiscale retinal representation of visual data is "scale-invariant" and there is an adaptation of the receptive field sizes to textural elements inside our field of view [12,20,33].

Kuffler [17] demonstrated that the receptive fields are roughly concentric, with the excitatory center and inhibitory surround or vice versa. Rodieck and Stone [32] and Enroth-Cugell and Robson [6] showed that the signals from the center and surround regions can be modeled by a weighted sum of photoreceptor outputs by two concentric Gaussians, with the different radii determining the ganglion cell response [19,31].

Computational studies of early visual processing and modeling of the physiological mechanisms of simple cells were reported by Marr and Ullman [23], inspired by the simple cell's directional sensitivity in the primate visual cortex. A CRF model for retinal GCs responses [25,26] is used to explain the Café Wall illusion and is the basis for quantitative measurement of tilt angle.

3.1   Formal Description and Parameters

The main stage in our experiment uses the DoG model (Fig. 1) to extract a bioplausible representation for the image (sample Café Wall image and feature maps are shown in Fig. 2). This feature representation is interpretable as an edge map in which the edges clearly reflect the perceived tilt. The RFs have a bioplausible implementation base on DoG. We simulate the DoG model on an image (illusion), and complete a quantitative analysis of the angles of the tilted lines in the feature map (Fig. 3).

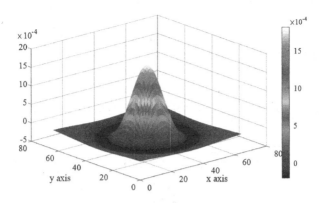

Fig. 1. DoG filter displayed in jetwhite colormap (downloadable from MathWorks central file exchange), with $\sigma_c = 8$, $\sigma_s = 16$, $h = 8$.

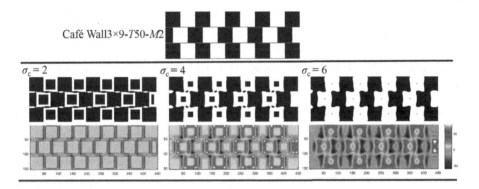

Fig. 2. Top: 50px Tile, 2px Mortar stimulus. Middle: Binary representation of the edge map with DoG scale (σ_c) ranges from 2 to 6. The other two parameters are constant: $s = 2$, $h = 8$. Bottom: Jetwhite colormap representation of above multiple scale edge map. (Reproduced by permission from [27])

Applying a Gaussian filter to an image creates a smoothed or blurred version. The DoG output of an image is the difference between such two blurred versions, and DoG processing acts as a band pass filter (Fig. 1). For a 2D signal such as image I, the DoG output of our retinal GCs model with antagonistic centre and surround organization is given by:

$$\Gamma_{\sigma,s\sigma}(x,y) = I * \frac{1}{2\pi\sigma^2}e^{-(x^2+y^2)/(2\sigma^2)} - I * \frac{1}{2\pi s\sigma^2}e^{-(x^2+y^2)/(2s^2\sigma^2)} \quad (1)$$

$$s = \sigma_{\text{surround}}/\sigma_{\text{center}} = \sigma_s/\sigma_c \quad (2)$$

where x and y represent the distance from the origin in the horizontal and vertical axes respectively and σ indicates the sigma of the centre Gaussian (σ_c). $s\sigma$ represents the sigma of the surround Gaussian (σ_s). This concentric representation

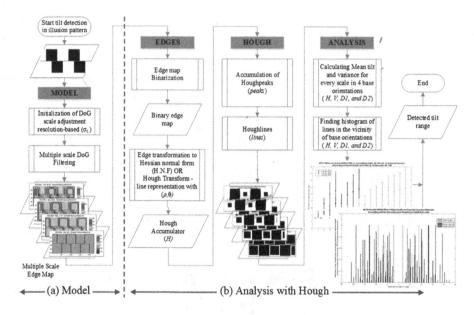

Fig. 3. Flowchart of experimental process. (Reproduced by permission from [27])

of the center and surround Gaussians, models the retinal PSF and LI inside layer two of the retina [6,19,32].

Taking the difference of two DoGs estimates the second derivative of the Gaussian, that is the Laplacian of Gaussian (LoG), and can provide more shades and shadows around the edges with more depth information. For modeling the receptive field of retinal GCs, DoG [19,31] is a good approximation of LoG, if the ratio of dispersion of centre to surround is about $s \approx 1.6$ ($\approx \phi$, the ubiquitous Golden Ratio) [22]. By increasing s, we achieve a wider area of surround suppression, although the height of the surround Gaussian declines. $s = 2$ is used in experimental runs.

In implementing the model, it is inconvenient and implausible to deal with Gaussians of unbounded extent, so we apply the DoG model only within a window of a size chosen so that the value of both Gaussians is negligible outside the window (e.g. less than 5% for the surround Gaussian) and we therefore control $windowSize$ as large windows have high computational cost. $windowSize$ is determined by h parameter defined as below:

$$windowSize = h \times \sigma_c + 1 \qquad (3)$$

where parameter h determines how much of the center Gaussian and the surround Gaussian are included in the filter.

Variations of DoGs have been tested with different values of s ($1.5 < s < 3.5$), while keeping σ_c and h constant. The result indicates that the final output of the model is not very sensitive to this modification range.

In particular, $h = 2$ corresponds to the diameter of the centre Gaussian, the part between the inflection points (68% centre, 31% surround); $h = 4$ corresponds to the diameter of outer Gaussian (95% centre, 68% surround); and $h = 8$ corresponds to the standard $p < 0.05$ significance for the outer Gaussian (99.94% centre, 95% surround). $h = 8$ is used in experimental runs.

3.2 DoG Representation of an Image

Figure 2 shows the output of the bioplausible model with DoG scales of 2, 4, and 6, on a Café Wall image with the size of 3×9 Tiles, where Tile sizes are 50×50px and Mortar size is 2 pixels. For easier naming of patterns, a few name tags are defined, representing 'tile size' with T and 'mortar size' with M followed by their values. So the pattern is referred to as Café Wall3×9-$T50$-$M2$.

We need to quantify the tilt of the lines induced by the DoG transformation so that we can compare with the tilt perceived by a human observer, and so embed the model in a processing pipeline involving multiple standard image processing transformations.

3.3 Image Processing Analysis Pipeline

We therefore turn to discuss the pipeline to extract edges and find and analyze their angles of tilt, as shown in Fig. 3.

MODEL - DoG Representation. For convenience, we focus on small cropped sections of the original image and this is illustrated in Fig. 4, with its DoG representation for a different range of σ_c varied from 1 to 4. The crop is selected from Café Wall3×9-$T50$-$M2$. The crop specifically selected in such a way to contain one single slanted line segment along a mortar line.

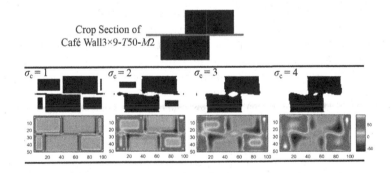

Fig. 4. Top: A crop section of Café Wall3×9-$T50$-$M2$. Middle: Binary representation of the edge map with the σ_c ranges from 1 to 4, $s = 2$, $h = 8$ are constant. Bottom: Jetwhite colormap representation of the above edge map. (Reproduced by permission from [27])

The most fundamental parameter in the model is the diameter of the center Gaussian (σ_c), and this value is highly correlated with the pattern elements and their characteristics, in this case the mortar size. To extract the tilted line segments along the mortar lines, we need σ_c to be of the same order as the mortar size. We undertook empirical test simulations to find an appropriate range for σ_c, testing range of $1/2M$ to $2M$ with incremental step of $1/2M$. Similar DoG representation is used by others in the deepest layer of a DNN [21].

The output in Figs. 2 and 4 show that by increasing the scale, the corner effect gets highlighted. This effect with the appearance of tilted line segments has a consequence on the appearance of square tiles to look similar to trapezoids, inducing convergent and divergent mortar lines. The trapezoid shape referred to as wedges in some previous studies on the Café Wall pattern and declared to be the explanation for the tilt effect in the Café Wall illusion [9]. On the other hand, the output of lower scale DoG filters preserve the outline and shape of the tiles and the connectivity of them by the mortar line. This local connectivity of tiles in adjacent rows by the mortar line, start to get disconnected when the scale of the filter is increased.

EDGES - Hesse Normal Form Representation - Hough Transform. The output of the model is an edge map based on DoGs. Such edge maps, often derived from multiple DoG scales, are common intermediate levels in a typical DNN [21]. In place of the usual higher level layers of the DNN, we introduce an analysis to characterize the tilted line segments present in this edge map. The first processing step for this extraction of line orientation and length information is to represent the edge pixels of the pattern in the *'Hesse normal form'* (HNF) [4]. First, the edge map should be converted to a binary map. The HNF and the voting procedure are sometimes referred to as Hough transform (HT).

In image processing, the Standard Hough Transform (SHT) [11,14] is for representing straight lines inside the image and uses a two-dimensional array called the accumulator with the dimensions equal to quantized values of ρ and θ in the pair (ρ, θ) where θ is in the range of $[0, \pi)$. In this representation, every possible line through the image is characterized by a (ρ, θ) pair that is the HNF, being a point in Hough space. All pixels that sit anywhere on a line are counted inside a corresponding bin in the accumulator matrix. Conversely, every edge pixel (x, y) in the image space, corresponds to a sinusoidal curve in (ρ, θ) space as given by (4), with θ as free parameter corresponding to the angle of the lines passing through the point (x, y) in image space:

$$\rho = x.\cos\theta + y.\sin\theta \tag{4}$$

where ρ represents the perpendicular distance between the line passing through that point with a specific θ and the origin, and θ is the counter-clockwise angle between the normal vector (ρ) and the positive direction of the x-axis. Therefore, two or more points that form a line in image space will produce sinusoids which cross at the point of (ρ, θ) in Hough space. So the output of the Hough transform is a two-dimensional accumulator matrix H, with the dimension of $\rho \times \theta$.

Each element of the matrix corresponds to the sum of the points or pixels that are located on the line represented by quantized parameters of (ρ_i, θ_i). For computing the Hough Transform of a binary image, the MATLAB hough function is used. The result of hough is all possible lines that could pass through every edge point in the edge map, and certainly not all of these lines are of interest.

HOUGH - Lines at Local Maxima of the Accumulator Matrix. In the EDGE-processing stage, all possible lines that could pass through every edge point in the edge map are extracted but we are just interested in the detection of tilt induction line segments inside the Café Wall image. The local maxima in the accumulator space show the most likely lines that can be extracted, and their approximate geometric definitions of (ρ, θ) values.

We have exploited two MATLAB functions called houghpeaks and houghlines that rely on further processing of the data in the accumulator matrix (H). The values of the parameters defined in these two functions should be selected based on the resolution of the Café Wall pattern and the scale of the bioplausible model in order to limit the line detection to the desired tilted line segments with the tilt inducing effect. More information about their parameters is explained below.

- The houghpeaks function finds the peaks in the Hough accumulator matrix H. Its parameters are *numPeaks* (maximum number of lines and line segments to be detected), *threshold* (threshold value of H for searching the peaks), and *NHoodSize* (the size of the suppression neighborhood that is set to zero after the peak is identified).
- The houghlines function extracts line segments from the edge image associated with particular bin in a Hough accumulator matrix (H). Its parameters are *FillGap* (the distance between two line segments associated with the same Hough bin. If the gap between them is shorter than this value, they are merged into a single line segment), and *MinLength* (specifies keeping or discarding the merged lines. Lines shorter than this value are discarded).

Figure 5 illustrates a sample output of HOUGH analysis stage in the model. The investigated pattern is a crop section of the Café Wall3 × 9-T800-M32 (Fig. 6-Right). The detected line segments are shown in green, displayed on a binarized edge map with four DoG scales (16, 32, 48, and 64). Blue lines indicate the longest detected line segment. Further explanations about the values of all parameters used in the model and the analysis are given in Sect. 4).

The absolute value of mean tilt and standard deviation are then calculated as a quantitative measurement for tilt deviation. The detected lines in the vicinity of four base orientations (Horizontal (H), Vertical (V), positive Diagonal $(+45°, D1)$, and negative Diagonal $(-45°, D2)$) are saved here for further processing. Applying Hough transform for line detection involves heuristics, rasterization and quantization [4] errors, so we present mean tilt and variance.

The parameters of houghpeaks and houghlines functions have been selected empirically for correct detection of slanted line segments. For instance, *MinLength* should be larger than T (TileSize) to not detect the outline of tiles.

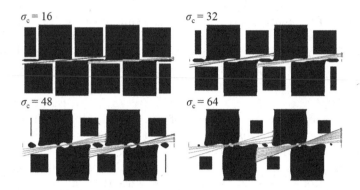

Fig. 5. HOUGH stage result. Detected tilted line segments displayed in green on a four scale edge map of cropped CW3 × 9-T800-M32 pattern. The DoG scales are 16, 32, 48, and 64. Blue lines have highlighted the longest lines. (Color figure online) (Reproduced by permission from [27])

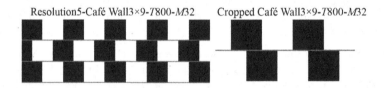

Fig. 6. Left: 800px Tile (T), 32px Mortar (M) stimulus. Right: Cropped section of the left pattern, with the size of $4.5 \times T$ (width), and $2 \times T + M$ (height), enlarged.

On the other hand FillGap parameter should fill the gap that starts to appear by DoG filtering and get larger in size when the scale of the DoG filter increases.

ANALYSIS. The information of the detected line segments including their orientations, are saved inside four matrices based on how close they are to one of the base orientations defined here for analysis. The four base orientations are horizontal (H), vertical (V), positive diagonal ($+45°$, $D1$), and negative diagonal ($-45°$, $D2$), and for the neighborhood around them, an interval of $[-22.5°, 22.5°)$ is chosen to cover the whole space. The parameter values defined in the implementation for tilt detection of the Café Wall, categorize the detected line segments in the horizontal class. Indeed, the perception of tilt in the Café Wall is near horizontal, except for the time when the distance is very far, and the model prediction should be the same. When the DoG scale is quite large compared to the mortar size, the near horizontal tilt is replaced by zigzag vertical joining of similar colored tiles [25,26].

The base orientations could be selected in a smaller angular interval such as in every 15° or 30° angles, but for the tile illusions, 45° increment for base orientations is enough as there are only horizontal and vertical edges. However the range of 15° or 30° might be well matched to other geometric illusions and better fit the way humans distinguish between angles.

The statistical analysis of the detected lines in the neighborhood of each base orientation is the output of this stage and includes the mean and standard deviation around the mean for each scale of the DoG. Also the histograms of line segments on their angular positions around base orientations can be graphed, representing the prominent tilt orientation. As DoG scale increases, both the mean tilt and standard deviation increases and the distribution of detected line segments gets more spread, covering more angular tilt range.

4 The Experimental Results

In this section, we explain two sample experiments for evaluating the model. The bioplausible model used here is similar to Robson's [31] in which finds retinal GCs response to the pattern (image). He claimed that a model of ganglion cell RF with antagonistic center and surround organization of DoG can predict the cell response to any arbitrary stimulus pattern. What is needed is finding the convolution between pattern's luminance function with the difference of Gaussian that is the weighting function of ganglion cells RFs. We now illustrate the details of two main experiments conducted.

4.1 Experiment One - Resolution Analysis

The aims of this experiment are to confirm that the resolution of an image does not affect the detected range of tilt, if their perceived sizes (visual angle) are the same and the DoG filtering is kept to scale. Viz. different resolutions of Café Wall pattern with the same perceived size either displayed on monitor or printed on paper, should have approximately the same tilt effect. From a practical perspective, this exploration of the role of resolution and subjective identification of the range of resolution that is reasonable to consider, is important to allow us to focus our quantitative experiments.

There is a close relationship between the resolution of pattern and the pattern's visual angle. The visual angle shows the person's subjective experience of how large an object or pattern's element looks. Stating the visual angle is a direct way to simply reference the size of the image on the retina rather than having to include both the size and distance parameters as it makes (apparent) size independent of distance.

It is possible that two distal stimuli subtend the same visual angle even though they have different extents. So, the visual angle as the pattern reference in retinal stimulation (or retinal image) is a more appropriate measure than the extent of the distal stimulus. We have not performed any psychophysical tests here therefore the image resolution is used as a reference not the visual angle.

For the resolution analysis experiment, five different resolutions of the Café Wall3 × 9 pattern have been generated. The characteristics of these variations of the Café wall3 × 9 patterns are given in Table 1. An identical cropped window

Table 1. Arithmatic Mean ± Standard Deviation of tilt for different resolutions of cropped Café Wall3 × 9 pattern, and the parameters of the model.

	Resolution1	Resolution2	Resolution3	Resolution4	Resolution5
Tilt and Mortar size (T, M)	50,2	100,4	200,8	400,16	800,32
$\sigma_c = (1/2 mortar)$	1	2	4	8	16
Tilt mean and variance	3.3 ± 1.2	3.6 ± 1.2	4 ± 1	4.5 ± 0.5	3.6 ± 1.24
$\sigma_c = (1 mortar)$	2	4	8	16	32
Tilt Mean and Variance	8.25 ± 1.3	8.25 ± 1.3	7.2 ± 1.6	7.2 ± 1.6	7.2 ± 1.6
$\sigma_c = (3/2 mortar)$	3	6	12	24	48
Tilt mean and variance	9.75 ± 3.6	9.87 ± 3.7	9.9 ± 3.5	9.75 ± 3.6	9.9 ± 3.5
$\sigma_c = (2 mortars)$	4	8	16	32	64
Tilt mean and variance	9.3 ± 7.5	10.44 ± 4.25	10.9 ± 5.7	11.5 ± 5.7	11.7 ± 5.7
houghpeaks & houghlines param					
numPeaks, Threshold	50,3	50,3	50,3	50,3	50,3
FillGap, MinLength	10,120	20,240	40,480	80,960	160,1920

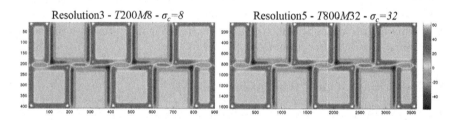

Resolution3 - *T200M8* - σ_c=8   Resolution5 - *T800M32* - σ_c=32

Fig. 7. Two samples of high resolution variations of cropped CW3 × 9 pattern. DoG scales are equal to the mortar size, and $s = 2$, $h = 8$ const. Left: $T = 200$, $M = 8$, $\sigma_c = 8$. Right: $T = 800$, $M = 32$, $\sigma_c = 32$. (Reproduced by permission from [27])

with the width of $4.5 \times T$, and the height of $2 \times T + M$ have been selected in all of these versions. The cropped sections have the same height above and below the mortar. Figure 6 shows the highest resolution of the Café Wall3 × 9 pattern that has been generated, with the $T = 800$, and $M = 32$ (Left), and the cropped section of the pattern (Right).

To make a distinction between the convergent and divergent mortar line, we talk about falling and rising, in which in the falling mortar, the right side of the mortar line is lower than its left side compared to the horizontal direction and for the rising mortar the vice versa. For instance in Fig. 6 (Left) the top mortar line is falling while the 2nd/bottom mortar line is rising.

Figure 7 illustrates the DoG output of two high resolution variations of cropped Café Wall3 × 9 pattern in jetwhite colors with their scales equal to their mortar size, Left: $(T, M) = (200, 8)$ with $\sigma_c = 8$, and Right: $(T, M) = (800, 32)$ with $\sigma_c = 32$. The result of parameters adjustment for five resolution versions of the Café wall3 × 9 pattern with the result of mean tilt and variance are provided in Table 1.

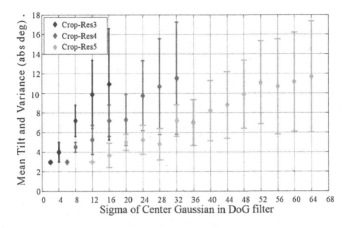

Fig. 8. Arithmetic Mean ± Standard Deviation for three resolutions of 3, 4, and 5 of Café Wall3 × 9-crop versions similar to Fig. 6-Right. The details of each resolution is given in Table 1. (Reproduced by permission from [27])

As shown in Table 1, for resolutions 3, 4, and 5 when the diameter of the center Gaussian is equal to the mortar size ($\sigma_c = M$), the detected mean tilt stay the same. There is one degree difference versus the lower resolution results. By increasing σ_c, there would be an increase in the mean tilt and variance. This is because the edge map gets thicker in the surrounding area in higher scales, and the detection of tilted lines with wider angular range is possible within these thick outlines. The `houghpeaks` and `houghlines` parameters have given in the last two rows of Table 1 for each resolution. The results of the mean tilt and variance with different DoG scales for three highest resolutions of the cropped Café Wall3 × 9 pattern (Resolutions 3, 4, 5) have been graphed in Fig. 8 with additional DoG scales compared to the values represented in Table 1.

4.2 Experiment Two - Cropping Parameterization

The aim of this experiment is to explore the effect of a focal or cropping region on tilt. To fix parameters not being investigated we restrict consideration initially to the Café wall3 × 9-$T400$-$M16$ pattern (resolution4). 50 samples are taken from the falling mortar (cropped as top mortar line) and 50 samples from the rising mortar (2nd/bottom mortar line). Sample dimensions are the same as the cropped section in the previous experiment (width of $4.5 \times T$ and height of $2 \times T + M$, with the same height above and below the mortar line). The sampling process starts from the left side of the pattern, with a resolution of 32 pixels between samples for both shift size and offset.

The result of Horizontal investigation of the mean tilt and variance of the detected lines, for all 50 samples of falling and 50 samples of rising mortar lines, have been analyzed for every DoG scale. The investigations show that as the scale of the bioplausible model (σ_c) increases, the variance also increases.

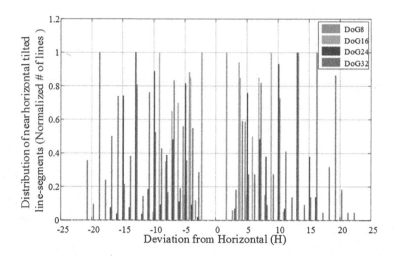

Fig. 9. The normalized of near horizontal detected lines for 4 different DoG scales of 8, 16, 24, and 32 of Café Wall3 × 9-T400-M16 with 100 samples in total. 50 Samples are taken along Falling mortar, and 50 samples from Rising mortar with 32px between samples as the offset/shift. (Reproduced by permission from [27])

Horizontal investigations of the detected lines based on their orientations is summarized in Fig. 9, indicating the normalized number of detected line segments from the 100 samples. The graph indicates the effect of the edge map scale on the range of detected tilts. This range covers a wider neighborhood area around horizontal axis when the DoG scale increases.

As shown, the scale of 32 has a high range of variations of tilt angle that is not reflected in our subjective perception of the pattern. Furthermore the angular values of detected slanted lines are very small in scale 8. So based on these results, the most informative parameter for the DoG scale in order to detect the convergence and divergence of the mortar lines and perceive consistent subjective tilt, is a center scale near the size of the mortar, here DoG16. The primary parameter of the DoG model is the scale factor (2σ is the diameter between the inflection points of the Gaussian and the range is smallest when this corresponds to the mortar size) with empirically insignificant parameters including the ratios between the Gaussians or to the window size used for filtering. The best way to confirm the prediction of this mortar-parameterized model as a plausible range for σ_c is by psychophysical experimental testing (future work).

A paired-samples t-test on 100 samples was conducted to compare the tilt mean value across the different scale DoGs. All comparisons were significant with $p < 0.01$.

5   Conclusions and Future Work

The optical illusion known as "Café Wall" is studied here, based on a bioplausible model of the earliest layers of our "deep neural network" as being retinal receptive fields (RF), modeled with ON-and OFF-center ganglion cells (GCs). The model is a classical RF (CRF) model [6,32], with circular center and surround organization based on difference of Gaussian, simulating a retinal response to the stimulus in low level vision [25,26], which we refer to as a multiple scale edge map of the pattern. The perceptual effect in the Café Wall illusion produces strong tilt effects, and our model explains the perception of divergent and convergent mortar lines in the pattern. The illustration of edge maps at multiple scales explains the concept of local and global (or foveal and peripheral) viewing of the pattern. This provides new insights into physiological models [3,7,8] as well as supporting Marr's theory of low level vision [22,23] and provides relevant insights for the construction of bioplausible DNNs.

The model's early stage output is investigated to quantify the degree of tilt using Hough transforms in place of the later DNN layers. Mean tilt and standard deviation of the detected tilt line segments are calculated for every scale of the edge map, as well as the histogram of near horizontal detected tilted lines in their positional angular values for a display of overall tilt, providing quantified predictions for both experiments with human subjects as well as features self-organized by DNNs.

More complex tile illusions having a diverse range of tilt will need fusion of multiscale (local and global) representations of the input pattern, considering the focus point (change of illusory effect with saccade) as well as a wholistic view of the pattern. The extension of the model to non-CRFs (nCRF) [28] based on elongated surrounds, might facilitate the directionality evaluation of a broader range of geometric illusions which will be investigated. A psychophysical assessment of the model predictions will also help the design of an analytical model for searching different visual clues in natural or illusion patterns similar to our visual processing.

The current models for geometric illusions are quite complicated and more research is needed to improve models of vision while directing them towards less sophisticated and more bioplausible detection of visual cues and clues. We believe that further exploration on the role of simple Gaussian-like models [21,22] in low level retinal processing, and Gaussian kernels in early stage DNNs, and its prediction or loss of perceptual illusion will lead to more accurate computer vision techniques and models and potentially steer computer vision towards or away from the features that humans detect. These effects can, in turn, be expected to contribute to higher-level models of depth and motion processing and generalized to computer understanding of natural images.

Acknowledgement. Nasim Nematzadeh was supported by an Australian Postgraduate Award (APA) scholorship for her Ph.D.

References

1. Bengio, Y.: Learning deep architectures for AI. Found. Trends® Mach. Learn. **2**(1), 1–127 (2009)

2. Blakeslee, B., McCourt, M.E.: A multiscale spatial filtering account of the White effect, simultaneous brightness contrast and grating induction. Vis. Res. **39**(26), 4361–4377 (1999)
3. Carandini, M.: Receptive fields and suppressive fields in the early visual system. Cogn. Neurosci. **3**, 313–326 (2004)
4. Duda, R.O., Hart, P.E.: Use of the Hough transformation to detect lines and curves in pictures. Commun. ACM **15**(1), 11–15 (1972)
5. Earle, D.C., Maskell, S.J.: Fraser cords and reversal of the Café Wall illusion. Perception **22**(4), 383–390 (1993)
6. Enroth-Cugell, C., Robson, J.G.: The contrast sensitivity of retinal ganglion cells of the cat. J. Physiol. **187**(3), 517–552 (1966)
7. Field, G.D., Chichilnisky, E.J.: Information processing in the primate retina: circuitry and coding. Annu. Rev. Neurosci. **30**, 1–30 (2007)
8. Gollisch, T., Meister, M.: Eye smarter than scientists believed: neural computations in circuits of the retina. Neuron **65**(2), 150–164 (2010)
9. Gregory, R.L., Heard, P.: Border locking and the Café Wall illusion. Perception **8**(4), 365–380 (1979)
10. Grossberg, S.: The link between brain learning, attention, and consciousness. Conscious. Cogn. **8**(1), 1–44 (1999)
11. Hough, P.V.: Method and means for recognizing complex patterns (1962)
12. Huang, J.Y., Protti, D.A.: The impact of inhibitory mechanisms in the inner retina on spatial tuning of RGCs. Sci. Rep. **6** (2016)
13. Hubel, D.H., Wiesel, T.N.: Receptive fields, binocular interaction and functional architecture in the cat's visual cortex. J. Physiol. **160**(1), 106–154 (1962)
14. Illingworth, J., Kittler, J.: A survey of the Hough transform. Comput. Vis. Graph. Image Process. **44**(1), 87–116 (1988)
15. Jameson, D., Hurvich, L.M.: Essay concerning color constancy. Annu. Rev. Psychol. **40**(1), 1–24 (1989)
16. Kitaoka, A., Pinna, B., Brelstaff, G.: Contrast polarities determine the direction of Café Wall tilts. Perception **33**(1), 11–20 (2004)
17. Kuffler, S.W.: Neurons in the retina: organization, inhibition and excitation problems. Cold Spring Harb. Symp. Quant. Biol. **17**, 281–292 (1952). Cold Spring Harbor Laboratory Press
18. LeCun, Y., Bengio, Y., Hinton, G.: Deep learning. Nature **521**(7553), 436–444 (2015)
19. Linsenmeier, R.A., Frishman, L.J., Jakiela, H.G., Enroth-Cugell, C.: Receptive field properties of X and Y cells in the cat retina derived from contrast sensitivity measurements. Vis. Res. **22**(9), 1173–1183 (1982)
20. Lourens, T.: Modeling retinal high and low contrast sensitivity filters. In: Mira, J., Sandoval, F. (eds.) IWANN 1995. LNCS, vol. 930, pp. 61–68. Springer, Heidelberg (1995). doi:10.1007/3-540-59497-3_157
21. Lv, Y., Jiang, G., Yu, M., Xu, H., Shao, F., Liu, S.: Difference of Gaussian statistical features based blind image quality assessment: a deep learning approach. In: IEEE International Conference on Image Processing (ICIP), pp. 2344–2348 (2015)
22. Marr, D., Hildreth, E.: Theory of edge detection. Proc. Roy. Soc. Lond. B: Biol. Sci. **207**(1167), 187–217 (1980)
23. Marr, D., Ullman, S.: Directional selectivity and its use in early visual processing. Proc. Roy. Soc. Lond. B: Biol. Sci. **211**(1183), 151–180 (1981)
24. Morgan, M.J., Moulden, B.: The Münsterberg figure and twisted cords. Vis. Res. **26**(11), 793–800 (1986)

25. Nematzadeh, N., Lewis, T.W., Powers, D.M.W.: Bioplausible multiscale filtering in retinal to cortical processing as a model of computer vision. In: ICAART 2015-International Conference on Agents and Artificial Intelligence. SCITEPRESS (2015)

26. Nematzadeh, N., Powers, D.M.W.: A quantitative analysis of tilt in the Café Wall illusion: a bioplausible model for foveal and peripheral vision. In: DICTA 2016-International Conference on Digital Image Computing: Techniques and Applications (2016)

27. Nematzadeh, N.: A neurophysiological model for geometric visual illusions. Ph.D. thesis, Flinders University (in preparation)

28. Passaglia, C.L., Enroth-Cugell, C., Troy, J.B.: Effects of remote stimulation on the mean firing rate of cat retinal ganglion cells. J. Neurosci. $21(15)$, 5794–5803 (2001)

29. Powers, D.M.W.: Lateral interaction behaviour derived from neural packing considerations. School of Electrical Engineering and Computer Science, University of New South Wales (1983)

30. Ratliff, F., Knight, B., Graham, N.: On tuning and amplification by lateral inhibition. Proc. Nat. Acad. Sci. $62(3)$, 733–740 (1969)

31. Robson, J.G.: Frequency domain visual processing. In: Braddick, O.J., Sleigh, A.C. (eds.) Physical and Biological Processing of Images. SSINF, vol. 11, pp. 73–87. Springer, Heidelberg (1983). doi:10.1007/978-3-642-68888-1_6

32. Rodieck, R.W., Stone, J.: Analysis of receptive fields of cat retinal ganglion cells. J. Neurophysiol. $28(5)$, 833–849 (1965)

33. Romeny, B.M.: Front-End Vision and Multiscale Image Analysis: Multiscale Computer Vision Theory and Applications, Written in Mathematica. Springer Science & Business Media, Heidelberg (2008)

34. Schmidhuber, J.: Deep learning in neural networks: an overview. Neural Netw. 61, 85–117 (2015)

35. Von der Malsburg, C.: Self-organization of orientation sensitive cells in the striate cortex. Kybernetik $14(2)$, 85–100 (1973)

36. Westheimer, G.: Irradiation, border location, and the shifted chessboard pattern. Perception $36(4)$, 483–494 (2007)

Image Patch Matching Using Convolutional Descriptors with Euclidean Distance

Iaroslav Melekhov[1(✉)], Juho Kannala[1], and Esa Rahtu[2]

[1] Department of Computer Science, Aalto University, Espoo, Finland
iaroslav.melekhov@aalto.fi
[2] Center for Machine Vision Research, University of Oulu, Oulu, Finland

Abstract. In this work we propose a neural network based image descriptor suitable for image patch matching, which is an important task in many computer vision applications. Our approach is influenced by recent success of deep convolutional neural networks (CNNs) in object detection and classification tasks. We develop a model which maps the raw input patch to a low dimensional feature vector so that the distance between representations is small for similar patches and large otherwise. As a distance metric we utilize L_2 norm, i.e. Euclidean distance, which is fast to evaluate and used in most popular hand-crafted descriptors, such as SIFT. According to the results, our approach outperforms state-of-the-art L_2-based descriptors and can be considered as a direct replacement of SIFT. In addition, we conducted experiments with batch normalization and histogram equalization as a preprocessing method of the input data. The results confirm that these techniques further improve the performance of the proposed descriptor. Finally, we show promising preliminary results by appending our CNNs with recently proposed *spatial transformer networks* and provide a visualisation and interpretation of their impact.

1   Introduction

Finding correspondences between image regions (patches) is a key factor in many computer vision applications. For example, structure-from-motion, multi-view reconstruction, image retrieval and object recognition require accurate computation of local image similarity. Due to importance of these problems various descriptors have been proposed for patch matching with the aim of improving accuracy and robustness. Many of the most widely used approaches, like SIFT [1] or DAISY [2] descriptors, are based on hand-crafted features and have limited ability to cope with negative factors (occlusions, variation in viewpoint etc.) making a search of similar patches more difficult. Recently, various methods based on supervised machine learning have been successfully applied for learning patch descriptors [3–6]. These methods significantly outperform hand-crafted approaches and inspire our research.

During recent years, neural networks have achieved great success in object classification [7] and other computer vision problems. Specifically, methods based

© Springer International Publishing AG 2017
C.-S. Chen et al. (Eds.): ACCV 2016 Workshops, Part III, LNCS 10118, pp. 638–653, 2017.
DOI: 10.1007/978-3-319-54526-4_46

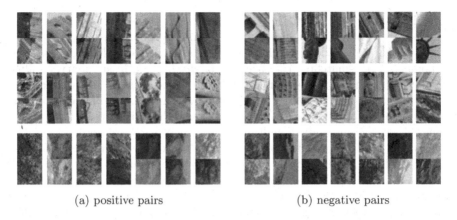

(a) positive pairs   (b) negative pairs

Fig. 1. Randomly picked matching (i.e. positive) and non-matching (i.e. negative) patch pairs of Multi-view Stereo Correspondence (MSC) dataset [3] which consists of three subsets: Liberty (top row), Notredame (middle row) and Yosemite (bottom row). The matching patches represent the same 3D structure so that their orientation, scale and location are roughly corresponding but there are still significant variations in viewpoint and illumination. The non-matching patches represent different 3D points and therefore they usually have quite different texture and appearance.

on Convolutional Neural Network (CNN) have showed significant improvements over previous state-of-the-art recognition and object detection approaches. Influenced by these works, we aim to create a CNN-based discriminative descriptor for patch matching task. In contrast to [8,9] where the representations of two patches are compared using a set of fully connected layers, we utilize *Euclidean distance* as a metric of similarity. The same metric is used in one of the most popular and applicable descriptor, SIFT. Therefore, our approach can be considered as a direct alternative to SIFT and similar techniques can be used for fast matching and indexing of descriptors as with SIFT. We utilize labeled patch pairs to learn the descriptor so that Euclidean distance (L_2 norm) between patches in the feature space is small for similar patches and large otherwise. This is analogous to face-verification problem where Siamese structure [10] has been utilized to predict whether the persons illustrated in an input image pair are the same or not.

For training and evaluation of the proposed descriptor we utilize Multi-view Stereo Correspondence (MSC) dataset [3], which is illustrated in Fig. 1 and consists of more than 1.5M grayscale patches. The dataset consists of pairs of matching and non-matching patches extracted from images of the Statue of Liberty, Notredame and Half Dome (Yosemite) by using Difference of Gaussian (DoG) interest point detector and matched by utilizing the respective 3D multi-view reconstructions computed from the images [3]. In detail, corresponding interest points were found by mapping between images using the dense stereo depth maps computed by the multi-view stereo algorithm of [11] based on the initial point cloud reconstructions by [12]. Pairs of patches corresponding to the same 3D point are defined to be matching (i.e. *positive* or *similar* pairs in our terminology) if they also originate from DoG interest points detected with sufficiently similar scale and orientation [3]. Pairs of patches sampled from different 3D points are non-matching (i.e. *negative* or *dissimilar*). In summary, as illustrated in Fig. 1, the matching pairs represent the same 3D structure with roughly correct geometric alignment so that their appearances are similar whereas the negative pairs typically have different texture and dissimilar appearance.

In this work, we conduct multiple experiments with preprocessing of raw patches and demonstrate that histogram equalization as well as batch normalization significantly improve the accuracy of the proposed descriptor.

We also explore different types of descriptor architectures evaluating their performance on MSC dataset. Our experimental evaluation shows that the proposed model outperforms recent state-of-the-art L_2-based approaches. In addition, we investigate the use of spatial transformer networks [13] in the patch matching problem.

The paper is organized as follows. Section 2 presents related work focusing on patch matching problem. Section 3 describes the proposed method of finding corresponding patches, discusses an architecture of the descriptor, objective function and details of data preprocessing. Section 4 presents the experimental pipeline and performance on the MSC dataset. In the end of this paper we summarize our results and point some directions of future work.

2   Related Work

Local image descriptors have been widely used in finding similar and dissimilar regions in images. Nowadays, the trend has changed from hand-crafted and carefully-designed methods (SIFT [1] or DAISY [2]) to a new generation of learned descriptors including unsupervised and supervised techniques like boosting [4], convex optimization [6] and Linear Discriminant Analysis (LDA) [3,14].

In our approach, however, we propose a descriptor based on deep convolutional neural networks (CNN) with batch normalization units accelerating learning and convergence. The first papers which utilized CNN based representations for finding matching image patches were [15,16]. More recently, Žbontar and LeCun [17] proposed a method for comparing image patches in order to extract stereo depth information. Their method is based on using convolutional networks minimizing a hinge loss function and showed the best performance on KITTI stereo evaluation dataset [18]. However, as that approach operates on very small patches (9×9 pixels), it restricts the area of applicability.

In addition, one recent related paper is [19], which utilizes Siamese network architecture for the challenging problem of matching street-level and aerial images. In contrast to our work, [19] concentrates on matching entire images in a specific application, i.e. ground-to-aerial geolocalization. Their approach is therefore not directly applicable in tasks where local features are currently used and it does not allow replacing or comparing with SIFT. Moreover, in their work the length of the proposed descriptor is significantly larger $(4,096)$ than that of SIFT and our representation (128).

Recent approaches [8,9,20] propose CNN descriptors trained with two-branch (Siamese) architecture which significantly exceed the accuracy of hand-crafted descriptors. However, in contrast to SIFT, in [8,9] the feature representations of input patches are compared by a set of fully connected layers (match network) that learns a complex comparison metric. Nevertheless, Zagoruyko and Komodakis [8] and Simo-Serra et al. [20] also conducted experiments in which the match network was replaced with Euclidean distance metric between the outputs of two branches and, hence, they are the closest works to ours. The implementation of [20] is not yet publicly available. Thus, in order to compare performance, we reproduced the network architecture of [20] and evaluated it using the standard protocol. The results show that our network architecture outperforms those of [8,20]. More detailed comparison is presented in Sect. 3.2.

3   Neural Descriptor

Our goal is to construct a system that efficiently distinguishes matching (similar) and non-matching (dissimilar) patches. To do this, we propose a method based on a deep convolutional neural network. As shown in Fig. 2, the model consists of two identical branches that share the same set of weights and parameters. Patches P_1 and P_2 are fed into branches and propagated through the model separately. The main objective of a proposed network is to map the raw patches to a low dimensional feature space so that the L_2 distance between pairs is small if the patches are similar and large otherwise. The same distance measure (L_2 distance) is usually applied also for matching hand-crafted descriptors.

The following section describes the proposed loss function and how it can be used in our approach.

3.1   Loss Function and Data Preprocessing

To optimize the proposed network, we have to use a loss function which is capable to distinguish similar (positive) and dissimilar (negative) pairs. More precisely, we train the weights of the network by using a loss function which encourages similar examples to be close, and dissimilar ones to have Euclidean distance larger or equal to a margin m from each other. In contrast to [8,20], which utilize hinge embedding loss [21], we use margin-based contrastive loss [22] defined as follows:

$$\mathcal{L}\left(P_1, P_2, l\right) = \frac{1}{2}lD^2 + \frac{1}{2}\left(1 - l\right)\left\{\max\left(0, m - D\right)\right\}^2 \tag{1}$$

where l is a binary label which selects whether the input pair consisting of patch P_1 and P_2 is a positive ($l = 1$) or negative ($l = 0$), $m > 0$ is the margin for negative pairs and $D = \|f(P_1) - f(P_2)\|_2$ is the Euclidean Distance between feature vectors $f(P_1)$ and $f(P_2)$ of input images P_1 and P_2.

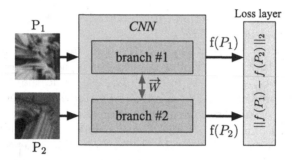

Fig. 2. Schematic illustration of the proposed descriptor based on Siamese architecture [10]. A pair of patches (P_1, P_2) is propagated through the network consisting of two identical branches and sharing the same set of weights (W). Feature representations of patches $(f(P_1), f(P_2))$ are extracted from the last layer of each branch separately and Euclidean distance is computed between them. Our objective is to learn a descriptor that minimizes the distance between similar pairs of patches and maximizes it for dissimilar pairs. It is important to note that at test time (i.e. after learning) the feature descriptor f can be computed independently for each individual patch since both branches are identical.

Dissimilar pairs contribute to the loss function only if their distance is smaller than the margin m. The idea of learning is schematically illustrated in Fig. 3. The loss function encourages matching patches (elements with the same color and shape) to be close in feature space while pushing non-matching pairs apart. Obviously, negative pairs with a distance larger than margin would not contribute to the loss (second part of (1)). Thus, setting margin to too small value would lead to optimizing the objective function only over the set of positive pairs and, as a result, would hamper learning.

To demonstrate what has been learned by our proposed descriptor, we illustrate the histogram of pairwise Euclidean distances of patch pairs of test set both before and after training in Fig. 4. The blue and brown bars represent pairwise distances of positive and negative pairs, respectively. It can clearly be seen that the training process of the descriptor on patch pairs effectively pushes

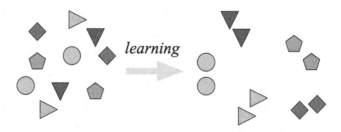

Fig. 3. Contrastive loss minimizes the distance between positive patch pairs (elements with the same color and shape) and maximizes otherwise. (Color figure online)

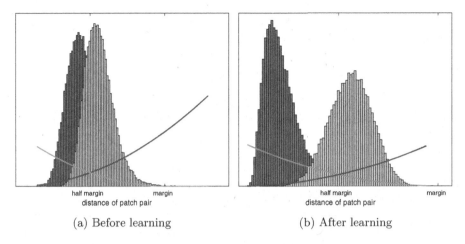

(a) Before learning   (b) After learning

Fig. 4. The distributions of feature distances D for positive (blue) and negative (brown) patch pairs of Notredame test dataset before (left) and after (right) training on Liberty patches of MSC dataset. Learning decreases distances of positive pairs and increases distances of negative pairs. The blue curve is the loss for positive pairs, i.e. $D^2/2$ in (1), and the red curve is the loss for negative pairs, i.e. $(\max(0, m - D))^2/2$. The curves intersect at $m/2$. (Color figure online)

non-matching pairs away and pulls matching pairs together. In the very beginning, the distributions of positive and negative pairs are grouped at the intersection of the blue (penalty for similar pairs (1)) and the red (penalty for dissimilar pairs) curves in Fig. 4a. We experimentally verified that for efficient training the margin value should be set to twice the average Euclidean distance between features of training patch pairs before learning.

Data Preprocessing and Augmentation. Data preprocessing plays an important role in machine learning algorithms. However, in practice it is hard to say in advance which preprocessing technique is helpful for achieving best performance. Here we calculate mean and standard deviation of pixel's intensities over the whole training dataset and use them to normalize intensity value of every pixel in the input grayscale patch. In addition, analysing raw patches in MSC dataset, we noticed that there are a lot of pairs where patches have significantly different contrast. To adjust patch intensities we apply histogram equalization before normalization. Histogram equalization is a technique that allows us to improve the contrast of images and it has been found to be a powerful technique in image enhancement. Equalized histogram of a discrete gray-level image represents the frequency of occurrence of all gray-levels in the image and well distributes the pixels intensity over the full intensity range. Finally, to prevent overfitting we used the same approach as [8] and augmented training data applying affine transformation by rotating both patches in pairs to 90, 180, 270 degrees and flipping them horizontally and vertically.

3.2   Network Architecture and Learning

The proposed network architecture for one branch of the Siamese network of Fig. 2 has following modules: convBlock[32,3,1,1]-convBlock[64,3,1,1]-pool[2]-convBlock[64,3,1,1]-convBlock[64,3,1,1]-pool[2]-convBlock[128,3,1,1]- convBlock [128,3,1,1]-pool[3]-convBlock[128,3,1,1]-L2norm. The shorthand notation: convBlock[N,w,s,p] consists of a convolution layer with N filters of size $\omega \times \omega$ with stride s and padding p, a regularisation layer (ReLU) and batch normalisation, pool[k] is a max-pooling layer of size $k \times k$ applied with stride k. This architecture dubbed *cnn7* was selected based on several experiments with different network structures having varying number of layers and involving also fully connected layers. We observed that convolutional networks without fully connected layers seemed to perform better than networks with fully connected layers, and *cnn7* had the best performance among the networks we experimented.

In our case, the benefit of applying batch normalization [23] and histogram equalization was verified experimentally, as is shown in Fig. 5 and described in Sect. 4. We also analyzed the network structure proposed by [20], titled *cnn3*, by re-implementing its architecture and utilizing contrastive loss objective function. As shown in Fig. 5 we noticed that our network architecture clearly outperforms *cnn3* even without histogram equalization of the input patches (blue and red

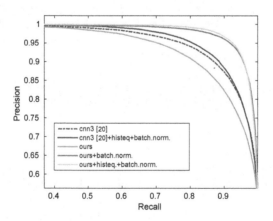

Fig. 5. Precision-recall curves for different descriptor architectures and data pre-processing approaches. To present the results more clearly we have zoomed on the recall axis. Here the performance is shown for 100k patch pairs of Notredame test data and the evaluated networks were trained in a Siamese architecture using 500k training pairs from the Liberty dataset.

curves respectively). Moreover, applying histogram equalization further improves the accuracy of the proposed method.

In contrast to *cnn3* [20] model and two models *siam-l₂*, *pseudo-siam-l₂* proposed by [8], we decomposed convolutional layers with a big kernel size into several filters with smaller kernels (3×3), separated by ReLU activations. According to [24], it increases nonlinearities of the whole network and makes the decision function more discriminative. Moreover, our model has only half the number of parameters compared to [8].

Learning. We minimize Contrastive loss function (1) over a training set using Stochastic Gradient Descent (SGD) with a standard back-propagation [25] and ADADELTA [26]. We train our descriptor in two stages. In the first stage, the training data has 500,000 patch pairs and it took about 1 day to finish 100,000 iterations of training, which is equal to 40 epochs of the training set. Weights are initialised randomly and the model is trained from scratch. In the second stage, we augmented the number of training samples up to 4M pairs by using also rotated and mirrored versions of the original patches, and then resumed training for another 20 epochs starting from pre-trained descriptor from the first stage. Learning rate (0.01), weight decay (0.001) as well as the size of mini-batch (100) remain constant during the training. The model[1] was trained using publicly available deep learning framework Caffe on one NVIDIA TITAN Z GPU.

4 Experiments

In this section, we present experimental results evaluating the proposed descriptor on MSC dataset. In order to compare results with previous work, we use exactly the same standard datasets for training and testing as used by e.g. [3,8]. That is, for each of the three subsets of MSC dataset (Liberty, Notredame, Yosemite) we use a test set of 100,000 pairs of patches originally provided by [3]. For training we utilize 500,000 pairs of patches from each subset (also provided by [3]). If we augment the training data by including rotated and mirrored versions of the original training patches, as described in Sect. 3, we get 4 million pairs from the original 0.5 million. We train three models by using training patches from the three different subsets, and evaluate each of the three models with test pairs of the two remaining subsets. In total we get six cases which are presented in Table 1.

Performance Metric. We follow the standard protocol of [3] and calculate ROC curves by thresholding the distance between feature pairs and determine the false positive rate at 95% recall. The numbers are shown in Table 1. As in [8], we also report the *mean* across all six combinations of training and test data. Like the original work [3], we also provide $mean_{[1:4]}$ metric which is the mean across the four cases obtained by training models only on Yosemite and Notredame.

---

[1] Source code and the model will be made available upon publication.

Table 1. Performance comparison of our descriptor and existing methods on Liberty (Lib), Notredame (ND) and Yosemite (Yos) image patches of MSC dataset. Numbers are false positive rate at 95% recall on each of the six combinations of training and test sets. **Bold** numbers are the best across all algorithms. The proposed architecture can outperform [8] in 4 cases out 6 and has the lowest average errors $mean$ and $mean_{[1:4]}$ (the average over the first four columns) with histogram equalization.

Method	Dim	Training test	Yos		ND		Lib		$mean$	$mean_{[1:4]}$
			Lib	ND	Lib	Yos	ND	Yos		
nSIFT + L2 (no training)	128d		29.84	22.53	29.84	27.29	22.53	27.29	26.55	27.38
nSIFT squared diff. linearSVM	128d		27.07	19.87	26.54	24.71	19.65	25.12	23.83	24.55
Brown et al. w/PCA	29d		18.27	11.98	16.85	**13.55**	-	-	-	15.16
Zagoruyko siam-L2 [8], 4M training pairs	256d		17.25	8.38	13.24	15.89	**6.01**	19.91	13.45	13.69
Zagoruyko pseudo-siam-L2 [8], 4M training pairs	256d		18.37	8.95	16.58	15.62	6.58	17.83	13.99	14.88
Ours, 500k training pairs	128d		**14.88**	9.47	16.57	19.50	9.01	17.21	14.44	15.11
Ours, 4M training pairs	128d		15.48	8.88	**11.84**	17.78	8.40	**15.07**	12.91	13.50
Ours, 500k training pairs, hist. eq.	128d		15.32	9.10	12.82	15.52	8.63	17.05	13.07	13.19
Ours, 4M training pairs, hist. eq	128d		15.19	**8.36**	12.20	14.72	6.93	15.86	**12.21**	**12.74**

The results in Table 1 confirm that the proposed model has better performance than [8] with the same number of training pairs. For instance, in Notredame-Liberty *siam-L2* outperforms hand-crafted descriptor nSIFT + L2 and nSIFT squared diff. by 16.6% and 13.3% respectively in absolute error. Our method with the same size of training data further improves accuracy by 1.4% in absolute error rate. Moreover, the length of our descriptor is significantly shorter than in [8]. The benefit of applying histogram equalization is presented in the last two rows of Table 1. The proposed model achieves 12.21% and 13.07% average error for with and without augmentation of training data, respectively. In general, it improves the performance of the proposed descriptor by 9.21% in relative units for average error and by 6.93% for $mean_{[1:4]}$ compared to [8].

Figure 6 shows top ranking false and correct matches of Notredame test dataset computed by our best model (the last row of Table 1). Specifically, we notice that some patches in false negative and false positive examples are so similar that even a human could make a mistake in interpretation. In fact, it seems that the top-ranking false positives (i.e. the pairs of negative patches whose descriptors are closest to each other) are probably originating from repeating texture patterns of the scene (i.e. similar texture appears in different 3D locations of the scene). Obviously, our descriptor or any other similar descriptor can not tell the difference here as it does not have access to multi-view information which was used to generate the ground truth labels. More interestingly, the top-ranking false negatives (i.e. the pairs of positive patches with descriptors furthest away from each other) seem to originate from patches where there is a perceived

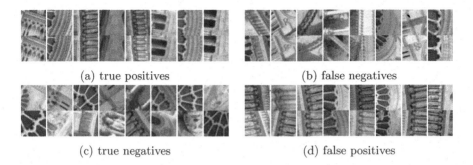

(a) true positives   (b) false negatives

(c) true negatives   (d) false positives

Fig. 6. Top-ranking true and false matches of Notredame patches by our best model.

Table 2. Performance (area under precision-recall curve) of our descriptor architecture and *cnn3* proposed by [20] on MSC dataset for 500k training pairs. Precision-recall curves corresponding to Liberty (Lib) training data and Notredame (ND) test data for considered descriptors are also illustrated in Fig. 5

Descriptor architecture	Yos		ND		Lib	
	Lib	ND	Lib	Yos	ND	Yos
cnn3	0.943	0.961	0.950	0.945	0.964	0.945
ours	**0.977**	**0.984**	**0.980**	**0.977**	**0.985**	**0.975**

dissimilarity because of inaccurate geometric alignment (due to non-planarity of the scene surface or due to inaccuracies in the orientation assignment or localization of the interest point). Thus, augmentation of training data and/or hard positive mining could bring further improvement and robustness to aforementioned factors in future. Nevertheless, Fig. 6 confirms the good behaviour of the proposed descriptor as the failure cases are intuitively understandable and hard to avoid in general without trade-offs.

Finally, we also calculated area under precision-recall curve for our method as this metric is used by [20] for comparing descriptor performance. The results presented in Table 2 show that our network architecture performs better than the cnn3 architecture of [20].

4.1 Spatial Transformer Networks

Our visualisation in Fig. 6 shows that the image patches in many of the false negative pairs have a slightly differing alignment. That is, the patches represent corresponding scene surfaces but the scales, orientations and locations assigned by the interest point detector do not match precisely. Thus, based on the visualisation and interpretation of our results in Fig. 6, we decided to further investigate that whether our descriptor could by made more robust to spatial misalignment by applying spatial transformer (ST) networks [13]. Specifically, the spatial transformer is a differentiable module performing explicit spatial transformations of

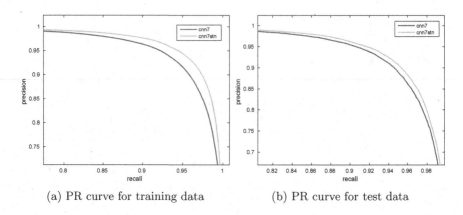

(a) PR curve for training data   (b) PR curve for test data

Fig. 7. Precision-recall curves for training and test data for *cnn7* and *cnn7stn*.

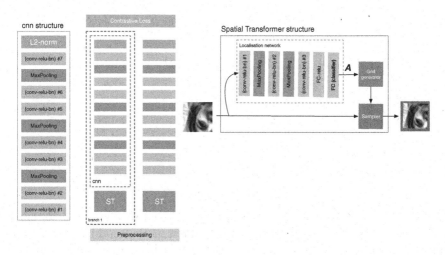

Fig. 8. Schematic representation of the pipeline incorporating Spatial Transformer layer in our experiments. ST layer consists of three different parts: *localisation network* predicts transformation parameters *A* that should be applied to the input feature map, i.e. a raw input patch after preprocessing (histogram equalization) procedure. *Grid generator* utilizes the predicted parameters *A* to construct a sampling grid which is used by *sampler* to produce the transformed output. The size of both the input and output of ST layer is 64 × 64. The warped output of the spatial transformer is fed to CNN model respectively.

input feature maps and can be placed at any part of a neural network easily. However, so far they have been mainly used in image classification problems [13] and, to the best of our knowledge, they have not been previously used for learning image similarity metrics with contrastive loss function.

Figure 8 schematically illustrates how we append our *cnn7* model (introduced in Sect. 3.2) by incorporating ST modules right after the preprocessing

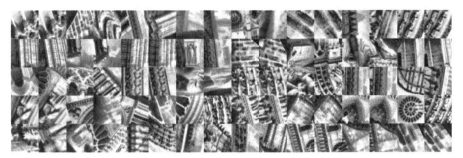

(a) false negatives by cnn7 which are true positives by cnn7stn (in total 688 pairs)

(b) false positives by cnn7 which are true negatives by cnn7stn (2488 pairs)

(c) false negatives by cnn7stn which are true positives by cnn7 (688 pairs)

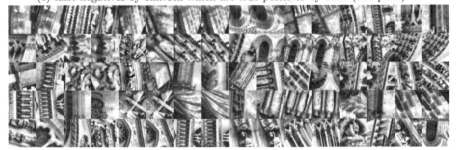

(d) false positives by cnn7stn which are true negatives by cnn7 (1414 pairs)

Fig. 9. Visualisation of some NotreDame test pairs which are classified differently by *cnn7* and *cnn7stn* when we set recall to 0.95 for both. As can be seen from Fig. 7 *cnn7stn* has higher precision. The total number of test pairs is 100k.

(a) false negatives by cnn7 which are true positives by cnn7stn (688 pairs)

(b) false positives by cnn7 which are true negatives by cnn7stn (2488 pairs)

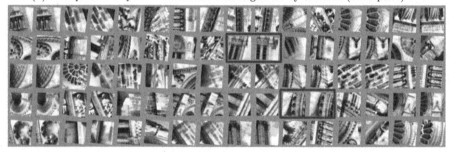

(c) false negatives by cnn7stn which are true positives by cnn7 (688 pairs)

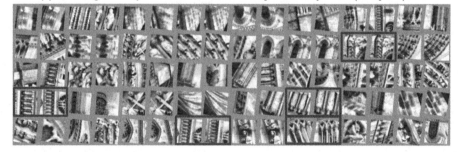

(d) false positives by cnn7stn which are true negatives by cnn7 (1414 pairs)

Fig. 10. The results of spatial transformations applied to the input image patches. Blue color represents cases where ST layer transforms patches so that mutual alignment is improved. (Color figure online)

layer. As we put ST module as the first layer in the network, it directly transforms the preprocessed input patches. The number of parameters A can vary and depends on the type of transformation used. Inspired by examples of Fig. 6, we aim to compensate errors caused by *rotation, translation* and *scaling*. Therefore, the number of estimated parameters by localisation network equals 4 (one for rotation, one for scaling and two for translation transformations). The architecture of the localisation network is as follows: convBlock[32,5,1,2]-pool[2]-convBlock[64,5,1,2]-pool[2]-convBlock[128,5,1,2]-fc[256]-fc[4] where fc[n] denotes a fully-connected layer with n outputs. The complete model with the ST layer is denoted as *cnn7stn*.

We train both *cnn7* and *cnn7stn* from random initialization using the histogram equalized pairs from the augmented Liberty training set (4M pairs). However, this time we did not use weight decay, and both models were trained using a smaller number of epochs than used for the results of Table 1 (due to a limited available training time). The models were evaluated with the NotreDame test set (100k pairs) and the results are shown in Fig. 7. We can see that *cnn7stn* gives better performance than *cnn7*.

In order to further visualize and analyse the difference Fig. 9 shows examples of pairs for which the two models give different classification result at 0.95 recall. Figure 10 shows the output of ST layer for the same patches. We can see that in most cases the ST layer transforms both patches of a pair quite similarly but in some cases (indicated with the blue color) the ST layer seems to improve the alignment which is probably the explanation for the better performance of *cnn7stn*. Hence, it seems that the ST layer has learnt the desirable behaviour to some extent. Still, there is probably room for further improvements since many misaligned pairs remain quite differently aligned after the ST layer (cf. Fig. 10).

5   Conclusion

In this paper, we use Siamese architecture to train a deep convolutional network for extracting descriptors from image patches. In training we utilized matching and non-matching pairs of image patches from MSC dataset. There are several conclusions that we can get from our experiments. First, we propose a descriptor with good performance, notably outperforming previous CNN-based L_2 norm descriptors on several datasets. We also show that utilizing histogram equalization for adjusting patch contrast improves the accuracy of the proposed model. In addition, we run preliminary experiments by appending our CNN architecture with spatial transformer layers and observe an improvement in the resulting descriptor. A potential future performance enhancement could be to investigate optimal structures of the localisation network of ST layers which could make the descriptor even more robust to geometric transformations.

References

1. Lowe, D.G.: Distinctive image features from scale-invariant keypoints. Int. J. Comput. Vis. **60**, 91–110 (2004)
2. Tola, E., Lepetit, V., Fua, P.: A fast local descriptor for dense matching. In: Proceedings of Computer Vision and Pattern Recognition (2008)
3. Hua, G., Brown, M., Winder, S.: Discriminant learning of local image descriptors. In: IEEE Transactions on Pattern Analysis and Machine Intelligence (2010)
4. Trzcinski, T., Christoudias, C.M., Lepetit, V., Fua, P.: Learning image descriptors with the boosting-trick. In: NIPS, pp. 278–286 (2012)
5. Trzcinski, T., Christoudias, M., Fua, P., Lepetit, V.: Boosting binary keypoint descriptors. In: Proceedings of the 2013 IEEE Conference on Computer Vision and Pattern Recognition, CVPR 2013, pp. 2874–2881. IEEE Computer Society (2013)
6. Simonyan, K., Vedaldi, A., Zisserman, A.: Learning local feature descriptors using convex optimisation. IEEE Trans. Pattern Anal. Mach. Intell. (2014)
7. Krizhevsky, A., Sutskever, I., Hinton, G.E.: Imagenet classification with deep convolutional neural networks. In: Pereira, F., Burges, C., Bottou, L., Weinberger, K. (eds.) Advances in Neural Information Processing Systems 25, pp. 1097–1105 (2012)
8. Zagoruyko, S., Komodakis, N.: Learning to compare image patches via convolutional neural networks. In: The IEEE Conference on Computer Vision and Pattern Recognition (CVPR) (2015)
9. Han, X., Leung, T., Jia, Y., Sukthankar, R., Berg, A.C.: Matchnet: unifying feature and metric learning for patch-based matching. In: The IEEE Conference on Computer Vision and Pattern Recognition (CVPR) (2015)
10. Chopra, S., Hadsell, R., LeCun, Y.: Learning a similarity metric discriminatively, with application to face verification. Comput. Vis. Pattern Recogn. **1**, 539–546 (2005)
11. Goesele, M., Snavely, N., Curless, B., Hoppe, H., Seitz, S.M.: Multi-view stereo for community photo collections. In: Proceedings of the 11th International Conference on Computer Vision (ICCV 2007), pp. 265–270. IEEE (2007)
12. Snavely, N., Seitz, S.M., Szeliski, R.: Modeling the world from internet photo collections. Int. J. Comput. Vis. **80**, 189–210 (2008)
13. Jaderberg, M., Simonyan, K., Zisserman, A., Kavukcuoglu, K.: Spatial transformer networks. Adv. Neural Inf. Process. Syst. **28**, 2017–2025 (2015)
14. Strecha, C., Bronstein, A., Bronstein, M., Fua, P.: LDAHash: improved matching with smaller descriptors. IEEE Trans. Pattern Anal. Mach. Intell. **34**, 66–78 (2012)
15. Jahrer, M., Grabner, M., Bischof, H.: Learned local descriptors for recognition and matching. In: Computer Vision Winter Workshop (2008)
16. Osendorfer, C., Bayer, J., Urban, S., Smagt, P.: Convolutional neural networks learn compact local image descriptors. In: Lee, M., Hirose, A., Hou, Z.-G., Kil, R.M. (eds.) ICONIP 2013. LNCS, vol. 8228, pp. 624–630. Springer, Heidelberg (2013). doi:10.1007/978-3-642-42051-1_77
17. Zbontar, J., LeCun, Y.: Stereo matching by training a convolutional neural network to compare image patches. In: The IEEE Conference on Computer Vision and Pattern Recognition (CVPR) (2015)
18. Geiger, A., Lenz, P., Stiller, C., Urtasun, R.: Vision meets robotics: the KITTI dataset. Int. J. Robot. Res. (IJRR) (2013)
19. Lin, T.Y., Cui, Y., Belongie, S., Hays, J.: Learning deep representations for ground-to-aerial geolocalization. In: 2015 IEEE Conference on Computer Vision and Pattern Recognition (CVPR) (2015)

20. Simo-Serra, E., Trulls, E., Ferraz, L., Kokkinos, I., Fua, P., Moreno-Noguer, F.: Discriminative learning of deep convolutional feature point descriptors. In: International Conference on Computer Vision (2015)
21. Mobahi, H., Collobert, R., Weston, J.: Deep learning from temporal coherence in video. In: Proceedings of the 26th Annual International Conference on Machine Learning, ICML 2009, pp. 737–744. ACM (2009)
22. Hadsell, R., Sumit, C., LeCun, Y.: Dimensionality reduction by learning an invariant mapping. Comput. Vis. Pattern Recogn. 2, 1735–6919 (2006)
23. Ioffe, S., Szegedy, C.: Batch normalization: accelerating deep network training by reducing internal covariate shift. CoRR abs/1502.03167 (2015)
24. Simonyan, K., Zisserman, A.: Very deep convolutional networks for large-scale image recognition. CoRR abs/1409.1556 (2014)
25. LeCun, Y., Boser, B., Denker, J.S., Henderson, D., Howard, R.E., Hubbard, W., Jackel, L.D.: Backpropagation applied to handwritten zip code recognition. Neural Comput. 1, 541–551 (1989)
26. Zeiler, M.D.: ADADELTA: an adaptive learning rate method. CoRR abs/1212.5701 (2012)

Author Index

Printed in the United States
By Bookmasters